INTERNATIONAL SERIES OF MONOGRAPHS IN
NATURAL PHILOSOPHY
General Editor: D. ter Haar

Volume 53

NUCLEAR PHYSICS

NUCLEAR PHYSICS

BY

M. G. BOWLER

Department of Nuclear Physics
Oxford University

PERGAMON PRESS

Oxford · New York · Toronto
Sydney · Braunschweig

Pergamon Press Ltd., Headington Hill Hall, Oxford

Pergamon Press Inc., Maxwell House, Fairview Park, Elmsford, New York 10523

Pergamon of Canada Ltd., 207 Queen's Quay West, Toronto 1

Pergamon Press (Aust.) Pty. Ltd., 19a Boundary Street, Ruschcutters Bay, N.S.W. 2011, Australia

Vieweg & Sohn GmbH, Burgplatz 1, Braunschweig

Copyright © 1973 Pergamon Press Ltd.

All Rights Reserved. No part of this publication may be reproduced, stored in a retrieval system, or transmitted, in any form or by any means, electronic, mechanical, photocopying, recording or otherwise, without the prior permission of Pergamon Press Ltd.

First edition 1973

Library of Congress Cataloging in Publication Data

Bowler, M   G
  Nuclear physics.

  (International series of monographs in natural philosophy, v. 53)
    1. Nuclear physics.
QC173.B673   1973   539.7   72-13262
ISBN 0-08-016983-X

Printed in Hungary

Contents

Preface		vii
Introduction		ix
Chapter 1. The Nuclear Periodic Table		1
1.1.	The periodic table, radioactivity and unstable nuclear states	1
1.2.	Nuclear mass	9
1.3.	Nuclear binding energy	13
1.4.	Nuclear size	15
1.5.	The semiempirical mass formula	34
1.6.	The independent particle model, magic numbers and shell structure within the nucleus	45
1.7.	Nuclear forces	56
Chapter 2. Angular Momentum and the Nucleus		60
2.1.	Central forces, orbital angular momentum and spin	60
2.2.	The quantum mechanical definition of angular momentum	63
2.3.	Parity	66
2.4.	Static electric moments	72
2.5.	Static magnetic moments—the magnetic dipole moment	82
2.6. ●	Diatomic molecular spectra and nuclear spin	97
Chapter 3. Nuclear Decay		101
3.1.	Introduction	101
3.2.	The Fermi Golden Rule	103
3.3.	Electromagnetic decay of nuclei	109
3.4.	Electric dipole transitions	114
3.5.	Transitions other than electric dipole terms	118
3.6. ●	Internal conversion	124
3.7. ●	$0 \to 0$ transitions	129
3.8.	Measurement of lifetimes in electromagnetic transitions	132
3.9.	Determination of multipolarities in electromagnetic decay	135
3.10.	Nuclear decay through the weak interactions—β-decay. Introduction	139
3.11.	The theory of β-decay	142
3.12.	Relativistic effects in β-decay	154
3.13.	Non-conservation of parity in the weak interactions	161
3.14.	The weak interactions and parity, charge conjugation and time reversal	172
3.15.	Odd remarks on β-decay	174
3.16. ●	α-decay	175
3.17. ●	α-decay and the Fermi Golden Rule	184
Chapter 4. Nuclear Reactions		188
4.1.	Introduction	188
4.2.	Qualitative features of nuclear reactions	189

Contents

4.3.	The concept of cross-section	192
4.4.	Nuclear reactions and the Fermi Golden Rule	198
4.5.	The partial wave analysis	206
4.6.	Cross-sections and spin	213
4.7.	Time-independent potentials: the optical model	218
4.8. ●	Time-independent potentials: the two-nucleon system	220
4.9. ●	Time-independent potentials: neutron capture by protons	237
4.10.	Resonances. Introduction	243
4.11.	Resonances in scattering from a potential hole	247
4.12.	Structure resonances: the compound nucleus and low-energy reactions	252
4.13. ●	The coulomb barrier	272
4.14.	Examples of direct reactions: peripheral scattering processes	287
4.15. ●	Examples of direct reactions: pickup and stripping reactions	292
4.16.	Examples of direct reactions: neutrino interactions	294

Chapter 5. Self-sustaining Nuclear Reactions and Nuclear Energy Sources — 300

5.1.	Introduction	300
5.2.	Nuclear fission	301
5.3.	Induced fission	308
5.4.	Secondary features of fission	309
5.5.	Physics of the fission reactor	311
5.6.	Time constant of a fission reactor	318
5.7.	A note on breeding and fast neutron reactors	321
5.8.	Thermonuclear reactions and stellar evolution. Introduction	321
5.9.	Hydrogen-burning processes in the stars	324
5.10.	Stellar evolution during hydrogen burning	334
5.11.	Helium burning	338
5.12.	Stellar evolution after helium burning	342
5.13.	Thermonuclear power plants? Introduction	351
5.14.	Nuclear physics of a fusion reactor	352
5.15.	Optimal conditions for fusion reactor operation	355
5.16.	Magnetic containment of charged particles	357

Chapter 6. ● Isospin — 362

6.1.	Introduction	362
6.2.	Exchange symmetry: identical and non-identical nucleons	362
6.3.	The nucleon and isospin	365
6.4.	Isospin and nuclear structure	372
6.5.	Isospin and nuclear decay: electromagnetic decay	379
6.6.	Isospin and nuclear decay: β-decay	386
6.7.	The pion and isospin	389
6.8.	Strange particles	398
6.9.	Concluding remarks	406

Appendix 1. Energy Units and Constants — 408

Appendix 2. Angular Momentum Coupling — 410

Index — 413

Other Titles in the Series — 419

Preface

IN TEACHING nuclear physics over the last ten years I have felt the need of a nuclear physics text aimed primarily at the undergraduate, and so not too detailed, but recognizing that nowadays undergraduates towards the end of their second year and in their third year have considerable mathematical and conceptual sophistication. I have therefore tried to write a book setting out the principles of the subject, and acceptable to the undergraduate who has taken courses in quantum mechanics to the level of solving the Schrödinger equation for the hydrogen atom and involving some perturbation theory, atomic physics including the amount of nuclear physics usually contained in atomic physics texts, and perhaps electromagnetic theory. The book is based on lectures I have recently given to undergraduates at Oxford at the end of their second year, but considerably elaborated. I would hope that an undergraduate could begin to use this book in the second year and master most of the material by the end of the third year if so inclined: the more advanced material might even be useful to graduate students in the first year.

This book is concerned primarily with low-energy nuclear physics rather than high-energy, or elementary particle physics. I did not want to make the book too long, and particle physics is a subject which is both sufficiently distinct from nuclear physics to merit a separate treatment, and changing so rapidly that only very basic concepts remain long undated. Much of the material of Chapters 3, 4 and 6 is relevant to beginning a study of particle physics, however, and I have introduced examples from particle physics where appropriate.

This book contains little on experimental techniques (which become rapidly obsolete and can only be learned by doing experiments) and nothing on the subject of the interaction of radiation with matter, which is very well documented. The subject matter reflects my own view of the basic and important material, and I have discussed in most detail those points that I have found my own pupils to have most difficulty with, although occasionally in preparing this book I got too interested in something to leave it out.

A thread running through the whole book is the Fermi Golden Rule and the general scheme is as follows. Chapter 1 is primarily concerned with nuclear structure at a fairly basic level; I have not attempted any discussion of the collective model of nuclear structure. Chapter 2 deals with spin and static electric and magnetic moments, but I have approached these topics from quantum mechanics rather than via the vector model of angular momentum. Chapter 3 is on nuclear decay and, because there exists no classical theory of β-decay, I have treated both electromagnetic and β-decay using quantum mechanics, but without formal field theory. Chapter 4 is devoted to the theory of nuclear reactions. I have discussed the concepts of cross-section and of resonance in considerable detail, having found that they

Preface

can give rise to much confusion. Chapter 5 is concerned with self-sustaining nuclear reactions. The first part deals with the nuclear physics of the fission reactor and is included not only because of practical importance in a world where fission reactors are proliferating, but also because the subject affords some nice illustrations of much of the preceding material. Secondly, I have discussed the nuclear aspects of stellar physics. This again affords many illustrations of material in the preceding chapters, and is to my mind the most interesting and exciting thing about nuclear physics. Nuclear physics has not taught us anything revolutionary, as electromagnetism taught us about relativity and atomic physics about quantum mechanics, but it has allowed us to understand more or less how the stars work. Chapter 5 ends with a further application of the theory of thermonuclear reactions: the problems of devising a thermonuclear power plant.

While I have introduced isospin here and there in the rest of the text, this subject is discussed in detail in Chapter 6. The first part of this chapter is concerned with charge independence and isospin in low-energy nuclear physics, and in the last part the extension of the original concept to pion physics and the physics of strange particles should afford an ascent to a convenient pass in the watershed separating nuclear physics from the physics of elementary particles.

Any long and involved section which is not essential to the main scheme and may consequently be skipped if desired has been tipped with the black spot ●.

My research interests lie at a tangent to the field covered in this book: I must apologize in advance to anyone whose subject I have inadvertently misrepresented, and extend my thanks to those of my colleagues who have kindly read and commented on various parts of this material: Professor D. H. Perkins, Professor K. W. Allen, Dr. I. J. R. Aitchison and in particular Dr. G. A. Jones whose help and advice on many aspects of low-energy nuclear physics have been invaluable.

I have learned a lot and had enormous fun in writing this book, and I hope that anyone who slogs through it will enjoy the same pleasant experiences.

Oxford

Introduction

THE atom is a structure governed by quantum mechanics in which electrons are confined with a region of space of diameter $\sim 10^{-8}$ cm by the attractive coulomb field generated by the atomic nucleus. If an electron is confined within a region of space of dimensions $\sim r$, then even in the ground state it possesses a minimum momentum p given by the Heisenberg Uncertainty Principle

$$pr \sim \hbar \quad \text{(where } \hbar \text{ is Planck's constant divided by } 2\pi\text{)}$$

and consequently a kinetic energy given by

$$T = \frac{p^2}{2m}$$

for non-relativistic electrons. The potential energy of the outermost electrons in an atom is given by

$$V \sim -\frac{e^2}{r}$$

and since for a bound structure

$$V + T < 0$$

$$\frac{e^2}{r} > \frac{p^2}{2m}$$

or

$$\frac{e^2}{r} \gtrsim \frac{\hbar^2}{2mr^2}$$

whence

$$r \gtrsim \frac{\hbar^2}{2me^2} \simeq 2.5 \times 10^{-9} \text{ cm.}$$

Taking this value of r

$$T = \frac{p^2}{2m} \simeq 2 \times 10^{-11} \text{ ergs} \simeq 50 \text{ eV}^\dagger$$

† Numerical value of constants and an explanation of the energy units are given in Appendix 1.

Introduction

while if alternatively we take a value $r = 10^{-8}$ cm, $T \simeq 13$ eV. We consequently expect that the energy needed to extract an electron from the outer regions of an atom to be ~ 10 eV and we expect the spacing of the excited levels of such an electron to range from ~ 10 eV down.

The source of the confining coulomb field, the nucleus, is very much smaller. The nucleus is an assembly of nucleons (that is, protons and neutrons) which are held together by the mutual interactions of the nucleons. Nucleons have mass ~ 2000 electron masses and a spatial extent $\sim 10^{-13}$ cm: in the nucleus they are spaced approximately as close-packed spheres and consequently the nucleus occupies a volume of space of diameter 10^{-13}–10^{-12} cm, depending on the number of nucleons present.

The momentum of a nucleon confined in a region of space $\sim 10^{-13}$ cm across is

$$p \sim \frac{\hbar}{10^{-13}} = 10^{-14} \quad \text{cgs units}$$

and is thus $\sim 10^5$ greater than the momenta of outlying electrons in the atom. The corresponding kinetic energy is

$$T = \frac{p^2}{2M} \sim 3 \times 10^{-5} \text{ ergs} \sim 20 \text{ MeV}.$$

The average potential experienced by a nucleon must therefore be of depth $\gtrsim 20$ MeV. At a separation of $\sim 10^{-13}$ cm the interaction energy between two particles of charge e due to the coulomb field is $e^2/r \sim 1$ MeV. Now protons carry a charge $+e$ and so this interaction is repulsive. Neutrons, as their name implies, are electrically neutral and so do not interact with the coulomb field of the protons. From these crude but rather general considerations, we can see that the nucleus must be held together by interactions which are not electromagnetic in origin and which are ~ 10–100 times the strength of the coulomb interaction between two protons in the nucleus. These are the strong interactions of nuclear and particle physics.

Nuclear physics is concerned with the properties of these clusters of nucleons orbiting under the influence of their mutual strong interactions, with the effect of the structure of nuclei on atomic structure, and at a deeper level with the properties and explanation of the strong interactions themselves.

Nuclear physics differs from classical physics and most areas of modern physics in that there exists no well-understood theoretical apparatus which may be applied with confidence to the physics of the nucleus. The very general principles of quantum mechanics and, where appropriate, special relativity seem to apply, but provide only a framework in which to calculate. The reason for the difficulty is twofold. First, the potential acting between two nucleons is still not understood in terms of a sound theoretical model. Indeed, it is not yet known whether or not three body forces (in which the force acting on one particle would depend on the relative positions of two others) are important. Secondly, even if the nucleon–nucleon forces were understood in detail the computation of any single nuclear property lands us at once in the many-body problem. We may bring out the point more clearly by contrasting the nucleus, an assembly of between one and over 200 nucleons, with the atom which is an assembly of between one and ~ 100 electrons. The electronic structure of the

Introduction

atom is dominated by the coulomb field of the positively charged nucleus. Any given electron interacts electromagnetically with both the nucleus and with all electrons, and the interaction with the nucleus is dominant. With the development of the Schrödinger equation, modified by Pauli to include electron spin and magnetic moment, the determination of atomic structure became merely a computational problem.† This problem is vastly simplified by the dominant role of the nuclear coulomb field, in that the electron–electron interactions may be regarded as perturbations acting on electron levels determined by the coulomb field of the nucleus alone, and a given problem may be attacked iteratively with fairly rapid convergence towards a solution.

In nuclear physics, the nucleon–nucleon force is not well understood and, as we shall see later, there is no dominant force which principally determines the nuclear wave functions. To make an analogy, if the calculation of atomic structure is likened to computation of planetary orbits, which is done readily using the gravitational interaction between the planets as a small perturbation on the dominant field of the sun, then the calculation of nuclear structure is comparable with the problem of determining stellar motions within a dense globular cluster—the perturbations are likely to be comparable in magnitude with the assumed starting potential. Thus the efforts of theoretical nuclear physicists are directed to attempting to solve this many-body problem, using plausible potentials and approximations to make the problem tractable. The simplest calculations, such as the binding energies of ^{3}H or ^{3}He, are therefore research problems of great complexity which can only be discussed at a specialist level. It is this feature of nuclear physics which makes quantitative discussion for the non-specialist difficult and sometimes gives the beginner the impresssion that the subject is vague and qualitative. It is not—and the disappointed embryo theorist should remember that the state of affairs discussed above is characteristic of a subject which is still open and growing. What would atomic physics have seemed like at the turn of the century?

---

† If relativistic effects are to be considered, the Dirac equation and the full apparatus of quantum electrodynamics must be applied.

CHAPTER 1

The Nuclear Periodic Table

1.1. The periodic table, radioactivity, and unstable nuclear states

The nuclear periodic table is intimately related to the periodic table of the elements, but contains an additional degree of freedom. A nucleus belonging to a chemical element X is specified by three numbers (one of which is redundant). The first is the mass number A which is the number of nucleons in the nucleus. The second is the charge on the nucleus Z, measured in units of e, and is the atomic number, equal to the number of protons in the nucleus. The third is the number of neutrons in the nucleus, N, and $A = Z+N$. The specification of a nucleus is thus

$$^{A}_{Z}X_N$$

where X is the symbol for the chemical element, e.g.

$$^{208}_{82}Pb_{126}.$$

The charge on the nucleus determines completely the electronic structure and hence the chemical properties of the corresponding element. Thus not only the neutron number N but also the atomic number Z are redundant and a nucleus is completely specified by

$$^{A}X, \quad \text{e.g.} \quad ^{208}Pb.$$

A chemical element is completely specified by the atomic number, but in the nucleus a given atomic number Z does not specify the neutron number. While the number of protons and neutrons in the nucleus is very approximately equal, a given element may have a number of *isotopes* of different mass number, which may be either stable or unstable against radioactive decay. It is in this sense that there is an extra degree of freedom in the nuclear periodic table, although this extra degree of freedom is not saturated in nature: we shall see that for a given mass number a nucleus becomes unstable if the ratio of protons to neutrons is unbalanced too far in either direction.

We may take the event marking the birth of nuclear physics to have been the discovery of radioactivity. In 1896 Becquerel found that photographic plates, properly isolated from light, were blackened when brought close to uranium salts: the inference being that these salts were emitting radiations capable of penetrating black paper and being absorbed by the

Nuclear Physics

photographic emulsion. In the next few years the Curies isolated the much stronger activities of polonium and radium by chemical and physical analysis of the uranium-bearing mineral pitchblende and Rutherford found evidence for two kinds of radiation, called by him α- and β-radiation. Further experiments, principally those of Rutherford and his colleagues, showed in the years following 1899 that the radiations have three components, α-, β- and γ-radiation and elucidated their properties. Deflection of the charged α- and β-rays in both electric and magnetic fields allowed the determination of the charge to mass ratios of the α- and β-particles, identifying the β-particle with the electron and making plausible the identification of the α-particle with the doubly ionized helium atom, that is, the nucleus ^{4}He. Direct measurement of the charge of the α-particle supported this identification (Rutherford and Geiger, 1908) and it was confirmed by allowing α-particles to penetrate a thin window into a discharge tube and subsequently observing the spectrum of helium (Rutherford and Royds, 1909). The electromagnetic nature of γ-radiation was established through studies of γ-ray diffraction by a crystal lattice (Rutherford and Andrade, 1914).

The relation of nuclear properties to the chemical periodic table is intimately connected with these early studies of radioactivity. In the early 1900s it was already apparent that emission of both α- and β-radiation changes the chemical properties of the emitting atom. The α-particle scattering experiments of Geiger and Marsden (1909, 1913) provided the evidence on which the nuclear model of the atom was founded, and the identification of the chemical atomic number with the nuclear charge Z was achieved by the X-ray studies of Moseley (1913) in conjunction with the Bohr model of the atom. Into this framework fitted the Displacement Law of radioactive decay, which states that in α-decay an atom is moved downward by two places in the chemical periodic table, in β-decay an atom is moved upwards by one place. This result was obtained (Soddy, 1911) from the chemistry of the sources of specified activities and the same work also showed that while in general different activities are associated with different chemical properties, in some cases different activities are associated with the same chemical properties, thus providing the first evidence for the existence of isotopes.

In 1912 two isotopes of neon, ^{20}Ne and ^{22}Ne, were separated by electric and magnetic deflection (J. J. Thomson) and the subsequent work of Aston firmly established the existence of stable isotopes throughout the chemical periodic table. The complete understanding of the extra degree of freedom in the nuclear periodic table was, however, only finally reached with the discovery of the neutron (Chadwick, 1932).

Let us now consider the properties of the particles emitted in radioactive decay. The energies of α's, β's and γ's are all $\sim$ MeV. The spectra of α- and γ- emission are discreet, while the spectra of β-rays are continuous, up to a fixed maximum energy for a particular decay process. The spectra are discreet in the sense that the energy of the emitted particle has one (or more) unique value and the width of the line is less than the experimental resolution (at least in the few MeV region). All three decay processes correspond to a quantum mechanical transition between two states of the nucleus of well-defined energy. The α- and γ-spectra are discreet because the final state is a two-particle state: the initial nucleus decays to the product nucleus and an α or γ. In the centre of mass of the decaying nucleus—which is the laboratory frame apart from thermal motion—conservation of momentum requires a unique energy for the α or γ. β-decay leads to a three-particle final state, the third particle being the neutrino,

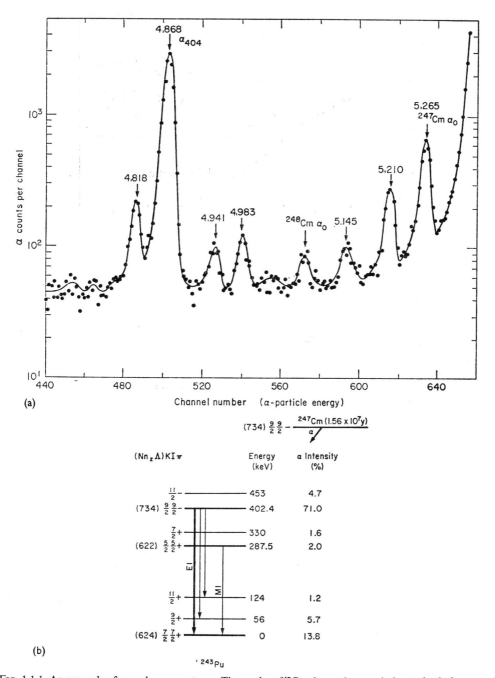

FIG. 1.1.1. An example of an α-decay spectrum. The nucleus ^{247}Cm decays by α-emission to both the ground and excited states of ^{243}Pu, giving rise to a complicated line spectrum, a portion of which is shown in (a). The 5.265 MeV line corresponds to decay to the ground state of ^{243}Pu; the most intense line at 4.868 MeV to decay to the excited state at 402.5 keV. (A line from the decay of ^{248}Cm to the ground state of ^{244}Pu is also present.) The decay scheme of ^{247}Cm is shown in (b). The data were taken with an Au–Si surface barrier detector: the line width reflects the experimental resolution. [From P. R. Fields *et al.*, *Nucl. Phys.* A **160**, 460 (1971).]

Nuclear Physics

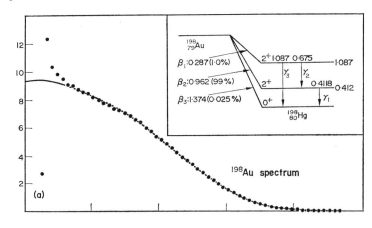

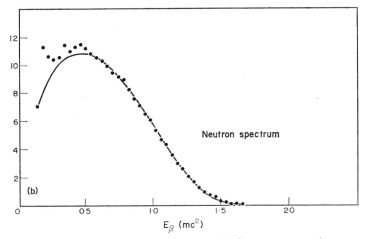

FIG. 1.1.2. The β-decay spectra of ^{198}Au and of the neutron. The β-spectra are continuous and in the case of ^{198}Au there are small contributions from two transitions other than the principal transition to the 2$^+$ excited state of ^{198}Hg at 412 keV. The data were taken with a magnetic spectrometer specially designed for the study of neutron β-decay. The solid lines were generated by folding theoretical spectra with the spectrometer resolution function. [From D. J. Christensen *et al.*, *Free Neutron Beta-decay Half-life*, Risö Report No. 226 (1971).]

the electron does not have a unique momentum and so does not have a unique energy. (The maximum energy of the electron measures the energy difference between initial and final nuclear states.)

Thus the qualitative study of the decay spectra has told us that nuclei exist in unstable states, which are semidiscreet and which transfer to a lower state by emission of radiation. While an atom transfers to a lower energy state by emission of a photon, of characteristic energy corresponding to optical photons ($\sim$ 10 eV) up to X-rays ($\sim$ keV) nuclei transfer to a lower state by emission of α-, β- and γ-radiation, with energies characteristically $\sim$ MeV.

The basic law of radioactive decay was formulated by Rutherford and Soddy (1903) in connection with their transformation hypothesis. The activity of a particular species of

The Nuclear Periodic Table

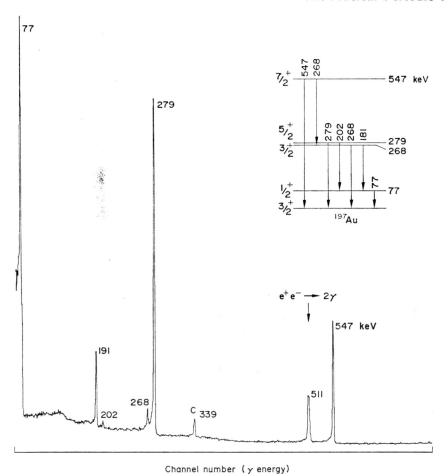

Fig. 1.1.3. An example of a γ-decay spectrum. The spectrum represents the decay of various excited states of ^{197}Au which were populated by coulomb excitation with a 11.8-MeV α-particle beam. The data were taken with a lithium-drifted germanium detector of resolution $\sim$ 2 keV. The decay scheme is shown as an insert and the γ energies given in keV. The line at 511 keV is due to positron annihilation following β^+ decay of proton rich nuclei also formed in the bombardment, and the line at 339 keV is due to an unidentified contaminant. [C. W. Cottrell, Nuclear Physics Laboratory, Oxford University.]

radioactive atom is specified by a decay rate which falls off exponentially in time with a characteristic lifetime.

$$\frac{dN}{dt} = \frac{dN}{dt}\bigg|_{t=0} e^{-t/\tau}$$

whence

$$N(t) = N_0 e^{-t/\tau} \qquad (1.1.1)$$

and

$$\frac{dN}{dt} = -\frac{N}{\tau}.$$

Nuclear Physics

The quantity τ is the mean lifetime of the nucleus

$$\frac{\int_0^\infty tN(t)\,dt}{\int_0^\infty N(t)\,dt} = \tau.$$

The decay rate $\lambda = 1/\tau$ and the half-life $T_{1/2}$ is defined by

$$N(T_{1/2}) = \tfrac{1}{2}N_0 = N_0 e^{-T_{1/2}/\tau}$$

$$T_{1/2} = \tau \ln 2.$$

The number of nuclei decaying is proportional to the number available for decay and any nucleus of a given species has the same probability of decaying as any other.

A sequence of radioactive decays is described by a set of coupled first order differential equations, the solution of which is straightforward. We shall not discuss the phenomenological treatment of radioactivity further except to introduce the branching ratio between two alternative decay modes. An example in the *Thorium series* of radioactive elements is the decay

$$^{212}_{83}\text{Bi} \xrightarrow{\beta} {}^{212}_{84}\text{Po}$$
$$\xrightarrow{\alpha} {}^{208}_{81}\text{Tl}$$

$$\frac{dN_\alpha}{dt} = -\lambda_\alpha N \qquad \frac{dN_\beta}{dt} = -\lambda_\beta N$$

λ_α measures the probability of α decay, λ_β of β-decay. Then

$$\frac{dN}{dt} = \frac{dN_\alpha}{dt} + \frac{dN_\beta}{dt} = -(\lambda_\alpha + \lambda_\beta)N = -\lambda N.$$

Therefore

$$\frac{dN_\alpha}{dt} = -\lambda_\alpha N_0 e^{-\lambda t} \qquad \frac{dN_\beta}{dt} = -\lambda_\beta N_0 e^{-\lambda t}.$$

Both decay modes exhibit the same half-life (1 hour for $^{212}_{83}\text{Bi}$), which is a characteristic of the decaying nucleus rather than of the decay modes. In this case the decay rates are less directly related to the lifetime. The branching ratio for, say, α decay is λ_α/λ (0.36 for $^{212}_{83}\text{Bi}$).

Radioactivity provided the first evidence that the energy levels of the nucleus are discreet and showed that such excited quantum mechanical levels decay exponentially in time. However, a characteristic of the wave function Ψ describing a state of unique energy is that $|\Psi|^2$ is time independent. An excited state which is capable of decay must therefore be described rather as a superposition of wave functions, each of which has almost identical spatial structure but different energy—a wave packet in time. The data on radioactive decay show that each nucleus in a sample has the same probability of decay in a given time interval. The probability that a specified nucleus has not decayed is thus $e^{-t/\tau}$ and since in quantum mechanics we identify the probability function of a state as the square of the wave function, we must write for such a state $|\Psi|^2 = e^{-t/\tau}$. Any state of definite energy E oscillates for ever with frequency $\omega = E/\hbar$ and so a decaying state cannot have a definite energy but must be represented as a superposition of states with the appropriate amplitudes and phases.

$$\Psi = \int a(\omega) e^{-i\omega t} d\omega.$$

As $\tau \to \infty$ the state Ψ approximates better and better to a state of definite energy E_0, and so we write

$$\Psi \simeq e^{-i\omega_0 t} e^{-t/2\tau} = \int a(\omega) e^{-i\omega t} d\omega$$

where we have suppressed the spatial parts of the wave functions. The coefficients $a(\omega)$ give the energy distribution in a decaying state and may be extracted as follows. Multiply by $e^{i\Omega t}$ and integrate over all time. The time independent component states are defined for all time and so for the term involving $a(\omega)$ the limits of integration are $t = \pm \infty$. The decaying state is by definition started off at $t = 0$ and so

$$\int_0^\infty e^{i(\Omega-\omega_0)t} e^{-t/2\tau} dt = \int_{-\infty}^\infty \int a(\omega) e^{i(\Omega-\omega)t} d\omega\, dt.$$

The term $e^{i(\Omega-\omega)t}$ oscillates indefinitely for $\Omega \ne \omega$ and so at once we find

$$a(\Omega) \sim \int_0^\infty e^{i(\Omega-\omega_0)t} e^{-t/2\tau} dt$$

we may write

$$\int_{-\infty}^\infty \int a(\omega) e^{i(\Omega-\omega)t} d\omega\, dt$$

$$= \lim_{T \to \infty} \int_{-T}^T \int a(\omega) e^{i(\Omega-\omega)t} d\omega\, dt$$

$$= \lim_{T \to \infty} 2 \int a(\omega) \frac{\sin(\Omega-\omega)T}{\Omega-\omega} d\omega.$$

On setting $(\Omega-\omega)T = y$ this becomes

$$\lim_{T \to \infty} 2 \int a(\Omega - y/T) \frac{\sin y}{y} dy$$

$$= a(\Omega) 2 \int_{-\infty}^\infty \frac{\sin y}{y} dy = 2\pi a(\Omega).$$

Thus

$$a(\Omega) = \frac{1}{2\pi} \int_0^\infty e^{i(\Omega-\omega_0)t} e^{-t/2\tau} dt$$

Nuclear Physics

and is the Fourier transform of Ψ with respect to Ω.

$$a(\Omega) = \frac{1}{2\pi} \frac{1}{i(\Omega-\omega_0) - \frac{1}{2\tau}}.$$

Setting $\hbar\omega = E$, $\hbar\omega_0 = E_0$ and $1/\tau = \Gamma/\hbar$ gives

$$a(E) = \frac{1}{2\pi i} \frac{\hbar}{E-E_0+i\Gamma/2}$$

and

$$|a(E)|^2 \sim \frac{1}{(E-E_0)^2+\Gamma^2/4}. \quad (1.1.2)$$

The relation $\Gamma\tau = \hbar$ is a precise statement of the Heisenberg Uncertainty Principle for energy, $\Delta E \Delta t \gtrsim \hbar$. For example, the shortest lifetime in the Thorium series of radioactive elements occurs for the α-decay of ^{212}Po, which has a half-life of 3×10^{-7} sec. The natural width of the corresponding line in the α-spectrum is thus 1.4×10^{-7} eV, to be compared with an α-energy of 8.9 MeV. The direct measurement of line widths is not in most cases feasible for lifetimes $\gtrsim 10^{-15}$ sec.

Thus a state which decays does not correspond to a discrete energy, but to a continuous distribution of energies with the relative intensities given by the formula (1.1.2), and the line is narrow if the lifetime is long. This expression is the famous Breit–Wigner formula for a single unstable state. While only an approximation, it applies to all narrow unstable quantum mechanical systems—to excited atoms as much as to excited nuclei. A rather different derivation may be found in Section 3.2.

It is interesting to note that very similar behaviour is exhibited by classical oscillators: consider an L–C–R series circuit. The response to forced oscillations of frequency ω is given by

$$I = \frac{V_0 e^{i\omega t}}{R+i\omega L + 1/i\omega C}; \quad |I|^2 = \frac{V_0^2}{R^2+(\omega L - 1/\omega C)^2}$$

$$= \frac{\omega^2 V_0^2/L}{\omega^2 R^2/L^2 + (\omega^2 - \omega_0^2)^2} \quad \text{where} \quad \omega_0^2 = \frac{1}{LC}. \quad (1.1.3)$$

Free oscillations (if the circuit is less than critically damped) satisfy

$$I = I_0 e^{-(R/2L)t} e^{i\{\omega_0^2 - R^2/4L^2\}^{\frac{1}{2}} t}$$

$$\sim I_0 e^{-(R/2L)t} e^{i\omega_0 t}$$

for small R, narrow width and hence high Q. The frequency breakdown is given by (1.1.3) and is essentially the same as that contained in (1.1.2). A version of the Breit–Wigner formula exactly analogous to our formula for $|I|^2$ is used a great deal in high-energy physics, but for narrow resonances it reduces to the form we are using, which is used most often in nuclear physics.

The Nuclear Periodic Table

1.2. Nuclear mass

It should be clear at this point that measurement of the decay spectra of nuclei provides one method of determining nuclear mass differences and hence nuclear mass, provided that the mass of one nucleus in a decay sequence is known. This method would be barren indeed if it were confined to the study of the naturally occurring radioactive nuclei, but with the discovery that the products formed in bombarding stable nuclei with beams of protons (or other light nuclei) at energies of several MeV are unstable and exhibit both radiative decay and β-decay (both positive and negative electrons are emitted, by appropriate products) the study of β-decay spectra became a powerful tool for the investigation of nuclear masses away from the *stability curve* of Z versus N, and the study of γ-spectra for the investigation of the excited states of a given nucleus: since the study of α-decay gives information only about the energy states of massive highly charged nuclei the most information is obtained from β- and γ-ray spectroscopy. It is worth digressing here to consider the principles and some of the difficulties of these studies.

The energies of charged particles in the 1 MeV (kinetic) energy region may be obtained in two ways. The first is by totally absorbing that energy in a detector which yields an optical or electrical output proportional to that energy. Such detectors are proportional gas counters, scintillation counters (plastic, liquid or NaI crystals) and solid-state counters such as lithium drifted germanium crystals. The latter have been recently developed and have a resolution of ~ 1 keV, much better than other energy detectors. The energy is dumped in the detector by ionization which incidentally is the way a charged particle makes a grain of silver bromide in a photographic emulsion developable.

The second method is magnetic spectroscopy, capable of much better resolution ($\lesssim 1\%$) than the previous group of techniques except for the solid-state detector. The radius of curvature of a charged particle in a magnetic field is given by

$$\frac{Bev}{c} = \frac{mv^2}{r} \quad \text{or} \quad p = \frac{Ber}{c}$$

in gaussian units, whence the energy, given the mass of the particle.

Total absorption detectors can be placed very close to a source. The main problem is to reduce the proportion of particles which escape without losing all their energy, thus smearing out the signal towards the low-energy end of the spectrum. Magnetic spectrometers do not have this problem, but in general accept only particles in a small solid angle and so suffer more from lack of intensity. Design of magnetic spectrometers consists largely of attempting to balance the requirements of a good acceptance against the requirement that the magnetic optics form as sharp a focus at the detector as possible in order to give good energy resolution.

γ-ray energies are nowadays beautifully measured by absorption of the energy in solid-state counters which have almost superseded absorption in NaI crystals. There are two other methods: Bragg reflection of low-energy γ-rays from either a plane or a curved crystal, and magnetic deflection of electron pairs produced when γ-rays (of energy >1.02 MeV) are passed through a thin converter.

Nuclear Physics

Both studies are affected by source problems: the source must be sufficiently intense to give a useful counting rate, but must be so thin that the electrons do not lose appreciable energy by ionization in passing through the target, nor γ-rays by the photoelectric or compton effects.

We shall not discuss these experimental techniques further in this book: for details we refer you to K. Siegbahn, *α-, β- and γ-Ray Spectroscopy*, Vol. 1 (North Holland, 1965).

The masses of stable nuclei have been determined primarily by the techniques of mass spectrometry. The force exerted on a charged particle of charge q esu is $\mathbf{F} = q(\mathbf{E} + \mathbf{v}/c \times \mathbf{B})$. The angular deflection of a charged particle in passing through an electric field depends on the charge and the velocity of the particle. The angular deflection of a particle in a magnetic field depends on the charge and the momentum. Thus, if a beam of particles of the same charge but having different masses and velocities is passed through a combination of electric and magnetic fields (either crossed fields or one following the other) a separation depending only on mass is obtained. The earliest example of this technique was the discovery of ^{20}Ne and ^{22}Ne among the "positive rays" of a discharge tube by J. J. Thompson. High-resolution mass spectrometry was initiated by Aston, some 50 years ago. Consider the principle of optical spectroscopy. Any optical spectrometer consists of a source, a focusing system, a dispersive element (diffraction grating, Fabry–Perot etalon, etc.) and a detector (photographic plate, eye, photomultiplier). Only one dispersive element is necessary because there is only one variable—the wavelength of the light. It is possible also to combine the dispersive element with part of the focusing system—for example, by using a curved reflection grating as the dispersive element.

There is an almost exact analogy in the principles of mass spectrometry. Here, however, there are two independent variables—the energy, which may be continuously distributed, and the mass, which is not continuously distributed but where it may be desired to separate two atoms very close in mass. The charge is a third variable, but is discreet and comes in units of e and is not a problem. Therefore we require two independent dispersive elements rather than one. (In β-spectroscopy the mass is known and it is desired to find the energy—one element, a magnetic deflection, suffices.)

The principles are simple. The problem once again is to obtain high resolution compatible with an adequate source strength (a single run had better not take a year!), and this is a function of the focusing properties of the electric and magnetic fields applied. It is usual for the dispersive elements to do the focusing, just as a curved diffraction grating does in optical spectroscopy.

Single focusing instruments bring to a focus particles of given mass and velocity emitted into a finite solid angle from the source. Their resolution depends therefore on the velocity selection properties of the spectrograph.

Double focusing instruments bring to a focus particles of given mass emitted into a finite solid angle and with a spread in velocity. Direction focusing focuses particles of given velocity with a spread in direction, and velocity focusing brings to a focus particles with a given direction and a spread in velocity. (The latter trick may be accomplished by balancing a magnetic dispersion against an electric dispersion.)

The Nuclear Periodic Table

Two kinds of electric filter have been used:

(i) Accelerate ions from the source to provide a beam of well-defined velocity. (The source may be a hot oven or gas discharge, for example.)

A potential difference of V volts is applied between the source and a grid. An ion with charge Ne, mass M, gains energy $\frac{1}{2}Mv^2 = NeV$ (NV electron volts).

We may apply a few kV to the grid, so that the final energy is $\sim$keV. Room-temperature velocities correspond to an energy $\sim 1/40$ eV so we may expect a spread in energy of $\lesssim$ eV. Then $\Delta E/E \sim 10^{-3}$–10^{-4} after the acceleration is complete.

After passing through this filter the ions have a velocity depending on the square root of the mass and subsequent magnetic deflection separates different masses in space, for example by using 180° deflection in a uniform field. The electric accelerator need have no special focusing properties, for a bend of 180° in a uniform field provides direction focusing. The velocity in the field is

$$v = \sqrt{2NeV/M}$$

whence

$$r = \frac{c}{B}\sqrt{\frac{2MV}{Ne}}.$$

(Note: e, V are in esu, B is in emu.)

$$\frac{\Delta r}{r} = \frac{1}{2}\frac{\Delta M}{M},$$

$M/\Delta M$ is the resolution of the instrument and may be $\sim 10^4$.

(ii) Electric fields may be used in another mode. Suppose we have a cylindrical condenser. The field is always at right angles to the path of an equilibrium orbit and we have $F = NeE = mv^2/r$ for such an orbit. Both sector magnetic fields and cylindrical electric fields have focusing properties, but both give rise to dispersion in velocity. For high-resolution mass spectrometry an electric and a magnetic element are combined so that both velocity and position focusing may be achieved. The principle problem in mass spectrometry is thus the design and production of electric and magnetic filters with the appropriate focusing properties, so that the analogues of spherical and chromatic aberration are removed.

The standard of atomic mass is now taken to be the ^{12}C atom, the mass of which is defined to be 12 atomic mass units. The relative masses of (ionized) atoms may be determined to $\sim 1/10^6$ by measuring the separation at the detector of two ions having the same charge and mass number, but different masses, for example: $(^{12}C\ ^1H_4)^+ - (^{16}O)^+$. The difference in mass may be determined to $\lesssim 1/10^3$ which yields a relative mass accurate to $\lesssim 1/10^6$, for nuclear binding energies are ~ 8 MeV/nucleon. Thus the mass difference between ^{16}O and ^{12}C ^{1}H$_4$ is ~ 30 MeV/c^2 in $\sim 16{,}000$ MeV/c^2, since the nucleon mass is ~ 939 MeV/c^2.

Mass spectrometry is nowadays a tool of considerable importance in the analysis of very low concentrations of atoms and may be used in the bulk separation of isotopes.[†] The principles are still of interest to the high-energy physicist, however. The beams of particles produced from proton synchrotrons or other accelerators must be transmitted from

---

[†] See, for example, H. Hinterberger, High sensitivity mass spectrometry in nuclear studies, *Ann. Rev. Nucl. Sci.* **12**, 435 (1962); *Modern Aspects of Mass Spectrometry*, Ed. R. I. Reed (Plenum Press, 1968).

Nuclear Physics

the target in which they are produced to the detecting equipment, and for many experiments, particularly those involving bubble chambers, it is necessary to purify the beam so that only one kind of particle enters the detector—for example, a positively charged beam may be selected by magnetic deflection but will still consist of protons, π-mesons and K-mesons. Quadrupole magnetic fields are used to focus the beam, magnetic deflection and collimation to provide momentum dispersion and subsequent electric deflection to separate particles of the same momentum and different mass.[†] The deflecting magnetic fields are ~ 10 k gauss, the gradients in the quadrupoles ~ 1000 gauss/cm and the potential difference applied across the beam is ~ 300 kV. Such beams have satisfactorily separated K-mesons (mass 495 MeV/c^2) from protons and from π-mesons (mass 139 MeV/c^2) at energies of up to 5 GeV: you should remember that at these energies particles are highly relativistic and all velocities are close to c.

A further method of studying nuclear masses is provided by the kinematics of nuclear reactions. We take this opportunity to distinguish between the *kinematics* and the *dynamics* of a nuclear reaction. The kinematics are those properties which depend only on conservation of energy and momentum: for example, if an elastic scattering process is studied, observation of the direction of the scattered beam determines through the kinematic equations of constraint the direction of the recoil and the energies of both the scattered particle and the recoil. These equations in themselves, however, give no information about the angular distribution of the scattered particles, which one must attempt to obtain from a dynamical theory of nuclear reactions. Consider a process in which a projectile nucleus of mass M_a reacts with a target nucleus of mass M_b to produce two nuclei in the final state of masses M_c and M_d.

Conservation of energy requires:

$$E_a + E_b = E_c + E_d$$

where E_a is the total energy of nucleus a, etc.

Now

$$E_a = T_a + M_a c^2 \quad E_b = T_b + M_b c^2,$$
$$E_c = T_c + M_c c^2 \quad E_d = T_d + M_d c^2$$

where T_a is the kinetic energy of nucleus a, etc., and T_b is usually zero.

Then

$$\begin{aligned} T_c + T_d &= T_a + M_a c^2 + M_b c^2 - M_c c^2 - M_d c^2 \\ &= T_a + Q \end{aligned} \quad (1.2.1)$$

where the energy released in the reaction is known as the Q value of the reaction. (The Q value is not an exclusive property of nuclear reactions—an exploding shell has a Q value which could be defined in exactly the same way.)

Q is zero for an elastic scattering;

Q is positive for an exothermic reaction; an example is ^{235}U$+n \rightarrow$ two fission fragments;

Q is negative for an endothermic reaction; an example is inelastic scattering, such as
$$p + {}^{14}\text{N} \rightarrow {}^{14}\text{N}^* + p,$$

[†] See, for example, O. Chamberlain, Optics of high energy beams, *Ann. Rev. Nucl. Sci.* **10**, 161 (1960).

where $^{14}N^*$ represents an excited state of ^{14}N and may subsequently decay electromagnetically $^{14}N^* \to {}^{14}N + \gamma$ or, if the excitation of $^{14}N^*$ is high enough, by particle emission. It is not, of course, necessary to measure both kinetic energies T_c and T_d. If the masses of particles a, b and c are known, measurement of T_c at a fixed angle to the incident beam is sufficient to determine the Q value and M_d by application of the four equations of conservation of momentum and energy.

In nuclear physics experiments T_a is usually $\sim$ MeV. In this energy region the most convenient accelerator is an electrostatic generator (fluxes $\sim 10^{15}$ protons/sec, currents ~ 0.1 mamp), and the voltage applied across the beam is stable to $\sim$ keV. It is necessary to use thin foil targets or, where this is not possible, the target material must be evaporated in a thin layer on a thin backing of relatively inert material. A 10 MeV proton loses all its energy through ionization[†] in traversing a thickness of copper that presents a mass of 0.2 g/cm² to the beam—that is, that thickness for which a cylinder of unit cross-section has mass 0.2 g (~ 0.2 mm). Target thickness is measured in units of g/cm² and it is clear that for good energy resolution we require targets in the mg/cm² region, foils of thickness $\sim$ microns. Some Q values have been measured to an accuracy of $\sim 0.1\%$.

The principles of mass determination through the study of kinematics are thus straightforward. An accuracy $\sim 10^{-5}$ can be achieved relative to ^{12}C and the technique is useful in checking the results of mass spectrometry on light atoms, in checking results from decay spectra of low-lying excited states, and is almost the only technique for studying the masses of states which are so excited that they decay by particle emission with lifetimes $\lesssim 10^{-20}$ sec.[‡]

1.3. Nuclear binding energy

Let us now consider the results of these studies, and begin by defining some terms. Consider Z protons and $A - Z$ neutrons, each separated from all others by such a distance that their interaction energy is negligible. The total energy of this system is the mass energy and the mass is

$$ZM_p + (A-Z)M_n.$$

If a stable nucleus is now assembled out of these nucleons the total energy will be less than the initial energy by an amount B and the mass of the nucleus is given by

$$M(Z, A) = ZM_p + (A-Z)M_n - B(Z, A)/c^2.$$

$B(Z, A)$ is the *binding energy* of a nucleus with mass (or nucleon) number A and atomic number Z. $B(Z, A)/c^2$ is known as the *mass defect* of the nucleus. $\bar{B} = B(Z, A)/A$ is the *mean binding energy*, or the binding energy per nucleon. Since nuclei are seldom studied stripped of all electrons, usually the *atomic mass* is specified:

$$M(Z, A) = ZM_H + (A-Z)M_n - B(Z, A)/c^2$$

[†] For detailed discussions of the energy loss of charged particles moving through matter see, for example, E. Fermi, *Nuclear Physics*, chap. II (Chicago, 1950); R. D. Evans, *The Atomic Nucleus*, chap. 18 (McGraw-Hill, 1955); E. Segre, *Nuclei and Particles*, chap. II (Benjamin, 1964); B. Rossi, *High Energy Particles*, chap. II (Prentice-Hall, 1952).
[‡] See Section 4.2.

Nuclear Physics

where M_H is the mass of the hydrogen atom. The mass is NOT uniquely specified by the mass number A.

The binding energy of the electrons is ignored to first approximation. The Thomas–Fermi model of the atom† gives the binding energy B_e of electrons in an atom of atomic number Z as $B_e(Z) = 15.73\, Z^{7/3}$ electron volts. For uranium $Z = 92$ and $B_e(92) = 6$ MeV; about 12 electron masses in 92. The mean binding energy of a nucleus in the uranium region is ~ 7 MeV/nucleon and there are ~ 240 nucleons in uranium. Thus, neglect of atomic binding energy yields an error of the order of only a few parts in a thousand even in the heaviest atoms available for study. (The absolute mass is, however, often of less significance than the relative masses of neighbouring nuclei. In a given region of the periodic table

$$\frac{dB_e(Z)}{dZ} = \frac{7B_e(Z)}{3Z}$$

$$\simeq \frac{1}{40} B_e(Z) \simeq 150 \text{ keV for uranium.}$$

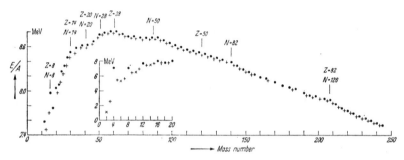

FIG. 1.3.1. Binding energy per nucleon E/A of the most β-stable isobars as a function of mass number. Points refer to even–even nuclides, crosses to such of odd mass; each symbol represents an average value for a few neighbouring nuclides of the type indicated. The positions of magic numbers are indicated. Insert: detailed picture for the lowest masses; in this part the four known stable odd-odd nuclides are indicated by oblique crosses. [From A. H. Wapstra, Atomic masses of nuclides, *Handbuch der Physik*, XXXVIII/1 (Springer-Verlag, 1958).]

Such an energy shift is easily measureable and will contribute to the Q value of nuclear reactions with heavy atoms and to the energetics of β-decay. This is another reason for using atomic masses rather than bare nuclear masses: the latter have more theoretical significance but it is the former that are directly measured.)

$$\bar{B}(Z, A) = \left\{ \frac{ZM_H + (A-Z)M_n - M(Z, A)}{A} \right\} c^2.$$

The mean binding energy for stable nuclei is shown in Fig. 1.3.1 as a function of A.

The greater the mean binding energy, the more stable a nucleus is. The greatest values of $\bar{B}$ are found for mass numbers between 50 and 60, a fact which is reflected in the astronomical abundance of the elements.‡ It is clear from the curve that the assembly of up to fifty

---

† See, for example, L. D. Landau and E. M. Lifshitz, *Quantum Mechanics*, 2nd ed., Section 70 (Pergamon Press, 1965).
‡ See H. E. Suess and H. C. Urey, *Rev. Mod. Phys.* **28**, 53 (1956).

The Nuclear Periodic Table

nucleons to form a stable nucleus will release energy if suitable mechanisms exist. Such reactions provide the stellar energy sources. Energy is also released if heavy nuclei in the region of $A \sim 250$ can be persuaded to undergo fission, while to assemble nuclei in the region of iron into more massive nuclei requires a net input of energy, such as gravitational energy in the collapse of a supernova.[†]

The principal features of the binding energy curve for stable nuclei are thus:

1. A rapid rise between $A = 1$ and $A \sim 30$ with a superimposed oscillation (representing extra stability at ^{4}He, ^{8}Be, ^{12}C, ^{16}O, ^{20}Ne, ^{24}Mg).
2. $\bar{B} \sim 8$ MeV/nucleon between $A \sim 30$ and $A \sim 200$.
3. A systematic decrease in $\bar{B}$ with increasing A, becoming severe for $A \sim 200$.

We shall defer discussion of the interpretation of nuclear mass studies until we have reviewed the next topic, nuclear size.

1.4. Nuclear size

The nucleus has dimensions $\sim 10^{-13}$–10^{-12} cm to be compared with atomic dimensions $\sim 10^{-8}$ cm. The structure of the nucleus may be studied through specifically nuclear reactions, or probed with beams of electrons and muons which are sensitive only to the charge and magnetic moment structure of the nucleus. The charge structure is of considerable importance in atomic physics, for the calculation of electronic structure in an atom proceeds, to a first approximation, under the assumption that the electrostatic field of the nucleus is that due to a point charge. The distribution of nuclear charge throughout a finite volume of space modifies this simple picture and affects the energy levels of the atom.

An upper limit on the dimensions of the distribution of positive charge in an atom was obtained from the observations of Geiger and Marsden, who studied the angular distribution of α-particles scattered by thin foils of gold and platinum. The angular distribution was found to be consistent with coulomb scattering by a point charge of magnitude Ze.[‡] The analysis employed classical mechanics: it is easy to show that the scattering angle θ of a particle with *impact parameter* b is given by[§]

$$b = \frac{Zze^2}{2E} \cot \frac{\theta}{2}.$$

If b is randomly distributed then the number of particles with impact parameter $<b$ is πb^2 (for unit flux) and the *scattering cross-section*

$$\sigma(>\theta) = \pi b^2 = \pi \left(\frac{Zze^2}{2E}\right)^2 \cot^2 \frac{\theta}{2}$$

[†] See Section 5.12.

[‡] Rutherford (1911), see *Foundations of Nuclear Physics* (Ed. Robert T. Beyer, Dover, 1949) or *The Collected Papers of Lord Rutherford* (George Allen & Unwin, 1962).

[§] This is shown in almost any elementary book on Atomic Physics.

Nuclear Physics

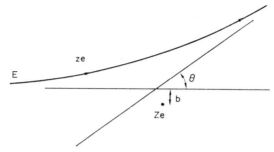

FIG. 1 4.1.

while the *differential cross-section*

$$\frac{d\sigma}{d\Omega} = \frac{d\sigma}{2\pi \sin\theta\, d\theta} = \left(\frac{Zze^2}{4E}\right)^2 \operatorname{cosec}^4 \frac{\theta}{2}.$$

We will not take up space by reproducing the whole argument here.

The concept of impact parameter is classical and has no analogy in wave mechanics. If we make an approximation and regard the incident α-particle as having a well-defined momentum **p**, then the wave function describing it is given by

$$\psi(\mathbf{r}) = e^{i\mathbf{p}\cdot\mathbf{r}/\hbar}$$

and $|\psi(\mathbf{r})|^2 = 1$. Thus the particle has the same probability of being *anywhere* in space and it is only possible to specify an impact parameter at the cost of complete uncertainty in momentum. It is a unique property of an inverse square field that a complete wave mechanical treatment yields the same result for the differential scattering cross-section as classical mechanics, but the mathematics of an exact treatment is too unfamiliar to be reproduced here.[†] An approximate derivation using quantum mechanics is given in Chapter 4.

We will indulge ourselves in one simple classical calculation. It is obvious that for an impact parameter of $b = 0$, $\theta = 180°$. The closest approach to the nucleus in an inverse square field will be given by

$$\frac{1}{2} M_\alpha v_0^2 = E = \frac{Zze^2}{r_{\min}},$$

$$r_{\min} = \frac{Zze^2}{E}.$$

Thus, if scattering does not depart from that expected for an inverse square field even at angles ~180° an upper limit to the nuclear radius is given by Zze^2/E—which for α-particles of 10 MeV on, for example lead, yields

$$r_{\text{nucleus}} \lesssim 2\times 10^{-12} \text{ cm}.$$

[†] If you require convincing of this, consult either N. F. Mott and H. S. W. Massey, *Theory of Atomic Collisions*, 3rd ed., chap. 3 (Oxford, 1965) or L. I. Schiff, *Quantum Mechanics*, 3rd ed., Sect. 21 (McGraw-Hill, 1968).

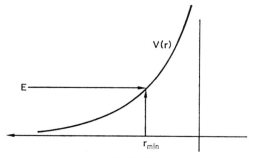

FIG. 1.4.2.

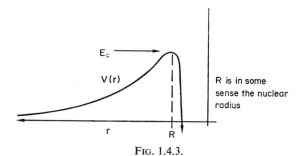

FIG. 1.4.3.

If the energy of the α-particle is sufficiently high (or if Z is sufficiently low) that the nucleus is penetrated, then coulomb scattering will be reduced—there is no electric field inside a spherically symmetric shell of charge due to that shell of charge. However, under these circumstances the α-particle is exposed to the short-range nuclear forces and the two scattering amplitudes may interfere. Thus, for α-particles (or any charged *strongly interacting* particle) we cannot conclude that large angle scattering will be suppressed by nuclear penetration, only that it will depart from the predicted for an inverse square field.

The first indications of *anomalous* α-particle scattering were obtained when Geiger and Marsden's work extended to light nuclei.

Consider the potential as a function of radius for a charged α-particle or proton approaching a nucleus.

The nucleus is only penetrated classically by a particle of energy $> E_c$. In quantum mechanics a particle of *any* energy $E > 0$ has a finite probability of penetrating the *coulomb potential barrier* or, in more appropriate language, the wave function of any particle of kinetic energy > 0 has a finite value inside the potential barrier. This is the phenomenon of tunnelling through a potential barrier—and is applied in reverse in the theory of α-particle decay. Under these circumstances the wave functions inside and outside may match well so that the transmission is large at a particular energy—this is *resonant* scattering—and then the scattering amplitude is given by

$$a_c + a_R + a_N$$

where a_c is the coulomb scattering amplitude, a_R the resonant scattering amplitude, and a_N a non-resonant scattering term. (The physics will be discussed further in Section 4.12.)

Nuclear Physics

The scattering cross-section is proportional to the square of this amplitude which is:

$$|a_c|^2 + |a_R|^2 + |a_N|^2 + 2Rea_c^*a_R$$
$$+ 2Rea_c^*a_N + 2Rea_R^*a_N$$

where the cross terms may be either positive *or* negative. (We should point out here that a full treatment of scattering through an inverse square field shows that it is inadequate to simply add a_c to the nuclear amplitudes—the incoming and outgoing waves are distorted by the coulomb potential which thus affects directly a_N and a_R.)

Because of barrier penetration, interpretation of anomalous α-particle scattering is complicated and it is necessary to compare the predictions of a model of the potential barrier and nuclear scattering with observation. It is clear that in charged particle scattering application of a classical argument will over-estimate the nuclear radius and if the projectile is a nucleus itself, as in α-scattering, a further correction is necessary for the finite charge distribution of the projectile.

The results of all determinations of nuclear radius are summed up in the approximate formula:

$$R = R_0 A^{1/3}.$$

It should be made very clear that attributing a radius R to the nucleus does not imply a sharp edge. R is a convenient parameter and to attach a precise meaning to it, it is necessary to explore the detailed distribution of nucleons in space, as seen by the experimental technique employed.

You should note at this point that a radius varying as $A^{1/3}$ corresponds to a uniform nuclear density—the packing of nucleons is largely independent of their number. The spatial distribution of the nucleus has been beautifully studied by high-energy electron scattering which has made it possible not only to determine a radius parameter but also to map out in detail the distribution of charge in the nucleus. Let us consider the conditions under which this may be done. We approximate the incident and final electron wave functions as plane waves and for simplicity suppose the nucleus to be infinitely heavy so that although momentum is transferred by the action of the nuclear electric field from the incident to the scattered electron, no energy is transferred. (This is always the case for elastic scattering in the centre of mass and by assuming that the nucleus is infinitely heavy all we are doing is identifying the laboratory frame with the centre of mass frame.)

The scattering will only be sensitive to structure if the wavelength of the incident electron is approximately the scale of the structure. Thus, if we want only to find a radius parameter, $\lambda \sim 10^{-12}$ cm,

so $$\lambda = \frac{h}{p} \sim 10^{-12} \quad pc = \frac{hc}{\lambda} = 1.8 \times 10^{-4} \text{ ergs},$$

so $$pc \sim 100 \text{ MeV} \quad p \sim 100 \text{ MeV}/c.$$

If it is desired to study nuclear structure down to $\sim 10^{-13}$ cm then the electron momentum must be ~ 1 GeV/c and at these energies it is possible to probe the electromagnetic structure of the proton and neutron.

We may easily make this discussion rather more quantitative. The scattering will be

The Nuclear Periodic Table

controlled by a matrix element which connects the initial and final states, and we will use first-order perturbation theory and consider the potential acting only once. Then the matrix element is

$$M(\mathbf{p}_f, \mathbf{p}_i) = \int \psi_f^* V(r) \psi_i \, d^3r$$

where

$$\psi_f = e^{i\mathbf{p}_f \cdot \mathbf{r}/\hbar} \quad \psi_i = e^{i\mathbf{p}_i \cdot \mathbf{r}/\hbar}.$$

The potential $V(r)$ acting on the initial plane wave state converts it into a superposition of outgoing scattered waves. If we neglect *magnetic moment scattering* (which we certainly cannot do at very large energies and scattering angles when nucleon structure is being probed) then if $V(r)$ is a pure coulomb potential

$$M(\mathbf{p}_f, \mathbf{p}_i) = \int e^{i\mathbf{q} \cdot \mathbf{r}/\hbar} \frac{Ze^2}{r} r^2 \, dr \, d\cos\theta \, d\phi$$

where

$$\mathbf{q} = \mathbf{p}_i - \mathbf{p}_f,$$

$$q^2 = 2p^2(1 - \cos\theta) = 4p^2 \sin^2 \frac{\theta}{2} \approx p^2 \theta^2 \quad \text{for small } \theta,$$

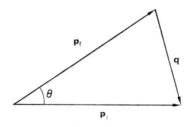

FIG. 1.4.4.

changing variables:

$$M(\mathbf{p}_f, \mathbf{p}_i) = \int e^{iqr \cos\chi/\hbar} \frac{Ze^2}{r} 2\pi r^2 \, dr \, d\cos\chi$$

where χ is the angle between $\mathbf{p}_i$ and $\mathbf{q}$

$$M(\mathbf{p}_f, \mathbf{p}_i) = \int_{-1}^{+1} \int_0^{\infty} r e^{iqr \cos\chi/\hbar} 2\pi Ze^2 \, dr \, d\cos\chi$$

$$= 2\pi Ze^2 \int_0^{\infty} \left[\frac{\hbar}{iq} e^{iqr \cos\chi/\hbar} \right]_{-1}^{1} dr.$$

$$= 2\pi Ze^2 \int_0^{\infty} \frac{\hbar}{iq} \{ e^{iqr/\hbar} - e^{-iqr/\hbar} \} \, dr$$

$$= 4\pi Ze^2 \left(\frac{\hbar}{q} \right)^2 \int_0^{\infty} \sin x \, dx \quad x = \frac{qr}{\hbar}$$

Nuclear Physics

so

$$M(\mathbf{p}_f, \mathbf{p}_i) = 4\pi Z e^2 \left(\frac{\hbar}{q}\right)^2. \tag{1.4.1}$$

Note that this is the three-dimensional Fourier transform of the coulomb field, and may be interpreted as giving the probability of a virtual photon with momentum q, and no energy, acting. A more detailed discussion is given in Section 4.4.

If the source of the coulomb field is distributed, we have instead

$$V(\mathbf{r}) = Ze^2 \int \frac{\varrho(\mathbf{r'})}{|\mathbf{r}-\mathbf{r'}|} d^3r'; \quad \int \varrho(\mathbf{r'}) d^3r' = 1$$

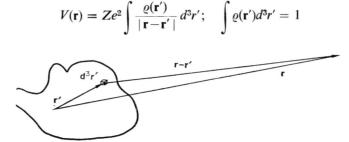

FIG. 1.4.5.

when

$$M(\mathbf{p}_f, \mathbf{p}_i) = Ze^2 \int e^{i\mathbf{q}\cdot\mathbf{r}/\hbar} \int \frac{\varrho(\mathbf{r'})}{|\mathbf{r}-\mathbf{r'}|} d^3r' \, d^3r.$$

If we perform the double integral first over $\mathbf{r}$, holding $\mathbf{r'}$ constant, we may set $\mathbf{r}-\mathbf{r'} = \mathbf{R}$ and then

$$M(\mathbf{p}_f, \mathbf{p}_i) = \iint e^{i\mathbf{q}\cdot\mathbf{R}/\hbar} e^{i\mathbf{q}\cdot\mathbf{r'}/\hbar} Ze^2 \frac{\varrho(\mathbf{r'})}{R} d^3r' d^3R$$

$$= 4\pi Ze^2 \left(\frac{\hbar}{q}\right)^2 \int e^{i\mathbf{q}\cdot\mathbf{r'}/\hbar} \varrho(\mathbf{r'}) d^3r'.$$

The quantity

$$F(q^2) = \int e^{i\mathbf{q}\cdot\mathbf{r'}/\hbar} \varrho(\mathbf{r'}) d^3r' \tag{1.4.2}$$

is the Fourier transform of the source and is called the *electric form factor* of the nucleus. This quantity is precisely analogous to the atomic form factor encountered in X-ray scattering theory.[†]

Since the transition rate between the two states depends on $|M|^2$, the ratio of the scattering measured at a fixed energy and angle (i.e. fixed $\mathbf{q}$) to the scattering predicted for a pure coulomb field yields $F^2(q^2)$. As $\varrho(\mathbf{r'})$ contains no odd electric moments, $F(q^2)$ is a real number and so may be calculated directly from $F^2(q^2)$. An inverse Fourier transformation yields $\varrho(\mathbf{r'})$. (In high-energy physics the form factor itself is regarded as being of more physical significance, since it is a relativistic invariant.)

[†] See, for example, C. Kittel, *Introduction to Solid State Physics*, chap. II (Wiley, 1956).

The Nuclear Periodic Table

Let us consider $F(q^2)$ in a little more detail. If we take $\varrho(\mathbf{r}')$ as spherically symmetric we have

$$F(q^2) = \int e^{iqr'\cos\chi'/\hbar} \varrho(\mathbf{r}') 2\pi r'^2 \, dr' \, d\cos\chi'$$

$$= 4\pi \frac{\hbar}{q} \int_0^\infty \sin\frac{qr'}{\hbar} \varrho(r') r' \, dr'.$$

If $qr'/\hbar$ is small then

$$\sin\left(\frac{qr'}{\hbar}\right) \simeq \frac{qr'}{\hbar} - \left(\frac{qr'}{\hbar}\right)^3 \frac{1}{3!} \cdots$$

and

$$F(q^2) = 1 - \left(\frac{q}{\hbar}\right)^2 \frac{\langle R^2 \rangle}{6}$$

where

$$\langle R^2 \rangle = \int_0^\infty r'^2 \varrho(r') \, d^3r'.$$

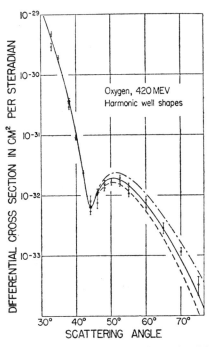

FIG. 1.4.6a. The figure shows the differential scattering cross-section for 420 MeV electrons on oxygen. The theoretical curves are calculated from the harmonic well version of the shell model: the dashed curve without correction, the dotted curve after correction for finite proton size and the full curve contain this correction and a correction for centre of mass effects. [From D. G. Ravenhall, *Rev. Mod. Phys.* **30**, 430 (1958).]

The expansion shows that F is a function of q^2 and demonstrates our assertion that we are sensitive to the nuclear size when $\lambda \lesssim$ nuclear radius—the maximum value of q^2 being $4p^2$.

Nuclear Physics

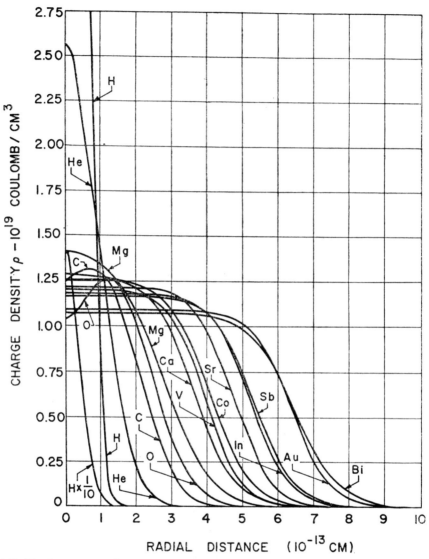

FIG. 1.4.6b. The figure summarizes the charge distributions found for a variety of nuclei by using electron scattering. [From R. Hofstadter, Nuclear and nucleon scattering of high-energy electrons, *Ann. Rev. Nucl. Sci.* **7**, 231 (1957).]

(If you are well acquainted with the special theory of relativity you will not be surprised to learn that if we allow the nucleus to recoil the form factor remains the same but its argument is now

$$\Delta^2 = (\mathbf{p}_i - \mathbf{p}_f)^2 - (E_i - E_f)^2/c^2$$

which is clearly a Lorentz invariant equal to q^2 if q^2 is defined in the centre of mass.)

These studies of high-energy electron scattering only became possible around 1950 with the development of hardware for handling centimetric electromagnetic waves at high-power

The Nuclear Periodic Table

densities. The major program was carried out by Hofstadter and his colleagues at Stanford[†] with an electron linear accelerator which was boosted steadily in energy until around 1960 it reached the GeV region and studies of the electromagnetic structure of nucleons were initiated. Figure 1.4.6b shows typical results for low A and large A nuclei.

We must emphasize that electron scattering is sensitive only to the charge distribution in the nucleus, which may be different from the density of nucleons. For example, ^{40}Ca (a doubly magic nucleus) has a charge radius $\sim 0.3\%$ greater than that of the unstable isotope ^{48}Ca—the additional eight neutrons seem to be added outside the core of ^{40}Ca and effect some compression.[‡] Electron beams of energies ~ 10 GeV are now available in several laboratories. The highest energy has been achieved at the Stanford Linear Accelerator Center where the energy is currently ~ 20 GeV. These beams allow the probing of the proton and neutron down to distances $\sim 10^{-15}$ cm and have revealed that the charge of the proton and the intrinsic magnetic moments of both proton and neutron are exponentially distributed with a mean radius $\sim 0.8 \times 10^{-13}$ cm. The proton and neutron can hardly be considered as elementary particles in the sense of the electron, which at such distances still behaves as a point charge.

The electron scattering results now available make most other methods of studying the charge distribution in nuclei redundant except in rather special cases, but we will discuss briefly some of the effects of the finite nuclear radius on atomic spectra.

Outside the nucleus the electric field due to the nucleus is that of a point charge Ze (apart from a possible small quadrupole component) while inside it flattens off so that the potential reaches a finite value as $r \to 0$. An atomic electron occupies an orbit with a radius parameter large compared with the nuclear radius: its wave function is therefore determined primarily by the electric field outside the nucleus. Consider the lowest energy s wave electron. Its wave function will be approximately that computed under the assumption of a point nucleus and we may calculate the change in energy of this state by computing

$$\Delta E = \int_0^\infty \psi_s^*(r)\, \Delta V(r)\, \psi_s(r)\, 4\pi r^2\, dr$$

where $\Delta V(r)$ is the difference between the potential due to a distribution of charge and the potential due to a point charge.

$$\psi_s(r) = \frac{1}{\sqrt{\pi}} \left(\frac{Z}{\alpha_B}\right)^{\frac{3}{2}} e^{-Zr/\alpha_B}$$

where α_B is the Bohr radius.

If we ignore the effect on this wave function of the atomic electrons we may calculate the energy shift of the s state due to a finite nuclear radius. Let us assume that the nucleus is uniformly charged. Then inside the nucleus

$$V(r) = -\frac{Ze^2}{R}\left\{\frac{3}{2} - \frac{1}{2}\left(\frac{r}{R}\right)^2\right\} \qquad r \leqslant R$$

[†] See *Nucleon and Nuclear Structure*, Ed. R. Hofstadter (Benjamin, 1963).
[‡] R. Frosch et al., *Phys. Rev.* **174**, 1380 (1968).

Nuclear Physics

and outside

$$V(r) = -\frac{Ze^2}{r} \qquad r > R$$

(see Fig. 1.4.7).

$$\Delta E = \int_0^R \psi_s^*(r) \left\{ \frac{Ze^2}{r} - \frac{Ze^2}{R}\left[\frac{3}{2} - \frac{1}{2}\left(\frac{r}{R}\right)^2\right] \right\} \psi_s(r) 4\pi r^2 \, dr$$

$$= \left(\frac{Z}{\alpha_B}\right)^3 4Ze^2 \int_0^R e^{-2Zr/\alpha_B} \left\{\frac{1}{r} - \frac{3}{2R} + \frac{1}{2}\frac{r^2}{R^3}\right\} r^2 \, dr,$$

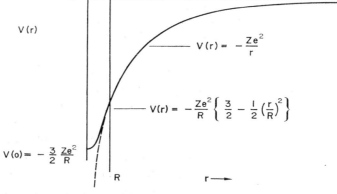

FIG. 1.4.7. This shows the coulomb potential experienced by an electron in the field of a uniformly charged sphere of radius R. (If the charge were distributed instead in a shell at R the potential at $r \leq R$ would be constant.)

to be compared with a total potential energy E of

$$E \simeq -\left(\frac{Z}{\alpha_B}\right)^3 4Ze^2 \int_0^\infty e^{-2Zr/\alpha_B} r \, dr = -\frac{Z^2 e^2}{\alpha_B},$$

$$\alpha_B = \frac{\hbar^2}{m_e^2 e^2} = 0.529 \times 10^{-8} \text{ cm}.$$

If $R \ll \alpha_B$,

$$\Delta E \simeq \left(\frac{Z}{\alpha_B}\right)^3 4Ze^2 \int_0^R \left\{r - \frac{3}{2}\frac{r^2}{R} + \frac{1}{2}\frac{r^4}{R^3}\right\} dr$$

$$= \left(\frac{Z}{\alpha_B}\right)^3 \frac{2}{5} Ze^2 R^2.$$

The fractional shift of the energy of the s-electron is

$$\left|\frac{\Delta E}{\frac{1}{2}E}\right| = \frac{4}{5} Z^2 \left(\frac{R}{\alpha_B}\right)^2$$

and this approximation is adequate provided $R \ll \alpha_B$. The first s state in lead (say) is reached from the next p state by emission of a K X-ray. Since the p wave function is not concentrated in the nucleus and falls to zero as $r \to 0$ the shift in p state energy is small in comparison with the shift in s state energy and the shift in the energy of the K X-ray $\Delta E_K/E_K$ ($p \to s$ transition) is just

$$\left|\frac{\Delta E_K}{E_K}\right| = \frac{4}{3}\left|\frac{\Delta E}{E}\right| \simeq \frac{16}{15} Z^2 \left(\frac{R}{\alpha_B}\right)^2,$$

$R \simeq 8 \times 10^{-13}$ cm, $Z = 82$ for Pb, so

$$\left|\frac{\Delta E_K}{E_K}\right| \simeq 10^{-4}.$$

This is a small effect and the quantity that has been studied is the splitting between the $2p_{\frac{1}{2}}$ and $2p_{\frac{3}{2}}$ levels which is affected much more strongly.

It is not at all obvious that this is going to be the case—you may be tempted to try and calculate the fractional shift in the splitting by evaluating[†]

$$\frac{\int_0^R \psi_p^* \frac{1}{r} \frac{d}{dr}\{\Delta V(r)\} \psi_p \, d^3r}{\int_0^R \psi_p^* \frac{1}{r} \frac{d}{dr}\left\{\frac{Ze^2}{r}\right\} \psi_p \, d^3r}$$

which is again $\approx Z^2(R/\alpha_B)^2$ and gives completely the wrong answer. The reason is that the approach we have followed is deficient in several respects. We have used a non-relativistic formalism and fed into our calculations Schrödinger wave functions for the electrons. While this would be all right for light nuclei in which the electrons are not highly relativistic, for heavy nuclei with high Z the low-lying states correspond to electron velocities so high ($\sim 0.5c$) that it is essential to use the Dirac equation in solving for the wave functions and energy levels. The spin–orbit coupling then, of course, appears naturally and does not have to be botched on to a Schrödinger equation treatment. Thus our first calculation of the shift in the energy of the $2p \to 1s$ transition is only approximate even if we do not count the effect of the atomic electrons. Now the solutions of the Dirac equation for a single electron have the property that neither the orbital angular momentum **L** nor the electron spin angular momentum **S** are conserved although the sum **J** is conserved.

The more relativistic the electron, the poorer an approximation in which a state is specified by **L**, **S** becomes. A correct calculation shows that the $p_{\frac{1}{2}}$ state takes on some of the characteristics of an $s_{\frac{1}{2}}$ state, in particular the wave function is NOT $\propto r$ near the origin but is large (see Fig. 1.4.8) while the $p_{\frac{3}{2}}$ wave function still behaves as r near the origin. The result is that splitting between $p_{\frac{1}{2}}$ and $p_{\frac{3}{2}}$ depends strongly on the nuclear radius and cannot be computed

[†] See L. I. Schiff, *Quantum Mechanics*, 3rd ed., sect. 53 (McGraw-Hill, 1968).

Nuclear Physics

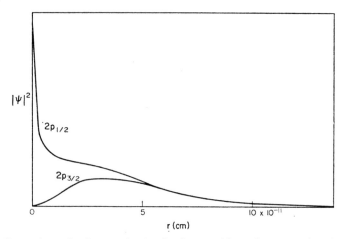

FIG. 1.4.8. This figure shows the electron density for $2p_{1/2}$ and $2p_{3/2}$ electrons when $Z = 85$. Relativistic effects concentrate the $2p_{1/2}$ wave function near the nucleus in a way quite uncharacteristic of the non-relativistic $2p$ wave functions. [From Schwalow and Townes, *Phys. Rev.* **100**, 1273 (1955).]

without using the Dirac equation, for a heavy atom. The effect of the nuclear radius on the splitting is $\sim 0.3\%$ for nuclei in the region of lead which is an order of magnitude greater than $Z^2(R/\alpha_B)^2$.

In passing we should also remark that for these highly relativistic electrons in strong fields even a treatment using the Dirac equation is not entirely adequate, and in order to obtain a reliable answer a treatment using the full apparatus of quantum electrodynamics is necessary. The reason is that the presence of virtual pairs of electrons in the strong field modifies the potential acting on the electron by an amount comparable to the effect of the nuclear charge distribution. These effects are known as Lamb shift effects since they are caused by the same physical phenomenon as the Lamb shift in the radio-frequency spectrum of atomic hydrogen. When these effects are taken into account the data are consistent with $R_0 \sim 1.1$–1.2×10^{-13} cm.

The finite radius of the nucleus also affects optical spectra in transitions involving s states of high principal quantum number. In this case it is practically impossible to make a calculation of the transition energy (although quite possible in principle) and what is observed is the difference in transition energy between two isotopes which have a different mass number although Z, the atomic number, is the same. Two even–even isotopes (even Z, even A–Z) are preferred because such nuclei have zero spin and so zero magnetic moment (see Chapter 2). Let us make a very crude calculation of this isotopic shift Δ for the X-ray spectra of ^{204}Pb and ^{208}Pb:

$$\Delta \approx \frac{Z^2}{\alpha_B^2} \{\langle R_{208}^2 \rangle - \langle R_{204}^2 \rangle\}.$$

If $R_0 \sim 1.1 \times 10^{-13}$ cm, $\Delta \sim 10^{-6}$ of the transition energy and we expect similar effects for the *optical* isotopic shift. Such fractional differences in energy (and hence wavelength) can be measured with multiple-beam interferometry and the difficulties lie in calculating the expected effect. Again, the data are consistent with $R_0 \sim 1.1 \times 10^{-13}$ cm. While isotopic shifts can

The Nuclear Periodic Table

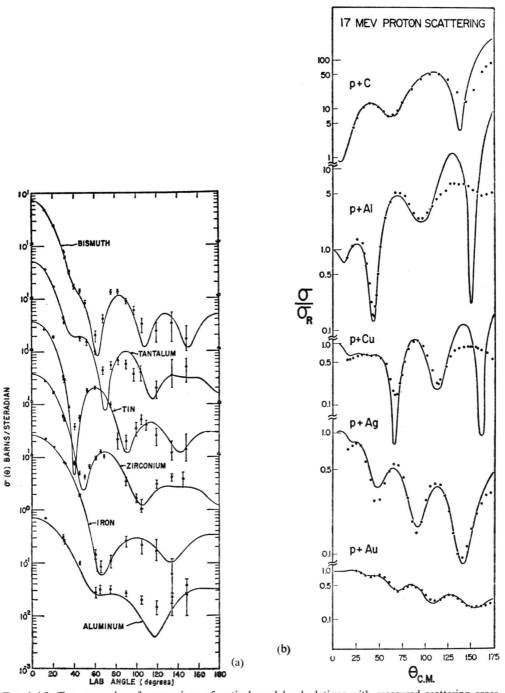

FIG. 1.4.9. Two examples of comparison of optical model calculations with measured scattering cross-sections. (a) shows experimental and theoretical differential cross-sections for 7-MeV neutrons on a variety of elements. [From S. Fernbach, *Rev. Mod. Phys.* **30**, 414 (1958).] (b) shows experimental and theoretical cross-sections for 17-MeV protons, normalized to the Rutherford scattering cross-section. [From A. E. Glassgold, *Rev. Mod. Phys.* **30**, 419 (1958).]

Nuclear Physics

be measured with great accuracy, the uncertainties in the theoretical interpretation preclude their being a sensitive measure of a gross radius parameter. They do, however, provide a very sensitive probe of the changes in nuclear charge distribution resulting from the addition of neutrons to a nucleus of specified Z, and yield information on the coupling of protons to neutrons in the nucleus and on the general subject of nuclear compressibility.

The X-ray spectra of muonic atoms provide another probe of the electromagnetic structure of the nucleus. The muon is a particle with only electromagnetic and weak interactions, with a mass 206 electron masses, behaving in every respect as a heavy electron. (Muons are produced in the decay of π-mesons. $\pi^{\pm} \to \mu^{\pm} + \nu$ and π-mesons are produced in nuclear interactions at several hundred MeV. Muons may also be produced in pairs by high-energy photons.)

The Bohr radius for a muonic orbit is 206 times smaller than for an electron and transition energies are correspondingly greater. A muon in a low-lying orbit about, say, a lead nucleus is thus well inside the electron cloud surrounding the nucleus and the energy levels can be calculated more easily than for a corresponding electron which is shielded from the nucleus by other electrons. The principles of the calculation are the same; but we cannot make even a pretence at a calculation here. In discussing the X-ray spectra and the isotopic shift we made the assumption that the wave functions were determined by a pure coulomb potential and then treated the departure of the potential from a pure coulomb form as a perturbation, evaluating the energy shift of a level in first-order perturbation theory.

Since the mass of the muon is 206 electron masses the first state in a pure $1/r$ potential would be

$$\sim e^{-Zr/\alpha_\mu} \qquad \alpha_\mu = \alpha_B/206.$$

The wave function itself would therefore be drastically modified and a first order perturbation calculation is useless. All we can do here is quote the results: the $2p_{\frac{3}{2}} \to 1s_{\frac{1}{2}}$ muonic transition in lead would give an X-ray of 16.41 MeV for a $1/r$ potential and is found to give a 6.02 MeV X-ray.[†] Both the gross spectra and the relativistic splitting have been studied—for a point lead nucleus the $2p_{\frac{3}{2}} - 2p_{\frac{1}{2}}$ splitting would be 0.55 MeV and is found to be 0.2 MeV. (The latter effect was at one time suggested as a possible method of determining the muon magnetic moment, but has been completely superseded by studies of the precession rate of the μ in a magnetic field.[‡])

With the development of lithium drifted germanium detectors, which have a resolution $\sim$ keV, studies of muonic X-rays have yielded information on the nuclear charge distribution which for $Z > 10$ is more accurate than that from electron scattering. In light atoms the data is sufficient only to determine the root mean square charge radius: in heavy atoms two parameters, for example, a radius and a surface thickness, have been derived. The results from electron scattering and muonic X-rays are in good agreement. The resolution of small-scale structure is of course limited by the scale of the muon orbits, while the wavelength of electron beams can be made almost arbitrarily small from the point of view of the nucleus. On the other hand, the muon, making many orbits before shifting levels, is sensitive to other

[†] V. L. Fitch and E. J. Rainwater, *Phys. Rev.* **92**, 789 (1953) (the first paper on this topic).
[‡] J. Bailey et al., *Phys. Lett.* **28** B, 287 (1968).

The Nuclear Periodic Table

effects scarcely accessible in electron scattering: the hyperfine structure of muonic X-ray transitions yields information on the quadrupole moment of the charge distribution: in particular the orbiting muon can excite quadrupole distortions of the nucleus about which it orbits.

The general subject of muonic atoms has been recently reviewed by Wu and Wilets.[†]

There is one last effect of the charge distribution that we may mention here. A uniformly charged sphere has a coulomb energy of $\frac{3}{5}(q^2/R)$ where R is the radius, and the sphere carries charge q. Consider two nuclei having the same mass number but differing in charge by one unit. If the specifically nuclear forces are identical for both these nuclei then the one with lower charge is more stable and a β transition will link the two:

$$^A_Z X \rightarrow {}^A_{Z-1}X + e^+ + \nu \quad \text{plus a lost atomic electron,}$$

e.g.

$$^{15}O \rightarrow {}^{15}N + e^+ + \nu$$

and the (maximum) electron energy provides a measure of the difference in coulomb energy. Consider

$$M(Z, A) = (Z-1)M_H + (A-Z)M_n + M_H \quad \text{— nuclear binding + coulomb energy,}$$
$$M(Z-1, A) = (Z-1)M_H + (A-Z)M_n + M_n \quad \text{— nuclear binding + coulomb energy.}$$

If the nuclear terms are the same, the maximum electron energy is given by

$$2m_e c^2 + E_{max} = E_{coulomb}(Z) - E_{coulomb}(Z-1) + (M_H - M_n)c^2$$
$$E_{max} = E_{coulomb}(Z) - E_{coulomb}(Z-1) - 1.8 \text{ MeV}$$
$$E_{coulomb}(Z) \simeq \tfrac{3}{5} Z^2 e^2 / R.$$

Thus

$$E_{max} \simeq 6/5 \frac{Ze^2}{R} - 1.8 \text{ MeV}$$

if R is assumed to be the same for both nuclei.

^{15}O and ^{15}N are an example of such a pair of *mirror nuclei*—replacing all protons by neutrons and all neutrons by protons in one of them yields the other. Both these nuclei may be regarded as an even–even core with one odd nucleon in orbit about the core: if specifically nuclear forces are charge independent, then clearly the energy released in a β transition between two mirror nuclei will be just the difference in coulomb energy minus the neutron–proton mass difference. (Such a pair of nuclei forms an *isotopic doublet*.)

If this crude model is used, the data yield a value of R_0 of $\sim 1.45 \times 10^{-13}$ cm. This value is high because the odd nucleon is often concentrated near the nuclear surface and the assumption of a uniform distribution of charge is wrong—the distribution is tailing off and so the coulomb energy of the decaying proton is less than the average value. In the framework of an independent particle model of this kind there are also further corrections due to the Pauli exclusion principle, which tends to keep the protons apart, and a complete treatment leads to the conclusion that $R_0 \sim 1.3 \times 10^{-13}$ cm. The agreement between the value obtained for mirror nuclei of mass numbers through the range 10–50 and the further agreement of the

[†] C. S. Wu and L. Wilets, Muonic atoms and nuclear structure, *Ann. Rev. Nucl. Sci.* **19**, 527 (1969).

Nuclear Physics

radius parameter obtained in this way with the results of electron-scattering experiments thus demonstrates that nuclear forces are substantially charge independent. (Strictly speaking, the results on mirror nuclei only demonstrate charge symmetry. Similar results may be obtained from *isotopic triplets* which indicate charge independence).

We may best summarize this section by saying that electron-scattering experiments yield the detailed charge distribution of nuclei and the result may be crudely represented by allotting to the charge distribution a radius R where $R = R_0 A^{1/3}$ with $R_0 = 1.1$–1.2×10^{-13} cm. These results are in agreement with the atomic effects of the nuclear size which are: fine structure of X-ray spectra, optical isotopic shift and X-ray spectra of muonic atoms. They also provide an explanation of the energy difference between mirror nuclei which are members of an isotopic doublet and the same parametrization works for all these phenomena.

The phenomena we have been discussing depend only on the electromagnetic interaction of electrons and muons, which are not coupled to the *strong interactions*, with the nuclear charge distribution. The interaction of neutrons with the nucleus will depend only on the distribution of nuclear matter within the nucleus, while the interaction of charged and strongly interacting particles (such as protons, deuterons, ^{4}He or π-mesons) with the nucleus will depend on both the charge distribution and the nuclear mass distribution, until such energies are reached that the coulomb barrier becomes insignificant.

The size the nucleus presents to strongly interacting particles is obtained from the total cross-section presented to a beam of particles and from the angular distribution in elastic scattering. The measured strong interaction nuclear radius is the radius parameter of an *optical model* nuclear potential. This parameter is extracted primarily by measurement of elastic scattering of either neutrons or charged particles. The analysis of the experimental data proceeds in the following way. It is assumed that a nucleon (or π-meson or α-particle) experiences, on entering the nucleus, a smoothly varying time independent potential which represents an average over the interactions with the individual nucleons in the nucleus, and this potential determines the scattering. This model can only be applied at relatively high energies (10's of MeV and above) even for neutrons which do not have to surmount a coulomb barrier in order to penetrate the nucleus—at energies in the MeV region and below the scattering mechanism involves an excited compound nucleus which may exist for perhaps 10^{-15} sec before decaying into the original nucleus and the scattered particle.

It should be quite clear that the scattering process can only take $\sim 10^{-22}$ sec if the assumption of a single-particle potential is to work. At such high energies it is easy to disrupt the nucleus when the incident particle, if it emerges at all, will have reduced energy. Thus the potential must have absorbtive properties which are included by allotting to it an imaginary part. Once a trial potential has been chosen, calculation of the scattering is only a matter of computation and the problem of fitting data has two parts: first, it is necessary to devise a plausible form for the potential and, secondly, the parameters which appear in that form must be varied until the calculated scattering provides a good representation of the measured scattering. The calculation presents no problem of principle. As a trivial example we will calculate the matrix element for elastic scattering of neutrons by a shallow square well potential, for which

$$\left. \begin{array}{ll} V(r) = -V_0 & r \leqslant R \\ V(r) = 0 & r > R \end{array} \right\} V(r) \text{ is real; there is no absorption.}$$

The Nuclear Periodic Table

For a potential small compared with the neutron energy we may use first order perturbation theory and evaluate

$$M_V = \int_0^\infty e^{i\mathbf{q}\cdot\mathbf{r}} V(r) r^2 \, dr \, d\phi \, d\cos\chi.$$

$$q = 2p \sin\frac{\theta}{2}$$

where p is the neutron momentum and θ the scattering angle and $\mathbf{q}\cdot\mathbf{r} = qr\cos\chi$.

$$M_V = -\int_0^R \int_0^{2\pi} \int_{-1}^{+1} e^{iqr\cos\chi} V_0 r^2 \, dr \, d\phi \, d\cos\chi$$

$$= -4\pi V_0 \int_0^R \frac{\sin qr}{qr} r^2 \, dr$$

$$= -\frac{4\pi}{q^3} V_0 [\sin qR - qR \cos qR].$$

The variation of the scattered intensity with angle is proportional to $|M_V|^2$ and so has zeros when $\sin qR - qR \cos qR$ is zero, and a principal maximum at $q = 0$. The position of the maxima and minima in this case depend only on the momentum and radius, not on the depth of the well. The magnitude of the scattering clearly depends on both the radius and the depth.

This trivial model illustrates the use made of an optical potential and also shows that maxima and minima in a scattering distribution do not *necessarily* imply diffraction scattering due to absorption of the beam—rapid oscillations in the scattering as a function of angle only indicate a rapid variation in the potential. You will recall that a $1/r$ potential (eq. (1.4.1)) did *not* produce oscillations. We could extend the calculation to describe the scattering of charged particles by replacing the potential $V(r)$ by $V(r) + \phi(r)$ where

$$\phi(r) = Zze^2/r \qquad r > R',$$

$$\phi(r) = \frac{Zze^2}{R'}\left\{\frac{3}{2} - \frac{1}{2}\left(\frac{r}{R'}\right)^2\right\} \quad r < R' \quad \text{or some other plausible form.}$$

(Note that the radius parameters R and R' do not have to be the same.) Optical model calculations are performed for charged particles by adding a plausible electrostatic potential to the assumed purely nuclear form.

Having made this first-order calculation, its complete inapplicability to real nuclear scattering must be pointed out. It should be clear that for an arbitrarily deep potential hole it is always possible to limit the discussion to energies sufficiently high that $E \gg V_0$. The difficulty arises from absorption effects. Suppose that all particles penetrating the nucleus are absorbed. It is clear that under such circumstances the potential is having an effect upon the wave functions that can hardly be classed as a small perturbation! The proper optical model treatment is discussed in Chapter 4.

Nuclear Physics

A given potential predicts the differential elastic cross-section, the elastic cross-section, and the absorption cross-section, and all three may be compared with the data to yield a radius parameter. The potentials used have both real and imaginary components, and a popular form for the radial dependence is the Woods–Saxon potential:

$$V(r) \propto [1+e^{(r-R)/a}]^{-1}$$

where R is the radius parameter and a measures the thickness of the nuclear surface. This form fits the charge distribution in heavy nuclei but the values of R obtained differ from the results of electron scattering. If we take $R = R_0 A^{1/3}$ then $R_0 \sim 1.1 \times 10^{-13}$ cm from electron scattering, but $R_0 \sim 1.25 \times 10^{-13}$ cm from optical model calculations. It should be remembered that the electron-scattering form factor measures the charge distribution, while the optical model parameter describes the spatial extent of the nuclear potential: it is not unreasonable that this potential extends beyond its source appreciably.

As implied by the name, this phenomenological treatment of scattering is closely analogous to certain optical phenomena. Consider a particle of given kinetic energy E. In a region of space where a potential V is acting, the wavelength of this particle is altered, either up or down depending on the sign of the potential, and if the potential is repulsive and greater than the kinetic energy ($V > E$) the wavelength becomes imaginary, corresponding to an exponential behaviour of the wave function rather than an oscillatory behaviour. These effects are analogous to the effect on light of entering a region of different refractive index— the exponential attenuation *without absorption* of light in a medium of imaginary refractive index (such as a gas of free electrons) providing the analogue to the case where $V > E$. (Another example is the attenuation in the electromagnetic field in a wave guide operated below the critical frequency, where again no power is absorbed.)

Consider how the reflection of light from a surface is calculated. The refractive indices on both sides of the surface are fed in and general propagating fields set up on both sides. On the input side the field is a sum of the incident and reflected fields, on the other side, the refracted field. The reflected and refracted fields are then uniquely calculated from two boundary conditions: the continuity of the tangential components of E and H. The calculation is trivial for plane waves and a plane surface with real refractive indices and more complicated for a complex refractive index. The scattering of light from a sphere may be calculated by first expanding the incident plane wave as a series of spherical waves and again using the boundary conditions.

The scattering of particles by a potential is treated in just this way. If the nuclear potential is assumed to have a sharp boundary, ψ and $\partial \psi / \partial r$ are assumed continuous across the boundary. Again, outside the nucleus ψ represents the total wave function and the scattered wave is determined from the propagation vectors inside and outside the nucleus and the boundary conditions. If the potential is not assumed to be sharp, continuity of ψ and $\partial \psi / \partial r$ is sufficient to allow the total wave function to be calculated. Subtraction of the incident wave then yields the scattered wave.

There is one last example of the relevance of optical analogies. Consider the problem of the scattering of light by a black disc. In elementary diffraction theory the problem is solved for a plane wave by supposing that on the immediate back side of the disc there is no field, applying Babinet's principle and integrating over the (infinite) plane in which the disc lies. This is only valid when the diameter of the disc is many wavelengths, since there is always some field near the edges of the back side due to diffraction—which is the very phenomenon we are trying to calculate!

Now if every particle in a beam that enters the nucleus induces an inelastic reaction, the nucleus will behave just like such a black disc and if the wavelength of the particles is $\ll R$, the nuclear radius, the boundary conditions may be approximated in the same way—no wave function on the back side of the disc. Thus a reaction cross-section of πR^2 will be accompanied by a differential scattering cross-section typical of diffraction by an absorbing disc if the nuclear edge is sharp, which integrates to a scattering cross-section of πR^2 again. It is not necessary to integrate the differential cross-section—the effect of the nucleus (or disc) is just to generate a wave over an area πR^2 of opposite phase to the incident one so that the net result behind the absorber is zero. (Incidentally, this is not just a way of seeing the result; this is what actually happens. In elementary diffraction theory the calculation is turned into an integral from R to ∞ by applying Babinet's principle.) The total cross-section is thus $2\pi R^2$. Neutron scattering at energies $\gtrsim 50$ MeV is very like this and it is of interest to note that the reaction cross-section for high-energy particles (π-mesons as well as protons and neutrons) on nuclei is approximately the geometric cross-section $\pi R_0^2 \, A^{2/3}$ at all energies from a few hundred MeV to those energies ~ 1000 GeV and above found in the cosmic radiation although at present inaccessible to accelerators.

Studies of the spatial extent of the nucleus may be summarized as follows: The radius of the nucleus must be operationally defined and there are two classes of phenomena which lend themselves to the measurement of such a quantity. The charge distribution of nuclei is obtained through the electron-scattering form factor, while the shape of the effective potential for high-energy reactions[†] is obtained through optical model calculations. For heavy nuclei ($A \gtrsim 40$) distributions of the Woods–Saxon form

$$f(r) = \frac{1}{1+e^{(r-R)/a}}$$

provide good fits with parameters for the two sets of data:

Electron scattering:

$R = 1.07 \times A^{1/3} \times 10^{-13}$ cm
$a = 0.55 \times 10^{-13}$ cm
These parameters describe the charge distribution

p and n scattering:

$R = 1.25 \times A^{1/3} \times 10^{-13}$ cm
$a = 0.65 \times 10^{-13}$ cm
These parameters describe optical model potentials

These are average values and fluctuations of $\sim 10\%$ may occur.

An excellent series of review articles dealing with nuclear size may be found in *Rev. Mod. Phys.* **30,** 412 (1958).

[†] Although we have not discussed it here we shall use the idea of nuclear radius in a discussion of low-energy reactions as well.

Nuclear Physics

1.5. The semiempirical mass formula

The semiempirical mass formula correlates the results we have already discussed, and provides some explanation of them and this leads us naturally into an elementary discussion of some models of the nucleus. We begin the discussion by noting the following gross features of the nuclear periodic table:

1. The nucleus has a radius R given by $R \simeq R_0 A^{1/3}$.
2. The binding energy per nucleon is approximately constant at ~ 8 MeV/nucleon from $A \sim 20$ to $A \sim 250$.
3. Between $A = 1$ and $A = 20$ the mean binding energy rises rapidly with a superimposed oscillation giving maxima at ^{4}He, ^{8}Be, ^{12}C, ^{16}O and ^{20}Ne.
4. The maximum mean binding energy is found in the region of $A = 60$ (Ni, Fe) and the mean binding energy decreases steadily as A increases beyond this region.
5. The most stable isobars of a family of given A are those for which $Z \sim A-Z$ (a proton or neutron excess is corrected via β-decay). (See Fig. 1.5.1.)

The first of these points implies that nuclear matter has approximate constant density (under terrestrial conditions at least) while the second suggests that nucleons only interact strongly with their near neighbours in the nucleus. Let us contrast this situation with coulomb repulsion. If we assume the protons in the nucleus to be approximately uniformly distributed, then we approximate the coulomb energy by $\frac{3}{5}(Z^2 e^2/R)$; each proton interacts through the coulomb field with all others and the contribution to the nuclear energy varies approximately as the square of the number of protons. The approximately constant mean binding energy suggests that an individual nucleon only interacts with a given number of other nucleons, regardless of the total number present in the nucleus. The constant density also supports this conclusion. Then our first approximation to a formula for the mass of a neutral atom characterized by Z, A is (neglecting electron binding energy):

$$M(Z, A) = ZM_H + (A-Z)M_n - \alpha A + \frac{3}{5}\frac{Z^2 e^2}{R}. \quad (1.5.1)$$

The first two terms merely give the mass of the constituents of the atom and do not include the binding energy at all. The third term represents the binding energy due to the strong interactions (and because nuclei *are* bound it is negative). The fourth represents crudely the disruptive effect of the coulomb energy and is responsible for massive nuclei becoming unstable against α-emission and eventually spontaneous fission (^{254}Cf has a half-life of 55 days against spontaneous fission, for example). The magnitude of this term can be calculated from scratch, given the electromagnetic radius, but α is a parameter which must be fitted.

Now let us suppose that a nucleon deep within the nucleus interacts strongly with an effective number n of its neighbours. Nucleons with less than n neighbours are bound less strongly and a correction must be made for this effect, *reducing* the binding:

$$M(Z, A) = ZM_H + (A-Z)M_n - \alpha A + \beta A^{2/3} + \frac{3}{5}\frac{Z^2 e^2}{R}. \quad (1.5.2)$$

The Nuclear Periodic Table

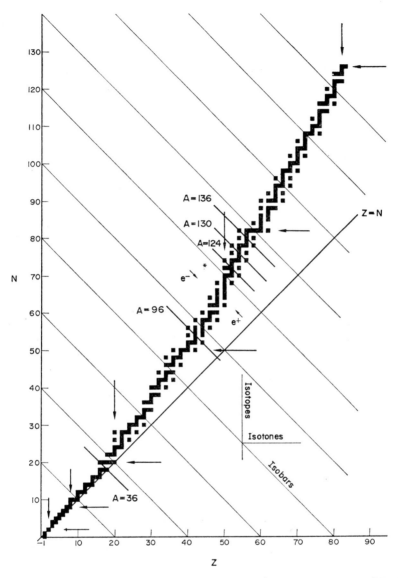

FIG. 1.5.1. Nuclei stable against β-decay processes are shown on an N versus Z plot. Beyond ^{40}Ca the increasing coulomb repulsion generates a neutron excess. Pairs of stable isobars first appear for even–even nuclei at $A = 36$ and triplets occur at $A = 96, 124, 130$ and 136. Magic numbers are indicated by arrows and finally a star indicates the fragments resulting from symmetric fission of ^{236}U, illustrating the inevitably neutron-rich character of fission fragments.

The correction term is taken as being proportional to the radius of the nucleus squared; that is, to the surface area of the nucleus. The reason is that those nucleons near the surface regions will have fewer neighbours within a given radius than those lying deep within the nucleus. The terms we have so far written down constitute a *liquid-drop model* of the nuclear ground state and the correction term corresponds to the surface energy of a liquid drop—two globules of mercury in contact congeal, thus reducing the total surface energy.

Nuclear Physics

The parameter β, of course, must be fitted to the data and cannot at present be predicted. Thus the rise from zero mean binding energy (for hydrogen) can be explained in terms of the surface energy effect—the smaller the nucleus the larger the ratio of surface area to volume and so the contribution of surface nucleons is relatively more important. (You should note at this point that so far all we are doing is to construct a formula which has qualitatively the right features—our arguments are shaky and the *only* justification is that the eventual formula works quantitatively.)

This formula still has no hope of working, and the reasons are obvious. It should be quite clear that for given A the most stable nucleus represented by the formula (1.5.2) would consist almost entirely of neutrons—both terms due to nuclear forces in the binding energy are the same for either protons or neutrons and if $Z = 0$ (or 1) there is no coulomb repulsion. Now nuclei of low A are most stable when they contain an equal number of protons and neutrons, while an increasing neutron excess sets in as A increases. The reason for the latter phenomenon is clear—because of the growing coulomb repulsion as Z increases, a stable nucleus of large Z, A needs relatively more neutrons than one of small Z, A in order to hold it together. But why is it that nuclei do not solve this problem thoroughly by consisting *only* of neutrons?

It is necessary to consider the purely quantum phenomenon of the Pauli exclusion principle in order to answer this question. Nucleons are particles of spin $\frac{1}{2}$ and so obey Fermi statistics—no two nucleons can have identical quantum numbers. This is the reason why nuclei have roughly equal numbers of protons and neutrons! In order to see how it works, let us construct a very simple model of the nucleus. We suppose that the effect of the nuclear forces on any individual nucleon may be represented by a short-range potential. Let us suppose initially that the well has simple harmonic form

$$V(r) = -V_0 + V_1 r^2$$
$$= -V_0 + V_1\{x^2 + y^2 + z^2\}.$$

The energy levels of such a particle are given by the energy levels of three independent one-dimensional simple harmonic oscillators and so are:

$$E_n = -V_0 + \hbar\omega\{n_1 + n_2 + n_3 + \tfrac{3}{2}\}$$

where ω measures the separation of the energy levels, and a level is specified by the three numbers.

The ground state corresponds to $n_1 = n_2 = n_3 = 0$ and the energy of the nucleon is $-V_0$ plus the zero point energy associated with the potential. Thus there is one single nucleon state labelled by n_1, n_2, n_3 with $E = -V_0 + \tfrac{3}{2}\hbar\omega$. There are three such states with $E = -V_0 + \tfrac{5}{2}\hbar\omega$ ($n_1 = 1$ or $n_2 = 1$ or $n_3 = 1$ with the other two zero). There are six with $E = -V_0 + \tfrac{7}{2}\hbar\omega$ and so on. (If we label these states by their orbital angular momenta and principal quantum numbers, the lowest state corresponds to the first s state, the three degenerate states with $E = -V_0 + \tfrac{5}{2}\hbar\omega$ to the three $1p$ orbitals, while for $\tfrac{7}{2}\hbar\omega$ the six independent degenerate states correspond to the $2s$ state and the five $1d$ orbitals, and so on.) Now two protons will go into any one of these states—so long as their spins are opposite. So will two neutrons—protons and neutrons are distinguishable! Suppose we take four nucleons and consider how to choose them among protons and neutrons so that we minimize

The Nuclear Periodic Table

the energy of the system (the interactions among the four are supposed to give the potential well). We have the following possibilities:

1. all four protons,
2. three protons, one neutron,
3. two protons, two neutrons,

and two more, the mirror images of (1) and (2). The results may be represented as in Fig. 1.5.2 and if our hypothetical nucleus with $A = 4$ was stable against nucleon emission then β-decay would put things right (Fig. 1.5.3).

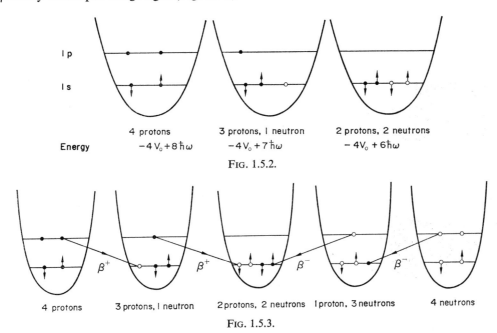

Fig. 1.5.2.

Fig. 1.5.3.

Thus the Pauli exclusion principle ensures that an equal number of protons and neutrons are, on average, present. Of course, odd A nuclei *cannot* have an equal number of protons and neutrons present, and *in our crude model* appreciable imbalances could occur—suppose we have filled the *s*-shell completely—two protons and two neutrons. Now start building up the *p*-shell. Clearly we could add up to six neutrons or six protons before the Pauli exclusion principle would start to bite, *if* the six states are degenerate.

These simple considerations tell us *what* is happening to fix $Z \sim A-Z$, but we need to make a guess at the form of the *asymmetry energy*. We suppose it to be, on average, zero when $Z = A/2$ and to vary as $(Z-A/2)^2$ for given A. (This is just the first non-zero term in an expansion about $Z = A/2$ if $Z = A/2$ represents a minimum in the asymmetry energy.) But presumably the effect will also vary with A and we must try to estimate this. Clearly the effect depends on the separation of the levels—$\hbar\omega$ in our simple harmonic oscillator model. The separation of levels will depend on both the radius and depth of the potential. As the radius of the nucleus expands with increasing A the difference in wavelength between two adjacent levels drops and so the spacing decreases. Our model clearly cannot be a pure

Nuclear Physics

simple harmonic potential and the potential must fall rapidly to zero outside the nucleus. We are examining the variation of binding energy of an isobar of given A with the asymmetry between the number of protons and the number of neutrons and so need the spacing of levels near the top of the potential well (lower-lying ones being already filled), and these are just the levels which depart most strongly from simple harmonic behaviour. We will therefore examine the effect of the Pauli exclusion principle using a square well potential rather than a simple harmonic oscillator—neither is very realistic and the square well is more tractable. So:

$$V(r) = -V_0 \quad r \leq R,$$
$$V(r) = 0 \quad r > R.$$

We make the additional assumption that the wave functions are zero for $r > R$ (although in reality they fall off exponentially outside the well). Thus the energy levels we are considering have energies $-V_0 + \varepsilon_n$ where ε_n are the energy levels of a particle confined within a spherical box. Now the number of single particle levels corresponding to a particle with momentum between p and $p + dp$ is well known to be[†]

$$\frac{4\pi p^2 \, dp}{(2\pi \hbar)^3} V$$

where V is the volume of the box. (You will notice the resemblance of this formula to that giving the number of normal modes of oscillation between ν and $\nu + d\nu$ of the electromagnetic field in a cavity—this is not surprising, the problem is almost identical!) Then the total number of levels corresponding to $p < p_{max}$ is

$$N = \frac{4\pi}{3} \frac{p_{max}^3}{(2\pi \hbar)^3} \frac{4\pi}{3} R_0^3 A$$

where because of the two different spin orientations of a given nucleon this number must be $\sim A/4$. Then p_{max} is approximately independent of A, and the density of levels is:

$$\left.\frac{dN}{d\varepsilon}\right|_{p_{max}} = \left(\frac{4\pi}{3}\right)^2 \frac{3 p_{max}^2 \, dp}{(2\pi\hbar)^3 \, d\varepsilon}\bigg|_{p_{max}} R_0^3 A$$

$$= \frac{3}{4} \frac{A}{p_{max}} \frac{dp}{d\varepsilon}\bigg|_{p_{max}}.$$

Now

$$\frac{1}{p}\frac{dp}{d\varepsilon} = \frac{1}{pv} = \frac{1}{2\varepsilon},$$

so

$$\left.\frac{dN}{d\varepsilon}\right|_{p_{max}} = \frac{3}{4}\frac{A}{2\varepsilon_{max}}$$

and since ε_{max} is approximately independent of A, the level spacing at the top of the potential

---

[†] This result is derived, in a different context, in Section 3.3.

The Nuclear Periodic Table

well is approximately proportional to $1/A$.† With the *asymmetry energy* added to the semi-empirical mass formula it now reads:

$$M(Z, A) = ZM_H + (A-Z)M_n - \alpha A + \beta A^{2/3}$$
$$+ \frac{3}{5}\frac{Z^2 e^2}{R} + \gamma \frac{(Z-A/2)^2}{A}. \quad (1.5.3)$$

The asymmetry energy sets $Z \simeq A/2$ for the most stable isobar of a given A, but may be overcome by the need to add neutrons so as to prevent disruption of the nucleus by the coulomb repulsion. Competition between these two terms determines the value of Z as a function of A on the stability line.

There is one more effect that must be included. It is found that nuclei of even A are more stable than those of odd A over most of the periodic table, and that for odd A nuclei there exists only one isobar stable against β-decay, while for even A nuclei there are two β-stable isobars in the majority of cases, and with four exceptions these stable isobars are all even–even nuclei (see Fig. 1.5.1). The four exceptions are ^{2}H (1,1), ^{6}Li (3,3), ^{10}B (5,5) and ^{14}N (7,7). Figure 1.3.1 shows that the mean binding energy of the most stable even A isobar is greater than that of the most stable isobar of odd A. Another manifestation is found in the very obvious peaks in the mean binding energy at ^{4}He, ^{8}Be, ^{12}C, ^{16}O and ^{20}Ne. There is so far no term in the mass formula which includes this information and so a final term is added: $\pm \delta(A_{\text{even}})$ where the $+$ sign is taken for odd–odd nuclei and the $-$ sign for even–even nuclei.

This final term is called the pairing energy and is a real effect over and above the discontinuities introduced in binding energy by the separation of single-particle states. It is to be understood in terms of the effect of residual interactions between nucleons which are left over after the major part of the nucleon–nucleon interactions have been approximated by an average time-independent potential. We must expect to find such residual interactions, for an average time-independent potential extending over the whole nucleus can hardly contain the effect of very close approaches between two nucleons, and they are needed to specify the way in which several nucleons couple together to give a nuclear ground state of definite spin. Consider two nucleons in p states in a potential generated by a spin zero core. Two protons or two neutrons can couple to make angular momenta 0 or 2, and a proton and neutron can couple to make 0, 1, 2 or 3; in the absence of residual interactions all these states would have equal energy. The presence of residual interactions removes such degeneracy.

A residual interaction which is attractive and of very short range in comparison with the nuclear size may be approximated by a δ function

$$V(\mathbf{r}_1 - \mathbf{r}_2) = -c\delta(\mathbf{r}_1 - \mathbf{r}_2).$$

Spatial parts of the wave functions of two nucleons will be either symmetric or antisymmetric:

$$\frac{1}{\sqrt{2}}\{\psi_1(\mathbf{r}_1)\psi_2(\mathbf{r}_2) \pm \psi_2(\mathbf{r}_1)\psi_1(\mathbf{r}_2)\}$$

† See E. Fermi, *Nuclear Physics*, p. 22 (Chicago, 1950) for a further discussion of this term.

Nuclear Physics

where ψ_1 and ψ_2 are two single-particle wave functions and where $\mathbf{r}_1$ refers to nucleon 1, $\mathbf{r}_2$ to nucleon 2. The shift in the energy of this term of the two-nucleon state is given in first-order perturbation theory by

$$-\frac{c}{2}\int\int \{\psi_1^*(\mathbf{r}_1)\psi_2^*(\mathbf{r}_2) \pm \psi_2^*(\mathbf{r}_1)\psi_1^*(\mathbf{r}_2)\}\,\delta(\mathbf{r}_1-\mathbf{r}_2)$$
$$\{\psi_1(\mathbf{r}_1)\psi_2(\mathbf{r}_2) \pm \psi_2(\mathbf{r}_1)\psi_1(\mathbf{r}_2)\}\,d^3\mathbf{r}_1\,d^3\mathbf{r}_2.$$

The δ function fixes $\mathbf{r}_1 = \mathbf{r}_2$ and the energy shift is zero for antisymmetric spatial parts of the two-particle wave function. For identical particles, antisymmetric spatial terms must be multiplied by triplet spin functions and symmetric terms by singlet spin functions. Physically the Pauli principle is preventing two identical nucleons with their spins parallel from getting close enough for the residual interactions to get a grip. Thus singlets of identical particles are displaced below triplets, and the lowest overall state will be that in which the statistical weight of the singlet combination, weighted by the overlap of the density functions, is greatest. Thus for two nucleons in p states with $L \cdot S$ coupling direct calculation shows the $L = 0$, $J = 0$ state to lie below the $J = 2$ state. Such effects will also exist for neutron and proton. Again an antisymmetric spatial part has no weight, but because of the charge independence of nuclear forces the residual interaction has as strong an effect on the symmetric spatial part with a singlet spin as for two protons or two neutrons. If there were no force in the pn triplet state, we would have the situation shown in Fig. 1.5.4, but it is a reasonable guess that the triplet pn state will lie even lower, because the deuteron, a triplet pn state, lies lower than the (unstable) singlet. ^{6}He, ^{6}Li, ^{6}Be are two p-wave nucleons added to a spin zero (^{4}He) core: the ground states of ^{6}He and ^{6}Be are spin zero, the ground state of ^{6}Li lies lower and has spin one.

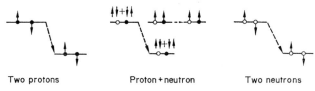

Two protons   Proton + neutron   Two neutrons

The broken lines indicate the effect of turning on a residual δ interaction in the singlet state

Fig. 1.5.4.

To trace the effect of even the simplest residual interaction when we have many nucleons with potentially good overlap of their density functions is very technical. Let us go over to the j–j coupling which is essential for the shell model (which we shall be discussing later). Then the greatest overlap of density functions is provided by coupling two nucleons in opposite magnetic substates to spin zero, and it is this fact that gives rise to the pairing term. For two identical nucleons where only singlet forces can contribute, short-range residual interactions may be said to pick out spin singlets and so force the total $\mathbf{j}$ vectors opposite in the lowest energy state. With many identical nucleons in a level the ground state has the maximum number of such pairs and for each pair the short-range residual interaction contri-

The Nuclear Periodic Table

butes extra binding. A pronounced odd–even effect results, even–even nuclei have spin zero and odd A nuclei the angular momentum of the last unpaired nucleon. The density overlap increases with j and so the pairing energy increases with j.[†]

With both protons and neutrons in a level of given j, the situation is much less clear, for it is quite improper to neglect proton–neutron interactions. However, for odd A and even–even nuclei the lowest energy state is still that which has the maximum number of pairs of nucleons coupling to spin zero because the overlap of densities is then greatest, and the pn triplet interaction is not quite strong enough to break up this scheme. Proton–neutron pairs contribute in the same way as pairs of identical nucleons—the forces are charge independent.[‡] An even–even nucleus thus has spin zero in the ground state and an odd A nucleus has the angular momentum of the last unpaired nucleon. The odd–even effect is still present, there being a contribution to the binding energy proportional to the number of pairs of nucleons with spin zero.

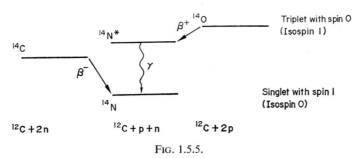

FIG. 1.5.5.

The ground states of odd–odd nuclei in most cases have two nucleons which are not coupled to spin zero. This is an effect of the symmetry term, which owes its bite to the effect of residual interactions removing degeneracies in filling a given set of single-particle levels, and is also naturally the place where the pn triplet force appears. Thus after writing in a symmetry term explicitly, the pairing effect is left. ^{14}N is the stable isobar of mass 14. The spin 0 excited state has as much pairing as ^{14}C or ^{14}O, but the ground state has two unpaired nucleons with spin 1 (Fig. 1.5.5). The same pattern is repeated for ^{18}F, but now the coulomb repulsion has raised the ground state of ^{18}F above the ground state of ^{18}O, and ^{19}F is the stable fluorine isotope (Fig. 1.5.6).

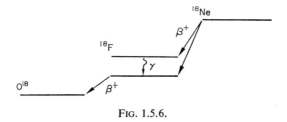

FIG. 1.5.6.

[†] M. G. Mayer, *Phys. Rev.* **78**, 22 (1950).
[‡] See Chapter 6.

Nuclear Physics

Residual interactions have further effects which we may as well mention here. Because the pairing term $\approx 2j+1$ the residual interactions can change the order in which levels are filled (thus an $h_{\frac{11}{2}}$ nucleon can steal a paired nucleon from a level of lower j). Also residual interactions can mix up different single—particle levels so that, for example, two nucleons in a $1d$ level with their spins coupling to zero may be contaminated by an admixture of two nucleons in a $2s$ level with spins coupling to zero—this sort of thing is called configuration mixing.

The semiempirical mass formula now becomes:

$$M(Z, A) = ZM_H + (A-Z)M_n - \alpha A + \beta A^{2/3}$$
$$+ \varepsilon Z^2 A^{-1/3} + \gamma \frac{(Z-A/2)^2}{A} \pm \delta(A_{\text{even}}). \quad (1.5.4)$$

With the observed dependence of R on A and three parameters α, β, γ, and a function δ† which must be fitted to the data, the masses and binding energies of several hundred nuclei may be fitted to $\sim$ a few per cent in the binding energy. The greatest departure from this average behaviour occurs for light nuclei, as we would expect, but it is interesting to note how the qualitative features, and in particular the low mass peaks in the binding energy for even–even nuclei, are reproduced by a formula which was only designed to fit nuclei with $A \gg 20$ (Fig. 1.5.7).

We must now put some numbers into the semiempirical mass formula and quote the form given by Wapstra in the *Handbuch der Physik*, XXXVIII/1: Wapstra gives for the binding energy of a nucleus $B(Z, A)$:

$$B(Z, A) = 15.835A - 18.33A^{2/3} - 0.1785(A-I)^2/A^{1/3}$$
$$- 23.2I^2/A \pm \delta(A_{\text{even}}) \text{ MeV} \quad (1.5.5)$$

where $I = N-Z$, $A = N+Z$, $\delta = 11.2A^{-1/2}$. You should note that the fitted parameters have varied appreciably in the past with who has done the fitting and, since we are concerned here only with an elementary discussion of the principles of nuclear physics, we refer you to Wapstra's article for a discussion. These numbers were fitted to nuclei in the regions *between* the *magic* peaks in the binding energy.

The semiempirical mass formula (originally due to Bethe and von Weizsäcker) provides a remarkably accurate fit to the masses of stable nuclei. It has, however, little predictive value—if it is desired to attempt a prediction of the binding energy of a hitherto unobserved nucleus the formula is too rough to be valuable and for such work formulae expected only to hold in a localized region, possibly with a larger number of fitted parameters, are used for interpolation or extrapolation from nuclei of known mass. The semiempirical mass formula gives a good picture of the gross features of the binding energy, however, and provides, through the simplicity of the terms, some insight into what is going on.

Let us now use the semiempirical mass formula to look at the behaviour of the binding energy as a function of the number of nucleons. Since the formula was developed primarily

---

† δ is always taken to have a simple variation with A so that only one extra parameter is involved here.

The Nuclear Periodic Table

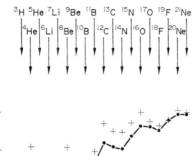

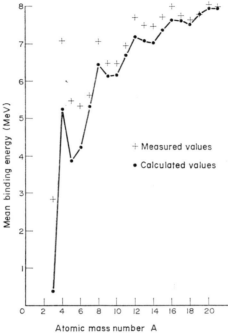

FIG. 1.5.7. The mean binding energies of nuclei with $A \leq 20$ are compared with the values calculated from the semiempirical mass formula, eq. (1.5.5). The pairing and asymmetry terms succeed in reproducing the peaks at ^{4}He, ^{8}Be, ^{12}C, ^{16}O and ^{20}Ne quite surprisingly well. [The measured values are taken from the compilation given by Wapstra, Atomic masses of nuclides, Handbuch der Physik, XXXVIII/1 (Springer-Verlag, 1958).]

to describe stable nuclei, we will first calculate the equation of the maximum stability curve, and then examine small excursions about this line. We write:

$$B(Z, A) = \alpha A - \beta A^{2/3} - \varepsilon Z^2/A^{1/3}$$
$$-\gamma (Z-A/2)^2/A \pm \delta(A_{\text{even}})$$

and for the time being we will drop the pairing term. Then if we treat Z as a continuous variable, the value of Z for which $B(Z, A)$ is a maximum for fixed A is given by:

$$\frac{\partial B(Z, A)}{\partial Z} = 0.$$

The maximum binding energy will not, however, give the line of maximum stability! The proton and neutron have different masses ($n \rightarrow p + e^- + \nu$) and so the equation we must

Nuclear Physics

use is:

$$\frac{\partial M(Z, A)}{\partial Z} = 0$$

or

$$M_H - M_n + 2Z_s \varepsilon A^{-1/3} + \gamma(2Z_s - A)/A = 0,$$

$$Z_s = \frac{M_n - M_H + \gamma}{2[\varepsilon A^{-1/3} + \gamma A^{-1}]}.$$

(The value of Z_s given by this expression is not integral—a consequence of treating Z as a continuous variable.)

Note that the volume energy $(-\alpha A)$ and the surface energy $(\beta A^{2/3})$ do not contribute to the equation of the stability curve, depending only on A and not on how A is divided between protons and neutrons. In the limit of very small A, $Z_s \to A/2$ while in the limit of very large A, $Z_s \to A^{1/3}\gamma/2\varepsilon$. The position of the most stable nucleus of given A is determined by the competing effects of the coulomb repulsion (reduced by reducing Z) and the asymmetry energy (increased by reducing Z). You should note that the minimum mass does *not*

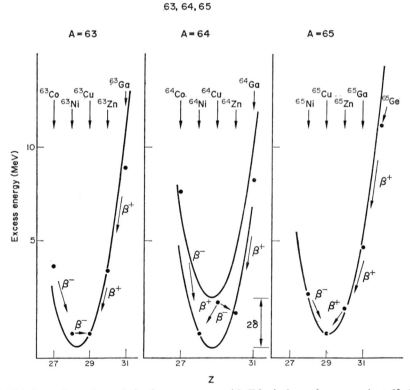

FIG. 1.5.8. This figure shows the variation in mass energy with Z for isobars of mass numbers 63, 64 and 65. The two odd A families are each represented by one parabola and have only one stable isobar, $Z = 29$ in each case. The even A family is represented by two parabolae and has two stable isobars, $Z = 28$ and 30. The smooth curves were calculated from eq. (1.5.5) and normalized to the most stable isobar in each case.

The Nuclear Periodic Table

correspond to a minimum in the parabolic asymmetry energy term alone, because of the coulomb energy.

We may now consider excursions about the point of minimum mass, for fixed A. We expect in general only one stable nucleus of given odd A, a neutron or proton excess being corrected by β-decay which changes the charge by one unit.

The situation is a little more complicated for even A nuclei. The pairing energy splits the parabola into two separated by 2δ (A even). We may have two stable isobars of given A (not to be confused with isotopes which have constant Z, varying A) and it is possible to have nuclei which decay via both β^+ and β^- emission. Pairs of stable isobars first appear at $^{36}_{16}$S, $^{36}_{18}$A. All these points are illustrated in Fig. 1.5.8.

At this point we leave the semiempirical mass formula, although we shall return to use it in other places. We built it up by appealing first to the *liquid drop model of the nucleus* for the first three terms—the volume energy, the surface energy, and the coulomb energy. The asymmetry energy was justified by using an independent particle model. In this model we considered single particle wave functions as constants of the motion, a point of view drastically opposed to the approach used for the first three terms where we did not consider single particle wave functions at all! Finally, in our discussion of the pairing energy, we started with the independent particle model and then began to destroy it by introducing correlations between nucleons through the residual interaction. This rapid changing of ground should impress upon you the horrible complexity of the nuclear many body problem—we find ourselves using violently opposed approximations simultaneously.

We have now reached an appropriate point to discuss the nuclear independent particle model.

1.6. The independent particle model, magic numbers and shell structure within the nucleus

We have already introduced ideas from the independent particle model of the nucleus to help fix up the semiempirical mass formula. We shall now go over the ground again, being more explicit about the physics involved. The basic assumption of the (extreme) independent particle model is that the potential acting on a nucleon as a result of all the two nucleon interactions may be treated as an average *time-independent* potential, so that the individual single particle bound states are eigenstates of the motion. A similar assumption is made in the optical model for nuclear reactions, which we mentioned while discussing nuclear radii. At first sight this assumption seems entirely ridiculous—here are two reasons why it seems ridiculous. The first is experimental: at relatively low energies ($\lesssim$ a few MeV) nuclear reactions proceed via a long-lived compound nucleus. An extreme example is provided by elastic scattering and (n, γ) reactions of thermal and epithermal neutrons, in which the cross-sections exhibit narrow peaks, which for heavy elements (such as silver, cadmium, uranium) have widths Γ and spacing $\sim eV$. Writing $\Gamma\tau = \hbar$ a width of 1 eV corresponds to a lifetime $\tau \sim 10^{-15}$ sec, while the time taken to cross a nucleus is $\sim 10^{-21}$ sec. The neutrons drop into the nucleus, forming an excited state with an excitation of ~ 8 MeV. This energy is shared among a large number of nucleons and the state is stable until either one nucleon gets all the

Nuclear Physics

energy through a lucky collision or the state decays electromagnetically. This compound nucleus formation and decay is discussed in more detail in Chapter 4. All we wish to do here is to point out that these excited nucleons do *not* sit in states which are eigenstates of the motion—if they did you would not find these very narrow resonances. Secondly, we will give a physical argument. Suppose we confine an assembly of nucleons within a box (representing an average potential). If there exists no residual interaction between them then collisions between nucleons do not take place—they just pass through one another. Under these circumstances the single particle states are eigenstates, and a particle deposited in one remains there—there is nothing to kick it out. Now suppose that we turn on a residual short-range potential between pairs of nucleons in the box. They can now scatter off each other—which means changing their states, and this is exactly what happens when we drop a slow neutron into a nucleus.

This argument, which would seem to underline the hopelessness of an independent particle model, can, however, be turned to our advantage to provide some physical justification for attempting this task. Suppose that we start with no residual interaction and the nucleons confined in a box. The single particle states are eigenstates of the motion. Let the assembly be in its *ground state*, and switch on the short-range residual interaction. The nucleons may now interact. If two nucleons are to scatter, they change their states *but* all energetically accessible states are already occupied in this overall ground state of the assembly. The Pauli exclusion principle thus *prevents* a change in state, except for two identical nucleons exchanging their roles, a situation which is not distinguishable from the initial set up. Thus, in the *ground state* of such an assembly we might expect that the single particle states are eigenstates of the motion. When this Fermi gas of nucleons is heated (by dropping in a slow neutron, say) nucleons may be lifted by collision into empty states above the top of the Fermi sea, emptying lower-lying states and allowing compound nucleus behaviour.

A crude independent particle model may be set up in the following way. We assume that the time-independent average potential felt by the individual nucleons can be represented by a square well potential

$$V(r) = -V_0 \quad r \leq R,$$
$$V(r) = 0 \quad r > R,$$

and that the wave functions of these nucleons are those of nucleons confined in a spherical box. On p. 38 we mentioned that the number of levels in such a box with momentum between p and $p+dp$ is

$$\frac{4\pi p^2 \, dp}{(2\pi\hbar)^3} V$$

where V is the volume of the box, and the total number of levels for which $p < p_{\max}$ is

$$N = \frac{4}{3}\pi \frac{p_{\max}^3}{(2\pi\hbar)^3} V.$$

Two nucleons of each kind can be fitted into each of such a sequence of levels. If we take

The Nuclear Periodic Table

$R = R_0 A^{1/3}$ then

$$V = \frac{4\pi}{3} R_0^3 A$$

and

$$N = \frac{4\pi}{3} \frac{p_{max}^3}{(2\pi\hbar)^3} \frac{4\pi}{3} R_0^3 A.$$

If p_{max} is the momentum in the topmost level, then ignoring the coulomb energy we expect $Z \simeq N$ and so

$$\frac{A}{4} = \frac{4}{3}\pi \frac{p_{max}^3}{(2\pi\hbar)^3} \frac{4\pi}{3} R_0^3 A$$

and hence p_{max} is independent of A.

Now from the nucleon separation energies we know that the total energy of a nucleon in he highest levels is between 0 and -8 MeV and so we can put

$$E_{max} - V_0 \sim -8 \text{ MeV}$$

where

$$E_{max} = \frac{p_{max}^2}{2M}$$

and hence calculate V_0, which turns out to be $V_0 \simeq 30$ MeV. Now the mean kinetic energy in the nucleus is easily found from the level density to be $\frac{3}{5}E_{max}$ and so the mean binding energy, $\bar{B}$, in the absence of coulomb effects and for $Z \simeq N$ is

$$\tfrac{3}{5}E_{max} - V_0 \simeq \tfrac{2}{5}E_{max} = 12 \text{ MeV}$$

and $\bar{B}$ must lie between 12 and 20 MeV, which is significantly in excess of the value 7–8 MeV found through much of the periodic table. We should remember, however, that we assumed the nucleons to be confined as though in a box of radius R with perfectly reflecting walls. For the potential we have assumed the walls have a height of only V_0, and so the higher-lying states will have wave functions penetrating exponentially into the region of zero potential, and consequently their potential energy has been overestimated. This is how the surface energy effect arises in an independent particle model.

Thus provided that we introduce as an additional postulate the variation of radius with A as $R = R_0 A^{1/3}$, an independent particle model accounts qualitatively for all the principle features of the semiempirical mass formula.

Our arguments have now swung too far from reality in the opposite direction. We know that collective motion exists within the nucleus. Some examples are: (i) α decay and fission —try treating these in terms of single nucleon states!; (ii) large nuclear electric quadrupole moments (see Chap. 2); (iii) the rotational excited states of aspherical nuclei; and (iv) the giant dipole resonances, in which an electric dipole field causes all protons in the nucleus to vibrate against all neutrons. Having argued that the single particle states are eigenstates of the motion, how can we accommodate these phenomena?

If $\psi_1, \psi_2, \ldots, \psi_N$ are single particle wave functions in an independent particle model, the total nuclear wave function will be

$$\Psi = \psi_1 \psi_2 \ldots \psi_N \quad \text{properly antisymmetrized.}$$

Nuclear Physics

The total energy of the nucleus will be the sum of the single particle energies

$$E = E_1 + E_2 + \ldots + E_N.$$

If we now introduce residual interactions written as $\sum_{ij} V(\mathbf{r}_i - \mathbf{r}_j)$ then the energy of the nuclear state is modified by a term

$$\int \Psi^* \sum_{ij} V(\mathbf{r}_i - \mathbf{r}_j) \Psi \, d^3r_1 \, d^3r_2 \ldots d^3r_n$$

in first order. In second order however the residual interactions mix together the independent particle states, including those above the top of the Fermi sea which are unfilled in the independent particle ground state. The ground state of the nucleus is now represented by

$$\Psi_g \rightarrow \Psi_g + \sum_k \Psi_k \int \Psi_k^* \sum_{ij} \frac{V(\mathbf{r}_i - \mathbf{r}_j)}{E_k - E_g} \Psi_g \, d^3r_1 \, d^3r_2 \ldots d^3r_n$$

where Ψ_k are all the other possible states of the whole nucleus constructed from the independent particle approximation. This new wave function does not factor into a product of single particle states (properly antisymmetrized), a result which is obvious because the new single particle states are no longer pure. In the ground state the Pauli principle inhibits this effect but does not prevent it and the new wave function contains correlations between the motion of individual nucleons and so collective behaviour on a scale determined by the relative importance of the time—independent potential and the residual interactions that must be superimposed.

Thus if we say that we have an independent particle model *with residual interactions* all possibilities are open to us. If the residual interactions are relatively weak, we expect an approximation to the extreme independent particle model in which the single particles sit in eigenstates. If the "residual" interactions are predominant, we expect this approximation to be useless and move in the direction of the liquid drop model in which we never even mention individual particle states, or with very strong and specific residual interactions a cluster model like the old α-particle model of nuclei. *A priori* we have no way of knowing which limit is right—we might make a pessimistic guess and suppose the truth to be somewhere in between. It is, and is treated in the framework of the collective model developed by Aage Bohr and Mottelson,[†] but the starting point is the independent particle model rather than the liquid-drop model.

After all these preliminary haverings, let us turn to the evidence which stimulated interest in the independent particle model and discuss its development.

The result which primarily raised interest in the independent particle model was the observation that nuclei containing 2, 8, 20, 50, 82 or 126 nucleons of one kind are extra stable, having binding energies greater than predicted by the semiempirical mass formula (including the asymmetry and pairing energy terms). This is illustrated in Fig. 1.6.1. 2, 8, 20, 50, 82, 126 are the famous magic numbers. A nucleus in which either the number of protons or the

---

[†] A. Bohr and B. R. Mottelson, *Nuclear Structure*, vol. I (Benjamin, 1969); vol. II and vol. III in preparation.

The Nuclear Periodic Table

number of neutrons is magic is called a magic nucleus, a nucleus in which both numbers are magic is said to be doubly magic (an example is $^{208}_{82}Pb_{126}$).

The extra stability manifests itself in various ways:

1. Magic nuclei are more tightly bound than their neighbours.
2. Nuclei with a magic number of protons have rather more stable isotopes than their neighbours, nuclei with a magic number of neutrons have more stable isotones.
3. The separation energies for nucleons show a rapid variation at the magic numbers: one nucleon over a magic number is less tightly bound than the preceding nucleons building up to a magic number and so the magic numbers are indicated by a discontinuity in the separation energy of the appropriate kind of nucleon.
4. Magic nuclei tend to be more abundant than their neighbours in the nuclear periodic table.

There are other indicators of magicity which we will discuss later. (Islands of isomerism, nuclear quadrupole moments.) This behaviour is reminiscent of the extra stability of the inert gases (and the extra reactivity of the halides and alkali metals) which is explained in terms of closure of a shell of electrons. Our interest is further roused by the seductive ease with which an extreme independent particle model explains the first three magic numbers.

Consider once more nucleons being fed into a simple harmonic oscillator potential. The single-particle levels fall in degenerate groups separated by $\hbar\omega$, with the lowest state lying $\frac{3}{2}\hbar\omega$ above the bottom of the potential. (In an independent particle model the radius increases with each particle added and ω will change too.) As we discussed earlier, the single particle energies depend on the quantity:

$$\hbar\omega[n_1+n_2+n_3+3/2]:$$

$n_1+n_2+n_3$	Energy	Max. occupation no.	Sum
0	$\frac{3}{2}\hbar\omega$	2	2
1	$\frac{5}{2}\hbar\omega$	6	8
2	$\frac{7}{2}\hbar\omega$	12	20
3	$\frac{9}{2}\hbar\omega$	20	40
4	$\frac{11}{2}\hbar\omega$	30	70

The maximum occupation number is just the number of independent ways of choosing n_1, n_2 and n_3 for a given value of the sum, multiplied by 2 for the two independent spin states.

We thus seek an explanation of the magic numbers in terms of closure of shells of nucleons. A shell is a group of independent particle levels which are either degenerate or lie very much closer in energy than the separation between different shells.

For a simple harmonic oscillator potential, we have the following shells:

$$(1s), (1p), (2s, 1d), (1f, 2p), (1g, 2d, 3s), (1h, 2f, 3p)$$

Nuclear Physics

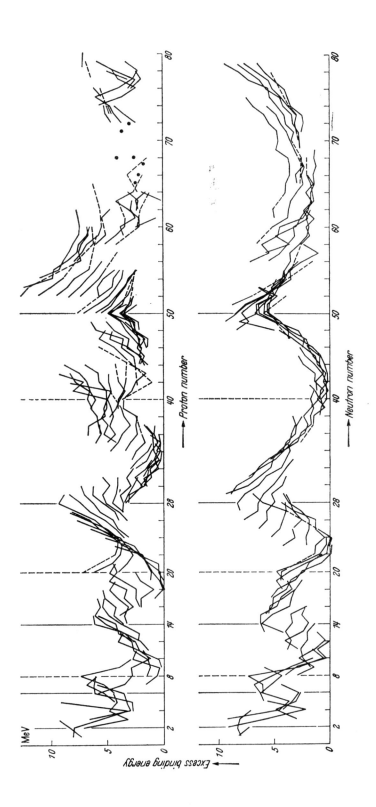

The Nuclear Periodic Table

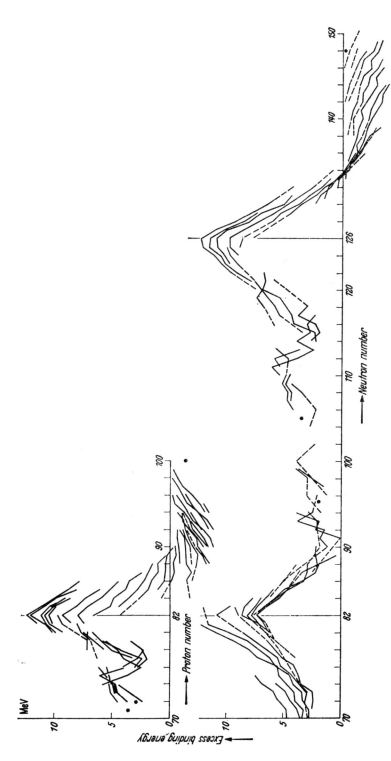

FIG. 1.6.1. Excess of experimental binding energies over the semiempirical mass formula, eq. (1.5.5). The upper parts show this quantity for isotones (constant neutron number) as a function of Z and the lower parts as a function of N for isotopes. [From A. H. Wapstra, Atomic masses of nuclides, *Handbuch der Physik*, XXXVIII/1 (Springer-Verlag, 1958).]

Nuclear Physics

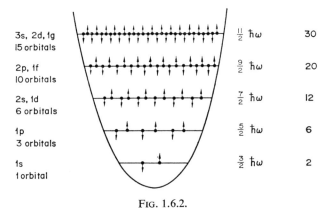

FIG. 1.6.2.

closing at 2, 8, 20, 40, 70, 112 nucleons of one kind in the nucleus. You should note that the independent particle levels do *not* have to have the same angular momentum or the same principle quantum number—the crucial property is the near degeneracy within a shell (exact for a pure SHO potential). The discontinuity introduced in the binding energy by the closure of a shell can alone explain most of the indicators of magicity. The total binding energy increases approximately linearly as nucleons are added to the nucleus, suffers a discontinuity as on passing a shell closure the next nucleon has to go into a much higher energy level, and then starts increasing again. Thus a local maximum develops in the total binding energy and this is reflected in the abundances of magic nuclei. The greater number of stable isotopes (for magic Z) or isotones (for magic N) may be attributed to the occurrence of a big gap in the spacing of the proton levels (at magic Z) while the neutron levels do not have such a gap, unless N is also magic. Thus many neutrons may be added to a magic Z nucleus before the most energetic neutron is raised above an accessible proton level (Sn, $Z = 50$, has 10 stable isotopes) and the mirror argument applies to isotones with magic N.

The separation energies show a discontinuity, but do not have a maximum on shell closure. Thus as a shell of neutrons is filled up, leaving the number of protons constant, the mean binding energy remains approximately constant until one over a closed shell is added. The closed shells do not exhibit an abnormal stability over the same partially filled shells. (The separation energy does have maxima relative to an average smooth curve such as that provided by the semiempirical mass formula.)

The effect of shell closure on nuclear quadrupole moments is discussed in Chapter 2.

If we wish to pursue this philosophy beyond the magic numbers at 2, 8, 20, we have to find some explanation for the absence of magic numbers at 40, 70, 112 and the presence of magic numbers at 50, 82 and 126. It is not necessary to stick to a simple harmonic oscillator potential in our model, but the potential we use must be of relatively short range—something like a square well with rounded corners. However, while a square well potential removes the degeneracy between states of different orbital angular momentum in a given SHO shell, it does not split up these levels sufficiently to produce a new scheme of shell closure, and hence the magic numbers 50, 82 and 126. (The splitting reduces the energy of states with higher orbital angular momentum relative to those with lower orbital angular momentum. The reason is that the centrifugal barrier term $\{(l(l+1)\}/r^2$ which occurs in the

The Nuclear Periodic Table

radial wave equation pushes wave functions of higher l away from the centre and makes them more sensitive to the edge of the potential which drops faster for a square well than for a SHO well.) The explanation for these values was found by Maria Mayer[†] and independently by Haxel, Jensen and Suess.[‡] These physicists postulated a strong spin–orbit coupling in individual single particle states, so that the degeneracy between states with the same orbital angular momentum but different total angular momentum be removed. For example, the six degenerate $1p$ states now split into a $1p_{\frac{3}{2}}$ level (occupation number 4) and a $1p_{\frac{1}{2}}$ level (occupation number 2 identical nucleons). With the assumption that the levels with the highest angular momentum are depressed in energy, and also assuming that the depression increases in magnitude with the nucleon orbital angular momentum, an entirely new level sequence may be generated, which reproduces the observed magic numbers (Fig. 1.6.3).

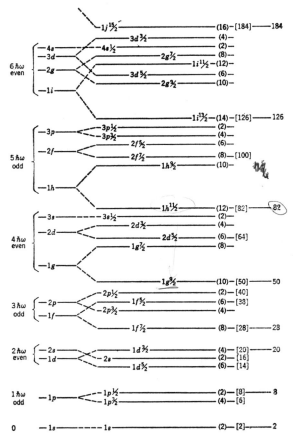

FIG. 1.6.3. Sequence of single-particle energy levels. The initial splitting from the SHO sequence is due to anharmonic terms in the central potential and the final splitting to the spin–orbit coupling. The crucial point about the spin–orbit splitting is that it brings the lowest level of a shell with little or no spin–orbit coupling down into the shell below. [From M. G. Mayer and J. H. D. Jensen, *Elementary Theory of Nuclear Shell Structure* (Wiley, 1955).]

† *Phys. Rev.* **75**, 1969 (1949); *Phys. Rev.* **78**, 16 (1950).
‡ *Phys. Rev.* **75**, 1766 (1949).

Nuclear Physics

This scheme succeeds not only in reproducing the magic peaks in the binding energy. You will remember that we indicated how a short-range residual interaction is capable, with j–j coupling, of reproducing the observed result that all even–even nuclear ground states have spin zero. Here is a scheme that allows us to see why j–j coupling should be important—the single particle levels are eigenstates of j and different j's are not degenerate even for the same orbital angular momentum. We can deduce more: if even–even cores have angular momentum zero, odd A nuclei will have the angular momentum of the extra nucleon, namely the unpaired nucleon in the highest level. This angular momentum may be read off at once from the single particle level sequence of Fig. 1.6.2. If the spin–orbit coupling is chosen so as to reproduce the magic numbers, then the spins of almost all odd A ground states, with a very few exceptions, are correctly predicted by this scheme. Finally, it is possible to determine not only the angular momentum but also the orbital angular momentum of the odd nucleon, thus distinguishing between, say, a $p_{\frac{3}{2}}$ and a $d_{\frac{3}{2}}$ state. The magnetic moment of the nucleus is sensitive to l as well as j and measurement of the magnetic moment allows a distinction between, say, $p_{\frac{3}{2}}$ and $d_{\frac{3}{2}}$ if we assume that the magnetic moment is almost entirely due to the odd nucleon. This will be discussed further in the next chapter: here we merely state that not only does the strong spin–orbit coupling scheme give the correct j value for odd A nuclear ground states, it also gives the correct orbital angular momentum.

Spin–orbit coupling means that we replace the assumed central potential in which all the nucleons are supposed to move with a potential

$$V(r) + \xi(r)\,\mathbf{l}\cdot\mathbf{s}$$

and the value of $\mathbf{l}\cdot\mathbf{s}$ for a state of given l and s is given by

$$2\mathbf{l}\cdot\mathbf{s} = j(j+1) - l(l+1) - s(s+1).$$

For $l \neq 0$ such a term splits states of given l and s into states characterized by $j = l+\frac{1}{2}$ and $j = l-\frac{1}{2}$, the splitting in energy being proportional to $2l+1$. With the sign of $\xi(r)$ such that the states with $j = l+\frac{1}{2}$ are lower than the states with $j = l-\frac{1}{2}$, we can generate the level sequence for nuclei. A spin–orbit potential is present in atomic structure, where it appears as a result of the interaction of the magnetic moment of the electron with the magnetic field generated by its orbital motion. In this case the splitting is such that states with $j = l-\frac{1}{2}$ are found at lower energies than a states with $j = l+\frac{1}{2}$, and is correctly given for single electron states by the Dirac equation. The splitting for the $1p$ level in hydrogen is $\sim 10^{-4}$ eV.

Electromagnetic effects are completely inadequate to account for the observed spin–orbit splitting in nuclei and this property must be ascribed to the strong interactions. Another manifestation of spin–orbit coupling is found in transverse polarization phenomena in nucleon–nucleus scattering. Suppose that a potential $\propto -\mathbf{l}\cdot\mathbf{s}$ exists when a nucleon is scattering off a nucleus ($\mathbf{l}$ and $\mathbf{s}$ both belong to the nucleon). Then the crude picture shown in Fig. 1.6.4 makes it plausible that nucleons with one polarization will be scattered preferentially to the left, nucleons with the other, to the right.

We emphasize that the coupling is between the $\mathbf{l}$ and $\mathbf{s}$ of the nucleon, and the term has no direct connection with the spin of the nucleus—^{12}C with spin 0 is an excellent polarizer: see Fig. 1.6.5.

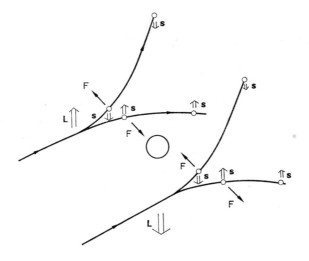

$F(r) = K(r) L \cdot s$

In this simple picture, spin up is scattered preferentially to the right, spin down to the left

FIG. 1.6.4. A semiclassical illustration of how a spin–orbit coupling in the interaction between a nucleon and a nucleus gives rise to polarization in scattering.

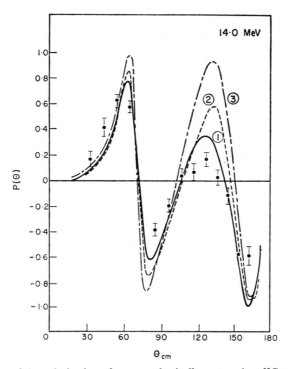

FIG. 1.6.5. The variation of the polarization of protons elastically scattered on ^{12}C at an energy of 14 MeV. The dots are experimental points and the curves represent various optical model predictions with a spin–orbit potential included in the optical potential. [From Nodvik, Duke and Melkanoff, *Phys. Rev.* **125**, 975 (1962).]

Nuclear Physics

Let us sum up the successes of the shell model of the nucleus with j–j coupling:

(i) It predicts correctly the magic numbers in binding energy.
(ii) It predicts correctly spin and magnetic moments for odd A nuclei (even–even nuclei are trivial, odd–odd nuclei require additional assumptions).
(iii) It makes correct predictions of the existence of "islands of isomerism". Isomers are long-lived unstable states of nuclei which decay electromagnetically (long-lived means $\tau \gtrsim$ sec in this context), and the level scheme provides an explanation: if the isomeric states are essentially single particle excited states, then as the shells at 82 and 126 near closure, the possibility of transition between two levels very close in energy, but with very different angular momenta, occurs. These are just the circumstances that give rise to anomalously long-lived states decaying electromagnetically (see Chap. 3). (This means that nuclear excited states are correctly predicted. A nearly magic core will be reluctant to exhibit collective states at low excitation, and some of the first few excited states may be given by the shell sequence.)
(iv) The level scheme makes predictions about β-decay spectra (see Chap. 3).

We shall return to the shell model in the next two chapters, where appropriate.

1.7. Nuclear forces

We conclude this chapter with a brief discussion of the nuclear forces that we deduce from the properties of the nuclear periodic table.

The variation of the principle term in the binding energy, the volume term, with A implies that the nucleon–nucleon forces are of short range—each nucleon interacts with a finite number only of other nucleons in the nucleus. In contrast, the coulomb energy of the nucleus varies as $\sim Z^2$: the electrostatic potential varies as $1/r$ and is a long-range force. The variation of the nuclear radius with $A^{1/3}$ implies a constant density for nuclei which means that the nucleons cannot get too close to each other—a conclusion which is also reached from the dependence of the binding energy on A rather than on A^2. We have no alternative but to conclude that the nucleon-nucleon interaction is NOT wholly attractive between all pairs of nucleons, even though it is short range. The reason is simple: let us suppose that the two nucleon potential is a simple attractive square well. Then the nuclear forces in the nucleus would be attractive between all pairs of nucleons separated by less than the range of the potential. Suppose that *all* nucleons are within range of each other. Then the depth of the well experienced by any one nucleon $\propto A-1$, and there are A nucleons, the binding energy, even for a short-range potential, would vary $\sim A^2$ unless there is something to keep the nucleons apart, so that only a finite number are within range of any one nucleon.

The Pauli exclusion principle modifies this simple picture but does not affect the conclusion. The binding energy of a given nucleon is the difference between the (negative) potential energy and the kinetic energy. Successive nucleons have to go into higher and higher energy states because of the exclusion principle. Let us make a simple calculation: the number of states dN between momentum p and momentum $p+dp$ for a given radius is $\propto p^2\,dp$. Then given a potential well of depth V_0 each nucleon has a binding energy of $V_0 - p^2/2M$ and the

The Nuclear Periodic Table

total binding energy is:

$$B = AV_0 - 4 \int_{E \sim 0}^{E \sim V_0} E \, dN$$

$$= AV_0 - 4 \frac{C}{5} \frac{(2MV_0)^{5/2}}{2M}$$

where C is a constant which we can get by noting that

$$\int_{E \sim 0}^{E_0 \sim V} dN \simeq A/4$$

and if

$$dN = Cp^2 \, dp,$$
$$A/4 = \tfrac{1}{3} C(2MV_0)^{3/2}$$

or

$$C = \tfrac{3}{4} \frac{A}{(2MV_0)^{3/2}}$$

when the binding energy becomes $B = AV_0(1-\tfrac{3}{5})$ which is positive and $\propto AV_0$. Now if all nucleons are within range of each other, $V_0 \propto (A-1)$ and so the binding varies as A^2, not as A, even after including the Pauli exclusion principle.[†] So if we had only a short-range attractive interaction between *all pairs*, all nuclei would have approximately the same radius (the "radius" of the two nucleon potential) and nuclear binding energies would have a dominant term A^2.

We can get out of this difficulty very easily by postulating a short-range attractive interaction, superseded at very short ranges by a repulsive force, so that the two nucleon potential has a minimum at $\sim 10^{-13}$ cm separation. Another way out is to introduce forces which are not attractive between *all* pairs, but are attractive or repulsive depending on the relative states of the two nucleons. These are *exchange forces*.

This term can cause some confusion. The force between any two given particles is mediated by exchange of an appropriate virtual particle (i.e. a particle for which $p^2c^2 - E^2 \neq -m^2c^4$). The static coulomb interaction between two charges is mediated by the exchange of longitudinally polarized virtual photons. (It is easy to see that they are longitudinally polarized: two charges attract or repel and momentum is exchanged between the two along the line of centres. The electric field is in this direction also, so $\mathbf{p} \cdot \mathbf{E} \neq 0$ in contrast to the free electromagnetic field.) The static coulomb interaction is a conservative central field, depending only on the radial separation of the two charges and not on their relative angular momenta, spin states or any other property.

The interaction between two nucleons is believed to be mediated by the exchange of mesons: the π-meson[‡] (postulated by Yukawa), and *many others*. Forces mediated by the exchange of virtual particles may or may not be of the "exchange" kind and this term is

---

[†] For a more complete discussion, see J. M. Blatt and V. Weisskopf, *Theoretical Nuclear Physics*, chap. III (Wiley, 1952); E. Fermi, *Nuclear Physics*, chap. VI, p. 111 (Chicago, 1950).
[‡] See Chapter 6.

Nuclear Physics

reserved in the jargon of nuclear physics for forces which depend on the relative states of two particles and not just on their separation.

An example of such a force is one which depends on the relative orientations of the spins of the two nucleons. Suppose we have a potential which contains a term $\propto \sigma_1 \cdot \sigma_2$. The two nucleons may be in a triplet or a singlet state, of total spin s. Now

$$\sigma_1 + \sigma_2 = s \qquad (\sigma_1+\sigma_2)^2 = s(s+1)$$

so

$$\sigma_1 \cdot \sigma_2 = \tfrac{1}{2}\{s(s+1) - 2\sigma(\sigma+1)\}$$

where $\sigma = \tfrac{1}{2}$. Then the operator $\sigma_1 \cdot \sigma_2$ has eigenvalues $\tfrac{1}{4}$ for $s=1$ (triplet) and $-\tfrac{3}{4}$ for $s=0$ (singlet). A potential dependent on $\sigma_1 \cdot \sigma_2$ will thus be attractive in the singlet state and repulsive in the triplet state or vice versa. It is plausible that such forces exist—some of the mesons which may be exchanged have spin 1 (and in any case the deuteron is the only bound state of two nucleons). This is just a spin–spin dependent potential (a classical example is the *magnetic* interaction of two electrons), but is also known as a spin exchange (or Bartlett) potential. What has it to do with exchange? Consider the operator $\tfrac{1}{4} + \sigma_1 \cdot \sigma_2$. Let it operate, for example, on an antisymmetric spin wave function of two spin $\tfrac{1}{2}$ particles which we write as

$$\chi(1,2) = \left[\binom{1}{0}_1 \binom{0}{1}_2 - \binom{1}{0}_2 \binom{0}{1}_1\right].$$

Each σ has three components (the Pauli spin matrices)

$$\sigma_x = \frac{1}{2}\begin{pmatrix}0&1\\1&0\end{pmatrix} \quad \sigma_y = \frac{1}{2}\begin{pmatrix}0&-i\\i&0\end{pmatrix} \quad \sigma_z = \frac{1}{2}\begin{pmatrix}1&0\\0&-1\end{pmatrix}$$

and σ_1 only operates on the wave function of particle 1, σ_2 only on the wave function of particle 2.

$$\left(\frac{1}{4}+\sigma_1\cdot\sigma_2\right)\chi(1,2) = \frac{1}{4}\chi(1,2)$$

$$+ 1/4\left\{\binom{0}{1}_1\binom{1}{0}_2 - \binom{0}{1}_2\binom{1}{0}_1\right\} \qquad x\text{ term}$$

$$+ 1/4\left\{(i)\binom{0}{1}_1(-i)\binom{1}{0}_2 - (-i)\binom{0}{1}_2(i)\binom{1}{0}_1\right\} \qquad y\text{ term}$$

$$+ 1/4\left\{-\binom{1}{0}_1\binom{0}{1}_2 + \binom{1}{0}_2\binom{0}{1}_1\right\} \qquad z\text{ term}$$

$$= 1/2\left\{\binom{0}{1}_1\binom{1}{0}_2 - \binom{0}{1}_2\binom{1}{0}_1\right\} = 1/2\,\chi(2,1).$$

So (apart from a numerical factor) $\tfrac{1}{4} + \sigma_1 \cdot \sigma_2$ is an operator which *exchanges* the spins of the two particles. Thus a potential which contains $\sigma_1 \cdot \sigma_2$ is called a spin exchange potential. Charge exchange can be similarly treated, most easily in the framework of the isotopic spin formalism (Heisenberg potentials), while the exchange of both charge *and* spin is equivalent to spatial coordinate exchange (Majorana forces). The term Wigner potential is applied to potentials which generate non-exchange forces.

For further details you may consult Blatt and Weisskopf, *Theoretical Nuclear Physics*, chap. III, where it is shown that a mixture of Wigner and Majorana forces, with the latter predominating, could account for the constant nuclear density and the saturation of the forces. We will not pursue the matter further.

The role of exchange forces (which certainly exist) has become less critical with the discovery that the nucleon–nucleon interaction has a hard (repulsive) core at short distances with a radius of about 0.5×10^{-13} cm. This repulsion at very short range would serve to stop nuclear collapse even in the absence of exchange forces.

The study of *pp* and *np* scattering at energies up to 300 MeV has resulted in the construction of potentials which are capable of reproducing the scattering data—and are genuine potentials in the sense that they are not energy dependent. Some theoretical effort has been devoted to attempts to reproduce the binding energy of three nucleon systems with such potentials as an input, but this is as far as attempts to calculate nuclear structure from the fitted two nucleon potentials have gone.

We conclude this chapter with a quotation from the abstract at the beginning of the paper discussing one of these potentials:[†] "The potential employs a hard core and is different for singlet-even, singlet-odd, triplet-even and triplet-odd states. It consists of central, tensor, spin–orbit and quadratic spin–orbit parts...."[‡]

[†] K. E. Lassila *et al.*, *Phys. Rev.* **126**, 881 (1962).

[‡] "Singlet" and "triplet" refer to the spin states of the two nucleons. "Even" and "odd" refer to the relative orbital angular momenta. A tensor force varies with the angular coordinates as well as the radial coordinate.

CHAPTER 2

Angular Momentum and the Nucleus

2.1. Central forces, orbital angular momentum and spin

In the last chapter we pointed out that an assembly of nucleons may be described in terms of an independent particle model with residual interactions, the kind of picture of the nucleus which emerges depending on the strength of the residual interactions. We may therefore begin a discussion of angular momentum and the nucleus with the properties of a single-particle state. To begin with we consider only orbital angular momentum and treat a (spinless) particle moving in a potential.

It is well known that if the potential is spherically symmetric, then the orbital angular momentum is a constant of the motion. Here is a rather easy proof:

The vector operator for orbital angular momentum is

$$\mathbf{L} = \mathbf{r} \times \mathbf{p} \tag{2.1.1}$$

when the angular momentum of a particle with wave function ψ is given by $\int \psi^* \mathbf{L} \psi \, dV$. The rate of change of orbital angular momentum with time is therefore

$$\frac{d}{dt}\left(\int \psi^* \mathbf{L} \psi \, dV\right) = \int \left[\left(\frac{\partial \psi^*}{\partial t}\right) \mathbf{L} \psi + \psi^* \left(\frac{\partial \mathbf{L}}{\partial t}\right) \psi + \psi^* \mathbf{L} \left(\frac{\partial \psi}{\partial t}\right)\right] dV. \tag{2.1.2}$$

Now

$$H\psi = i\hbar \frac{\partial \psi}{\partial t}.$$

So

$$\frac{d}{dt}\left(\int \psi^* \mathbf{L} \psi \, dV\right) = \int \left[-\frac{1}{i\hbar} \psi^* H \mathbf{L} \psi + \psi^* \frac{\partial \mathbf{L}}{\partial t} \psi + \frac{1}{i\hbar} \psi^* \mathbf{L} H \psi\right] dV.$$

$\mathbf{L}$ has no explicit time dependence ($\mathbf{r}$, $\mathbf{p}$ are defined by ψ) and so if the orbital angular momentum does not change with time then $(\mathbf{L}H - H\mathbf{L}) = 0$, written

$$[\mathbf{L}, H] = 0, \tag{2.1.3}$$

Angular Momentum and the Nucleus

or the angular momentum operator *commutes* with the Hamiltonian H.†

$$H = \frac{p^2}{2m} + V \qquad \mathbf{L} = \mathbf{r} \times \mathbf{p} = \frac{\hbar}{i}(\mathbf{r} \times \nabla).$$

So
$$[\mathbf{L}, H] = \frac{\hbar}{i}\{(\mathbf{r} \times \nabla)V - V(\mathbf{r} \times \nabla)\} \qquad (2.1.4)$$

which is zero if and *only if* V is spherically symmetric. Consider, for example, L_x:

$$L_x = yp_z - zp_y = \frac{\hbar}{i}\left\{y\frac{\partial}{\partial z} - z\frac{\partial}{\partial y}\right\},$$

$$\left\{y\frac{\partial}{\partial z} - z\frac{\partial}{\partial y}\right\}V(r) = y\frac{\partial}{\partial z}V(r) - z\frac{\partial}{\partial y}V(r)$$

$$+ yV(r)\frac{\partial}{\partial z} - zV(r)\frac{\partial}{\partial y} \qquad (2.1.5)$$

and
$$\frac{\partial}{\partial z}V(r) = \frac{\partial V}{\partial r}\frac{\partial r}{\partial z} = \frac{z}{r}\frac{\partial V}{\partial r}.$$

So that if V is a function only of r—a *central potential*—then the orbital angular momentum is a constant of the motion.

If we take the Schrödinger equation

$$\nabla^2 \psi + \frac{2m}{\hbar^2}(E - V)\psi = 0$$

where V is a function of r only, then we may separate the equation by writing

$$\psi = R(r)\Theta(\theta)\Phi(\phi)$$

in spherical polar coordinates r, θ, ϕ and expect to find that the angular part, for a central field, does not depend on the form of V. The result is very well known:

$$\Theta(\theta) = P_l(\cos\theta)$$
$$\Phi(\phi) = e^{im\phi} \qquad m \leq l$$

and
$$\frac{1}{r^2}\frac{d}{dr}\left(r^2\frac{dR}{dr}\right) + \frac{2m}{\hbar^2}\left[E - V(r) + \frac{\hbar^2}{2m}\frac{l(l+1)}{r^2}\right]R = 0 \qquad (2.1.6)$$

where l is the angular momentum, m is its third component. l^2 and m are *constants of the motion* and have eigenvalues $l(l+1)$ and m respectively, where l and m are integers. But it must be emphasized that this separation is possible *only* for a central field.

Let us suppose that we have a particle moving in a predominantly central field, to which a small non-central component is added: $V = V(r) + U(r, \theta, \phi)$. The orbital angular momen-

† For another way of getting this result, see L. D. Landau and E. M. Lifshitz, *Quantum Mechanics*, 2nd ed., sect. 26 (Pergamon Press, 1965).

Nuclear Physics

tum is no longer a constant of the motion. We can find out what the new wave function looks like by applying perturbation theory. (This is why we required $U(r, \theta, \phi) \ll V(r)$.) To first order the energy levels are changed by an amount $\int \psi^* U(r, \theta, \phi) \psi \, dV$ where ψ is a solution for $V = V(r)$ and to compute the new wave functions we must go to second-order perturbation theory. The *correction* to the wave functions is given by adding a term

$$\sum_m \psi_m \int \frac{\psi_m^* U \psi_n}{E_n - E_m} dV$$

to ψ_n for all terms *except* $m = n$. If U is a function *only* of r then the integral is non-zero *only* when ψ_m and ψ_n have the *same* values of both l and m because of the orthogonality properties of the angular functions. If U is a function only of r, θ then the integral is non-zero only when ψ_m and ψ_n have the same value of m while if U is a function of r, θ, ϕ the integral may be non-zero for different l and m. Thus the effect of a non-central potential is to mix states of different orbital angular momentum L or of different L_z: the eigenstates of the potential are mixtures of different orbital angular momentum states and orbital angular momentum is no longer a constant of the motion. The time derivative of the orbital angular momentum is no longer zero because the potential is no longer spherically symmetric.

Alternatively, if we start a particle off in a state of given orbital angular momentum —which is not an eigenstate of the motion any more—then we may solve a time-dependent problem and find as t increases that the perturbation mixes in more and more of states of different orbital angular momentum: orbital angular momentum is not conserved by a non-central interaction.

An example in nuclear physics is provided by the deuteron, which is basically a bound s-wave state of a proton and a neutron. A non-central force is also present however, and mixes into the deuteron about 4% of a d-wave function.[†]

Examples of non-central fields abound in nature—an obvious classical example is the magnetic interaction between two magnetic dipoles. In general, then, we have no reason to expect that the orbital angular momentum of a particle will be a conserved quantity. If we wish to retain the principle of conservation of angular momentum, then the orbital angular momentum of our particle plus some other angular momentum quantity must be a constant of the motion. These are several possibilities:

1. If the motion of the particle is in an *external field* then the orbital angular momentum of the particle need not be conserved since the external field can transfer angular momentum to the external system generating the field. For example, if we study the motion of a particle in an electromagnetic field established by laboratory apparatus, then the apparatus, and hence the earth, can absorb angular momentum from the particle and the angular momentum of the particle alone would not necessarily be conserved.

2. Consider a system consisting of a π^--meson (spin zero) and a nucleus. Suppose that the field in which the π^- moves is generated only by the nucleus and is non-central. Then if total angular momentum is to be conserved, the nucleus must be capable of absorbing angular momentum from the π and so must have an intrinsic angular momentum. This may be due in principle just to the orbital motion of the constituent nucleons, in which case neither

---

[†] See Section 4.8.

Angular Momentum and the Nucleus

of the two orbital angular momenta would be individually conserved, but the (vector) sum would be.

3. Suppose we consider two particles in orbit about each other. If the interaction between them is non-central and we are to preserve conservation of angular momentum, then at least one of the two particles must have an intrinsic angular momentum which can couple to the orbital angular momentum so that the sum is conserved. So far this is just the situation in case (2) but suppose *both* particles are elementary. They then, by definition, cannot have any substructure so that their intrinsic angular momentum cannot be orbital angular momentum and we must enlarge our concept of angular momentum to include the *intrinsic spin* of a particle.

A note on terminology. The term spin is used in two different ways, not always clearly differentiated, in nuclear physics. First, it is used for the intrinsic angular momentum of an elementary particle—such as an electron or a nucleon. Secondly, it is also used to describe the total angular momentum of a nucleus (or of an atom). In the latter case, of course, this spin may be made up from both orbital angular momentum and the intrinsic spins of the constituent particles. Thus we say that the spin of the electron is $\frac{1}{2}$ (intrinsic spin) and that the spin of ^{17}O is $\frac{5}{2}$ which is made up of the vector sum of the orbital and intrinsic angular momenta of the odd (unpaired) nucleon (2 and $\frac{1}{2}$ respectively).

Here are two examples of non-central forces:

1. We have already mentioned the deuteron. In the absence of any non-central force the deuteron would be a bound state of a proton and a neutron in an *s*-state, with the nucleon (intrinsic) spins summing to 1 (parallel spins, a triplet state). The non-central forces mix in 4% *d*-state, spins still parallel with the sum of the spins and the orbital angular momentum still adding up to 1, so that the total angular momentum is well defined, although the orbital angular momentum is not.

2. Consider a single electron in orbit about a ^{4}He nucleus (chosen because it has no spin). There is a spin–orbit coupling acting on the electron, which results in a non-central term in the potential, and only the vector sum of the electron orbital and spin angular momenta is conserved. In this case different eigenstates of L do not mix (because of parity conservation) but states of different L_z, S_z do mix.

We may now reverse the argument. Consider a π^--meson in orbit around a ^{4}He nucleus. Both the π^- and the ^{4}He have zero intrinsic angular momentum. Therefore, if angular momentum is conserved, the orbital angular momentum is conserved and so there can be no *non-central force* between the π^- and ^{4}He because their spins are zero.

Thus there is an intimate connection between the possible existence of non-central forces and spin (used in the wider sense). Indeed we can now see that non-central forces can only arise through a coupling to a spin. This is very important for understanding many internal and external properties of nuclei and we shall bring this point out again, in the next section.

2.2. The quantum mechanical definition of angular momentum

This is a very complicated subject which we cannot ignore if we wish to reach a proper understanding of many properties of the nucleus. We might hope to proceed by taking the

Nuclear Physics

classical definition of angular momentum $\mathbf{r}\times\mathbf{p}$ and obtain an angular momentum operator by replacing $\mathbf{p}$ by $(\hbar/i)\nabla$. This hope is destroyed by noting that elementary particles have spins, and the concept of intrinsic spin is only encountered in quantum mechanics and not in classical mechanics. One of the ways of passing from quantum mechanics to classical mechanics is to let $\hbar \to 0$. Now the orbital angular momentum L of a system is given by $l\hbar$ where l is an integer, and so any values of orbital angular momentum may be retained classically by allowing $l \to \infty$ as $\hbar \to 0$ so as to keep L constant. Now consider the spin of an elementary particle, value $S = s\hbar$ (where s is integral or $\frac{1}{2}$ integral). The number s characterizes the particle and so must be held constant as $\hbar \to 0$ so that $S \to 0$. Intrinsic spin has no classical analogue and so we must establish a concept of angular momentum which is defined quantum mechanically and which contains orbital angular momentum *and* spin.

This is done by considering the properties of a system when viewed from different coordinate systems. Consider a system isolated in space. We set up coordinates with which to describe this system: suppose we use cartesian coordinates. Then our choice for the direction of the Z-axis is arbitrary. The X-axis must be chosen at right angles to the Z-axis but is otherwise arbitrary, and $\mathbf{Y} = \mathbf{Z}\times\mathbf{X}$. We do not suppose that the internal properties of the system will depend on which direction we choose for the Z-axis (or the X-axis) so that if we rotate the coordinate system by some arbitrary amount we would still expect to see the same energy levels of the system, for example. That is, the Hamiltonian is expected to be invariant under a rotation, or H is a scalar operator. An obvious example: take the Schrödinger equation Hamiltonian $p^2/2m + V(r)$. p^2 is the square of a vector and the length of a vector does not change under a rotation of the coordinates. The length of the radius vector is r and does not change under a rotation of the coordinates. If V is a function of $\mathbf{r}$ then $\mathbf{r}$ *does* change under a rotation and then H would change. This is just the condition for which orbital angular momentum is not conserved: we are establishing a connection between angular momentum and the behaviour of a system, when the coordinate system is rotated.

Now if the operator H is not changed by rotating the coordinate system then $R(H\psi) = H(R\psi)$ where R stands for an operator which rotates the coordinates. That is $HR - RH = 0$; R commutes with the Hamiltonian. R just rotates the coordinate system of course, but if an operator generating a physical quantity is proportional to R, then this quantity is conserved. If $J = \alpha R$ then

$$\frac{dJ}{dt} = [JH - HJ] = \alpha[RH - HR] = 0. \tag{2.2.1}$$

It should be obvious by now that we are going to define the angular momentum operator in terms of the rotation operator. This gives a more general definition than $\mathbf{r}\times\mathbf{p}$ because a rotation is concerned *only* with angles and makes no reference to $\mathbf{r}$ or $\mathbf{p}$ whatsoever.

In general, the angular momentum operators are defined in terms of rotation operators and the behaviour of a state under a rotation of the coordinates is used to define the angular momentum, either orbital or intrinsic.

Let us consider some simple examples. An s-state wave function has no angular dependence and is given by some function $f_s(r)$. r is invariant under rotations so $f_s(r)$ is a scalar. A wave function with $J = 0$ behaves under rotations like a scalar and this is the definition in quantum mechanics of spin zero.

Angular Momentum and the Nucleus

The wave function for a state with angular momentum 1 has three components

$$\left.\begin{array}{l}\psi(1,+1) \propto f_p(r) \sin\theta e^{i\phi} \\ \psi(1,\ \ 0) \propto f_p(r) \cos\theta \\ \psi(1,-1) \propto f_p(r) \sin\theta e^{-i\phi}\end{array}\right\} \quad (2.2.2)$$

if $\hat{r}$ is a unit vector in the direction of r we may rewrite this as

$$\left.\begin{array}{l}\psi(1,+1) \propto f_p(r)\{\hat{x}+i\hat{y}\} \\ \psi(1,\ \ 0) \propto f_p(r)\{\hat{z}\} \\ \psi(1,-1) \propto f_p(r)\{\hat{x}-i\hat{y}\}.\end{array}\right\} \quad (2.2.3)$$

Any mixtures of magnetic substates can be rewritten as a sum of

$$f_p(r)\hat{x}, \quad f_p(r)\hat{y} \quad \text{and} \quad f_p(r)\hat{z}$$

with complex coefficients. A system with angular momentum 1 then transforms under rotations like a vector, and this is the definition in quantum mechanics of spin 1.

Angular momentum 2 has five components. A symmetric, traceless second-rank tensor has five independent components and angular momentum 2 transforms under rotations like such a tensor. Finally a system with spin $\frac{1}{2}$ has two components which transform into each other under rotations.

We can now see another reason why non-central forces involve coupling to spins. Suppose a nucleus generates a potential which is not spherically symmetric. When viewed from a rotated coordinate system the potential looks different—and the definition of a system with spin zero is that the wave function looks the same from a rotated coordinate system.

A more detailed discussion of the connection between rotations and angular momentum would at this stage result in a loss of contact with the physics. Admirable discussions have been given by Feynman[†] to which we refer you if you are interested. We merely make the following points. The rotation operator is more general than the orbital angular momentum operator. The various magnetic substates J_z of an angular momentum J range from $+J$ to $-J$ by integral steps and the integral steps are a general property unassociated with the values of J.[‡] Then $2J$ is integral so $J = 0, \frac{1}{2}, 1$, etc. If J is just the orbital angular momentum L, we can calculate the values of $J^2 = L^2 = L(L+1)$ from the operator $r \times \nabla$ and find that J is integral. If we generalize to a definition of angular momentum in terms of rotations, then it is no longer necessary for J to be restricted to integral values: there is no reason now to suppose $\frac{1}{2}$ integral values to be impossible and all we can say is that if they do exist, they do not correspond to orbital angular momentum. This generalization of our definition of angular momentum is justified by the existence in nature of particles of spin $\frac{1}{2}$ (fermions).

This concludes our introductory discussion of the angular momentum of a nucleus and we move on to a property which has no classical analogue, but which is related to the concepts of left- and right-handedness, and to the concepts of even and odd functions.

[†] *The Feynman Lectures on Physics*, vol. III (Addison-Wesley, 1963); *The Theory of Fundamental Processes* (Benjamin, 1962).
[‡] See, for example, L. I. Schiff, *Quantum Mechanics*, 3rd ed., sect. 27 (McGraw-Hill, 1968).

Nuclear Physics

2.3. Parity

In the last section we set up a system of cartesian coordinates and chose the direction of the Z-axis arbitrarily and then the direction of the X-axis to be normal to Z but otherwise arbitrary. We completed the set by defining $\mathbf{Y} = \mathbf{Z} \times \mathbf{X}$ and in setting up those axes we assumed it to be impossible for the physics of any studied system to depend on the orientation of our axes in space. This is an assumption familiar throughout classical physics—as an example consider vector operators. We can make all sorts of first-order differential operators, like $i\partial/\partial x - 5j\partial/\partial y$ but we do not give them special names and build physical laws out of them. For three-dimensional problems we define the operator

$$\nabla = \mathbf{i}\frac{\partial}{\partial x} + \mathbf{j}\frac{\partial}{\partial y} + \mathbf{k}\frac{\partial}{\partial z}$$

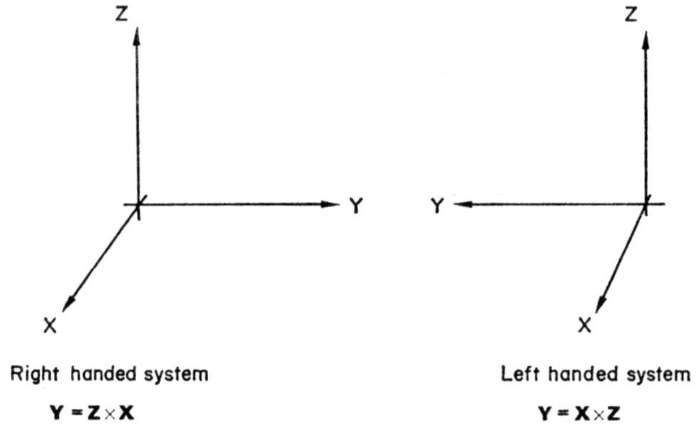

Right handed system
$\mathbf{Y} = \mathbf{Z} \times \mathbf{X}$

Left handed system
$\mathbf{Y} = \mathbf{X} \times \mathbf{Z}$

Fig. 2.3.1.

and use it (to first order) in three ways: we form $\nabla \phi$ (where ϕ is a scalar), $\nabla \cdot \mathbf{A}$ (where $\mathbf{A}$ is a vector) and $\nabla \times \mathbf{A}$ and this is all. The reason is simple: $\nabla \phi$ and $\nabla \times \mathbf{A}$ have three components which transform into one another when the coordinate system is rotated just like the three components of a vector transform into one another, while $\nabla \cdot \mathbf{A}$ remains the same under a rotation of the coordinate systems. Then a physical law like $\mathbf{E} = -\nabla \phi$ or $\mathbf{B} = \nabla \times \mathbf{A}$ is true, independent of the choice of the coordinate system. This is the way we construct physical laws and so far we have not been wrong about rotational invariance.

We chose two axes arbitrarily and then fixed the third $\mathbf{Y} = \mathbf{Z} \times \mathbf{X}$, thus defining a right-handed system of coordinates. This was arbitrary too—while Y must be perpendicular to X and Z, why should we choose $\mathbf{Y} = \mathbf{Z} \times \mathbf{X}$, rather than $\mathbf{Y} = \mathbf{X} \times \mathbf{Z}$ which defines a left-handed system of coordinates?

How can we transform the first system into the second? Not by rotations alone (try it!) but simply by *reflecting* the Y-axis through the X–Z-plane, or by reflecting all coordinates and then making a rotation.

If we suppose that our choice of right-handed or left-handed coordinate systems is indeed arbitrary, then the laws of physics cannot depend on whether we use (1) or (2) above—the laws of physics must be invariant under a reflection. Let us define an operator P which has

Angular Momentum and the Nucleus

the property of reflecting all coordinates, i.e. which takes us from a right-handed system to a left-handed system. Consider the equation

$$H\psi = E\psi.$$

Then under a reflection

$$P(H\psi) = P(E\psi) = E(P\psi) \tag{2.3.1}$$

since E is a number. If $P\psi$ is an equally good physical state with the same energy then

$$H(P\psi) = E(P\psi) \quad \text{and so} \quad PH - HP = 0. \tag{2.3.2}$$

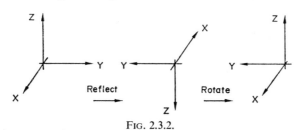

FIG. 2.3.2.

The operator P has eigenvalues. Suppose we reflect once: $\psi \to P\psi$ and again $P\psi \to P(P\psi)$ which takes us back to the original state. Then $P^2\psi = \psi$ and so if ψ is an eigenstate of the *Parity Operator* P its eigenvalues are ± 1 and the state is said to be *even* or *odd* under parity. Then since $PH - HP = 0$ the parity of a state does not vary with time: if the laws of physics are invariant under reflection, then no interaction is capable of mixing states with opposite parity (we can, of course, have an accidental mixture of two degenerate states—for example, in the first excited state of atomic hydrogen). Consider

$$H = \frac{p^2}{2m} + V(r).$$

Then under a reflection $\mathbf{p} \to -\mathbf{p}$ but p^2 is the same. r as a length has no direction: $r = (x^2 + y^2 + z^2)^{1/2}$. Under reflection the coordinates x, y, z of a particle in the right-handed frame become $-x, -y, -z$ in the new (left-handed) frame but r is still the same. Then clearly for such a Hamiltonian $PH - HP = 0$. Thus if $PH - HP = 0$, H is *even* under reflection. If we may classify states according to their P eigenvalues, the states are eigenstates of a Hamiltonian H which is even under reflection. We may then add a small perturbation H' to H and see what mixtures of states result:

$$\psi_n \to \psi_n + \sum_{m \neq n} \psi_m \int \frac{\psi_m^* H' \psi_n}{E_n - E_m} dV. \tag{2.3.3}$$

Suppose ψ_m and ψ_n have the *same* parity. Then if H' is even under reflection, the integrand is even and H' can mix ψ_m and ψ_n. If ψ_m and ψ_n have opposite parity the integrand is odd and the integral vanishes. If, on the other hand, H' is odd under reflections, then the integral vanishes if ψ_m and ψ_n have the same parity, but when ψ_m and ψ_n have opposite parity the integrand is even and so the integral may have a finite value. Thus an interaction which is odd under reflection has the property of mixing states of opposite parity together when it is

Nuclear Physics

switched on. In the classical limit the idea of parity goes away. If in the classical limit a system is described by a wave function ψ, then the observed probability density is $|\psi|^2$, which contains no information about parity. If in general

$$\psi = \psi_{\text{even}} + \psi_{\text{odd}}$$

then

$$|\psi|^2 = |\psi_{\text{even}}|^2 + |\psi_{\text{odd}}|^2 + 2\text{Re}\psi^*_{\text{even}}\psi_{\text{odd}}. \qquad (2.3.4)$$

Only the interference term changes sign when we go from a right-handed system of coordinates to a left-handed system and the presence of such a term is indicative of both parities being present in the original wave function ψ. If ψ is an eigenstate of parity, then $|\psi|^2$ contains no information on the parity of the state.

Consider how we pass from the quantum mechanical description of a particle in orbit in a central field to the classical description. In the classical limit the particle is localized and executing periodic motion. This must be represented quantum mechanically as a wave packet oscillating periodically and the frequency of the motion will emerge as the beat frequency between the components of the wave packet:

$$\Psi = \sum_n a_n \psi_n$$

where the set ψ_n are eigenstates of the motion. Now since the particle is localized at a particular value of (r, θ, ϕ) the components ψ_n must be both even and odd with respect to inversion through the origin of the field in order to interfere destructively on one side of the origin and constructively on the other. Positive and negative parity eigenstates must therefore be present with essentially equal weight.

Then the classical values of the components of angular momentum are given by

$$\int \Psi^* L_x \Psi \, dV,$$
$$\int \Psi^* L_y \Psi \, dV,$$
$$\int \Psi^* L_y \Psi \, dV$$

and all may have finite values well defined in the classical limit—the wave packet has a narrow spread in angular momentum about mean values of L_x, L_y and L_z. But if we try to define the value of the parity in a similar way

$$\int \Psi^* P \Psi \, dV = \sum_{m,n} a^*_m a_n \int \psi^*_m P \psi_n \, dV \simeq 0$$

because both parities are present with equal weight. In classical mechanics parity goes away.

We now have two ways of seeing what is meant by non-conservation of parity. An interaction, part of which is even and part odd under reflection, means that $PH - HP \neq 0$ and so the average value of P is not time independent. Alternatively, starting with states of definite parity, the action of an odd perturbation is to mix in states with the opposite parity: a mixture of even and odd perturbation connects an initial state which has a well-defined parity with a final state which does not. The strong and electromagnetic interactions conserve parity, the weak interaction is parity violating to the maximal extent: as far as the weak interactions are concerned, our assumption that the choice of a left-handed or right-

Angular Momentum and the Nucleus

handed coordinate system is arbitrary appears to be utterly wrong. This subject will be dealt with in detail in the section on nuclear decay (Chapter 3). In this section we are concerned primarily with static properties of nuclei. The states of nuclear matter are determined predominantly by the strong interactions, secondarily by electromagnetic interactions and weak interactions are of negligible importance. If only strong and electromagnetic interactions were to be included in the nuclear Hamiltonian, then the eigenstates would be eigenstates of parity—the mixing between states of opposite parity caused by the weak interactions may be ignored in most cases.[†]

Let us now consider the properties of a single nucleon state. The nucleon moves in a potential V, which is even under reflection. If the potential is central, then the orbital angular momentum is conserved and the nucleon is described by a wave function $f_l(r)\, P_l^m(\cos\theta)e^{im\phi}$. Let us find the parity of this function. Consider a particular point in space and reflect the coordinate system (see Fig. 2.3.3).

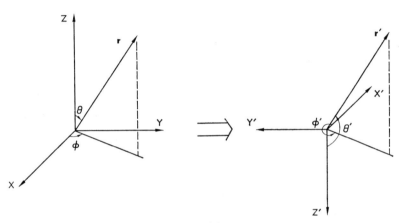

Fig. 2.3.3.

r is a length and is unchanged, so

$$f(r') = f(r) \tag{2.3.5}$$
$$\theta' = \pi - \theta$$

so

$$P_l^m(\cos\theta) \to P_l^m(\cos\theta') = P_l^m(\cos(\pi-\theta))$$
$$= P_l^m(-\cos\theta), \tag{2.3.6}$$
$$\phi' = \pi + \phi,$$

so

$$e^{im\phi} \to e^{im\pi + im\phi'} = (-1)^m\, e^{im\phi'}. \tag{2.3.7}$$

From the properties of the associated Legendre polynomials[‡]

$$P_l^m(-\cos\theta) = (-1)^{l+m}\, P_l^m(\cos\theta) \tag{2.3.8}$$

[†] For an exception see K. S. Krane et al., Phys. Rev. Lett. **26**, 1579 (1971).
[‡] See, for example, L. I. Schiff, Quantum Mechanics, 3rd ed., sect. 14 (McGraw-Hill, 1968).

Nuclear Physics

so

$$f(r) P_l^m(\cos\theta) e^{im\phi} \to f(r') P_l^m(\cos\theta') e^{im\phi'}$$
$$= f(r)(-1)^l P_l^m(\cos\theta) e^{im\phi}. \quad (2.3.9)$$

The *parity* of such a state is $(-1)^l$—the state is *even* (positive parity) if l is even and *odd* (negative parity) if l is odd.

If we now switch on a non-central potential as a perturbation, we mix together states with different values of l but if such a potential is still even under reflection then all states mixed together must have the same parity. This is the reason why in the deuteron the *s*- and *d*-triplet states are mixed together and the *p*-state is not present—a *p*-state has opposite parity and the strong interactions *conserve* parity.

If we want to characterize the state of a nucleon, we may take the orbital angular momentum which is the sum of the orbital and spin angular momenta, and label a state, say $p_{3/2}$ or $p_{1/2}$. In general a non-central interaction is present and while the total angular momentum is conserved, the orbital angular momentum is not, and states of different orbital angular momentum but the same total angular momentum are mixed in. If the state is still predominantly one orbital angular momentum—like 96% *s*, 4% *d* for the deuteron, we still refer loosely to the state by the orbital angular momentum, but if the mixing is severe this becomes meaningless. However, if the whole potential is even under a reflection, we may label states by their parity. If in a central field we have states $s_{1/2}$, $p_{1/2}$, $p_{3/2}$, $d_{3/2}$, $d_{5/2}$... these become respectively $1/2^+$, $1/2^-$, $3/2^-$, $3/2^+$, $5/2^+$... . This is an alternative description for motion in a central field. In a strongly non-central field, and when a single particle description is inadequate, it is the only one to use. (The deuteron is 1^+.)

We may conclude this section with something rather more qualitative. Conservation of parity is frequently "explained" by saying that nature does not distinguish between right and left; conversely the non-conservation of parity which occurs in β-decay is sometimes described by saying that nature is weakly left handed. What meaning can we attach to these very woolly statements which explain nothing until the physics underlying them has been understood? We can construct an example which brings out the principles very clearly. Suppose we have a spinless nucleus which decays into two (distinguishable) particles which recoil away in opposite directions. The original state has a well-defined parity. Suppose we look at one of these particles, and always find that its angular momentum is directed along its direction of motion. Then by conservation of angular momentum the other particle must also always have its angular momentum along its direction of motion (Fig. 2.3.4).

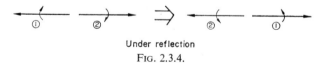

Under reflection

FIG. 2.3.4.

Now if a wave function ψ is an eigenstate of the parity operator P, $|\psi|^2 = |P\psi|^2$. The system we have just described does not correspond to this; on reflection right-handed particles become left-handed. We could take

$$\psi \propto 1 + \frac{\mathbf{p}\cdot\boldsymbol{\sigma}}{|p||\sigma|}$$

when
$$P\psi \propto P\left\{1+\frac{\mathbf{p}\cdot\mathbf{\sigma}}{|p||\sigma|}\right\} = 1-\frac{\mathbf{p}\cdot\mathbf{\sigma}}{|p||\sigma|}$$

because angular momentum is an axial vector and does not change direction under reflection. Then

$$|\psi|^2 = 1+\left[\frac{\mathbf{p}\cdot\mathbf{\sigma}}{|p||\sigma|}\right]^2 + 2\frac{\mathbf{p}\cdot\mathbf{\sigma}}{|p||\sigma|} \tag{2.3.10}$$

while

$$|P\psi|^2 = 1+\left[\frac{\mathbf{p}\cdot\mathbf{\sigma}}{|p||\sigma|}\right]^2 - 2\frac{\mathbf{p}\cdot\mathbf{\sigma}}{|p||\sigma|}. \tag{2.3.11}$$

The odd part of $|\psi|^2$ is the interference term between an even and odd part to the wave function. If interactions connected states only of opposite parity, we would never know the difference since we observe $|\psi|^2$ not ψ.

An observation of one circularly polarized particle from our system is not enough to allow us to infer that the final state wave function has even and odd parts. The wave function is a probability amplitude and $|\psi|^2$ gives us the average behaviour taken over a lot of observations. Thus for 1 event if we observe such a right-handed circular polarization, this does not require parity non-conservation, provided that we find left-handed circular polarization is also a possible physical state and is produced with the same frequency. If we found only *one* polarization when we averaged over a lot of observations, then *provided* we knew that the initial state had definite parity, we could then infer that parity was violated and *as a consequence* we could use the sense of polarization to *define* "handedness", in a way which would be reproducible in any laboratory in the world as a standard. It is in this sense that parity non-conservation may be interpreted by saying that a handedness is defined or parity conservation interpreted by saying that nature does not distinguish between left and right. Our thought experiment is duplicated by nature: in the decay $\pi^+ \to \mu^+ + \nu$ the ν is 100% polarized left-handedly. This is a weak interaction—"nature is weakly left-handed".

In ordinary experience electromagnetic effects predominate and electromagnetic interactions conserve parity. We do not observe (as far as we know) the effects of weak interactions in the macroscopic world. This may be the reason that some children have immense difficulty in learning which is their right hand! The label is arbitrary since π-mesons cannot be studied in the nursery. A child may decide (or be told) that he uses his pencil with his right hand. This works fine until he meets someone who uses his pencil with the other hand —a perfectly possible physical state. Or he may define his right by saying he combs his hair from the left—until he meets someone who does it the other way. Even if he knows his heart is on the left-hand side of his body, this is not adequate—a few people exist whose heart is on the other side. There is no obvious reason why most people are right-handed— this must be regarded as a biological accident and while it may be used to establish a *convention*, it does not establish anything *absolute*. Someone who cannot understand which is his right hand and which is his left is not necessarily very stupid—perhaps he really understands the implication of parity conservation in the everyday world. The situation might be really serious for a pair of mirror image twins!

Nuclear Physics

2.4. Static electric moments

We may now tackle the static electric and magnetic moments of a nucleus. We will start with the electric field of a nucleus since this is a slightly simpler case. Consider a classical distribution of charge of arbitrary shape, and density $\varrho(\mathbf{r}')$ and calculate the potential at a point $\mathbf{r}$ (Fig. 2.4.1).

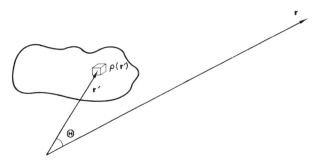

Fig. 2.4.1.

$$V(\mathbf{r}) = \int \frac{\varrho(\mathbf{r}')\, d^3r'}{|\mathbf{r}-\mathbf{r}'|} \tag{2.4.1}$$

$$|\mathbf{r}-\mathbf{r}'| = \sqrt{r^2+r'^2-2\mathbf{r}\cdot\mathbf{r}'}$$

and if $r \gg r'$ so that $|\mathbf{r}-\mathbf{r}'| \sim r$, then since

$$|\mathbf{r}-\mathbf{r}'| = r\sqrt{1+\frac{r'^2}{r^2}-\frac{2\mathbf{r}\cdot\mathbf{r}'}{r^2}} = r\sqrt{1+\frac{r'^2}{r^2}-2\frac{rr'}{r^2}\cos\Theta} \tag{2.4.2}$$

we may expand $V(\mathbf{r})$ and obtain

$$\begin{aligned}V(\mathbf{r}) &= \frac{1}{r}\int \varrho(\mathbf{r}')\left\{1-\frac{1}{2}\left[\frac{r'^2}{r^2}-2\frac{r'}{r}\cos\Theta\right]+\frac{3}{8}\left[\frac{r'^2}{r}-2\frac{r'}{r}\cos\Theta\right]^2\ldots\right\}d^3r' \\ &= \frac{1}{r}\int \varrho(\mathbf{r}')\left\{1+\frac{r'}{r}\cos\Theta+\frac{1}{2}\left(\frac{r'}{r}\right)^2[3\cos^2\Theta-1]\ldots\right\}d^3r'.\end{aligned} \tag{2.4.3}$$

The nth *multipole moment* is given by

$$\int \varrho(\mathbf{r}')r'^n P_n(\cos\Theta)\, d^3r' \tag{2.4.4}$$

so that the field due to the zero moment is

$$\frac{1}{r}\int \varrho(\mathbf{r}')\, d^3r'$$

which is just the coulomb field of a point change of magnitude $\int \varrho(\mathbf{r}')d^3r'$. The second term is a dipole field generated by a dipole moment $\int r'\varrho(\mathbf{r}')\cos\Theta\, d^3r'$, and an electrical quadrupole field is generated by a quadrupole moment

$$\int r'^2 \varrho(\mathbf{r}') P_2(\cos\Theta)\, d^3r'.$$

Angular Momentum and the Nucleus

You should note that these moments all refer to a measurement with respect to a specific direction r in space.

Electric dipole and quadrupole moments are the only static electric moments we will be concerned with (apart, of course, from the total charge!) and so let us consider them one at a time, starting with the dipole moment. The square of the wave function of a system in a well-defined state of J, M is invariant under rotations about the M-axis. Let us therefore define a new set of coordinates, with the Z-axis coinciding with M. Then because of the rotational symmetry, it does not matter how we choose the X-axis.

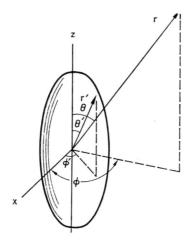

Fig. 2.4.2.

The potential due to the dipole moment is

$$V(\mathbf{r}) = \frac{1}{r^2} \int \varrho(\mathbf{r}')r' \cos\Theta \, d^3r' = \frac{1}{r^3} \int \varrho(\mathbf{r}')\mathbf{r}' \cdot \mathbf{r} \, d^3r'.$$

If a point at $\mathbf{r}'$ is specified by (r', θ', ϕ') and at $\mathbf{r}$ by (r, θ, ϕ) then

$$\mathbf{r} \cdot \mathbf{r}' = x'x + y'y + z'z$$
$$= r'r\{\sin\theta' \cos\phi' \sin\theta \cos\phi + \sin\theta' \sin\phi' \sin\theta \sin\phi + \cos\theta' \cos\theta\}. \quad (2.4.5)$$

We integrate with respect to $d^3r' = r'^2 \, d\cos\theta' \, d\phi'$ and using the rotational invariance of the squared wave functions about Z (i.e. no ϕ' dependence) we are left with only

$$V(\mathbf{r}) = \frac{\cos\theta}{r^2} \int \varrho(\mathbf{r}')r' \cos\theta' \, d^3r'. \quad (2.4.6)$$

If the state has a definite parity then the square of the wave function is even under the transformation $Z \to -Z$ and so we have two results: (1) the potential at some angle θ to the z-axis is $d \cos\theta/r^2$ where $d = \int \varrho(\mathbf{r}') z' d^3r'$ and is the dipole moment defined with respect to the z axis of the charge distribution and (2) a state of unique parity *cannot* have an electric dipole moment. Consider an example from atomic physics. In the ground state of atomic hydrogen, only a second-order Stark effect is observed. When the applied electric field is

73

Nuclear Physics

zero, then the atom is in the 1s ground state. Application of an electric field **E** mixes in the excited states of opposite parity in first-order perturbation theory (**E** is odd) and the mixed state, being no longer of unique parity, has a dipole moment, the strength of which increases with the mixing and hence with the applied field. This gives an interaction energy $\propto E^2$ and so a second-order Stark effect. In the first excited state, however, we have a degeneracy between the 1s and 2p levels. The degeneracy is quite accidental and specific to the coulomb field, and results in an electric dipole moment even in zero field. The interaction energy is therefore proportional to the applied field E and a first-order Stark effect results.

Thus observation of an electric dipole moment in a system either implies a parity violating interaction *or* implies that the state is not pure because of an accidental degeneracy between the two states of opposite parity. Thus while it would be interesting if a nucleus was observed to have an electric dipole moment, it would not require a parity violating interaction. If on the other hand an elementary particle—like the neutron—was observed to have an electric dipole moment, then the implication would be that it was caused by a parity violating interaction, unless we decided that even a neutron is not elementary.[†]

We may define an internal electric quadrupole moment with respect to the symmetry axis of the charge distribution in a similar way.

$$V_Q(\mathbf{r}) = \frac{1}{r^3} \int \varrho(\mathbf{r'}) r'^2 \left\{ \frac{3}{2} \cos^2 \Theta - \frac{1}{2} \right\} d^3 r' \tag{2.4.7}$$

and

$$\cos^2 \Theta = [\sin \theta' \cos \phi' \sin \theta \cos \phi + \sin \theta' \sin \phi' \sin \theta \sin \phi + \cos \theta' \cos \theta]^2. \tag{2.4.8}$$

Because of the assumed rotational symmetry, $\varrho(\mathbf{r'})$ is independent of ϕ' and we may integrate directly over ϕ'. The cross terms in $\cos^2 \Theta$ vanish and we are left with

$$\begin{aligned}
&\int \varrho(\mathbf{r'}) \cos^2 \Theta \, d\phi' \, d\cos \theta' \\
&= \int \varrho(\mathbf{r'}) [\sin^2 \theta \sin^2 \theta' \cos^2 \phi \cos^2 \phi' + \sin^2 \theta \sin^2 \theta' \sin^2 \phi \sin^2 \phi' \\
&\quad + \cos^2 \theta' \cos^2 \theta] \, d\phi' \, d\cos \theta' \\
&= 2\pi \int \varrho(\mathbf{r'}) [\tfrac{1}{2} \sin^2 \theta \sin^2 \theta' \cos^2 \phi + \tfrac{1}{2} \sin^2 \theta \sin^2 \theta' \sin^2 \phi \\
&\quad + \cos^2 \theta' \cos^2 \theta] \, d\cos \theta' \\
&= 2\pi \int \varrho(\mathbf{r'}) \{\tfrac{1}{2} \sin^2 \theta \sin^2 \theta' + \cos^2 \theta \cos^2 \theta'\} d\cos \theta', \\
V_Q(\mathbf{r}) &= \frac{2\pi}{r^3} \int \varrho(\mathbf{r'}) r'^2 \left\{ \frac{3}{4} \sin^2 \theta \sin^2 \theta' + \frac{3}{2} \cos^2 \theta' \cos^2 \theta - \frac{1}{2} \right\} \\
&\quad \times r'^2 \, d\cos \theta' \, dr'.
\end{aligned} \tag{2.4.9}$$

Now

$$\begin{aligned}
&\tfrac{3}{4} \sin^2 \theta \sin^2 \theta' + \tfrac{3}{2} \cos^2 \theta' \cos^2 \theta - \tfrac{1}{2} \\
&= (\tfrac{3}{2} \cos^2 \theta - \tfrac{1}{2})(\tfrac{3}{2} \cos^2 \theta' - \tfrac{1}{2})
\end{aligned} \tag{2.4.10}$$

[†] An electric dipole moment for a pure state would also imply time reversal non-invariance, see Section 3.14.

Angular Momentum and the Nucleus

so that
$$V_Q(\mathbf{r}) = \frac{1}{r^3}\left(\frac{3}{2}\cos^2\theta - \frac{1}{2}\right)\int \frac{r'^2}{2}(3\cos^2\theta' - 1)\varrho(\mathbf{r}')\,d^3r'$$
$$= \frac{1}{2}\frac{1}{r^3}\left(\frac{3}{2}\cos^2\theta - \frac{1}{2}\right)Q \qquad (2.4.11)$$

where Q is the quadrupole moment referred to our axis of symmetry. θ is the angle between the spin axis of the system and the radius vector of the point of observation.

This serves as an introduction to the quadrupole field of the nucleus. A quadrupole moment is detected by the interaction of the quadrupole moment with an electrostatic field. Suppose that the electrostatic potential due to sources other than the nucleus is $V(x, y, z)$. Then the energy of the nucleus in such a field is

$$-\int \varrho(x, y, z)\, V(x, y, z)\, dx\, dy\, dz.$$

Expanding $V(x, y, z)$ about 0, we obtain for the term involving second derivatives

$$-\frac{1}{2}\sum_{i,j}\int \varrho(\mathbf{x}) x_i x_j \frac{\partial^2 V}{\partial x_i \partial x_j}\Big|_0 \qquad (2.4.12)$$

where i, j run from 1 to 3, and the second derivatives of the potential are evaluated at the origin.

Then there are nine terms of form

$$-\frac{1}{2}Q_{ij}\frac{\partial^2 V}{\partial x_i \partial x_j}\Big|_0$$

where the nine Q_{ij} are the nine components of the quadrupole moment tensor, referred to our arbitrary set of axes x, y, z. V must clearly be second order or greater in the coordinates, and only second-order terms survive in $\partial^2 V/\partial x_i \partial x_j$. Let us further assume that V is cylindrically symmetric about the z-axis. Then since V is a solution of the Laplace equation $\nabla^2 V = 0$ and is finite at the origin, the solutions are of the form

$$A_n r^n P_n(\cos\theta) \qquad (2.4.13)$$

which leaves us only with the term

$$\tfrac{1}{2}Kr^2(3\cos^2\theta - 1) = \tfrac{1}{2}K(3z^2 - r^2). \qquad (2.4.14)$$

All cross terms vanish and

$$\frac{\partial^2 V}{\partial z^2} = 2K \qquad \frac{\partial^2 V}{\partial x^2} = \frac{\partial^2 V}{\partial y^2} = -K.$$

The interaction energy is then

$$-\tfrac{1}{4}\{2Q_{zz} - Q_{xx} - Q_{yy}\}\frac{\partial^2 V}{\partial z^2}\Big|_0$$
$$= -\tfrac{1}{4}\left[\int \varrho(x, y, z)(3z^2 - r^2)\, dx\, dy\, dz\,\frac{\partial^2 V}{\partial z^2}\Big|_0\right]$$
$$= -\tfrac{1}{4}\left\{\frac{3}{2}\cos^2\theta - \frac{1}{2}\right\}\frac{\partial^2 V}{\partial z^2}\Big|_0 Q \qquad (2.4.15)$$

Nuclear Physics

where Q is the internal quadrupole moment of our nucleus measured with respect to its rotational symmetry axis, and θ is the angle between the axis of symmetry of the field and of the nucleus.

We will now give an argument to show that a nucleus with angular momentum < 1 (i.e. $\frac{1}{2}$ or 0) cannot have a static quadrupole moment! This is purely a quantum mechanical result. Consider a charged particle (for simplicity suppose it to have spin 0) in orbit around a nucleus which in addition to a $1/r$ potential also possesses a potential generated by a quadrupole moment. If the wave functions without the quadrupole terms are ψ, then treating the quadrupole potential as a small perturbation, the first-order energy change is proportional to

$$e \int \psi_n^* \frac{1}{r^2} P_2(\cos\theta) \psi_n \, dV. \qquad (2.4.16)$$

This integral is zero if ψ_n does not depend on θ but is finite if $\psi_n \propto P_l(\cos\theta)$ where $l \geq 1$. $\int \psi_n^* P_2(\cos\theta) \psi_n$ is clearly the quadrupole moment of the orbiting charge distribution and is zero for a spinless particle in an s-wave. This is obvious—such a charge distribution is spherically symmetric. A p-wave spinless particle has a charge distribution like $\cos^2\theta$ or $\sin^2\theta$ and so has a quadrupole moment. The complications of an argument like this arise only if we consider half integral spins. If we do it the hard way, we can use the angular momentum raising and lowering operators to find the combination of angular momentum 1 and angular momentum $\frac{1}{2}$ which yield $p_{\frac{1}{2}}$ and $p_{\frac{3}{2}}$ states. The $\frac{1}{2}$ states are†

$$|\tfrac{1}{2}, +\tfrac{1}{2}\rangle = \sqrt{\tfrac{2}{3}} Y_1^1(\theta,\phi)|-\tfrac{1}{2}\rangle - \sqrt{\tfrac{1}{3}} Y_1^0(\theta,\phi)|+\tfrac{1}{2}\rangle,$$
$$|\tfrac{1}{2}, -\tfrac{1}{2}\rangle = \sqrt{\tfrac{1}{3}} Y_1^0(\theta,\phi)|-\tfrac{1}{2}\rangle - \sqrt{\tfrac{2}{3}} Y_1^{-1}(\theta,\phi)|+\tfrac{1}{2}\rangle,$$

i.e.
$$|\tfrac{1}{2}, +\tfrac{1}{2}\rangle \propto -\sqrt{\tfrac{1}{3}} P_1^1(\theta,\phi)|-\tfrac{1}{2}\rangle - \sqrt{\tfrac{1}{3}} P_1^0(\theta,\phi)|+\tfrac{1}{2}\rangle, \qquad (2.4.17)$$
$$|\tfrac{1}{2}, -\tfrac{1}{2}\rangle \propto \sqrt{\tfrac{1}{3}} P_1^0(\theta,\phi)|-\tfrac{1}{2}\rangle - \sqrt{\tfrac{1}{3}} P_1^{-1}(\theta,\phi)|+\tfrac{1}{2}\rangle.$$

It should be obvious that these combinations have respectively $J_z = \pm\tfrac{1}{2}$. It is *not* immediately obvious that they have $J = \tfrac{1}{2}$. Suppose we take two combinations for $+\tfrac{1}{2}$:

$$AY_1^1|-\tfrac{1}{2}\rangle + BY_1^0|+\tfrac{1}{2}\rangle,$$
$$CY_1^1|-\tfrac{1}{2}\rangle + DY_1^0|+\tfrac{1}{2}\rangle, \qquad (2.4.18)$$

and require one to represent spin $\tfrac{1}{2}$ and the other spin $\tfrac{3}{2}$. First require normalization to 1. Then

$$A^2 + B^2 = 1,$$
$$C^2 + D^2 = 1.$$

Secondly, require orthogonality:

$$\int \{A^* Y_1^{1*}|-\tfrac{1}{2}\rangle^* + B^* Y_1^{0*}|+\tfrac{1}{2}\rangle\}\{CY_1^1|-\tfrac{1}{2}\rangle + DY_1^0|+\tfrac{1}{2}\rangle\} \, d\Omega = 0.$$

Then
$$A^*C + B^*D = 0.$$

† See Appendix 2.

Angular Momentum and the Nucleus

There are several solutions, differing only in an arbitrary phase, e.g.

$$A = \sqrt{\tfrac{2}{3}} \quad B = -\sqrt{\tfrac{1}{3}} \quad C = \sqrt{\tfrac{1}{3}} \quad D = \sqrt{\tfrac{2}{3}} \quad (2.4.19)$$

which is the one we took. Then operating with $J^+ = J_x + iJ_y = L_x + iL_y + S_x + iS_y$ yields zero on the first and a non-zero answer on the second. Since $J_x + iJ_y$ raises J_z by one unit, the first combination corresponds to $J = \tfrac{1}{2}$, the second to $J = \tfrac{3}{2}$.

$$\langle \tfrac{1}{2}, +\tfrac{1}{2} | P_2(\cos\theta) | \tfrac{1}{2}, +\tfrac{1}{2}\rangle \propto \int (\sin^2\theta + \cos^2\theta) P_2(\cos\theta)\, d\cos\theta = 0, \quad (2.4.20)$$

$\langle \tfrac{1}{2}, -\tfrac{1}{2} | P_2(\cos\theta) | \tfrac{1}{2}, -\tfrac{1}{2}\rangle$ is also equal to zero while the combinations for spin $\tfrac{3}{2}$ are non-zero. But it is not necessary to be so complicated.

$$\psi_n^{J*} P_2(\cos\theta)\, \psi_n^J = \psi_n^{J*}(P_2(\cos\theta)\, \psi_n^J) \quad (2.4.21)$$

and the term in brackets can be interpreted as the wave function for orbital angular momentum 2, multiplied by the wave function ψ_n^J. Such a product can be split up.

$P_2(\cos\theta)\psi^0$ is a state with angular momentum 2,

$P_2(\cos\theta)\psi^{1/2}$ is a state with angular momentum $\tfrac{3}{2}$ or $\tfrac{5}{2}$,

$P_2(\cos\theta)\psi^1$ is a state with angular momentum 1, 2, or 3.

Thus for $J < 1$ the term in brackets can be split up into terms all of which are orthogonal to ψ^J while for $J = 1$ or greater, one or more terms in brackets are not orthogonal to ψ^J. Thus the integral vanishes for $J < 1$ which means no quadrupole interaction, which means no quadrupole moment.

Now apply this to a nucleus instead of a single electron moving in a quadrupole field generated by a nucleus. We may detect a nuclear quadrupole moment by observing a shift in the energy levels of a system comprising a nucleus and an electric field which could in principle be external but is in practice an atomic field, a molecular field or a crystal field. A single proton orbiting under the influence of the strong interactions and having a wave function ψ has its energy changed by an amount proportional to

$$\frac{e}{r^2} \int \psi^* P_2(\cos\theta)\, \psi\, dV \quad (2.4.22)$$

where its quadrupole moment is given by the integral. For N-nucleons the nuclear wave function is

$$\psi(r_1, r_2, \ldots, r_N)$$

where the first p-coordinates are proton coordinates and the rest neutron coordinates. The function is normalized by

$$\int \psi^*(r_1, \ldots, r_N)\, \psi(r_1, \ldots, r_N)\, dV_1, dV_2, \ldots, dV_N = 1 \quad (2.4.23)$$

and the quadrupole moment is given by

$$\sum_{i=1}^{p} e \int \psi^*(r_1, \ldots, r_N) r_i^2 P_2(\cos\theta_i)\, \psi(r_1, \ldots, r_N)\, dV_1, \ldots, dV_N. \quad (2.4.24)$$

Nuclear Physics

$P_2(\cos\theta_i)$ acts only on the ith proton of the nuclear wave function but since the strong interactions couple together all protons and neutrons, it is the angular momentum of the nucleus, not that of the ith proton, in which we are interested. Again, the the rules of addition of angular momentum require that the nuclear angular momentum is ≥ 1 for a quadrupole interaction energy to result.

Whether this is interpreted by saying that it is not possible to observe a quadrupole moment of a state with angular momentum < 1 or by saying that no pure angular momentum state with angular momentum < 1 can have a quadrupole moment is a matter of taste—these two statements make identical physical predictions and so cannot be distinguished.

Here is a more physical argument. A state with spin zero looks the same from any orientation of the coordinate frame in space, and therefore cannot generate a field which does not look the same—any field is therefore spherically symmetric and a nucleus with spin zero cannot have an electric moment other than the zero nuclear moment.

A state with spin $\frac{1}{2}$ is rotationally invariant about the z-axis and has only two possible orientations in space with respect to the z-axis—which we may define by an electric field, say as in Fig. 2.4.3.

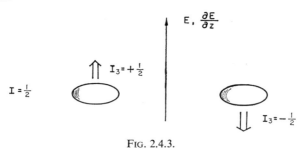

FIG. 2.4.3.

From the symmetry in the two cases, the interaction energy of the system is the same in the two cases and so no effect due to a quadrupole moment exists.

We may now go ahead and work out the energy levels of a nucleus (or an atom) in an electric field. We first note that the distribution of charge given by a single particle with a wave function $P_l^m(\theta)e^{im\phi}$ is given by $[P_l^m(\theta)]^2$ and so has a symmetry axis along the direction of the z-component of angular momentum. Because of the relation of angular momentum to transformation under rotations, this result carries over into a complex many-body system such as a nucleus. The energy levels of a nucleus in an external field are given by

$$-\tfrac{1}{4}\int \psi^*(3z^2-r^2)\,\psi\, dx\, dy\, dz\, \left.\frac{\partial^2 V}{\partial z^2}\right|_0 \tag{2.4.25}$$

and so we need to evaluate the integral. $3z^2-r^2$ is made up of vectors and the matrix element of any vector is proportional to the matrix element of the angular momentum J.[†] Thus

$$\int \psi^*(3z^2-r^2)\,\psi\, dV \propto \int \psi^*(3J_z^2-J^2)\,\psi\, dV \tag{2.4.26}$$

writing this as

$$\langle J, J_z | 3J_z^2 - J^2 | J, J_z \rangle$$

[†] See, for example, L. D. Landau and E. M. Lifshitz, *Quantum Mechanics*, 2nd ed., sect. 29 (Pergamon Press, 1965).

Angular Momentum and the Nucleus

to represent the angular momentum of the nucleus explicitly, the energy levels of a nucleus due to the quadrupole component of an external electric field are

$$-\tfrac{1}{4}C \langle J, J_z | 3J_z^2 - J^2 | J, J_z \rangle \frac{\partial^2 V}{\partial z^2}\bigg|_{z=0} \tag{2.4.27}$$

where C could be calculated if we knew all the properties of the wave functions.

We define

$$\langle J, J | 3z^2 - r^2 | J, J \rangle = Q_0 = C \langle J, J | 3J_z^2 - J^2 | J, J \rangle$$
$$= C[3J^2 - J(J+1)]. \tag{2.4.28}$$

The energy levels in an external field are

$$-\tfrac{1}{4}Q_0 \frac{3J_z^2 - J(J+1)}{3J^2 - J(J+1)} \frac{\partial^2 V}{\partial z^2}\bigg|_{z=0}. \tag{2.4.29}$$

This expression may be regarded as a definition of the quadrupole moment of a nucleus. The same result may of course be obtained by using the vector model of angular momentum. If we return to our expression (2.4.15) the energy levels are

$$-\tfrac{1}{4}(\tfrac{3}{2}\cos^2\theta - \tfrac{1}{2}) \frac{\partial^2 V}{\partial z^2} Q$$

where Q is the internal quadrupole moment. Then we write

$$\cos\theta = \frac{J_z}{\sqrt{J(J+1)}}$$

so that

$$3\cos^2\theta - 1 = \frac{3J_z^2 - J(J+1)}{J(J+1)},$$

Q_0 is the maximum measurable quadrupole moment in the vector model and is not equal to Q, rather

$$Q_0 = \frac{3J^2 - J(J+1)}{J(J+1)} Q$$

which then yields the result given above.

It should be fairly obvious that we cannot readily measure the quadrupole moment of a nucleus by placing it in an external field. However, the presence of a nuclear quadrupole moment will perturb the atomic energy levels (unless the electrons have total angular momentum zero or $\tfrac{1}{2}$) and so we need to find the splitting due to a nuclear quadrupole moment. We first require that both the electrons and the nucleus have quadrupole moments q_0, Q_0 respectively and then expand the field due to the atomic electrons at the nucleus

$$E = -\int_{V_e}\int_{V_n} \frac{\varrho_e \varrho_n}{r} dV_e \, dV_n \tag{2.4.30}$$

Nuclear Physics

where $\mathbf{r} = \mathbf{r}_e - \mathbf{r}_n$,

$\mathbf{r}_n \ll \mathbf{r}_e$,

$$\frac{1}{r} = \frac{1}{r_e} + \frac{r_n}{r_e^2}\cos\theta + \tfrac{1}{2}\frac{r_n^2}{r_e^2}(3\cos^2\theta - 1)\ldots$$

where θ is the angle between $\mathbf{r}_e$ and $\mathbf{r}_n$. Both the electrons and the nucleus have angular momenta and the evaluation of the integral requires a careful study of the couplings of angular momenta. This is discussed in great detail in Ramsay, *Nuclear Moments*[†] and the result is

$$E_{Qq} = \frac{e^2 q_0 Q_0 [3(\mathbf{I}\cdot\mathbf{J})^2 + \tfrac{3}{2}\mathbf{I}\cdot\mathbf{J} - I(I+1)J(J+1)]}{[3J^2 + J(J+1)][3I^2 + I(I+1)]}. \qquad (2.4.31)$$

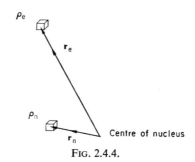

FIG. 2.4.4.

(In this expression I is the angular momentum of the nucleus and J that of the electrons.)

The vector model would lead us to suppose a form $3(\mathbf{I}\cdot\mathbf{J})^2 - I(I+1)J(J+1)$ and is wrong in this application.

If the total angular momentum of the atom is $\mathbf{F} = \mathbf{I} + \mathbf{J}$ then

$$F(F+1) = I(I+1) + J(J+1) + 2\mathbf{I}\cdot\mathbf{J} \qquad (2.4.32)$$

and substitution into the above expression yields the hyperfine splitting due to a quadrupole–quadrupole interaction. Measurement of this splitting then gives Q_0 if the electronic structure of the atom is sufficiently well known to allow a calculation of q_0.

Let us attempt to evaluate the quadrupole moment of ^{153}Eu, sixty-three protons and ninety neutrons. The angular momentum is that of the unpaired proton which is in a $2d$ state. If we neglect complications due to spin we may evaluate

$$\langle 2, +2 | 3z^2 - r^2 | 2, +2 \rangle$$

$$Y_2^2 = \tfrac{1}{4}\sqrt{\tfrac{15}{4\pi}}\sin^2\theta\, e^{2i\phi}$$

giving for the angular part

$$\int \tfrac{1}{16}\tfrac{15}{4\pi}\sin^4\theta\{3\cos^2\theta - 1\}\, d\cos\theta\, d\phi \simeq -\tfrac{1}{3}.$$

$\sin^2\theta$ yields an equatorial bulge, so we would expect a negative value.

[†] N. F. Ramsay, *Nuclear Moments* (Wiley, 1953).

Angular Momentum and the Nucleus

For the radial part, let us represent the radial wave function of the unpaired proton by a δ-function at the nuclear radius so $Q \sim -\frac{1}{3}R^2$ or $Q/R^2 \simeq -\frac{1}{3}$ so $Q/Z^2R \simeq -0.005$. If the even–even core makes no contribution to the quadrupole moment, we would thus expect a negative quadrupole moment characterized by

$$\frac{Q}{ZR^2} \sim -0.005.$$

The negative value would be a characteristic of even N–odd Z nuclei. In fact, ^{153}Eu has $Q/ZR^2 = +0.05$ and even N–odd Z nuclei have quadrupole moments which are positive just before shell closure and negative just after.

The magnitude of the quadrupole moment of ^{153}Eu indicates that the quadrupole moment must be generated by some coupled motion of nucleons. ^{153}Eu has 13 protons above shell closure at 50 and is thus half-way between 50 and 82. This, of course, is the reason why we chose to consider this nucleus. Collective motion in the incomplete shell above 50 must be invoked to explain the observed quadrupole moment and it is clear that the independent particle model falls down badly in an attempt to explain the quadrupole moments for nuclei in the middle of shells. This is further emphasized by quadrupole moments of nuclei which have an even number of protons and an odd number of neutrons, so that their spin is non-zero. On an independent particle model we would expect such nuclei to have no quadrupole moment, regardless of angular momentum, but they do have. Consider, for example, the nucleus ^{173}Yb which has 70 protons and 103 neutrons. It has a quadrupole moment given by $Q/ZR^2 = 0.08$. It is clear that not only are there strong collective motions within a given proton or neutron shell, but that there is also substantial coupling between the proton and neutron shells.

The behaviour of quadrupole moments of even N, odd Z near closed shells is now clear. The first proton over a closed shell will perturb the tightly bound core little and will contribute almost the whole of the quadrupole moment, which will be negative and $Q/ZR^2 \sim -0.005$ to -0.01. A shell which is one proton below closure must be supposed to behave like a closed shell with a hole—contributing a small positive quadrupole moment because the hole behaves like a negative charge. In the middle of a shell strong correlations exist between the nucleons, and the quadrupole moments have their largest values. It is this characteristic behaviour as a function of proton number which acts as one of the indicators of the magic numbers. Note that quadrupole moments close to zero are not by themselves an indication of shell closure. Quadrupole moments start negative just above a shell closure and are positive just before the shell closes—they must go back through zero somewhere in between. It is a *negative gradient* associated with crossing zero that indicates a shell closure. This behaviour is illustrated in Fig. 2.4.5.

The behaviour of the quadrupole moments between closed shells may be qualitatively understood in terms of the shell model as follows. The residual interactions can mix unfilled states in a shell in with the occupied states, and so collective phenomena will develop when we have plenty of nucleons in a shell and plenty of unfilled states. This situation obtains in the middle of a shell, but not just before or just after shell closure, when the only unfilled states allowing many nucleons to behave collectively are those in the next shell, further away in energy and so not strongly admixed.

Nuclear Physics

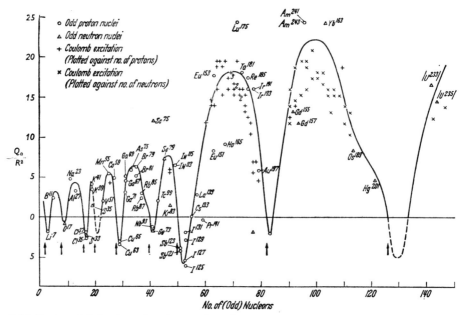

FIG. 2.4.5. Systematic behaviour of nuclear quadrupole moments.
$$Q_0 = \frac{(I+1)(2I+3)}{I(2I-1)} Q \quad \text{and} \quad R = 1.2 \times 10^{-13} A^{1/3}.$$
Arrows mark closed shells, the longer arrows indicating shells of major importance. [From C. H. Townes, Determination of nuclear quadrupole moments, *Handbuch der Physik*, XXXVIII/1, p. 377 (Springer-Verlag, 1958).]

This concludes our discussion of quadrupole moments and we now move on to magnetic multipole moments, of which the only static moment of importance is the *magnetic dipole moment*.

2.5. Static magnetic moments—the magnetic dipole moment

The magnetic dipole moment of the nucleus consists in general of two parts: a term due to the orbital motion of the nucleons, and a term from the intrinsic magnetic moments of the nucleons. The first term may be handled in the framework of non-relativistic quantum mechanics. If we are only concerned with the orbital motion of the nucleons, then we may either calculate the magnetic field due to an assembly of circulating currents by writing

$$\mathbf{A}(\mathbf{r}) = \int \frac{\mathbf{J}(\mathbf{r}')d^3r'}{|\mathbf{r}-\mathbf{r}'|} \tag{2.5.1}$$

where **A** is the magnetic potential and **J** the current density in the nucleus which generates it, or we may write the energy of such a current distribution in a magnetic field as

$$-\int \mathbf{A} \cdot \mathbf{J}(\mathbf{r}') \, d^3r'. \tag{2.5.2}$$

Angular Momentum and the Nucleus

Adopting the first approach, it is easily shown that[†]

$$\mathbf{A}(\mathbf{r}) = \oint \frac{I(\mathbf{r}')}{2} [\mathbf{r}' \times d\mathbf{r}'] \times \frac{\mathbf{r}}{r^3}; \qquad (2.5.3)$$

we take only the first term in the expansion which is written as

$$\mathbf{A}(\mathbf{r}) = \mathbf{M} \times \frac{\mathbf{r}}{r^3} \qquad (2.5.4)$$

where $\mathbf{M}$ is the magnetic dipole moment of the current distribution.

Now $I(\mathbf{r}') \, d\mathbf{r}' = \mathbf{v}' \frac{\varrho}{c} dV$ where ϱ/c is the charge density. Then

$$\mathbf{M} = \int \frac{\varrho}{2c} [\mathbf{r}' \times \mathbf{v}'] \, d^3r' \qquad (2.5.5)$$

which is parallel and proportional to the angular momentum of the system. For a quantum mechanical system we write (for a single particle of charge e)

$$\mathbf{M} = \frac{e}{2c} \int \psi^*(\mathbf{r}') [\mathbf{r}' \times \mathbf{v}'] \psi(\mathbf{r}') \, d^3r' \qquad (2.5.6)$$

$$= \frac{e}{2mc} \int \psi^*(\mathbf{r}') \mathbf{L} \psi(\mathbf{r}') \, d^3r' \qquad (2.5.7)$$

where m is the particle mass.

L_x and L_y are connected with the raising and lowering operators of angular momentum:

$$\begin{aligned}(L_x + iL_y) \psi(L, L_z) &\propto \psi(L, L_z+1), \\ (L_x - iL_y) \psi(L, L_z) &\propto \psi(L, L_z-1).\end{aligned} \qquad (2.5.8)$$

Since $\psi(L, L_z+1)$ and $\psi(L, L_z-1)$ are orthogonal to $\psi(L, L_z)$ only the z-component remains and

$$M = \frac{e}{2mc} \int \psi^*(\mathbf{r}') L_z \psi(\mathbf{r}') \, d^3r' = \frac{e\hbar}{2mc} l_z \qquad (2.5.9)$$

where l_z is an integer. If we define the maximum value of this quantity as the magnetic moment M_0 of the system, then the magnetic moment in any substate is

$$M_0 \frac{l_z}{L}.$$

The magnetic moment due to orbital motion is thus proportional to the third component of the angular momentum, and we have been able to work out the constant of proportionality without a detailed knowledge of the wave functions, in contrast to our study of the electrical quadrupole moment.

This only applies to orbital angular momenta of spinless particles. It is tempting to try to generalize by writing

$$M = \frac{e}{2mc} \int \psi^*(\mathbf{r}') J_z \psi(\mathbf{r}') \, d^3r' \qquad (2.5.10)$$

[†] See, for example, J. R. Reitz and F. J. Milford, *Foundations of Electromagnetic Theory*, chap. 8 (Addison-Wesley, 1960).

Nuclear Physics

for non-integral J but this is not valid. For non-integral J we must take account of the intrinsic magnetic moments of the proton and neutron, and, for applications to atomic physics, of the electron.

The electron has a spin of $\frac{1}{2}$ and a magnetic moment of $e\hbar/2mc$. If we write this as

$$g \frac{e\hbar}{2mc} \mathbf{S}$$

then $g = 2$, the gyromagnetic ratio of the electron.[†] This was observed first through the so-called anomalous Zeeman effect, and had to be added on *ad hoc* to non-relativistic quantum mechanics. However, the Dirac equation[‡] which is relativistically invariant describes in the non-relativistic limit a particle with spin $\frac{1}{2}$ and in an external magnetic field has a term in the Hamiltonian

$$\frac{e\hbar}{mc} \mathbf{S} \cdot \mathbf{B}$$

where $\mathbf{B}$ is the magnetic field. This is interpreted as giving to a particle of spin $\frac{1}{2}$ a magnetic moment of $e\hbar/2mc$ and hence a gyromagnetic ratio of 2. (Quantum electrodynamics predicts very accurately the departure of g from 2.) Thus there exists a theoretical framework within which the electron magnetic moment may be understood. The situation is different for the proton and neutron. Both have magnetic moments, of

$$2.79 \frac{e\hbar}{2M_p c} \quad \text{and} \quad -1.91 \frac{e\hbar}{2M_p c}$$

respectively. They are not described by an unmodified Dirac equation (in particular, a Dirac particle with zero charge would have zero magnetic moment) and while the anomalous magnetic moments are supposed to arise from circulating meson currents within the proton and the neutron, there exists no theoretical framework within which their values are given. Then for a single nucleon we must write

$$M = \frac{e\hbar}{2Mc} \int \psi^*(\mathbf{r}')[L_z + gS_z] \psi(\mathbf{r}') \, d^3r' \qquad (2.5.11)$$

where $g = 5.58$ for a proton and -3.82 for a neutron. We can use this formula to see that a nucleus of angular momentum zero cannot exhibit a static magnetic dipole moment. Suppose

$$M = \frac{e\hbar}{2Mc} \int \psi^*(\alpha L_z + \beta S_z) \psi \, dV. \qquad (2.5.12)$$

If $\alpha = \beta = 1$ then we would just have $\langle J_z \rangle$ which of course is zero. But what happens if $\alpha \neq \beta$ and we have several nucleons for which the orbital angular momenta and the spin sum are not separately zero but only the sum of the two is zero? If $\langle \mathbf{L} + \mathbf{S} \rangle = 0$ then

---

[†] g is not precisely equal to 2. Quantum electrodynamics predicts a small correction, measured as $g/2 = 1.001\ 159\ 549(30)$. B. N. Taylor, W. H. Parker and D. N. Langenberg, *Rev. Mod. Phys.* **41**, 375 (1969).

[‡] See, for example, L. I. Schiff, *Quantum Mechanics*, 3rd ed., sect. 52 (McGraw-Hill, 1968).

Angular Momentum and the Nucleus

$\langle L \rangle = -\langle S \rangle$. There is no intrinsic direction and the system viewed from any direction must therefore be represented by a function

$$\psi_0 = C\{a_L\psi_L(L, +L)\psi_S(S, -S) + a_{L-1}\psi_L(L, L-1)\psi_S(S, 1-S) \\ \ldots a_{-L}\psi_L(L, -L)\psi_S(S, +S)\} \quad (2.5.13)$$

where C is a normalization constant and the a's are phase factors of modulus 1, which need not concern us further here. Such a state is an equal mixture of all terms with $J_z = 0$ and with appropriate choice of the phase factors, is invariant under rotations.

[Consider the example of two p-wave states, each of angular momentum 1, coupling to make angular momentum 0.

Write

$$\begin{aligned} \psi_1(1, +1) &= \frac{1}{\sqrt{2}}(x_1+iy_1) & \psi_2(1, +1) &= \frac{1}{\sqrt{2}}(x_2+iy_2), \\ \psi_1(1, 0) &= z_1 & \psi_2(1, 0) &= z_2, \\ \psi_1(1, -1) &= \frac{1}{\sqrt{2}}(x_1-iy_1) & \psi_2(1, -1) &= \frac{1}{\sqrt{2}}(x_2-iy_2). \end{aligned} \quad (2.5.14)$$

Then the combination

$$\begin{aligned} \psi_0 &= \psi_1(1, +1)\psi_2(1, -1) + \psi_1(1, 0)\psi_2(1, 0) + \psi_1(1, -1)\psi_2(1, +1) \\ &= \tfrac{1}{2}(x_1+iy_1)(x_2-iy_2) + z_1z_2 + \tfrac{1}{2}(x_1-iy_1)(x_2+iy_2) \\ &= \mathbf{r}_1 \cdot \mathbf{r}_2 \end{aligned} \quad (2.5.15)$$

which is a scalar and so corresponds to spin 0.]

It is trivial to show that

$$J_z\psi_0 = 0 \quad (2.5.16)$$

and if we now evaluate

$$\int \psi_0^*(\alpha L_z + \beta S_z)\psi_0 \, d\Omega$$

we have, because each term is an eigenfunction of both L_z and S_z and successive terms are orthogonal:

$$\int \{\psi_L^*(L, +L)\psi_S^*(S, -S)[\alpha L - \beta S]\psi_L(L, +L)\psi_S(S, -S) \\ + \psi_L^*(L, L-1)\psi_S^*(S, 1-S)[\alpha(L-1) - \beta(S-1)]\psi_L(L, L-1)\psi_S(S, 1-S) \\ + \ldots \psi_L^*(L, -L)\psi_S^*(S, +S)[-\alpha L + \beta S]\psi_L(L, -L)\psi_S(S, +S)\} \, d\Omega \quad (2.5.17)$$

and pairing terms inwards from the ends yields zero (Fig. 2.5.1).

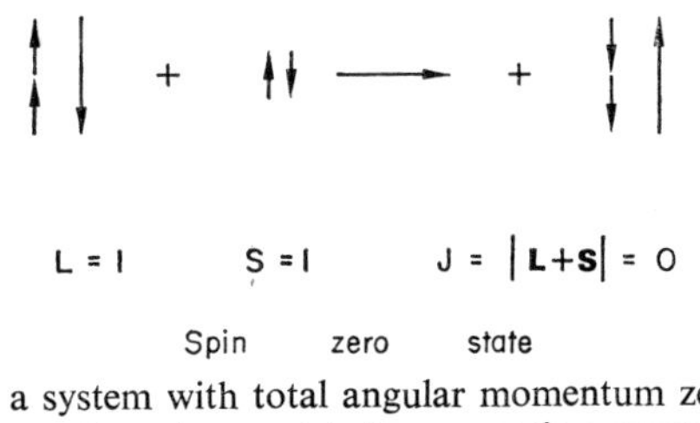

$$\uparrow\downarrow \quad + \quad \uparrow\downarrow \quad \longrightarrow \quad + \quad \uparrow\downarrow$$

$$L = 1 \qquad S = 1 \qquad J = |L+S| = 0$$

Spin zero state

FIG. 2.5.1. This illustrates how a system with total angular momentum zero has zero magnetic moment. With each of the amplitudes shown there is associated a magnetic moment, but the magnetic moment for spin zero is given by a sum over these three amplitudes and hence is zero.

Nuclear Physics

As a concrete example, consider an electron in a p state about a proton. The total angular momentum of this system can be 0, 1, or 2. The state with $J = 0$ is a triplet state so that $L = 1$ and $S = 1$. The magnetic moment of the proton is $g_p S_{zp}$ and of the electron $g_e S_{ze}$ while the orbital motion contributes L_z, all in units of $e\hbar/2mc$ where m is the appropriate mass.

The spins couple to make 1 so that the spin wave functions are

$$\begin{aligned} \psi(1, +1) &= \psi_p(+\tfrac{1}{2})\psi_e(+\tfrac{1}{2}) \\ \psi(1, \ \ 0) &= \frac{1}{\sqrt{2}}\{\psi_p(+\tfrac{1}{2})\psi_e(-\tfrac{1}{2})+\psi_p(-\tfrac{1}{2})\psi_e(+\tfrac{1}{2})\} \\ \psi(1, -1) &= \psi_p(-\tfrac{1}{2})\psi_e(-\tfrac{1}{2}) \end{aligned} \qquad (2.5.18)$$

and the state with $J = 0$ is

$$\Psi(0) = f(r)\sqrt{1/3}\{\sqrt{3/8\pi}\sin\theta e^{i\phi}\,\psi(1,-1) - \sqrt{3/4\pi}\cos\theta\,\psi(1,0) \\ + \sqrt{3/8\pi}\sin\theta e^{-i\phi}\,\psi(1,+1)\} \qquad (2.5.19)$$

where

$$\Psi^*(0)\Psi(0) = \frac{|f(r)|^2}{4\pi}$$

and (r, θ, ϕ) are the coordinates of the electron with respect to the proton.

$$\frac{e\hbar}{2m_e c}\int \Psi^*(0)\left\{L_z + g_e S_{ze} + g_p S_{zp}\frac{m_e}{M_p}\right\}\Psi(0)\,dV$$

$$= \frac{e\hbar}{2m_e c}\int \frac{|f(r)|^2}{4\pi}\left\{\frac{1}{\sqrt{2}}\sin\theta e^{-i\phi}\psi^*(1,-1) - \cos\theta\,\psi^*(1,0)\right.$$
$$\left. + \frac{1}{\sqrt{2}}\sin\theta e^{i\phi}\psi^*(1,+1)\right\}$$

$$\times\left\{\left(1 - \tfrac{1}{2}g_e - \tfrac{1}{2}g_p\frac{m_e}{M_p}\right)\frac{1}{\sqrt{2}}\sin\theta e^{i\phi}\psi(1,-1)\right.$$

$$-\left(-\tfrac{1}{2}g_e\left[\psi_p\left(+\tfrac{1}{2}\right)\psi_e\left(-\tfrac{1}{2}\right) - \psi_p\left(-\tfrac{1}{2}\right)\psi_e\left(+\tfrac{1}{2}\right)\right]\right.$$

$$\left.+\tfrac{1}{2}g_p\frac{m_e}{M_p}\left[\psi_p\left(+\tfrac{1}{2}\right)\psi_e\left(-\tfrac{1}{2}\right) - \psi_p\left(-\tfrac{1}{2}\right)\psi_e\left(+\tfrac{1}{2}\right)\right]\right)\cos\theta$$

$$\left. + \left(-1 + \tfrac{1}{2}g_e + \tfrac{1}{2}g_p\frac{m_e}{M_p}\right)\frac{1}{\sqrt{2}}\sin\theta e^{-i\phi}\psi(1,+1)\right\}\,r^2\,dr\,d\cos\theta\,d\phi$$

$$= \frac{e\hbar}{2m_e c}\int\frac{|f(r)|^2}{4\pi}\left\{\frac{1}{\sqrt{2}}\sin\theta\,e^{-i\phi}\psi^*(1,-1) - \cos\theta\,\psi^*(1,0)\right.$$
$$\left. + \frac{1}{\sqrt{2}}\sin\theta e^{i\phi}\psi^*(1,+1)\right\}$$

$$\times\left\{\left(1 - \tfrac{1}{2}g_p\frac{m_e}{M_p} - \tfrac{1}{2}g_e\right)\frac{1}{\sqrt{2}}\sin\theta e^{i\phi}\psi(1,-1)\right.$$

86

Angular Momentum and the Nucleus

$$-\frac{1}{2}\left(g_p\frac{m_e}{M_p}-g_e\right)\cos\theta\,\psi(0,0)$$

$$-\left(1-\frac{1}{2}g_p\frac{m_e}{M_p}-g_e\right)\frac{1}{\sqrt{2}}\sin\theta e^{-i\phi}\,\psi(1,+1)\Big\}\,r^2\,dr\,d\cos\theta\,d\phi$$

$$=0 \tag{2.5.20}$$

regardless of the values of g_p and g_e.

Thus while we might have expected on classical grounds that this state would exhibit a dipole moment due to lack of cancellation of the individual moments, it does not. Only a state with $J \geqslant \frac{1}{2}$ can exhibit a magnetic dipole moment, regardless of the details of its composition.

In general we write

$$M_J = \mu \int \psi^* g_J J_z \psi\, dV \tag{2.5.21}$$

where g_J is the gyromagnetic ratio of a state with total angular momentum J and μ is either the Bohr or nuclear magneton.

Let us now work out the effect of the nuclear magnetic moment on atomic spectra. Each atomic energy level will be split by the interaction between the magnetic dipole moment of the electrons and the magnetic dipole moment of the nucleus. If $\mathbf{r}$ describes the electron coordinates and $\mathbf{r}'$ the nuclear coordinates

$$\left.\begin{aligned}\Delta E &= -\int \mathbf{A}_n(\mathbf{r})\cdot\mathbf{J}_e(\mathbf{r})\,d^3r\\ &= -\iint \frac{\mathbf{J}_n(\mathbf{r})\cdot\mathbf{J}_e(\mathbf{r})}{|\mathbf{r}-\mathbf{r}'|}\,d^3r'\,d^3r = -\mathbf{M}_N\cdot\mathbf{B}_e\end{aligned}\right\} \tag{2.5.22}$$

where $\mathbf{J}_e$ and $\mathbf{J}_n$ are the current densities.

For $r' \ll r$

$$\left.\begin{aligned}\Delta E &= -\iint \mathbf{J}_e(\mathbf{r})\cdot\left[\frac{\mathbf{I}_N(\mathbf{r}')}{2}(\mathbf{r}'\times d\mathbf{r}')\times\frac{\mathbf{r}}{r^3}\right]d^3r\\ &= -\iint \frac{I_e(\mathbf{r})I_n(\mathbf{r}')}{2}\,d\mathbf{r}\cdot\frac{[(\mathbf{r}'\times d\mathbf{r}')\times\mathbf{r}]}{r^3}\end{aligned}\right\} \tag{2.5.23}$$

where I_e and I_n are the currents corresponding to $\mathbf{J}_e$ and $\mathbf{J}_n$.

$$\left.\begin{aligned}d\mathbf{r}\cdot[(\mathbf{r}'\times d\mathbf{r}')\times\mathbf{r}] &= -(\mathbf{r}'\times d\mathbf{r}')\cdot(d\mathbf{r}\times\mathbf{r})\\ &= (\mathbf{r}'\times d\mathbf{r}')\cdot(\mathbf{r}\times d\mathbf{r}).\end{aligned}\right\} \tag{2.5.24}$$

So the interaction energy is $\propto \mathbf{I}\cdot\mathbf{J}$. We have only shown this for orbital angular momentum but the result carries over to total angular momentum.

Now in obtaining eq. (2.5.9) we argued that only $\int \psi^* L_z \psi\, dV$ is non-zero. We can, however, easily see that the interaction energy given by $\int \psi^*(\mathbf{I}\cdot\mathbf{J})\psi\, dV$ is not the same as $\int \psi^*(I_z J_z)\psi\, dV$. Consider a state of the atom with angular momentum F, F_z.

Nuclear Physics

Then
$$\psi(F, F_z) = A(I_z)\psi_n(I, I_z)\psi_e(J, F_z - I_z)$$
$$+ A(I_z - 1)\psi_n(I, I_z - 1)\psi_e(J, F_z - I_z + 1) \ldots A(F_z - J_z)\psi_n(I, F_z - J_z)\psi_e(J, J_z) \quad (2.5.25)$$

where the A's are numbers. Then $I_x J_x$ and $I_y J_y$ are also non-zero:
$$\begin{aligned}
I_x &= \tfrac{1}{2}(I_x + iI_y) + \tfrac{1}{2}(I_x - iI_y) = \tfrac{1}{2}(I^+ + I^-), \\
I_y &= +\frac{i}{2}(I^- - I^+), \\
I_x J_x &= \tfrac{1}{4}(I^+ + I^-)(J^+ + J^-) = \tfrac{1}{4}\{I^+J^+ + I^-J^- + I^-J^+ + I^+J^-\}, \\
I_y J_y &= -\tfrac{1}{4}(I^- - I^+)(J^- - J^+) = -\tfrac{1}{4}\{I^+J^+ + I^-J^- - I^-J^+ - I^+J^-\}.
\end{aligned} \quad (2.5.26)$$

I^+J^+, I^-J^- move $\psi(F, F_z)$ to a state of different F_z. I^-J^+ and I^+J^- terms keep F_z the same and the mean value of these terms is non-zero.
$$\int \psi^*(F, F_z) \mathbf{I}\cdot\mathbf{J}\, \psi(F, F_z)\, dV$$
is proportional to the splitting of an energy level. Since $\mathbf{F} = \mathbf{I} + \mathbf{J}$,
$$\mathbf{I}\cdot\mathbf{J} = \frac{F^2 - I^2 - J^2}{2}$$

and
$$\mathbf{I}\cdot\mathbf{J}\,\psi(F, F_z) = \frac{F(F+1) - I(I+1) - J(J+1)}{2}\psi(F, F_z). \quad (2.5.27)$$

So the splitting is
$$\propto \frac{F(F+1) - I(I+1) - J(J+1)}{2}$$

where the constant of proportionality depends on the magnetic moment of the nucleus and on the magnetic field due to the atom at the nucleus.

The splitting depends only on F and not F_z: without the magnetic interaction states of different F and F_z are degenerate. The magnetic interaction is purely internal and so affects only states of different F.

Calculation of the magnetic field due to the electrons which acts at the nucleus is easy only for hydrogen-like atoms. Let us consider atomic hydrogen in the ground S-state. There is no orbital angular momentum and the magnetic field $\mathbf{B}$ at the nucleus is just $8\pi\,\mathbf{M}(0)/3$ where $\mathbf{M}(0)$ is the magnetization at the nucleus. This result will be familiar to you as the field at the centre of a uniformly magnetized sphere. The hydrogen electron cloud is not uniformly magnetized, the magnetization is
$$\psi^*(r)\,\frac{e\hbar}{2mc}\,g_e \mathbf{S}_e\,\psi(r)$$

which falls off exponentially with the radius r, but the direction is constant and the spherical symmetry alone yields the above result. The interaction energy is then
$$E = -\frac{8\pi}{3} g_e g_p \frac{e\hbar}{2mc}\,\mathbf{S}_e\cdot\mathbf{S}_p\,\frac{e\hbar}{2Mc}\,\psi^*(0)\,\psi(0); \quad \psi(r) = \sqrt{\frac{1}{\pi a^3}}\,e^{-r/a} \quad (2.5.28)$$

where a is the Bohr radius, so $\psi^*(0)\,\psi(0) = 1/\pi a^3$.

$$E = -\frac{8}{3a^3} g_e g_p \, \mathbf{S}_e \cdot \mathbf{S}_p \, \mu_B \mu_N$$

where μ_B is the Bohr magneton and μ_N the nucleus magneton.

$$g_e = 2 \quad \text{and} \quad g_p = 5.58,$$
$$a = 5.29 \times 10^{-9} \text{ cm},$$
$$\mu_B = 0.927 \times 10^{-20} \text{ erg gauss}^{-1},$$
$$\mu_N = 0.505 \times 10^{-23} \text{ erg gauss}^{-1}.$$

The dimensions of the right-hand side are $\text{erg}^2 \text{ gauss}^{-2} \text{ cm}^{-3}$ and since $\text{gauss}^2 \text{ cm}^3$ has the dimensions of energy (in ergs) the whole expression is dimensionally correct.

$$S_e(S_e+1) = S_p(S_p+1) = \tfrac{3}{4} \qquad S_e + S_p = 0 \quad \text{or} \quad 1.$$

Therefore

$$\mathbf{S}_e \cdot \mathbf{S}_p = -\tfrac{3}{4} \quad \text{for the singlet,}$$
$$= +\tfrac{1}{4} \quad \text{for the triplet.}$$

The difference in energy between the singlet and triplet is just

$$\frac{8}{3a^3} g_e g_p \, \mu_B \mu_N = 9.6 \times 10^{-18} \text{ ergs} = 6 \times 10^{-6} \text{ eV}.$$

Compare this with the ionization potential of atomic hydrogen—14 eV.

The wavelength of radiation emitted in a transition between the singlet and triplet states of atomic hydrogen is 21 cm—this transition is responsible for the famous 21 cm line which is of great importance in radio astronomy.

Finally, note that the magnetic field B at the core of the hydrogen atom is 1.67×10^5 gauss.

Let us now calculate the magnetic moments of nuclei using the independent particle model, although we have already found this model to be inadequate in considering electric quadrupole moments. We have seen in Chapter 1 that the ground states of even–even nuclei have spin zero. Then these nuclei will exhibit no magnetic moment, and this result is independent of the model of the nucleus we use and depends only on their zero spin. In the independent particle model, even–odd and odd–even nuclei are regarded as having an even–even core as a result of pairing with angular momentum zero and a single odd nucleon in orbit about this core. Then in this model the magnetic moment of odd A nuclei will be due solely to the orbital motion and intrinsic magnetic moment of the odd nucleon. The resulting magnetic moment is easily calculated.

$$\mu = \frac{e\hbar}{2Mc} \int \psi^*(r', J_z = J)[L_z + g_p S_z] \psi(r', J_z = J) \, d^3r' \qquad (2.5.29)$$

Nuclear Physics

for an odd proton and

$$\mu = \frac{e\hbar}{2Mc} \int \psi^*(r', J_z = J)[g_n S_z] \psi(r', J_z = J) \, d^3r' \tag{2.5.30}$$

for an odd neutron which, being neutral, only contributes its intrinsic magnetic moment.

Since $\psi(J, J_z)$ contains a mixture of various values of L_z and S_z subject only to the restriction that $L_z + S_z = J_z$, evaluation of the integral is not completely trivial. We could work it out the long way, using the proper mixture of different states, or we can do it as follows: the average value of any vector must be along $\mathbf{J}$, the only well-defined angular momentum vector in the problem. Then

$$\int \psi^*(J) \mathbf{L} \psi(J) \, dV = R_L \int \psi^*(J) \mathbf{J} \psi(J) \, dV. \tag{2.5.31}$$

Then

$$\int \psi^*(J) \mathbf{L} \cdot \mathbf{J} \psi(J) \, dV = R_L \int \psi^*(J) J^2 \psi(J) \, dV = R_L J(J+1). \tag{2.5.32}$$

So

$$R_L = \int \frac{\psi^*(J) \mathbf{L} \cdot \mathbf{J} \psi(J) \, dV}{J(J+1)}, \tag{2.5.33}$$

$$\mathbf{L} \cdot \mathbf{J} = \frac{J^2 + L^2 - S^2}{2},$$

$$\therefore \int \psi_J^* (\mathbf{L} \cdot \mathbf{J}) \psi_J \, dV = \frac{J(J+1) + L(L+1) - S(S+1)}{2},$$

$$\therefore \int \psi_J^* L_z \psi_J \, dV = J_z \frac{J(J+1) + L(L+1) - S(S+1)}{2J(J+1)},$$

$$\mathbf{S} \cdot \mathbf{J} = \frac{J^2 + S^2 - L^2}{2},$$

$$\therefore \int \psi_J^* S_z \psi_J \, dV = J_z \frac{J(J+1) + S(S+1) - L(L+1)}{2J(J+1)},$$

$$\mu_{J_z} = \frac{e\hbar}{2Mc} \left\{ \frac{J(J+1) + L(L+1) - S(S+1)}{2(J+1)} + g_p \frac{J(J+1) + S(S+1) - L(L+1)}{2(J+1)} \right\} \frac{J_z}{J} \tag{2.5.34}$$

for a proton. (Note that if $S = 0$ then $J = L$ and $\mu_{J_z} = e\hbar L_z/2Mc$ as we defined things before.)

$$\mu_{J_z} = \frac{e\hbar}{2Mc} \left\{ g_n \frac{J(J+1) + S(S+1) - L(L+1)}{2J(J+1)} \right\} J_z. \tag{2.5.35}$$

Then

μ (odd proton)
$$= \frac{e\hbar}{2Mc} J \left\{ \frac{J(J+1) + L(L+1) - S(S+1)}{2J(J+1)} + g_p \frac{J(J+1) + S(S+1) - L(L+1)}{2J(J+1)} \right\},$$
$$\mu \text{ (odd neutron)} = \frac{e\hbar}{2Mc} J \left\{ g_n \frac{J(J+1) + S(S+1) - L(L+1)}{2J(J+1)} \right\}. \tag{2.5.36}$$

Angular Momentum and the Nucleus

We may write $\mu = g_J \mu_N J$ where μ_N is the nuclear magneton, in which case

$$g_J = \frac{J(J+1)+L(L+1)-S(S+1)}{2J(J+1)} g_L + \frac{J(J+1)+S(S+1)-L(L+1)}{2J(J+1)} g_S, \quad (2.5.37)$$

where $g_L = 1$ for an odd proton,
$ = 0$ for an odd neutron;
$g_S = 5.58$ for an odd proton,
$ = -3.82$ for an odd neutron.

Now
$$S = \tfrac{1}{2}, \quad L = J \pm \tfrac{1}{2},$$

$$\begin{aligned}
J(J+1)+L(L+1)-S(S+1) &= 2J^2+3J & L &= J+\tfrac{1}{2},\\
&= 2J^2+J-1 & L &= J-\tfrac{1}{2};\\
J(J+1)+S(S+1)-L(L+1) &= -J & L &= J+\tfrac{1}{2},\\
&= J+1 & L &= J-\tfrac{1}{2}.
\end{aligned}$$

Then for an odd proton this model gives

$$\left.\begin{aligned}
\mu &= (J-\tfrac{1}{2})\mu_N + g_p \mu_N & J &= L+\tfrac{1}{2},\\
\mu &= \frac{J^2+\tfrac{3}{2}J}{J+1}\mu_N - \tfrac{1}{2}\frac{J}{J+1} g_p \mu_N & J &= L-\tfrac{1}{2},
\end{aligned}\right\} \quad (2.5.38)$$

and for an odd neutron

$$\left.\begin{aligned}
\mu &= \tfrac{1}{2} g_n \mu_N & J &= L+\tfrac{1}{2},\\
\mu &= -\tfrac{1}{2} \frac{J}{J+1} \mu_N g_n & J &= L-\tfrac{1}{2}.
\end{aligned}\right\} \quad (2.5.39)$$

These are the Schmidt limits, sometimes known as Schmidt lines. (Magnetic moments can also be calculated for odd–odd nuclei, but only if the specific coupling scheme is known.)

Finally, let us work out the magnetic moments expected for an odd proton in the $p_{\frac{3}{2}}$ and $p_{\frac{1}{2}}$ states first using the formulae derived above and then going back to (2.5.29):

$$p_{\frac{3}{2}} : J = \tfrac{3}{2} \quad \mu = \mu_N + \mu_p \quad \mu_p = \tfrac{1}{2} g_p \mu_N, \quad (2.5.40)$$
$$p_{\frac{1}{2}} : J = \tfrac{1}{2} \quad \mu = \tfrac{2}{3}\mu_N - \tfrac{1}{3}\mu_p. \quad (2.5.41)$$

In equation (2.5.29) we wrote

$$\mu = \mu_N \int \psi^*(J, J_z = J)\{L_z + g_p S_z\} \psi(J, J_z = J) \, dV.$$

Now
$$\psi(\tfrac{3}{2}, +\tfrac{3}{2}) = \psi(1, +1)\psi(\tfrac{1}{2}, +\tfrac{1}{2}),$$
$$L_z \psi(\tfrac{3}{2}, +\tfrac{3}{2}) = \psi(\tfrac{3}{2}, +\tfrac{3}{2}),$$
$$S_z \psi(\tfrac{3}{2}, +\tfrac{3}{2}) = \tfrac{1}{2}\psi(\tfrac{3}{2}, +\tfrac{3}{2}),$$

whence (2.5.40)

$$\psi(\tfrac{1}{2}, +\tfrac{1}{2}) = \sqrt{\tfrac{2}{3}}\psi(1, +1)\psi(\tfrac{1}{2}, -\tfrac{1}{2}) - \sqrt{\tfrac{1}{3}}\psi(1, 0)\psi(\tfrac{1}{2}, +\tfrac{1}{2}),$$
$$L_z \psi(\tfrac{1}{2}, +\tfrac{1}{2}) = \sqrt{\tfrac{2}{3}}\psi(1, +1)\psi(\tfrac{1}{2}, -\tfrac{1}{2}),$$
$$S_z \psi(\tfrac{1}{2}, +\tfrac{1}{2}) = -\tfrac{1}{2}\sqrt{\tfrac{2}{3}}\psi(1, +1)\psi(\tfrac{1}{2}, -\tfrac{1}{2}) - \tfrac{1}{2}\sqrt{\tfrac{1}{3}}\psi(1, 0)\psi(\tfrac{1}{2}, +\tfrac{1}{2}).$$

Nuclear Physics

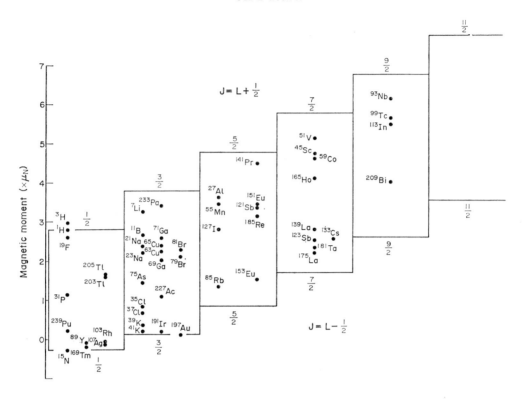
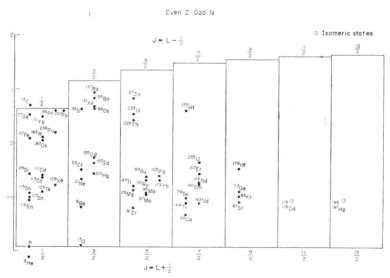

FIG. 2.5.2. In this figure the spins and magnetic moments of odd A nuclei are compared with the Schmidt limits. With a few exceptions the magnetic moments are significantly smaller in magnitude than the predictions of a single-particle model. [Data from G. H. Fuller and V. W. Cohen, Nuclear Moments, Appendix 1 to *Nuclear Data Sheets* (1965).]

Angular Momentum and the Nucleus

So the integral becomes

$$\tfrac{2}{3}(1-\tfrac{1}{2}g_p)+\tfrac{1}{2}\tfrac{1}{3}g_p$$

which of course gives (2.5.41) back to us.

The Schmidt limits enclose the observed magnetic moments very nicely: nuclear magnetic moments typically lie within these limits, but sufficiently close to them that, given the total angular momentum of the nucleus, J, the orbital angular momentum L may be inferred from a measurement of the magnetic moment. The actual values of the magnetic moments are correct to within a factor of 2 or so: the independent particle model is much more successful in explaining the magnetic dipole moments of nuclei than in explaining the values of the electric quadrupole moments which as we have already seen differ by an order of magnitude from those predicted by a single-particle model. The values of the magnetic moments sufficed to establish the single-particle levels of unpaired nucleons and were crucial to the development of the shell model.

We should not expect the magnetic moments to be predicted accurately by the elementary shell model. If we turn up the residual interactions, some of the unfilled states will be mixed in with the occupied single-particle states and the angular momentum of the last unpaired nucleon may be shared with other nucleons. Conservation of angular momentum ensures that the spin of the nucleus is not changed from the single-particle value, but the other nucleons will contribute to the magnetic moment, and in most cases reduce it in magnitude.

The determination of the spins, magnetic moments and quadrupole moments of nuclei may be regarded as a problem in experimental atomic physics rather than nuclear physics if you care to make a distinction. Almost all the information on these properties of nuclei has been obtained from a study of atomic and molecular spectra: both optical emission spectra and radiofrequency and microwave absorption spectra. No account of nuclear angular momentum and associated properties would be complete without some discussion of the way in which these quantities may be deduced.

Consider first an atom in which the nucleus has a spin $\geqslant \tfrac{1}{2}$ and the electrons also have a total angular momentum $\geqslant \tfrac{1}{2}$. The electrons generate a magnetic field at the nucleus, and the energy levels of the atom depend on the relative orientations of the field axis and the nuclear magnetic moment. The magnetic interaction splits levels which would otherwise be degenerate and contributes to the hyperfine structure of these levels. The splitting is proportional to the electron magnetic field at the nucleus, to the nuclear magnetic moment and to the factor

$$\mathbf{I}\cdot\mathbf{J} = \frac{F(F+1)-I(I+1)-J(J+1)}{2} \quad (2.5.27)$$

where I is the nuclear angular momentum, J is the electronic angular momentum and F is the total angular momentum of the atom. States with different F have different energies for the same I and J. The maximum value of F is $I+J$ and the minimum value is $|I-J|$:

$$F = I+J, I+J-1, I+J-2, \ldots |I-J|.$$

Then if $J < I$ there are $2J+1$ values of F for given values of I and J, while if $I < J$ there are $2I+1$ values of F. The number of different levels can thus yield I if J is known. For a very

Nuclear Physics

simple example consider atomic hydrogen again. In the ground state there is no orbital angular momentum and so no fine structure to the ground state. I and J are both $\frac{1}{2}$, $F = 1$ or 0. The ground $s_{\frac{1}{2}}$ level is split. A spin flip transition between these states is responsible for the 21 cm emission of galactic hydrogen.

Fig. 2.5.3.

The hyperfine splitting which is produced in optical spectra is mostly due to deeply penetrating $s_{\frac{1}{2}}$ electrons. $p_{\frac{1}{2}}$ electrons are also deeply penetrating because of relativistic effects, but electrons with $l \geqslant 2$ do not in general interact sufficiently strongly with the nuclear magnetic moment to produce optically observable hyperfine structure. Atomic transitions between a level with negligible hyperfine splitting, and a level which is split produce a spectral line which is split and the components can in many cases be resolved by multiple-beam interferometry.

A sample of excited atoms will contain atoms in many different states of J and of F. If J is less than I, then the level with given J has $2J+1$ components. If I is less than J then there are $2I+1$ components. Then the nuclear spin may be obtained from the maximum number of hyperfine components found among a number of levels with different J, and this number may be inferred from the optical spectrum.

The nuclear spin may also be obtained from the relative spacing of hyperfine levels. The interaction energy may be written as

$$\triangle w = \alpha[F(F+1) - I(I+1) - J(J+1)]. \tag{2.5.42}$$

The maximum value of F is $I+J$:

$$\Delta w_{I+J} = \alpha[2IJ],$$
$$\Delta w_{I+J-1} = \alpha[2IJ - 2(I+J)],$$
$$\Delta w_{I+J-n} = \alpha[2IJ - 2n(I+J) + n(n-1)],$$
$$\Delta w_{I+J-n-1} = \alpha[2IJ - 2(n-1)(I+J) + (n-1)(n-2)],$$
$$\Delta w_{I+J-n} - \Delta w_{I+J-n-1} = \alpha[-2(I+J) + 2(n-1)].$$

The relative spacing of the terms in the split level thus also gives I if J is known: for example,

$$\frac{\Delta w_{I-J} - \Delta w_{I+J-1}}{\Delta w_{I+J+n} - \Delta w_{I+J-N-1}} = \frac{I+J}{I+J-(n-1)}. \tag{2.5.43}$$

If the electronic structure of a level characterized by electronic angular momentum J is known well enough to allow a calculation of the magnetic field at the nucleus, then the magnetic dipole moment may also be obtained from the spacing of the hyperfine levels. This is rarely the case.

Angular Momentum and the Nucleus

Multiple-beam interferometry has adequate resolving power for study of the optical hyperfine splitting, but the width of the lines to be resolved must be smaller than their separation. Doppler broadening can destroy the resolution of these hyperfine components and sources with a relatively low spread of atomic velocities must be employed: for example, excited atoms may be injected into a cold gas or an atomic beam be formed with relatively uniform velocity and the atoms excited non-thermally, for example, by electron bombardment.

Atomic beams may also be studied with magnetic resonance methods. If the beam is passed through an inhomogeneous magnetic field, atoms are deflected according to the orientation of their magnetic moments, as in the original Stern–Gerlach experiment. They emerge from such a field in definite states of F, separated spatially according to F_z. The beam may be recombined by the inverse of the sorting field. If in between the two inhomogeneous fields they encounter an r.f. field, transitions may be induced between different hyperfine levels and the beam will no longer be completely recombined by the second inhomogeneous field if the frequency of the r.f. field is appropriate. The hyperfine structure of a level may be obtained from the variation of transmission through the system with applied radiofrequency. If a magnetic field is applied in the transition region, then for atoms in which there is no electron angular momentum, the frequency corresponds to the energy needed to change I_z in the applied field. If the atoms have electronic magnetic moments, then the hyperfine levels are themselves split by the action of the applied magnetic field and transitions may be induced between these components.

Let us briefly consider the effect of an applied magnetic field on an atomic level which possesses hyperfine structure and take as an example the ground state of atomic hydrogen once more. If we could switch off the proton magnetic moment we would have a single energy for the $S_{1/2}$ ground state, which is split into two components by the magnetic interaction between the proton and the electron.

We treat the magnetic interaction as a perturbation which removes the degeneracy between the two states of different F. Now suppose an external magnetic field is applied, which is very weak. This is an additional perturbation and to first order it leaves the wave functions unaltered, but removes the degeneracy between the magnetic substates F_z. The $F = 1$ term is split into three components and the $F = 0$ level is unperturbed.

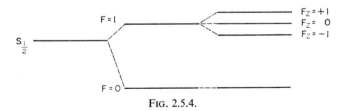

Fig. 2.5.4.

The splitting of the $F = 1$ term is linear in the applied field until first-order perturbation theory is no longer adequate, and the calculation of the level structure becomes very complicated. We can, however, easily see what happens in the limit of a very strong field, when the dominant perturbation is the applied field and the magnetic interaction between the proton and electron can be neglected. F, F_z are no longer good quantum numbers but I, I_z and J, J_z are—clearly if there is no magnetic coupling between the nucleus and the electron

Nuclear Physics

this will be the case, and we must regard the magnetic interaction in the strong field case as a perturbation on energy levels determined by the coulomb field and an applied magnetic field.

The electrons have energy $\pm \mu_e B$ and the protons have energy $\pm \mu_N B$ over and above the energy due to the s-wave orbital. Thus we have the sequence shown in Fig. 2.5.5.

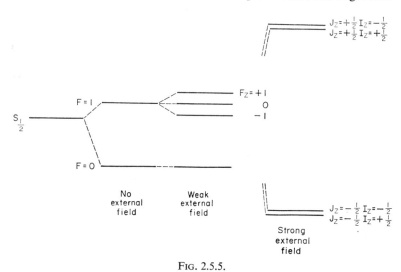

FIG. 2.5.5.

In the intermediate region the levels are eigenfunctions of neither F, F_z nor J, J_z and I, I_z. The definition of a strong field is that J, J_z and I, I_z are constants of the motion and it is reached when the shift of the levels becomes linear with magnetic field. The splitting of the hyperfine structure in a strong field or in the intermediate region when the Zeeman type splitting disappears is called the Paschen–Back effect and counting the number of terms for a given J gives nuclear spin, while the spacing gives the nuclear magnetic moment. This splitting may be studied in optical spectra or in atomic beam magnetic resonance experiments. In the latter case the nuclear magnetic moments are obtained with less accuracy than the transition frequency: while the applied magnetic field can be measured with great accuracy ($\sim$ one part in 10^5) the local field acting on the nucleus is in general less well known because of atomic diamagnetism, and this uncertainty may give rise to errors $\sim 10\%$ in the deduced magnetic moment. The energy levels of nuclei in an applied magnetic field may also be studied by nuclear resonance absorption or resonance induction techniques.

Nuclear electric quadrupole moments have mostly been studied through departures from the hyperfine splitting which would be expected through a magnetic interaction alone: the electric field due to the electrons produces a further splitting which has a different dependence on I and J and F. We should add that studies have been made not only on essentially free atoms in gases but also, by using resonance absorption or induction methods, in liquids and solids. Molecular spectra have yielded a lot of information: we shall not go into this, since it is a matter for the professional, except to discuss one relatively simple and elegant example: determination of nuclear spin through study of the band spectra of homonuclear diatomic molecules.

2.6. Diatomic molecular spectra and nuclear spin

Consider a diatomic molecule. Such a molecule will have an electronic structure which has excited states but it also has extra degrees of freedom which are absent in an atom. The two nuclei are separated by a finite distance, sitting at the bottom of a potential well generated by their coulomb interaction with each other and by the coupling of one nucleus to the other through the electronic structure of the molecule. The two nuclei may vibrate along the line joining them under the influence of this potential. Finally, the molecule may rotate about its centre of inertia since it is not spherically symmetric. If the nuclear vibrations have a small amplitude, so that the nuclear motion does not perturb significantly the electronic energy levels, then electronic and vibrational energies are independent. If the rotational motion is not sufficiently vigorous to appreciably change the separation between the nuclei, the rotational and vibrational energies will be independent (in general this independence of the terms does not obtain, but in most cases it is a good approximation). Then the energy of the molecule may be written as the sum of three independent terms

$$w = w_e + w_v + w_R$$

and the energy change in a transition is

$$\delta w = \delta w_e + \delta w_v + \delta w_R.$$

We shall not be concerned with the form of w_e or w_v. w_R may be written as

$$\frac{\hbar^2 K(K+1)}{2\mathscr{L}} \tag{2.6.1}$$

where K is the rotational angular momentum and $\mathscr{L}$ is the moment of inertia.

There is yet a further degree of freedom which is not present in an atom, if the two nuclei have spin: the relative orientation of the spins of the two nuclei. While the energy of the molecule is negligibly affected by different relative orientations, this extra degree of freedom plays a crucial role. If the two nuclei are identical, then we know that the entire molecular wave function must be either symmetric or antisymmetric under exchange of the two nuclei. The vibrational part depends only on the relative separation and is clearly symmetric.

The electronic part may be either (but is most often symmetric). The rotational part is symmetric for even K and antisymmetric for odd K. The nuclear spin part of the wave function may be either symmetric or antisymmetric under exchange of the two nuclei. If we specify the electronic state and the vibrational state then even K is associated with a symmetric spin wave function and odd K with an antisymmetric spin wave function or vice versa. We may divide the states of the molecule for given electronic and vibrational states into two classes depending on the nuclear spin symmetry and each class has rotational states separated by $\Delta K = 2$. In a transition between different excited states of the molecule, the probability of the transition being between the two classes is negligible because of the negligible interaction between the spins of the two nuclei. Transitions are therefore always within the same class.

Two charged objects of equal mass rotating around each other constitute a fluctuating electric quadrupole, as do two charged objects of equal mass vibrating along their line of

Nuclear Physics

centres. As we shall see in the next chapter, the coupling of a fluctuating quadrupole to the electromagnetic field is very small unless the wavelength of the radiation is comparable to the dimensions of the quadrupole, and so the probability of a change in only w_R (or w_v) is much less than the probability of a decay involving a change of *both* w_e and w_R: the lines due to electronic transitions in a molecule are split by the different rotational energy.

Suppose we restrict ourselves to transitions involving only the rotational motion and the electrons simultaneously, and consider one electronic transition of energy δw_e. K either does not change or changes by 2 units for a given nuclear spin symmetry.

$$\left.\begin{aligned}\delta w &= \delta w_e + \delta w_R, \\ \delta w_R &= \frac{\hbar^2}{2\mathcal{L}}[(K+2)(K+3) - K(K+1)] = \frac{\hbar^2}{2\mathcal{L}}(4K+6).\end{aligned}\right\} \quad (2.6.2)$$

$K = 0, 2, 4, \ldots$ belong to one spin symmetry class, while $K = 0, 3, 5, \ldots$ belong to the other. Then the *transition energies* are

$$\left.\begin{aligned}\delta w_R &= 6, 14, 22, 30 \ldots \frac{\hbar^2}{2\mathcal{L}} \quad (K \text{ even}), \\ & 10, 18, 26, 34 \ldots \frac{\hbar^2}{2\mathcal{L}} \quad (K \text{ odd}).\end{aligned}\right\} \quad (2.6.3)$$

The different rotational states split the lines due to specified electronic transitions. The spacing of the rotational structure is constant and alternate lines belong to different nuclear spin symmetry classes.

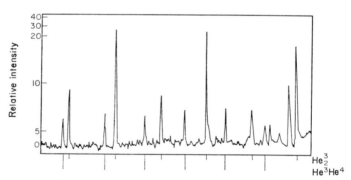

FIG. 2.6.1. Band spectra of ^{3}He–^{3}He and ^{3}He–^{4}He molecules. The ratio of successive lines in the ^{3}He–^{3}He spectrum is 3 : 1, thus demonstrating that the spin of ^{3}He is indeed $\frac{1}{2}$. [From Douglas and Herzberg, *Phys. Rev.* **76**, 1529 (1949).]

A study of this rotational structure in band spectra yields the nuclear spin in the following way. Transitions between states symmetric in the nuclear spin occur more frequently than transitions between states antisymmetric in the nuclear spin, because there are more symmetric states: the ratio between symmetric and antisymmetric states is $(I+1)/I$ where I is the nuclear spin. An antisymmetric state has the form

$$\sqrt{\tfrac{1}{2}}\left\{\chi(1)_{M_1}\chi(2)_{M_2} - \chi(1)_{M_2}\chi(2)_{M_1}\right\} \quad (2.6.4)$$

Angular Momentum and the Nucleus

where χ_M is the spin state of a nucleus with a spin I and $I_z = M$ and the argument labels the nucleus. There are $2I+1$ values of M_1, M_2 and the number of combinations of form $\chi_{M_1} \chi_{M_2}$ with $M_1 \neq M_2$ is easily found to be $I(2I+1)$. This, then, is the number of antisymmetric states. States like

$$\sqrt{\tfrac{1}{2}}\{\chi_{M_1}(1)\chi_{M_2}(2) + \chi_{M_2}(1)\chi_{M_1}(2)\} \tag{2.6.5}$$

are symmetric under interchange of the nuclear spins and there are again $I(2I+1)$ of them. To these must be added the $2I+1$ states with $M_1 = M_2$, which can only be symmetric. Thus there are $(I+1)(2I+1)$ symmetric states, $I(2I+1)$ antisymmetric states.

The symmetric and antisymmetric classes have different rotational angular momenta and so in general different total angular momenta. However, if $K \approx 10$ the factor $2K+1$ is very close to $2(K+1)+1$ and the relative statistical weights of the symmetric and antisymmetric states is approximately $(I+1)/I$.

The $(I+1)(2I+1)$ symmetric states for given K are highly degenerate, as are the $I(2I+1)$ antisymmetric states. At ordinary temperatures, each of these spin configurations has an equal probability and so the ratio of the number of molecules with symmetric spins to the number with antisymmetric spins is $I+1 : I$. Transitions between states with even spin symmetry are therefore $I+1/I$ more common than transitions between states with odd nuclear spin symmetry and the intensities (not amplitudes) of alternate lines in the band spectra of homonuclear diatomic molecules also are in this ratio, whence the nuclear spin. It is important to note that a knowledge of the value of K associated with a particular line is unnecessary: neither does the argument depend on the electronic wave function or on whether the nuclei are fermions or bosons.

Consider the special case when $I = 0$. Then there are no antisymmetric states present. This is obvious from the form of the antisymmetric states but can also be seen differently: if a rotation of π is applied to such a pair of nuclei, the nuclear configuration is the same after the rotation has been applied as before. Angular momentum is specified by the behaviour of a wave function under rotations—any odd angular momentum state changes sign under a rotation of π.

If the electronic structure of the molecule is known and the values of K associated with each line can be inferred, then it is immediately possible to associate a given nuclear spin symmetry with even or odd values of K as the case may be. It is in this way that the statistics obeyed by nuclei have been checked—no departure from the rule that half integral spins mean Fermi statistics and integral spins mean Bose statistics has ever been found.

This chapter has been concerned with the static mechanical, electric and magnetic moments of the nucleus. If the operator appropriate to a classical moment $\mathcal{M}$ is M, then

$$\mathcal{M} = \int \Psi^* M \Psi$$

and setting in the classical limit

$$\Psi \to \sum_n a_n \psi_n$$

we find that $\mathcal{M}$ is made up of terms like

$$\int \psi_n^* M \psi_n$$

Nuclear Physics

and

$$\int \psi_m^* M \psi_n \qquad m \neq n.$$

The former are the diagonal matrix elements of the moment operator M and correspond to the static moments we have been studying. The latter terms are the off-diagonal matrix elements of the operator M and in the quantum limit correspond to the amplitude for a transition between the initial state ψ_n and a different final state ψ_m. In Chapter 3 we shall be discussing these off-diagonal transition matrix elements.

CHAPTER 3

Nuclear Decay

3.1. Introduction

In Chapter 1 we already discussed briefly the phenomenology of radioactive decay. In this chapter we shall be concerned with the dynamics of nuclear decay: the detailed mechanisms through which transitions between nuclear states proceed. We shall distinguish three kinds of nuclear decay. The first is *electromagnetic*, in which a nucleus de-excites by emitting electromagnetic radiation, or by ejecting an atomic electron through a varying local electromagnetic field, or by emission of a pair of electrons in certain rare cases. The second is *β-decay*, or decay through the weak interactions, in which the nucleus emits an electron and a neutrino, or as an alternative to emission of a positive electron, absorbs an atomic electron. Finally, there are decay processes in which the nucleus *emits nucleons* or other nuclear fragments. This class may be divided into two categories: those processes where the driving forces are provided by the strong interactions, such as the boiling off of nucleons from a highly excited nucleus which is unstable over a time scale of $\lesssim 10^{-20}$ sec, or emission of neutrons from a moderately excited nucleus with a time scale of $\lesssim 10^{-15}$ sec, and those processes driven by coulomb repulsion, such as α-decay and spontaneous fission. The latter processes, although driven by the electromagnetic interaction, are not normally classified as electromagnetic decays. In this chapter we shall be concerned primarily with electromagnetic, β- and α-decay processes, leaving decay through strong interactions to Chapter 4.

You will remember from Chapter 1 that with a decay process which a given nuclear state may undergo we associate a decay rate; a quantity with the dimensions of sec^{-1} which measures the probability per unit time that the nucleus decays via the specified process. The total decay rate is the sum of the partial decay rates, and the mean lifetime of the nuclear state is the inverse of the total decay rate. We can easily construct idealized classical systems which behave in this way. Suppose that we take a box with a small hole in one wall and in the box we place one gas molecule with a specified energy but arbitrary direction of motion. If we know the direction of motion, then in principle we can calculate how long it will be before the molecule gets out through the hole, that is, the system decays. If we have a large number of such identical boxes, each with a molecule of the same energy inside but a random distribution of initial directions, the number of boxes containing a molecule will decrease exponentially, the characteristic time being the inverse of the probability per unit time for a

Nuclear Physics

randomly directed molecule to get out of one box. It is also clear that if we open up another decay channel, for example by having two holes in the boxes, that decay rate will go up: the probability for a molecule to leave its container will be the sum of the probabilities for escape through each hole.

It is instructive to compare such an idealized classical system with the decay of a quantum mechanical system, where we have a wave function Ψ_i describing the initial system, a wave function Ψ_f describing the final system and an interaction linking the two. Since the initial and final systems are different, Ψ_i and Ψ_f are orthogonal:

$$\int \Psi_f^* \Psi_i \, dV = 0. \tag{3.1.1}$$

The interaction operator I acts on Ψ_i so as to convert it into another state which is no longer orthogonal to the specified final state Ψ_f:

$$M_{fi} = \int \Psi_f^* (I\Psi_i) \, dV \neq 0. \tag{3.1.2}$$

This integral is the matrix element for the transition and the transition rate is governed by the strength of the interaction I and the spatial overlap of the state $(I\Psi_i)$ with the state Ψ_f. The transition rate per unit time from the state Ψ_i to the state Ψ_f is given by

$$\frac{2\pi}{\hbar} \left| \int \Psi_f^* I \Psi_i \, dV \right|^2 \tag{3.1.3}$$

and for a continuum of final states we obtain the total transition rate by summing over all the degenerate states with the same energy as the initial state. This gives the decay rate as

$$\frac{1}{\tau_i} = \frac{2\pi}{\hbar} |M_{fi}|^2 \frac{dn}{dE_i} \tag{3.1.4}$$

where dn/dE_i is the density of final states in the continuum per unit interval of the energy of the system. This expression is known in the trade as the Fermi Golden Rule. Before we go on to discuss it in more detail we will point out the difference between the quantum mechanical treatment of a decay and our classical system.

The first difference is that in the classical case the concept of a decay probability arises because of incomplete information—classically it is possible to calculate the state of one molecule in one box at any time, given the initial conditions. A quantum system is described by its wave function, and this is interpreted as a probability amplitude. A knowledge of the wave function allows us to predict the average behaviour of a large number of nuclei, but is not capable of telling us anything about the state of one nucleus as a function of time. It may be that quantum mechanics conceals a deeper theory in which hidden variables determine the behaviour of an individual state, but if so we do not know about them. The only way for us to determine whether an unstable system has decayed or not is by going and looking—we cannot find out by calculation and we have no theory which prescribes a way of calculating the evolution of a single state even in principle.

There is another difference. In the classical system we discussed, each molecule was given a definite energy and classically the whole system has a definite energy regardless of whether the molecule can escape or not. A quantum mechanical system which is unstable does not have a definite energy:

and
$$|\psi_i|^2 = |\psi_i(0)|^2 e^{-t/\tau},$$
$$\psi_i(t) = e^{-(i/\hbar)E_i t} e^{-t/\tau}. \qquad (3.1.5)$$

The energy structure is obtained by a Fourier transform with respect to time yielding [see eq. (1.1.2)]

$$\psi_i(E) \propto \frac{1}{E-E_i+i\Gamma/2} \qquad (3.1.6)$$

where $\Gamma\tau = \hbar$ and

$$|\psi_i(E)|^2 \propto \frac{1}{(E-E_i)^2+\Gamma^2/4}. \qquad (3.1.7)$$

An unstable state has a central energy E_i and a Breit–Wigner shape with a full width at half height of Γ.

This completes our introductory remarks to this chapter and we now go on to consider the physical content of the Fermi Golden Rule.

3.2. The Fermi Golden Rule

Consider a rather simple quantum mechanical system: a box containing an electron in a magnetic field B. The electron has two possible states, spin up and spin down, and these two states are separated in energy by

$$\frac{e\hbar}{mc} B.$$

In the absence of any other interaction, either state is absolutely stable, and an electron in the upper state will stay there. Now suppose we apply an oscillating electromagnetic field to this system. If the upper state contains initially a population with amplitude A^+ and the lower state a population with A^-, then power will be absorbed in increasing A^+ at the expense of A^- and be radiated in decreasing A^+ and increasing A^-. The applied field is a time-dependent perturbation which in first order merely alters the populations of the two states and does not shift their energies.

If the applied field is turned off, the population of the upper level will decrease, spontaneously rather than by stimulated emission because the electrons couple to a free electromagnetic field. Both the stimulating field and the radiated field may be considered as coupling with the electrons to provide a perturbation, and so the behaviour of the amplitudes A^+ and A^- with time is complicated. In the presence of an external stimulating field A^+ and A^- go up and down, subject to the constraint that $|A^+|^2+|A^-|^2 = $ constant and after the stimulating field is removed $|A^+|^2$ will decay feeding $|A^-|^2$. If such a system is started off with $A^+ \neq 0$ and with no externally applied stimulating field, then initially A^+ will decrease with the emission of radiation. If the system is enclosed in a box which reflects electromagnetic waves, then subsequently the radiation will be reflected back from the walls of the box and will raise electrons from the lower state to the upper state.

Nuclear Physics

If we wish to discuss the time evolution of such a system we must recognize that the frequencies of electromagnetic radiation in the box are restricted to the normal modes and that consequently the electromagnetic field is quantized. The general state, if we have only one electron in the box, will consist of an electron and photons.

Let us now imagine a very simple system of this kind, possessing only two possible states. The first state has a wave function ψ_1 and consists of an electron in the upper state. The second has a wave function ψ_2 and consists of an electron in the lower state and a photon. The total wave function of the whole system is ψ and satisfies the time-dependent Schrödinger equation

$$i\hbar \frac{\partial \psi}{\partial t} = H\psi \qquad (3.2.1)$$

where H contains the electron energy and the photon energy and the interaction energy between them.

The total wave function ψ is a linear superposition of ψ_1 and ψ_2

$$\psi = a_1(t)\psi_1 + a_2(t)\psi_2 = A_1(t)\phi_1 + A_2(t)\phi_2$$

where the time dependence of each state has been factored out so as to leave the orthonormal spatial parts ϕ_1 and ϕ_2. Thus for the whole system

$$i\hbar \frac{\partial}{\partial t}\{A_1(t)\phi_1 + A_2(t)\phi_2\} = H\{A_1(t)\phi_1 + A_2(t)\phi_2\}. \qquad (3.2.2)$$

We now multiply both sides by ϕ_1^* and integrate over all space (that is, the volume of the box)

$$\int \phi_1^* \phi_1 \, dV = 1 \quad \int \phi_2^* \phi_2 \, dV = 1 \quad \int \phi_1^* \phi_2 \, dV = 0,$$

$$i\hbar \frac{\partial A_1(t)}{\partial t} = A_1(t) \int \phi_1^* H \phi_1 \, dV + A_2(t) \int \phi_1^* H \phi_2 \, dV$$

and similarly multiplying through by ϕ_2^*

$$i\hbar \frac{\partial A_2(t)}{\partial t} = A_1(t) \int \phi_2^* H \phi_1 \, dV + A_2(t) \int \phi_2^* H \phi_2 \, dV$$

and we rewrite these two equations as

$$\left.\begin{aligned} i\hbar \frac{\partial A_1}{\partial t} &= A_1 \langle 1|H|1\rangle + A_2 \langle 1|H|2\rangle, \\ i\hbar \frac{\partial A_2}{\partial t} &= A_1 \langle 2|H|1\rangle + A_2 \langle 2|H|2\rangle. \end{aligned}\right\} \qquad (3.2.3)$$

These two equations describe completely the time evolution of our coupled states and are analogous to the equations describing two coupled oscillators in a.c. theory, and may be solved by the same methods.

Write

$$A_1 = A_1(0)e^{\lambda t}$$
$$A_2 = A_2(0)e^{\lambda t}$$

Nuclear Decay

whence

$$i\hbar\lambda A_1 = \langle 1|H|1\rangle A_1 + \langle 1|H|2\rangle A_2, \\ i\hbar\lambda A_2 = \langle 2|H|1\rangle A_1 + \langle 2|H|2\rangle A_2,\} \quad (3.2.4)$$

or

$$\{i\hbar\lambda - \langle 1|H|1\rangle\} A_1 - \langle 1|H|2\rangle A_2 = 0, \\ -\langle 2|H|1\rangle A_1 + \{i\hbar\lambda - \langle 2|H|2\rangle\} A_2 = 0$$

and solutions only exist if

$$\{i\hbar\lambda - \langle 1|H|1\rangle\}\{i\hbar\lambda - \langle 2|H|2\rangle\} - \langle 2|H|1\rangle\langle 1|H|2\rangle = 0. \quad (3.2.5)$$

If the free field photon did not interact with the electron, then the two states ψ_1 and ψ_2 would not be coupled and would be eigenstates of the energy with eigenvalues E_1, E_2. For a coupling that is not too strong we have solutions

$$i\hbar\lambda_1 \sim \langle 1|H|1\rangle \sim E_1, \\ i\hbar\lambda_2 \sim \langle 2|H|2\rangle \sim E_2.$$

This corresponds to a very weak mutual inductance in an a.c. analogue. If, on the other hand, the coupling is very strong then λ_1 and λ_2 will be very different from the decoupled values, as in the case of two oscillators coupled by a large mutual inductance.

Now we can solve for A_2 in terms of A_1 for either value of λ by writing

$$A_2 = \frac{A_1\langle 2|H|1\rangle}{i\hbar\lambda - \langle 2|H|2\rangle}.$$

Similarly

$$A_1 = \frac{A_2\langle 1|H|2\rangle}{i\hbar\lambda - \langle 1|H|1\rangle}.$$

If we start off with $A_2 = 0$ and $A_1 = 1$ then A_2 is fed from A_1 through a term governed by $\langle 2|H|1\rangle$. In a small box A_2 will then promptly start to feed A_1 through a term initially dominated by $\langle 1|H|2\rangle\langle 2|H|1\rangle$. Thus if we can ignore the repopulation of ψ_1 from ψ_2 this corresponds to ignoring the second-order matrix elements $\langle 1|H|2\rangle\langle 2|H|1\rangle$, and this is just the situation we have for a decay process. In terms of our electron system, the larger the box the smaller the amplitude of the photon at the position of the electron. If radiation is emitted in a transition, at very small times it clearly is not described by either a plane wave or a standing wave, but can be built up from a superposition of such waves with the appropriate amplitudes and relative phases. The larger the box, the longer it takes for radiation bouncing back from the walls to restimulate the system. Decay with emission of a free field thus corresponds to the limit where the size of the box containing the emitting system becomes infinite. The radiation escapes and never returns.

Thus our boundary conditions must be

$$A_1(0) = 1, \\ A_2(0) = 0$$

and we solve the equations in the limit $\langle 1|H|2\rangle\langle 2|H|1\rangle \to 0$.

Nuclear Physics

With the volume of the box tending to ∞ the photon energy levels get closer and closer and merge into a continuum of states of the kind described by ψ_2. These states can only feed each other through re-excitation and decay of the initial system and so under the conditions we are considering they are each fed only from the initial state. Considering one particular state ψ_2 we have

$$A_1(t) = A_1^1(0)\, e^{\lambda_1 t} + A_1^2(0)\, e^{\lambda_2 t},$$

$$A_2(t) = \left\{ \frac{A_1^1(0)\, e^{\lambda_1 t}}{i\hbar\lambda_1 - \langle 2|H|2\rangle} + \frac{A_1^2(0)\, e^{\lambda_2 t}}{i\hbar\lambda_2 - \langle 2|H|2\rangle} \right\} \langle 2|H|1\rangle \qquad (3.2.6)$$

and as $i\hbar\lambda_2 \to \langle 2|H|2\rangle$, $A_1^2(0)$ must $\to 0$ to keep A_2 finite, in such a way that $A_2(0) \to 0$. That is

$$\frac{A_1^2(0)}{i\hbar\lambda_2 - \langle 2|H|2\rangle} \to -\frac{A_1^1(0)}{i\hbar\lambda_1 - \langle 2|H|2\rangle}.$$

Then

$$A_1(t) = A_1^1(0)\, e^{\lambda_1 t}$$

and

$$A_2(t) = \frac{\langle 2|H|1\rangle\, A_1^1(0)}{i\hbar\lambda_1 - \langle 2|H|2\rangle}\, \{e^{\lambda_1 t} - e^{\lambda_2 t}\}. \qquad (3.2.7)$$

Since in the limit of an infinitely large box the initial state is never re-excited, λ_1 must have a real as well as an imaginary part and this part must be negative. Then as $t \to \infty$

$$A_1(t) \xrightarrow[t\to\infty]{} 0,$$

$$A_2(t) \longrightarrow \frac{\langle 2|H|1\rangle\, A_1^1(0)}{i\hbar\lambda_1 - \langle 2|H|2\rangle}\, \{-e^{\lambda_2 t}\}.$$

The state ψ_2 is a ground state and does not decay. Consequently λ_2 is pure imaginary and as $t \to \infty$

$$|A_2(t)|^2 \longrightarrow \left| \frac{\langle 2|H|1\rangle\, A_1^1(0)}{i\hbar\lambda_1 - \langle 2|H|2\rangle} \right|^2 \qquad (3.2.8)$$

it is now easy to generalize to the other states of the continuum ψ_n

$$A_n(t) \xrightarrow[t\to\infty]{} \frac{\langle n|H|1\rangle\, A_1^1(0)}{i\hbar\lambda_1 - \langle n|H|n\rangle}\, \{-e^{\lambda_n t}\},$$

$$|A_n(t)|^2 \xrightarrow[t\to\infty]{} \left| \frac{\langle n|H|1\rangle\, A_1^1(0)}{i\hbar\lambda_1 - \langle n|H|n\rangle} \right|^2. \qquad (3.2.9)$$

For our decaying state

$$i\hbar\lambda_1 = \langle 1|H|1\rangle$$

and we write

$$\langle 1|H|1\rangle = E_1 - i\Gamma_1/2 \qquad (3.2.10)$$

so that

$$\lambda_1 = \frac{E_1}{i\hbar} - \frac{\Gamma_1}{2\hbar}.$$

Nuclear Decay

Similarly we write for the continuum states

$$\langle n|H|n\rangle = E_n. \qquad (3.2.11)$$

Our derivation is now nearly complete. As $t \to \infty$ the population of the state ψ_1 diminishes to zero and the depopulation of this state feeds the continuum states. For a very large (but not infinite) box conservation of probability requires that

$$\lim_{t \to \infty} \sum_n |A_n(t)|^2 = |A_1^1(0)|^2 \qquad (3.2.12)$$

for the boundary conditions appropriate to decay.
Then

$$\sum_n \left| \frac{\langle n|H|1\rangle}{E_n - E_1 + i\Gamma_1/2} \right|^2 = 1.$$

As we pass to the limit of an infinitely large box the states blend into a continuum and so the sum must be replaced by an integral over the energy of the continuum

$$\sum_n \left| \frac{\langle n|H|1\rangle}{E_n - E_1 + i\Gamma_1/2} \right|^2 \longrightarrow \int_0^\infty \frac{|\langle f|H|1\rangle|^2}{(E-E_1)^2 + \Gamma_1^2/4} \frac{dn_f}{dE} dE, \qquad (3.2.13)$$

where ψ_f is a continuum state of energy E and the density of final continuum states is dn_f/dE. If Γ_1 is not too large in comparison with E_1 then the major part of the integral is contributed from values of E in the neighbourhood of E_1. We approximate by writing

$$\int_0^\infty \frac{|\langle f|H|1\rangle|^2}{(E-E_1)^2 + \Gamma_1^2/4} \frac{dn_f}{dE} dE \simeq \left\{ |\langle f|H|1\rangle|^2 \frac{dn_f}{dE} \right\}\bigg|_{E_1} \int_0^\infty \frac{dE}{(E-E_1)^2 + \Gamma_1^2/4}$$

$$= \left\{ |\langle f|H|1\rangle|^2 \frac{dn_f}{dE} \right\}\bigg|_{E_1} \frac{2\pi}{\Gamma_1} = 1.$$

Then

$$\Gamma_1 = 2\pi \left\{ |\langle f|H|1\rangle|^2 \frac{dn_f}{dE} \right\}\bigg|_{E_1} \qquad (3.2.14)$$

and

$$|A_1(t)|^2 = e^{-t/\tau}$$

where

$$\tau = \Gamma_1/\hbar$$

and the decay rate T_1 of ψ_1 is

$$T_1 = \frac{1}{\tau} = \frac{2\pi}{\hbar} |\langle f|H|1\rangle|^2 \frac{dn_f}{dE} \qquad (3.2.15)$$

which is the Fermi Golden Rule.

Nuclear Physics

$$\langle f|H|1\rangle = \int \phi_f^* H \phi_1 \, dV \qquad (3.2.16)$$

is the matrix element connecting the initial state with the continuum states, and dn_f/dE is the density of final states, the *phase space* for the decay.

This treatment has demonstrated that a decaying state does not have a well-defined energy, for the variation of intensity of the final states is given not by a spike at E_1 but by the intensity factor

$$\frac{|\langle f|H|1\rangle|^2}{(E-E_1)^2 + \Gamma_1^2/4} \qquad (3.2.17)$$

which is the famous Breit–Wigner formula for a single decaying state. Indeed, the density of states factor enters through an integration over the energy spread of the initial state.

The matrix elements $\langle 1|H|1\rangle$, $\langle n|H|n\rangle$ and $\langle n|H|1\rangle$ merit a little further consideration. Suppose we divide up the Hamiltonian H into two parts

$$H = H_0 + H_I$$

such that

$$H_0 \psi_1 = \varepsilon_1 \psi_1,$$
$$H_0 \psi_n = \varepsilon_n \psi_n.$$

Then

$$\langle 1|H|1\rangle = \langle 1|H_0|1\rangle + \langle 1|H_I|1\rangle$$
$$= \varepsilon_1 + \langle 1|H_I|1\rangle$$
$$= E_1 - i\frac{\Gamma_1}{2}.$$

$\langle 1|H_I|1\rangle$ thus shifts the energy of the unperturbed state ψ_1 by an amount $E_1 - \varepsilon_1$ as well as introducing the imaginary part $\Gamma_1/2$.

$$\langle n|H|n\rangle = \varepsilon_n + \langle n|H_I|n\rangle = E_n$$

and the second term shifts the energy of the unperturbed final states to the physically observed values.

$$\langle n|H|1\rangle = \langle n|H_0|1\rangle + \langle n|H_I|1\rangle$$
$$= \langle n|H_I|1\rangle$$

and it is the perturbing interaction that is responsible for the decay—turn off the coupling between the electron and the free radiation field and we get no decay at all: an electron in the upper state stays there.

We should point out one more feature of the Fermi Golden Rule. If Γ_1 is not small in comparison with E_1, then Γ_1 is dependent on the energy E and the state no longer decays exponentially. Exponential decay is a limiting case appropriate to narrow quantum mechanical levels.

To conclude this paragraph we note that there may be more than one set of continuum states available. For example, in nuclear physics a nucleus may in some circumstances decay via two different γ-transitions. Writing $\langle f|H|i\rangle \to M_{fi}$ and $dn_f/dE = \varrho_{fi}$ we would then have

$$\frac{\partial |A_1|^2}{\partial t} = -\frac{2\pi}{\hbar}\{|M_{21}|^2 \varrho_{21} + |M_{31}|^2 \varrho_{31}\} \qquad (3.2.18)$$

Nuclear Decay

corresponding to the summation of partial widths mentioned in Chapter 1. The matrix element M_{32} and the phase space factor ϱ_{32} do not enter into the decay rate unless the initial nuclear state and one of the final states are very close in comparison with the width of either. Re-excitation of the initial state cannot then be ignored, interference is possible and very complicated effects may occur (see Fig. 3.2.1).

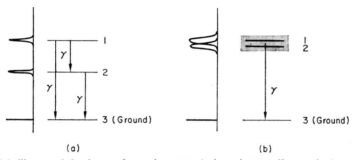

FIG. 3.2.1. In (a) is illustrated the decay of a nuclear state 1 via an intermediate excited state 2 and directly to the ground state. The population of state 2 does not affect the decay rate of state 1. In (b) are illustrated two broader overlapping states. If two such states have different quantum numbers a specific production process may prepare the nucleus in either one or the other: if they have the same quantum numbers it becomes meaningless to ask whether the excited system is in state 1 or state 2.

3.3. Electromagnetic decay of nuclei

Since this is a problem in which single photons are emitted, a proper treatment requires the use of quantum electrodynamics and is beyond the scope of this book. A possible semi-classical treatment consists of writing down the energy radiated by a classical distribution of currents, replacing the classical currents by the matrix element of the current operators taken between the initial and final states and then dividing by the energy of the emitted photon, $\hbar\omega$, in order to find the number of photons emitted per second, the transition rate. This method is well-documented[†] and we shall not use it here. Since there exists no classical theory of β-decay, and we are therefore forced to start from the Fermi Golden Rule in that case, we prefer to calculate the properties of electromagnetic transitions in the same way. For electromagnetic processes we must, however, be careful to define our photon states so that the final expressions reduce to the well-known classical forms in the appropriate limit. Thus in the Fermi Golden Rule (3.2.15)

$$T = \frac{2\pi}{\hbar} |\langle f|H|i\rangle|^2 \frac{dn_f}{dE} \qquad (3.3.1)$$

the initial state is described by $\Psi_a e^{-iE_a t/\hbar}$, and is a state of the atom or nucleus with no photon present. The final state is described by a product of a single photon wave function

$$\frac{\epsilon_0}{\sqrt{V}} e^{i(\mathbf{k}\cdot\mathbf{r}-\omega t)}$$

which is a normalized plane wave and by a final atomic or nuclear wave function $\Psi_b e^{-iE_b t/\hbar}$.

---

† E. Fermi, *Nuclear Physics*, chap. V (Chicago, 1950); L. I. Schiff, *Quantum Mechanics*, 3rd ed. (McGraw-Hill, 1968); J. M. Blatt and V. Weisskopf, *Theoretical Nuclear Physics*, chap. XII (Wiley, 1952).

Nuclear Physics

Then

$$\langle f|H|i\rangle = \int \Psi_b^* \frac{e^{-i\mathbf{k}\cdot\mathbf{r}}}{\sqrt{V}} \boldsymbol{\epsilon}_0 \cdot \mathbf{O}(\omega, E_a, E_b) \Psi_a \, d^3r. \tag{3.3.2}$$

Ψ_a and Ψ_b may be decomposed into contributions from different individual particles, and for each such particle the photon wave function enters only at the coordinate $\mathbf{r}$ of the particle because the interaction is local. The unit vector $\boldsymbol{\epsilon}_0$ gives the direction of the electric field associated with the photon and the vector quantity $\mathbf{O}$ creates a photon of energy ω and rearranges the wave function Ψ_a so that it is no longer orthogonal to Ψ_b. Our problem is to find the operator.

A classical external field interacting with a classical current density $\mathbf{J}$ has an interaction energy $1/c \int \mathbf{A} \cdot \mathbf{J}$ where $\mathbf{A}$ is the vector potential of the field, and for a radiation field we may choose the scalar potential ϕ to be zero. This at once suggests that we should write

$$\langle f|H|i\rangle = q \int \Psi_b^* \frac{e^{-i\mathbf{k}\cdot\mathbf{r}}}{\sqrt{V}} A\boldsymbol{\epsilon}_0 \cdot \frac{\mathbf{p}}{mc} \Psi_a \, d^3r \tag{3.3.3}$$

where A is a number, $\mathbf{p}$ is the momentum operator, $(\hbar/i)\nabla$ and q are the charge of the particle while $A\boldsymbol{\epsilon}_0 \, e^{-i(\mathbf{k}\cdot\mathbf{r}-\omega t)}/\sqrt{V}$ is interpreted as the vector potential. We can confirm this by writing the Schrodinger equation for a particle in the presence of an electromagnetic field. If there is no potential present

$$\left(\frac{p^2}{2m} - E\right)\psi = 0.$$

With an electromagnetic field present a particle of charge q is described by replacing $E \to E + q\phi$ and $\mathbf{p} \to \mathbf{p} + q\mathbf{A}/c$ where $\mathbf{A}$ is the vector potential of the classical field and ϕ the scalar potential. The latter result can be tediously derived in classical mechanics, but follows easily if it is remembered that the Schrödinger equation is the limit of a relativistic equation in which the three components of $\mathbf{p}$ and E each enter; either linearly or quadratically. On replacing $E \to E + q\phi$ we must replace $\mathbf{p} \to \mathbf{p} + q\mathbf{A}/c$ in order to maintain a relativistically invariant equation, for $(\mathbf{A}, i\phi)$ forms a 4-vector just as does $(\mathbf{p}, iE/c)$. Then for $\phi = 0$

$$\frac{1}{2m}\left(\mathbf{p} + \frac{q\mathbf{A}}{c}\right)^2 \psi = E\psi \tag{3.3.4}$$

and the interaction energy due to the non-zero value of q is

$$q\frac{\mathbf{p}\cdot\mathbf{A}}{mc} + \frac{q^2 A^2}{2mc^2}. \tag{3.3.5}$$

We may neglect the last term as being small, or on the grounds that, referring back to eq. (3.3.2), it should be interpreted as creating two photons. Then indeed

$$\langle f|H|i\rangle \to q \int \Psi_b^* \frac{\mathbf{A}\cdot\mathbf{p}}{mc} \Psi_a \, d^3r.$$

But a classical plane wave is real and we may represent it by

$$A = A_0 \cos(\mathbf{k}\cdot\mathbf{r}-\omega t+\delta)$$
$$= \frac{A_0}{2} e^{i(\mathbf{k}\cdot\mathbf{r}-\omega t)} + \frac{A_0^*}{2} e^{-i(\mathbf{k}\cdot\mathbf{r}-\omega t)}. \qquad (3.3.6)$$

If the field A contains in the quantum limit only one photon, then the energy

$$\int \left\{\frac{E^2+B^2}{8\pi}\right\} dV = \hbar\omega \qquad (3.3.7)$$

extended over the volume of the normalizing box.

$$E = -\frac{1}{c}\frac{\partial \mathbf{A}}{\partial t} \qquad \mathbf{B} = \nabla\times\mathbf{A}$$

so that

$$\hbar\omega = \frac{\omega^2 A_0^2 V}{8\pi c^2} \quad \text{or} \quad A_0^2 = \frac{8\pi\hbar c^2}{V\omega}. \qquad (3.3.8)$$

The implication of a comparison of (3.3.3) with the expansion of the classical field is that we should write

$$\frac{A}{\sqrt{V}} \to \frac{A_0^*}{2} = \sqrt{\frac{2\pi\hbar c^2}{V\omega}}. \qquad (3.3.9)$$

The significance of this factor of $\tfrac{1}{2}$ is realized by noting that while classically the interaction energy between an external field and a current is $\int \mathbf{A}\cdot\mathbf{J}\, dV$ the interaction energy between field $\mathbf{A}$ generated by a current density $\mathbf{J}$ and that current is only $\tfrac{1}{2}\int \mathbf{A}\cdot\mathbf{J}\, dV$. Thus the matrix element for the creation of a photon in a transition between two atomic or nuclear states is given by the sum over all participating particles of

$$\sqrt{\frac{2\pi\hbar c^2}{\omega V}}\,\frac{q}{mc}\int \Psi_f^* e^{-i\mathbf{k}\cdot\mathbf{r}}\,\boldsymbol{\epsilon}_0\cdot\mathbf{p}\,\Psi_i\, d^3r \qquad (3.3.10)$$

where $E_f = E_i - \hbar\omega$ for conservation of energy and the matrix element for the inverse process is the complex conjugate

$$\sqrt{\frac{2\pi\hbar c^2}{\omega V}}\,\frac{q}{mc}\int \Psi_i^* e^{i\mathbf{k}\cdot\mathbf{r}}\,\boldsymbol{\epsilon}_0\cdot\mathbf{p}\,\Psi_f\, d^3r.$$

We have thus obtained very plausibly an expression for the matrix elements to be fed into the Fermi Golden Rule for one photon processes, without plunging into the murky depths of quantum field theory. The justification of this result is that it works![†] Then for a particle of charge e transferring between states Ψ_a and Ψ_b

$$\frac{1}{\tau} = \frac{2\pi}{\hbar}\frac{e^2}{c^2}\left|\int \Psi_b^* \frac{e^{-i\mathbf{k}\cdot\mathbf{r}}}{\sqrt{V}}\,\boldsymbol{\epsilon}_0\cdot\mathbf{v}\,\Psi_a\, d^3r\right|^2 \frac{2\pi\hbar c^2}{\omega}\frac{dn}{dE} \qquad (3.3.11)$$

and this is our *first application of the Fermi Golden Rule*.

[†] A very beautiful demonstration of this is afforded by Einstein's treatment of spontaneous emission. See, for example, L. I. Schiff, *Quantum Mechanics*, 3rd ed., sect. 45 (McGraw-Hill, 1968).

Nuclear Physics

It remains to work out the phase space term dn/dE which we may do by counting the number of normal modes of radiation in a cavity of volume V—for radiation the evaluation of the phase space is no more than the counting of modes in a resonant cavity.

Any wave with propagation vector $\mathbf{k}$ can be broken up into three waves travelling in the x, y, z directions. Classically the tangential components of the electric field vanish at the walls of a perfectly reflecting box: in a photon picture the wave function must be continuous, and hence zero at the walls. In either picture the standing wave-boundary conditions imply that an integral number of half wavelengths must fit into each dimension of the box. For simplicity we consider a cubic box with side L, when

$$L = n_x \frac{\lambda_x}{2} = n_y \frac{\lambda_y}{2} = n_z \frac{\lambda_z}{2}$$

and

$$\lambda_x = \frac{2\pi}{k_x}, \quad \text{etc.}$$

Then for a given mode

$$k^2 = k_x^2 + k_y^2 + k_z^2 = \left(\frac{\pi}{L}\right)^2 \{n_x^2 + n_y^2 + n_z^2\},$$

$$\omega^2 = k^2 c^2 = \left(\frac{\pi c}{L}\right)^2 \{n_x^2 + n_y^2 + n_z^2\} = \left(\frac{\pi c}{L}\right)^2 n^2.$$

The number of modes with frequency $n\pi c/L$ is given by the number of different ways of choosing n_x, n_y, n_z so that $n_x^2 + n_y^2 + n_z^2 = n^2$. For example, if $n^2 = 1$ there are three independent modes of energy given by $\omega = \pi c/L$:

n_x^2	n_y^2	n_z^2
1	0	0
0	1	0
0	0	1

Similarly there are three modes for $n^2 = 2$, only one for $n^2 = 3$, three for $n^2 = 4$, six for $n^2 = 5$ and so on. We have not counted negative values of any n_i, for with standing waves no distinct wave function results.

It is easy to see that the number of ways of making up an allowed value of n^2 is given by the number of points with coordinates (n_x, n_y, n_z) lying on a sphere of radius n extended from the origin. The number of modes with frequency $< N\pi c/L$ is given by the number of points enclosed by a sphere of radius N (and which lie in the positive octant) and for large N this is given by

$$\frac{1}{8} \frac{4\pi}{3} N^3 \tag{3.3.12}$$

so that number of modes with wave number $< k$, frequency $< \omega$ is given by

$$n_k = \frac{1}{8} \cdot \frac{4\pi}{3} \left(\frac{Lk}{\pi}\right)^3$$

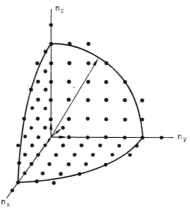

FIG. 3.3.1. The figure shows points in the n_x, n_y, n_z grid which also lie in the planes $n_x = 0$, $n_y = 0$ a $n_z = 0$. The intersection of a sphere of radius $n = 5$ with these planes is indicated, and radius vectors are drawn to the three points with $n = 1$ and the single point with $n = \sqrt{3}$.

and the number of modes between ω and $\omega + d\omega$ is then just

$$dn_k = \frac{V}{2\pi^2} k^2 \, dk$$

$$= \frac{V}{2\pi^2 c^3} \omega^2 \, d\omega.$$

Then

$$\frac{dn_k}{dE} = \frac{V}{2\pi^2 c^3} \omega^2 \frac{d\omega}{dE} = \frac{V}{2\pi^2 c^3} \frac{\omega^2}{\hbar}. \tag{3.3.13}$$

While this result has been obtained for standing waves, we can make running waves in either direction by a combination of standing waves of different phase. In order to work with running waves directly, we may adopt the device of periodic boundary conditions, requiring that a wave leaves one boundary with the same amplitude that it enters the opposite boundary. Then we require that an integral number of wavelengths fits into the normalization volume, but for running waves a negative value of n_i gives a different solution from a positive value of n_i. The expression for the density of final states remains the same. However, using running waves makes it clear that we can select on the direction of the wave vector and write

$$\frac{dn_k}{d\Omega} = V \frac{k^2 \, dk}{(2\pi)^3}$$

and the element of phase space for transitions of energy E into an element of solid angle $d\Omega$ is then

$$\frac{d^2 n_k}{dE \, d\Omega} = \frac{Vk^2}{(2\pi)^3} \frac{dk}{dE}$$

which in terms of the momentum p of a photon (or any other particle) becomes

$$\frac{d^2 n_p}{dE \, d\Omega} = \frac{Vp^2}{(2\pi\hbar)^3} \frac{dp}{dE}. \tag{3.3.14}$$

Nuclear Physics

We can now write our final expression for the rate at which a nucleus described by Ψ_a decays into the nucleus described by Ψ_b. For a single-particle transition

$$\frac{1}{\tau} = \frac{2\pi}{\hbar} \frac{e^2}{c^2} \left(\frac{2\pi \hbar c^2}{\omega}\right) \int_\Omega |\Psi_b^* e^{-i\mathbf{k}\cdot\mathbf{r}} \boldsymbol{\epsilon}_0 \cdot \mathbf{v}\Psi_a d^3 r|^2 \frac{\omega^2}{(2\pi c)^3 \hbar} d\Omega \qquad (3.3.15)$$

where $\mathbf{v}$ is the velocity operator for a single particle and the nuclear part of the matrix element must be summed over all charged particles participating. (The interpretation of the nuclear part of the matrix element is illustrated in the following sections.) This applies to a specific polarization $\boldsymbol{\epsilon}_0$ of the photon, and if we are not detecting the polarization the matrix element must, in addition, be summed over the two photon polarizations.

Equation (3.3.15) contains all the essential physics encountered in the problem of the radiative decay of nuclei, and much of the rest of this chapter is concerned only with the properties of the matrix element.

$$\int \Psi_b^* e^{-i\mathbf{k}\cdot\mathbf{r}} \boldsymbol{\epsilon}_0 \cdot \mathbf{v}\Psi_a d^3 r. \qquad (3.3.16)$$

3.4 Electric dipole transitions

First of all consider transitions in which the energy of the photon is sufficiently small that $e^{i\mathbf{k}\cdot\mathbf{r}}$ does not vary appreciably over the nuclear volume—that is, the wavelength of the radiation is very much greater than the nuclear radius. Then as a first approximation $e^{i\mathbf{k}\cdot\mathbf{r}} \simeq 1$ and we must evaluate

$$\int \Psi_b^* \boldsymbol{\epsilon}_0 \cdot \mathbf{v} \Psi_a dV.$$

Now

$$e \int \Psi_b^* \boldsymbol{\epsilon}_0 \cdot \mathbf{v} \Psi_a dV \qquad (3.4.1)$$

is the integral of the current density in the transition between a and b, and this is given by the time derivative of the dipole moment so that

$$|\int \Psi_b^* \boldsymbol{\epsilon}_0 \cdot \mathbf{v} \Psi_a dV|^2 = \omega^2 |\int \Psi_a^* \boldsymbol{\epsilon}_0 \cdot \mathbf{r} \Psi_a dV|^2$$

yielding

$$\begin{aligned}\frac{1}{\tau} &= \frac{2\pi}{\hbar} \frac{e^2}{c^2} \left(\frac{2\pi\hbar c^2}{\omega}\right) \omega^2 \int \left|\boldsymbol{\epsilon}_0 \cdot \int \Psi_b^* \mathbf{r}\Psi_a dV\right|^2 \frac{\omega^2 d\Omega}{\hbar(2\pi c)^3} \\ &= \frac{e^2}{\hbar c} \frac{\omega^3}{2\pi c^2} \int \left|\boldsymbol{\epsilon}_0 \cdot \int \Psi_b^* \mathbf{r}\Psi_a dV\right|^2 d\Omega.\end{aligned} \qquad (3.4.2)$$

Since $\mathbf{E} = -(1/c)\,\partial\mathbf{A}/\partial t$ the polarization vector $\boldsymbol{\epsilon}_0$ must be at right angles to the propagation vector $\mathbf{k}$. We are interested in the angular distribution of the radiation, the angle θ given by $\mathbf{k}\cdot\mathbf{r}$ rather than $\boldsymbol{\epsilon}_0\cdot\mathbf{r}$ and must obtain this for each polarization. Let us take a system of axes X, Y, Z, so that $\mathbf{k}$ has components k_x, k_y, k_z and $\mathbf{r}$ has components x, y, z (see Fig. 3.4.1). Define $\boldsymbol{\epsilon}_1$ in the kZ plane and $\boldsymbol{\epsilon}_2$ by $\mathbf{k}\times\boldsymbol{\epsilon}_2 = \boldsymbol{\epsilon}_1$. Then

$$\mathbf{r}\cdot\boldsymbol{\epsilon}_1 = x\cos\Theta\cos\Phi + y\cos\Theta\sin\Phi + z\sin\Theta$$

while

$$\mathbf{r}\cdot\boldsymbol{\epsilon}_2 = x\sin\Phi - y\cos\Phi.$$

Nuclear Decay

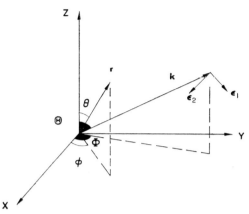

FIG. 3.4.1. Coordinate system for the evaluation of the quantity $\boldsymbol{\epsilon}_0 \cdot \mathbf{r}$ occurring in eq. (3.4.2). The wave vector **k** has coordinates (k, Θ, Φ) and the position vector **r** has coordinates (r, θ, ϕ).

If we now choose the Z-axis to be in the direction of $\int \Psi_b^* \mathbf{r} \Psi_a dV$ then the matrix elements of x and y are zero and we are left with no contribution from the final state with $\boldsymbol{\epsilon}_0 = \boldsymbol{\epsilon}_2$, while for $\boldsymbol{\epsilon}_0 = \boldsymbol{\epsilon}_1$ we have a matrix element

$$\boldsymbol{\epsilon}_0 \cdot \int \Psi_b^* \mathbf{r} \Psi_a^* dV = \sin \Theta \int \Psi_b r \cos \theta \Psi_a \, dV \\ = \sin \Theta \, \langle b | z | a \rangle. \quad (3.4.3)$$

Then

$$\frac{1}{\tau} = \frac{e^2}{\hbar c} \frac{\omega^3}{2\pi c^2} |\langle b | z | a \rangle|^2 \int_\Omega \sin^2 \Theta \, d\cos\Theta \, d\Phi$$

$$= \frac{4}{3} \frac{e^2}{\hbar c} \frac{\omega^3}{c^2} |\langle b | z | a \rangle|^2.$$

The energy radiated per unit time is

$$\hbar\omega/\tau = \frac{4}{3} e^2 \frac{\omega^4}{c^3} |\langle b | z | a \rangle|^2 = \frac{4}{3} \frac{\omega^4}{c^3} |D_0|^2 \quad (3.4.4)$$

where $D_0 = e \langle b | z | a \rangle$ and this energy has a $\sin^2 \Theta$ angular distribution with respect to the transitory electric current.

We may now consider just what sort of decay this approximation represents. From the classical theory of the scattering of light we know that when the wavelength is large in comparison with the size of the scattering object, induced currents in all parts of the object are in phase and we have an oscillating electric dipole, driven by the incident radiation. We have made this approximation and it corresponds in our case, where oscillations are spontaneous rather than induced, to the transitory current at all points in the nucleus having constant phase. The resultant is a linear current, oscillating along z and not in a closed curve, so we have electric dipole radiation and not magnetic dipole radiation. This is confirmed by a comparison of the properties of eq. (3.4.4) with the classical result.

The factor of $\frac{4}{3}$ in the expression (3.4.4) can cause some confusion when this result is compared with the classical result for the rate at which energy is radiated from an oscillating

Nuclear Physics

dipole. A classical one-dimensional dipole of moment

$$d = d_0 \cos \omega t$$

radiates at a rate

$$\tfrac{1}{3}(\omega^4/c^3)|d_0|^2.$$

It is in fact straightforward to derive this result from the expression (3.4.4) in passing to the classical limit.

In the classical limit the energy levels of a system are close together and the localization of a particle results from the wave function describing the particle being spread over a band of states of the system, yielding a wave packet the motion of which gives the classical behaviour of the system.† If this total wave function is Ψ then the value of any classical quantity is given by

$$F_c \rightarrow \int \Psi^* F_0 \Psi \, dV \qquad (3.4.5)$$

where F_0 is the operator appropriate to the classical quantity F_c. This is easy to see. Expand Ψ in a set of functions which are eigenfunctions of F_0:

$$\Psi = \sum_n a_n \Psi_n^F \qquad (3.4.6)$$

and

$$F_0 \Psi_n^F = F_n \Psi_n^F.$$

Then

$$\int \Psi^* F_0 \Psi \, dV = \sum_n |a_n|^2 F_n; \quad \sum_n |a_n|^2 = 1$$

which is the weighted average of F corresponding to the classical quantity. For example, if we take the position operator z_0 the classical value is

$$z_c = \sum_n |a_n|^2 z_n$$

and corresponds to the averaged position of the wave packet which has a spread in z about z_c.

Now we expand the function Ψ in terms of the states involved in the transitions (which are NOT eigenstates of z)

$$\Psi = \sum_n A_n \Psi_n \quad \sum_n |A_n|^2 = 1.$$

Then

$$\begin{aligned}
z &= \int \Psi^* z_0 \Psi \, dV \\
&= \sum_n \sum_m A_m^* A_n \Psi_m^* z_0 \Psi_n \, dV \\
&= \sum_n \sum_m A_m^* A_n \, e^{i(\omega_m - \omega_n)t} \langle m | z_0 | n \rangle.
\end{aligned} \qquad (3.4.7)$$

In the classical limit states in a narrow band are equally spaced, the amplitudes A_n vary only slowly with n and the matrix elements drop off very rapidly as $|m-n|$ increases. Thus we pick a particular difference $m-n = \alpha$ and write

$$z \rightarrow \sum_\alpha \sum_n A_n^* A_n \, e^{i\alpha\omega_0 t} \langle n+\alpha | z_0 | n \rangle$$

† For a nice discussion of the classical limit of a SHO see L. I. Schiff, *Quantum Mechanics*, 3rd ed., sect. 13 (McGraw-Hill, 1968).

where $\langle n+\alpha | z_0 | n \rangle$ varies rapidly with α but not with n. The real value of z oscillating with a fixed frequency ω is given by picking the value of α such that $\alpha\omega_0 = \omega$. The real value involves the matrix elements in both directions.

$$z(\omega) = \sum_n |A_n|^2 \{\langle n+\alpha | z_0 | n \rangle e^{i\omega t} + \langle n | z_0 | n+\alpha \rangle e^{-i\omega t}\}.$$

Since the matrix elements hardly vary with n we can write this as

$$\begin{aligned} z(\omega) &= \langle m | z_0 | n \rangle e^{i\omega t} + \langle n | z_0 | m \rangle e^{-i\omega t} \\ &= \langle m | z_0 | n \rangle e^{i\omega t} + \text{complex conjugate} \\ &= 2 |\langle m | z_0 | n \rangle| \cos(\omega t + \phi) \end{aligned} \quad (3.4.8)$$

where ϕ is an unimportant phase. Thus if

$$d = d_0 \cos(\omega t + \phi)$$

then

$$|d_0|^2 = 4 |\langle m | z_0 | n \rangle|^2$$

so that in the classical limit

$$\tfrac{4}{3} \frac{\omega^4}{c^3} |D_0|^2 \to \tfrac{1}{3} \frac{\omega^4}{c^3} |d_0|^2 \quad (3.4.9)$$

which is indeed the result from classical electromagnetic theory.

In order for the motion of a particle in the classical limit to be periodic, the inverse matrix elements enter with the same strength as the direct ones. We made the expansion of Ψ in terms of eigenstates of the energy of the system described, and so the population of any eigenstate ψ remains constant with time under these conditions. Thus the transition $n \to m$ induced by an operator such as z_0 is accompanied by the inverse transition $m \to n$ with a definite phase relationship between the two, and the energy "emitted" in the transition remains in the system by virtue of the inverse terms. Indeed, picking out one frequency as we have done means that no decay of the amplitude of the oscillations is possible, and turning on a coupling with the free electromagnetic field is necessary to generate such a decay, and at the same time introduces a spread of frequencies.

As we move further and further from the classical limit only transitions between a particular pair of states in general come in, and the initial state is now prepared pure in one of the ψ_i. Under these circumstances $\langle m | z_0 | n \rangle$ describes the transition downwards from n to m and $\langle n | z_0 | m \rangle$ is the matrix element involved in electric dipole absorbtion of radiation in the presence of a stimulating field.

The result we have calculated using the Fermi Golden Rule reduces in the classical limit to the result obtained classically from Maxwell's equations, thus providing some justification for the whole procedure. The angular distribution is also the same. Classically $\sin^2 \Theta$ gives the intensity of energy radiated (per unit solid angle) while in quantum theory it gives the probability of emission of the photon (per unit solid angle) at an angle Θ to the transitory electric dipole moment. Finally note that the direction of the electric field is also given correctly: for any $\mathbf{k}$ it lies in $k-z$ plane at right angles to $\mathbf{k}$.

Nuclear Physics

We now infer some of the properties of an electric dipole transition between two states from the matrix element

$$\int \Psi_b^* z \Psi_a \, dV.$$

The first is very simple. In Chapter 2 we pointed out that if $\Psi_b = \Psi_a$ then because z is an odd function such an integral vanishes, provided Ψ_a and Ψ_b have definite parity. The off-diagonal matrix elements also vanish if Ψ_b merely has the same parity as Ψ_a for the same reason. Thus our first *Selection Rule* for electric dipole transitions is provided by the answer to the question: Does the parity of the nucleus change? YES.

The state $(z\Psi_a)$ is a mixture of states with $J = J_a+1$, $J = J_a$ and $J = J_a-1$, because $\cos\theta$ is associated with angular momentum 1 and we are multiplying a state with angular momentum by a function equivalent to another state with angular momentum 1. The function $\cos\theta$ also corresponds to $m = 0$ so that, *referred to the axis of the dipole, m does not change*, and $J_b = J_a \pm 1, 0$. An inspection of the matrix elements of vectors[†] shows that on choosing an arbitrary set of axes for the quantization of angular momentum we can write

$$\mathbf{J}_b = \mathbf{J}_a \pm 1,$$

$$\mathbf{J}_b = \mathbf{J}_a \pm 1, 0.$$

(But NOT $0 \to 0$.)

Then we may summarize:

The Properties of Electric Dipole Transitions

$$\frac{1}{\tau_{E_1}} = \frac{4}{3} \frac{e^2}{\hbar c} \frac{\omega^3}{c^2} |\langle b | z | a \rangle|^2,$$

$\mathbf{J}_b = \mathbf{J}_a \pm 1.$ YES
$\mathbf{J}_b = \mathbf{J}_a \pm 1, 0.$ (NOT $0 \to 0$.)

3.5 Transitions other than electric dipole terms

If we set $e^{i\mathbf{k}\cdot\mathbf{r}} \simeq 1$ the charge elements in the radiating distribution are all moving with the same phase. Then in order to find decay rates for a magnetic dipole transition and transitions corresponding to higher magnetic and electric multipoles, this approximation must be dropped and we have to take higher terms in an expansion of $e^{i\mathbf{k}\cdot\mathbf{r}}$. A simple Taylor expansion gives

$$e^{i\mathbf{k}\cdot\mathbf{r}} = 1 + i\mathbf{k}\cdot\mathbf{r} - (\mathbf{k}\cdot\mathbf{r})^2/2 \ldots \quad (3.5.1)$$

and the successive matrix elements from this expression decrease by $\sim kR$ in each term. This expansion suffers from the lack of orthogonality of terms—for example, the first and third can interfere even after integration over angles. It is better to expand in spherical harmonics so that successive terms are orthogonal. This expansion is easily worked out,

---

[†] See, for example, L. D. Landau and E. M. Lifshitz, *Quantum Mechanics*, 2nd ed., sect. 29 (Pergamon Press, 1965).

taking the direction of **k** as the z-axis (see Section 4.5)

$$e^{i\mathbf{k}\cdot\mathbf{r}} = \frac{\sin kr}{kr} + 3\left(\frac{\cos kr}{ikr} - \frac{\sin kr}{ik^2r^2}\right)P_1(\cos\theta);$$
$$+5\left\{\frac{\sin kr}{kr} + \frac{3\cos kr}{(kr)^2} - \frac{3\sin kr}{(kr)^3}\right\}P_2(\cos\theta)\ldots\right\}$$
(3.5.2)

and for $kr \ll 1$ this yields

$$e^{i\mathbf{k}\cdot\mathbf{r}} \simeq 1 + ikr P_1(\cos\theta) - \tfrac{1}{3}(kr)^2 P_2(\cos\theta)\ldots \quad (3.5.3)$$

Again, successive terms decrease by $\sim kR$ if R is the nuclear radius.

The first two terms in each expansion are the same, and this is as far as we need to go here. So the next term to consider is

$$\int \Psi_b^*(i\mathbf{k}\cdot\mathbf{r})(\boldsymbol{\epsilon}_0\cdot\mathbf{v})\Psi_a\, dV. \quad (3.5.4)$$

In both classical theory and a quantum theory this term gives the radiation due to a magnetic dipole and an electric quadrupole. The dipole term corresponds classically to current oscillating in a closed loop, but without any buildup of charge anywhere, while the quadrupole term corresponds to two regions with positive charge and two with negative charge reversing their roles periodically. In both cases all current elements are not in phase and in the long-wavelength limit the matrix elements go to zero for this reason.

The standard treatment of this term is as follows:

$$\mathbf{v}(\mathbf{k}\cdot\mathbf{r}) - \mathbf{r}(\mathbf{k}\cdot\mathbf{v}) = \mathbf{k}\times(\mathbf{v}\times\mathbf{r}) = -\frac{\mathbf{k}\times\mathbf{L}}{m}$$

where $\mathbf{L} = (\mathbf{r}\times\mathbf{p})$ and is the angular momentum of a circulating particle.

Then

$$\mathbf{v}(\mathbf{k}\cdot\mathbf{r}) = -\frac{\mathbf{k}\times\mathbf{L}}{m} + \mathbf{r}(\mathbf{k}\cdot\mathbf{v}),$$

$$2\mathbf{v}(\mathbf{k}\cdot\mathbf{r}) = -\frac{\mathbf{k}\times\mathbf{L}}{m} + \mathbf{r}(\mathbf{k}\cdot\mathbf{v}) + \mathbf{v}(\mathbf{k}\cdot\mathbf{r}).$$

So

$$\mathbf{v}(\mathbf{k}\cdot\mathbf{r}) = -\frac{\mathbf{k}\times\mathbf{L}}{2m} + \tfrac{1}{2}\{\mathbf{r}(\mathbf{k}\cdot\mathbf{v}) + \mathbf{v}(\mathbf{k}\cdot\mathbf{r})\}$$
$$= -\frac{\mathbf{k}\times\mathbf{L}}{2m} + \tfrac{1}{2}\frac{d}{dt}\{(\mathbf{k}\cdot\mathbf{r})\mathbf{r}\}.\right\}$$
(3.5.5)

The time derivative gives us a factor of $i\omega$ for the second term, just as it did for the electric dipole case, while in the first term k is of course proportional to ω.

The first term clearly represents a circulating current, and the matrix element of such a circulating current gives the magnetic dipole transition. The diagonal matrix element of $e\mathbf{L}/2mc$ is a static magnetic moment: the off-diagonal element

$$\int \Psi_b^* \frac{e\mathbf{L}}{2mc} \Psi_a\, dV$$

Nuclear Physics

is the magnetic moment generated in the transition between Ψ_a and Ψ_b and we must evaluate

$$\epsilon_0 \cdot \int \Psi_b^* \frac{(\mathbf{k} \times \mathbf{L})}{2mc} \Psi_a \, dV. \tag{3.5.6}$$

We make an analysis similar to that for electric dipole radiation. $\mathbf{L}$ is normal to $\mathbf{r}$ (since $\mathbf{L} = \mathbf{r} \times \mathbf{p}$) and ϵ_1, ϵ_2 are normal to $\mathbf{k}$ and each other.

$$\epsilon_0 \cdot (\mathbf{k} \times \mathbf{L}) = \mathbf{L} \cdot (\epsilon_0 \times \mathbf{k})$$

so

$$\epsilon_1 \times \mathbf{k} = \epsilon_2 k,$$
$$\epsilon_2 \times \mathbf{k} = -\epsilon_1 k.$$

Then on defining the z-axis to be along the direction of the transitory magnetic moment (Fig. 3.5.1), we find that the electric field is in the direction of ϵ_2 and that

$$\frac{1}{\tau_{M_1}} = \frac{\omega^3}{2\pi\hbar c^3} |M_0|^2 \int_\Omega \sin^2 \Theta' \, d\Omega \tag{3.5.7}$$

where

$$M_0 = \int \Psi_b^* \frac{eL_z}{2mc} \Psi_a \, dV. \tag{3.5.8}$$

The energy radiated per unit time is

$$\frac{4}{3} \frac{\omega^4}{c^3} |M_0|^2 \tag{3.5.9}$$

and again our results are in accord with classical theory on setting

$$4|M_0|^2 = m_0^2.$$

So far we have neglected entirely the fact that nucleons have spin. The modification to M_0 is straightforward: to $eL_z/2mc$ we add $g_n(eS_z/2mc)$ where S_z is the nucleon spin operator and g_n the nucleon g factor. In general, transitions will correspond to more than one nucleon shifting its state and so we write

$$M_0 = e \int \Psi_b^* \frac{g_J J_z}{2mc} \Psi_a \, dV \tag{3.5.10}$$

[or $M_0 = (e\hbar/2mc) \int \Psi_b^* g_J J_z \Psi_a \, dV$ if J is measured in units of $\hbar$]. There is also a modification to be made to the electric dipole matrix element. The reason is fairly clear. Suppose a single proton changes its orbit so that $\Delta L = 1$ and at the same time, due perhaps to the strong coupling with all the other nucleons, flips its spin so $\Delta S = 1$. Then $\Delta J = 0, 1$ or 2 and there is a change of parity. Thus the $\Delta J = 1$ transition obeys the selection rules for an electric dipole transition, but the spin of the particle took part. Since a magnetic moment is generated by a current loop, we have to replace $\epsilon_0 \cdot \mathbf{j}$ by $\epsilon_0 \cdot (\mathbf{j} + c \nabla \times \mathbf{M})$, where $\mathbf{M}$ is the magnetization due to intrinsic magnetic moments, and this is the time derivative of the electric dipole moment. (Note that both $\mathbf{j}$ and $\nabla \times \mathbf{M}$ are vector operators.)

The magnetic moment operator is an axial vector (it has even parity where a vector has odd parity) and so in answer to the question: Does the parity of the nucleus change in a magnetic dipole transition? No.

Nuclear Decay

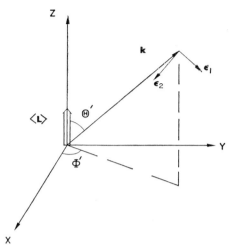

FIG. 3.5.1. The quantities $\mathbf{k}$, $\boldsymbol{\epsilon}_1$, and $\boldsymbol{\epsilon}_2$ for a magnetic dipole transition, the direction of the transition magnetic moment being taken along the z-axis.

In general, the magnetic moment operator is to be summed over all particles, but as its behaviour under rotations is that of a vector, we once again have the selection rule

$$\mathbf{J}_b = \mathbf{J}_a \pm 1$$

and summarize:

The Properties of Magnetic Dipole Transitions

$$\frac{1}{\tau_{M_1}} = \frac{4}{3} \frac{e^2}{\hbar c} \frac{\omega^3}{c^2} |\langle b|\mu_z|a\rangle|^2,$$

$\mathbf{J}_b = \mathbf{J}_a \pm 1$ NO
$\mathbf{J}_b = \mathbf{J}_a \pm 1, 0$ (but NOT $0 \to 0$).

The second term in eq. (3.5.5), $\frac{1}{2}(\mathbf{k}\cdot\mathbf{r})\mathbf{r}$ yields the radiation due to the quantum equivalent of a fluctuating electric quadrupole. This may easily be seen by considering the derivative $\mathbf{r}(\mathbf{k}\cdot\mathbf{v}) + \mathbf{v}(\mathbf{k}\cdot\mathbf{r})$ which describes the current distribution. The z-component of this vector is

$$z(k_x v_x + k_y v_y + k_z v_z) + v_z(k_x x + k_y y + k_z z)$$

and the term coupling to k_x is $zv_x + xv_z$ integrated over the wave function. For a particular value of x, z we may represent this in the x, z-plane as shown in Fig. 3.5.2. Thus charge is

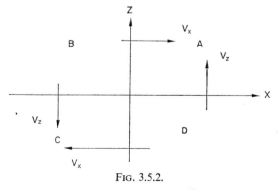

FIG. 3.5.2.

Nuclear Physics

building up in the regions A and C at the expense of charge in the regions B and D. This is reversed in the next half cycle.

To find a convenient expression for the matrix element, we choose a z-axis in a direction such that we have radial symmetry about the z-direction and reflection symmetry (required because nuclear states are parity eigenstates) about the plane $z = 0$. Then only the matrix elements of $r_i r_j$

$$\langle b| r_x^2 |a\rangle = \langle b| r_y^2 |a\rangle \quad \text{and} \quad \langle b| r_z^2 |a\rangle$$

survive.

With our usual definitions (see Fig. 3.4.1)

$$\begin{aligned}
\mathbf{r} &= (r\sin\theta\cos\phi,\ r\sin\theta\sin\phi,\ r\cos\theta), \\
\mathbf{k} &= (k\sin\Theta\cos\Phi,\ k\sin\Theta\sin\Phi,\ k\cos\Theta), \\
\boldsymbol{\epsilon}_1 &= (\varepsilon_1\cos\Theta\cos\Phi,\ \varepsilon_1\cos\Theta\sin\Phi,\ \varepsilon_1\sin\Theta), \\
\boldsymbol{\epsilon}_2 &= (\varepsilon_2\sin\Phi,\ -\varepsilon_2\cos\Phi,\ 0),
\end{aligned}$$

we have

$$\begin{aligned}
&\tfrac{1}{2}\langle b| (\boldsymbol{\epsilon}_1\cdot\mathbf{r})(\mathbf{r}\cdot\mathbf{k}) |a\rangle \to \\
&\tfrac{1}{2}\langle b| k\sin\Theta\cos\Theta\cos^2\Phi\, r_x^2 + k\sin\Theta\cos\Theta\sin^2\Phi\, r_y^2 \\
&\quad - k\sin\Theta\cos\Theta\, r_z^2 |a\rangle \\
&= -\tfrac{1}{2} k\sin\Theta\cos\Theta \langle b| r^2 P_2(\cos\theta) |a\rangle \\
&= -\tfrac{1}{2} k\sin\Theta\cos\Theta\, Q_0/e \quad \text{if} \quad Q_0 = e\langle b| r^2 P_2(\cos\theta) |a\rangle
\end{aligned} \qquad (3.5.11)$$

so

$$\begin{aligned}
\frac{1}{\tau_{E_1}} &= \frac{\omega^3 k^2}{8\pi\hbar^2 c^3} |Q_0|^2 \int_\Omega \sin^2\Theta\cos^2\Theta\, d\Omega \\
&= \frac{\omega^3 k^2}{15\hbar c} |Q_0|^2.
\end{aligned} \qquad (3.5.12)$$

$P_2(\cos\theta)$ is even under parity and so the selection rule on the parity of the nuclear states is NO. $P_2(\cos\theta)\Psi_a$ contains states with the same parity as Ψ_a and $J = J_a+2,\ J_a+1,\ J_a,\ J_a-1,\ J_a-2$.

A quadrupole field carries off angular momentum 2 with respect to the z-axis and so the angular momentum selection rule is

$$J_b = J_a + 2.$$

Once more our result for the energy radiated per unit time is in accord with the classical result.

Then for electric dipole radiation

$$\frac{1}{\tau_{E_1}} = \frac{4}{3}\frac{e^2}{\hbar c}\frac{\omega^3}{c^2}\left|\frac{D_0}{e}\right|^2$$

while for magnetic dipole radiation

$$\frac{1}{\tau_{M_1}} = \frac{4}{3}\frac{e^2}{\hbar c}\frac{\omega^3}{c^2}\left|\frac{M_0}{e}\right|^2$$

Nuclear Decay

and for higher multipoles the rate goes down *approximately* as k^2R^2 for each step, thus

$$1/\tau_{E_1} \sim \omega(kR)^2, \quad 1/\tau_{E_2} \sim \omega(kR)^4, \quad 1/\tau_{E_l} \sim \omega(kR)^{2l}.$$

The angular momentum carried off by the photon is l for both magnetic and electric transitions. For even l the parity selection rule is NO for electric transitions, YES for magnetic transitions, while for odd l the rule is YES for electric transitions and NO for magnetic transitions. The reason is that a current $\mathbf{j}$ which generates electric radiation is a vector, while the circulation of a current, which generates magnetic radiation, is an axial vector.

The whole subject of multipole radiation in nuclear decay is dealt with in great detail in Blatt and Weisskopf, *Theoretical Nuclear Physics*,† to which we refer you for any further details.

We may now make some numerical calculations. In atoms the coupling of electrons with electrons is relatively weak and transitions are mostly single electron transitions. We take as a typical electric dipole moment ea_0 where a_0 is the Bohr radius for hydrogen and as a typical magnetic moment $e\hbar/2mc$ where m is the electron mass. Then for radiation in the optical region, say 5000 Å, the wavelength is ~ 5000 times the atomic dimension and we expect only electric dipole radiation to be important.

If
$$\lambda \simeq 5 \times 10^{-5} \text{ cm}, \quad \omega \simeq 4 \times 10^{15} \text{ sec}^{-1}, \quad k \simeq 10^5 \text{ cm}^{-1},$$

$$\frac{1}{\tau_{E_1}} \simeq \frac{e^2}{\hbar c} \omega(ka_0)^2 = \frac{1}{137} \omega(ka_0)^2 \simeq 10^7 \text{ sec}^{-1}$$

so $\tau_{E_1} \simeq 10^{-7}$ sec for atoms radiating in the visible spectrum;

$$\frac{1}{\tau_{M_1}} \simeq \frac{k^3}{\hbar}\left(\frac{e\hbar}{2mc}\right)^2 \simeq 10^2 \text{ sec}^{-1},$$

so $\tau_{M_1} \simeq 10^{-2}$ sec for atoms radiating in the visible spectrum;

$$\frac{1}{\tau_{E_2}} \simeq \frac{(ka_0)^2}{\tau_{E_1}} \simeq 2.5 \text{ sec}^{-1},$$

so $\tau_{E_2} \simeq 0.5$ sec for atoms radiating in the visible spectrum.

It is clear why transitions other than electric dipole are known as forbidden transitions in atomic spectroscopy—radiation corresponding to such transitions will only be observed if the collision time of atoms is comparable with or in excess of the lifetimes of the states.

In nuclei the coupling among nucleons is strong and transitions corresponding to single particles changing their state are rare. In general the coupling of single-particle states into collective motion complicates the picture, particularly for $E2$ transitions. However, we repeat the calculations for nuclei, assuming a nuclear radius of 5×10^{-13} cm and a photon of energy 1 MeV.

† J. M. Blatt and V. Weisskopf, *Theoretical Nuclear Physics* (Wiley, 1952).

Nuclear Physics

A 1-MeV photon has a frequency $\omega = 1.6 \times 10^{21}$ sec^{-1} and $k \simeq 5 \times 10^{-10}$ cm^{-1}. Then $kR \simeq 0.025$ and for a single-particle transition

$$\frac{1}{\tau_{E_1}} \simeq \frac{e^2}{\hbar c} \omega (kR)^2 \simeq 10^{17} \text{ sec}^{-1},$$

$\tau_{E_1} \sim 10^{-17}$ sec for a single proton nuclear transition.

If we suppose the magnetic dipole amplitude for a nuclear transition to be $\simeq e\hbar/2Mc$ where M is the nucleon mass, then

$$\frac{1}{\tau_{M_1}} \simeq \frac{k^3}{\hbar} \left(\frac{e\hbar}{2Mc}\right)^2 \simeq 3 \times 10^{12} \text{ sec}^{-1},$$

$\tau_{M_1} \simeq 3 \times 10^{-13}$ sec in a nuclear transition, while

$$\frac{1}{\tau_{E_2}} \simeq \frac{(kR)^2}{\tau_{E_1}} \simeq 6 \times 10^{13} \text{ sec}^{-1},$$

$\tau_{E_2} \simeq 10^{-14}$ sec in a single proton nuclear transition.

The variation goes as k^{2l+1}. If we tabulate such rough estimates we find:

	1 MeV		0.01 MeV		0.01 MeV	
l	τ_{E_l}	τ_{M_l}	τ_{E_l}	τ_{M_l}	τ_{E_l}	τ_{M_l} (sec)
1	10^{-17}	3×10^{-13}	10^{-14}	3×10^{-10}	10^{-11}	3×10^{-7}
2	1.6×10^{-14}	5×10^{-10}	1.6×10^{-9}	5×10^{-5}	1.6×10^{-4}	5
3	3×10^{-11}	10^{-7}	3×10^{-4}	1	30	10^5
4	5×10^{-7}	1.5×10^{-3}	500	1.5×10^6	5×10^{11}	1.5×10^{15}

Thus a 0.1-MeV transition involving a single particle, an angular momentum change of 4 and a change of parity can easily have a lifetime of a year, despite the typical electromagnetic lifetime of $\sim 10^{-16}$ sec. This is the explanation of nuclear isomerism—and we can now see why nuclear isomers cluster in the region of shell closure. The recipe for a long-lived electromagnetic decay is (1) two closely spaced states and (2) a large angular momentum change. The single-particle states exhibit these properties just before high shells close (see Fig. 1.6.3).

The sorting out of these long-lived isomers is dependent on a knowledge of the multipolarity—the lifetimes depend too much on the details of nuclear structure to allow this to be deduced from the lifetime alone—and one of the ways of achieving this is by the study of internal conversion.

3.6. Internal conversion

This is the name given to a process in which the nucleus de-excites electromagnetically by ejecting an electron from the atom rather than by γ-emission. The name is a misnomer, implying as it does that the nucleus emits a photon which then ejects an atomic electron through the photoelectric effect. Such a process does exist, but is insignificant in comparison with the direct process in which the nuclear current interacts with the electron current via virtual rather than real photons. Real photons correspond to a radiation field—and the atomic electrons are well within the induction zone.

For internal conversion to occur, the transition energy must be greater than the binding energy of the electron ejected, which then emerges with the transition energy minus the electron binding energy. The electron spectrum is thus discreet—and internal conversion was first observed through the discrete electron spectrum superimposed on the continuous β-decay spectrum.

We will calculate the probability of this process as our *second example of the Fermi Golden Rule*. The initial state consists of a nucleus Ψ_a and a bound electron with a wave function ϕ_i. The final state contains the nucleus Ψ_b and a free electron with a wave function ϕ_f. We will assume that the free electron can be treated non-relativistically (energy $\lesssim$ 100 keV) and that we can neglect the effect of the nuclear charge on the free electron wave function and represent it as a plane wave $\phi_f = e^{-i\mathbf{k}\cdot\mathbf{r}}$ (normalized to unit volume). For this illustration we will also suppose the atomic electron to be in an s-state with a wave function

$$\phi_i = e^{-r/a}(\pi a^3)^{-1/2} \quad \text{where} \quad a = a_0/Z.$$

The process involves the interaction of two electromagnetic currents and may be represented by

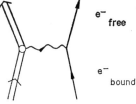

Fig. 3.6.1.

and the interaction energy is that of two currents coupling through the electromagnetic field:

$$(\mathbf{j}_N \cdot \mathbf{j}_e - \varrho_N \varrho_e)/|\mathbf{r}_N - \mathbf{r}_e|.$$

Since we are concerned with non-relativistic electrons, and nucleon velocities in the nucleus, we may neglect the term

$$\mathbf{j}_N \cdot \mathbf{j}_e \sim \frac{v_e v_N}{c^2} \varrho_N \varrho_e$$

if we are dealing with an electric transition. This corresponds to the electron being ejected by the pulsating electric moments of the nucleus. We cannot use this approximation for magnetic transitions, but since we can only illustrate the physical principles here, we will take the interaction energy as

$$\frac{\varrho_N \varrho_e}{|\mathbf{r}_N - \mathbf{r}_e|}$$

and for $r_N \ll r_e$ we may expand this expression and write it as

$$\frac{\varrho_e}{r_e}\left[\varrho_N\left\{\sum_l \left(\frac{r_N}{r_e}\right)^l P_l(\cos\theta)\right\}\right] \tag{3.6.1}$$

Nuclear Physics

where θ is the angle between r_N and r_e. Then for the transition matrix element we write

$$-\int\int \left\{\frac{\phi_f^* e \phi_i}{r_e^{l+1}}\right\} \{\Psi_b^* er_N^l P_l(\cos\theta)\Psi_a\} \, d^3r_N \, d^3r_e \tag{3.6.2}$$

which for an initial s-wave electron and a final plane wave becomes

$$-\int\int \left\{\frac{e^{i\mathbf{k}_1\cdot\mathbf{r}_e}}{r_e^{l+1}} \frac{e^{-r_e/a}}{(\pi a^3)^{1/2}} d^3r_e\right\} \{\Psi_b^* er_N^l P_l(\cos\theta)\Psi_n\} \, d^3r_N. \tag{3.6.3}$$

We can expand the plane wave in spherical waves—the coefficients go as powers of r_e and so remove divergence problems at the origin. The angle θ, being between $\mathbf{r}_e$ and $\mathbf{r}_N$, is not convenient as a variable of integration and it is easier to define a z-axis of symmetry for the nucleus and refer to this axis the nuclear radius vector $\mathbf{r}_N$, the electron radius vector $\mathbf{r}_e$ and the electron momentum vector $\mathbf{k}$.

Consider the case of a dipole transition: if the fluctuating dipole forms the z axis, then the only component of $\mathbf{r}_e \cdot \mathbf{r}_N$ which contributes is $r_{e_z} r_{N_z}$. Thus the expression becomes

$$-\langle b|ez|a\rangle \int \frac{e^{i\mathbf{k}\cdot\mathbf{r}_e} e^{-r_e/a}}{r_e^2(\pi a^3)^{1/2}} \frac{r_{e_z}}{r_e} d^3r_e. \tag{3.6.4}$$

It is clear that an angular integration uses r_{e_z} to pick out the $l=1$ component of the plane wave final state, as we expect from conservation of angular momentum and parity. Writing $d^3r_e = r_e^2 \, d\cos\Theta \, d\Phi$ and denoting $\mathbf{r}_e$ by (r_e, Θ, Φ), $\mathbf{k}$ by (k, α, β) the $l=1$ term in the expansion is

$$i\{\sin\Theta\cos\Phi\sin\alpha\cos\beta + \sin\Theta\sin\Phi\sin\alpha\sin\beta + \cos\Theta\cos\alpha\}f(kr_e)$$

so that the matrix element becomes, on integrating over Φ,

$$-\langle b|ez|a\rangle \int \frac{ef(kr_e)\cos^2\Theta e^{-r_e/a}\cos\alpha \, 2\pi r_e^2 \, dr_e \, d\cos\Theta}{r_e^2(\pi a^3)^{1/2}} \tag{3.6.5}$$

where

$$f(kr_e) = 3\left\{\frac{\cos kr_e}{ikr_e} - \frac{\sin kr_e}{ik^2r_e^2}\right\}.$$

Note that while kr_N is small, kr_e in general is NOT. The integral becomes

$$\frac{4\pi}{3} \frac{e}{(\pi a^3)^{1/2}} \int_0^\infty f(kr_e) e^{-r_e/a} \, dr_e \tag{3.6.6}$$

and since k corresponds to energies in the γ-ray region, the exponential varies slowly in comparison with $f(kr)$ and so may be set equal to unity, yielding

$$\frac{4\pi}{3} \frac{e}{(\pi a^3)^{1/2}} \frac{1}{k} F \cos\alpha$$

Nuclear Decay

where F is the dimensionless quantity $\int_0^\infty f(x)\, dx = 3i$. Then the transition rate becomes

$$T_{E_1}^K = \frac{2\pi}{\hbar} \left| 4\pi \frac{e^2}{(\pi a^3)^{1/2}} \frac{1}{k} \langle b | z | a \rangle \right|^2 \int \cos^2 \alpha \, d\Omega \, \frac{p^3 \, dp/dE}{(2\pi\hbar)^3}. \tag{3.6.7}$$

$T_{E_1}^K$ denotes a dipole transition from the K-shell where E is the transition energy, equal to the kinetic energy of the electron (plus the electron binding energy)

$$\frac{dp}{dE} = \frac{m_e}{p}$$

so for two K-shell electrons

$$T_{E_1}^K = 2 \times \frac{2\pi}{\hbar} \left\{ 4\pi \frac{e}{(\pi a^3)^{1/2}} \frac{1}{k} \right\}^2 |\langle b | ez | a \rangle|^2 \frac{4\pi}{3} \frac{m_e p}{(2\pi\hbar)^3}$$

$$= \frac{32}{3} \frac{e^2}{a^3} \frac{1}{k} \frac{m_e}{\hbar^3} |\langle b | ez | a \rangle|^2. \tag{3.6.8}$$

Having worked this example in some detail we may now return to eq. (3.6.3) and extract the physical dependence without paying attention to the numbers. The successive spherical harmonics have coefficients $f_l(kr)$ and to set the integral over the electron coordinates in a dimensionless form we have to write

$$\frac{d^3 r_e}{r_e^{l+1}} \sim k^{l-2} \frac{d^3(kr_e)}{(kr_e)^{l+1}}.$$

The square of the matrix element varies as k^{2l-4} and the transition rate as k^{2l-3}.

Thus the essential physical features of internal conversion are:

1. It contains an extra factor of e^2, coming from the coupling of the virtual photon to the electron.
2. It increases as the cube of the charge on the initial nucleus.
3. It varies with $k(k = p/\hbar)$ to the power $2l-3$.
4. The nuclear matrix element is identical with that for the corresponding radiative transition.

As a result of (4) the ratio of the internal conversion rate to the radiative decay rate can be calculated exactly (although we have not done it) and is independent of the nuclear matrix element. The ratio

$$\frac{T_l^K}{\tau_l} \sim \frac{k_e^{2l-3}}{k_\gamma^{2l+1}}. \tag{3.6.9}$$

The transition energy $E = \hbar\omega$ where $\omega = k_\gamma c$ and if the binding energy of the electron may be neglected then

$$\frac{p^2}{2m_e} = \hbar\omega = \hbar k_\gamma c$$

so that $k_e \propto k_\gamma^{1/2}$ for a non-relativistic electron.

Nuclear Physics

Thus the K-shell *internal conversion coefficient* is

$$\alpha_{K_l} = \frac{T_l^K}{\tau_l} \propto k_\gamma^{-l-5/2} = c_l k_\gamma^{-l-5/2} \qquad (3.6.10)$$

where the number c_l can be precisely calculated. A measurement of the coefficient α_K is capable of yielding a direct determination of the multipolarity of the transition. Armed with this information, a measurement of the lifetime of the state provides the nuclear matrix element, so that radiative decay processes constitute a probe of nuclear structure.

We have only worked out the form of α_K for electric transitions leading to non-relativistic final states, and must emphasize that the calculations must be done properly, using relativistic electron theory, in order to yield useful results. The calculations for magnetic transitions necessarily involve flipping the electron spin and are a bit more complicated in a non-relativistic calculation of the kind we have made above. However, it should be clear that the value of a conversion coefficient depends on both the multipolarity and on whether the nuclear transition is electric or magnetic. Consider for an example dipole transitions. The electromagnetic field must transfer one unit of angular momentum to the electron. This means that for an electric dipole transition the ejected electron is produced in a *p* state —since the nuclear parity changes and the overall parity of the whole system cannot change— and the electron ejected in a magnetic dipole transition will find itself still in an *s* state (with the spin flipped over) or in a $d_{3/2}$ state. The overlap of these final electron wave functions with the nucleus and the initial wave function depends on the angular momentum and so the magnetic transitions of order *l* will in general have internal conversion coefficients different from those for electric transitions of order *l*.

Finally we would point out that internal conversion from atomic states higher than the *s* state can occur—the ratios for the different coefficients also carry information about the nature of the transition.

The results of a systematic study of isomeric transitions was summarized in a classic paper by Goldhaber and Sunyar[†] which is very well worth your reading, not only for its intrinsic interest but also as an example of the convergence of scattered data to the physical reality.

In the regions of the "Islands of Isomerism" before shell closure, transitions are primarily of the E3 and M4 types, corresponding to the large spin changes expected from the independent particle model with spin–orbit coupling. The transition rates are approximately equal to those expected on the assumption that a single nucleon only is involved in the transition.

A strong contrast is offered by transitions known to be E2. Here the rates are an order of magnitude higher than expected on the basis of single nucleon transitions, just as static quadrupole moments are an order of magnitude greater than those calculated from a single unpaired nucleon. It is clear that collective transitions are involved—for example, transitions between different rotational states of an aspherical nucleus—which can take place even for an even–even nucleus (in the excited states) and which may involve rotation of an outer layer with respect to an inner core rather than rotation of the nucleus as a whole.

[†] M. Goldhaber and A. W. Sunyar, *Phys. Rev.* **83**, 906 (1951); see also *α-, β- and γ-Ray Spectroscopy*, Ed. K. Siegbahn, Vol. 2, chap. XVIII (North Holland, 1965).

Nuclear Decay

Another collective effect is demonstrated by the E1 transitions in massive nuclei: the rates are down by several orders of magnitude on single-particle rates. This is a manifestation of the strong coupling of protons with neutrons in collective motion (another example of which is the large electric quadrupole moments of nuclei with an odd number of neutrons). In order for a dynamic electric dipole moment to appear, the charge in the nucleus must fluctuate from end to end—the nucleus cannot just wiggle up and down since its centre of mass remains fixed. Therefore the charge must wiggle up and down with respect to the centre of mass—and if the protons were rigidly tied to the neutrons this could not happen. The strong coupling between the two kinds of nucleon does not suppress electric dipole transitions completely, but does cut down the dynamic dipole moment in complex nuclei by several orders of magnitude below that expected on a single-particle basis—in complete contrast to atomic physics where the electrons are free to wiggle with respect to the centre of mass, the nucleus. (The electric dipole transitions which are suppressed by the strong nuclear coupling are those in which the nuclear rearrangement involves neither a single nucleon nor the whole nucleus. In very light nuclei electric dipole transitions corresponding to single nucleons changing their orbits take place with about the expected rate, while it is also possible for all the protons in a nucleus to vibrate against all the neutrons. In the latter case the effect of the strong nuclear coupling is to drive the frequency corresponding to this mode up to photon energies ~ 20 MeV, with a line width $\sim$ MeV. It is the excitation of this mode of oscillation which gives rise to the giant dipole resonances in the scattering of photons by nuclei.)

Independent particle model transitions were straightened out in the early 1950s. The principal interest in electromagnetic transitions nowadays is to find numerical values for matrix elements which can be compared with calculations made with more complicated models of the nucleus. This topic is for professionals and we shall leave it alone.

3.7. 0→0 transitions

In Sections 3.5 and 3.6 we saw that the vector difference between the initial and final nuclear angular momenta is at least 1 (for dipole transitions). While this allows $1 \to 1$ and $\frac{1}{2} \to \frac{1}{2}$ transitions, clearly a $0 \to 0$ transition cannot go by emission of a single real photon. A $0 \to 0$ transition has both initial and final states spherically symmetric. The nuclear transition matrix element cannot behave like a vector and the field cannot carry off any angular momentum. In classical terms such a transition corresponds to a pulsating sphere of charge—and outside a spherically symmetric region of charge density the field is pure coulomb. Thus well away from the nucleus in a $0 \to 0$ transition, there is a constant coulomb field and so no radiation.

However, such transitions can occur, but not by emission of a single photon. One possibility is the emission of two correlated photons, but the probability of this is way down. The most common mechanism for electromagnetic de-excitation between two states of spin 0 is internal conversion. The *s*-state electrons are deeply penetrating and that part of the wave function within the nuclear region is exposed to the fluctuating coulomb field to be expected *within* a pulsating sphere of constant charge. Energy can thus be transferred between the

Nuclear Physics

excited nucleus and an atomic electron, the rate depending on the degree of penetration of s-wave electrons into the nucleus. The transition rate is given by insertion of the matrix element

$$\int \frac{(\phi_f^* e \phi_i)(\Psi_f^* e \Psi_i)}{|\mathbf{r}_N - \mathbf{r}_e|} d^3 r_e \, d^3 r_N \tag{3.7.1}$$

but the whole contribution to the integral now comes from a region where $r_e \sim r_N$. Since all transitions are here between states of 0 orbital angular momentum we do not need to distinguish any direction in space, the emitted electrons will be isotropic. If the angle between $\mathbf{r}_e$ and $\mathbf{r}_N$ is defined as θ, then the matrix element becomes

$$\int \frac{(\phi_f^*(r_e) \, e \phi_i(r_e))(\Psi_f^*(r_N) \, e \Psi_i(r_N)) \, d^3 r_N 2\pi r_e^2 \, dr_e \, d\cos\theta}{\sqrt{r_N^2 + r_e^2 - 2 r_e r_N \cos\theta}}. \tag{3.7.2}$$

The integration over $\cos\theta$ is straightforward, provided that we remember to take always the positive square root in the denominator:

$$\int_{-1}^{+1} \frac{d\cos\theta}{\sqrt{r_N^2 + r_e^2 - 2 r_e r_N \cos\theta}} = \left[\frac{\sqrt{r_N^2 + r_e^2 - 2 r_e r_N \cos\theta}}{r_e r_N} \right]_{-1}^{+1}$$

$$= \frac{1}{r_e r_N} \left\{ \sqrt{r_N^2 + r_e^2 - 2 r_e r_N} - \sqrt{r_N^2 + r_e^2 + 2 r_e r_N} \right\}$$

$$= \frac{1}{r_e r_N} \left\{ \sqrt{(r_N - r_e)^2} - \sqrt{(r_N + r_e)^2} \right\}$$

$$= \frac{1}{r_e r_N} \left\{ (r_N - r_e) - (r_N + r_e) \right\} \qquad r_N > r_e$$

$$= \frac{1}{r_e r_N} \left\{ (r_e - r_N) - (r_N + r_e) \right\} \qquad r_N < r_e.$$

The integral over r_e becomes

$$\int_0^{r_N} \phi_f^*(r_e) \frac{e}{r_N} \phi_i(r_e) 4\pi r_e^2 \, dr_e + \int_{r_N}^{\infty} \phi_f^*(r_e) \frac{e}{r_e} \phi_i(r_e) 4\pi r_e^2 \, dr_e \tag{3.7.3}$$

where there is a subsequent integration to be made over the variable r_N. However, for $r_e > r_N$ the charge within r_N contributes nothing to the fluctuating field, only to the static energy of the system. Since we want that part of the integral which yields transitions in a fluctuating coulomb field, we discard the integral for $r_e > r_N$ and just have

$$\int_0^{r_N} \phi_f^*(r_e) \frac{e}{r_N} \phi_i(r_e) 4\pi r^2 \, dr_e$$

$$\simeq \phi_f^*(0) \, e \phi_i(0) \frac{4\pi}{3} r_N^2$$

Nuclear Decay

and the whole matrix element becomes

$$\phi_f^*(0)\, e\phi_i(0)\, \frac{4\pi}{3} \int \Psi_f^*\, er_N^2 \Psi_i\, d^3r_N; \tag{3.7.4}$$

here again the nuclear integral is understood to be summed over as many nucleons as is necessary.

There is another process through which a nucleus may make a $0 \to 0$ electromagnetic transition, again involving a single virtual photon. Consider the diagram for internal conversion:

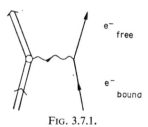

FIG. 3.7.1.

This is clearly a special case of the electron scattering diagram

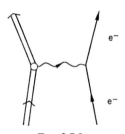

FIG. 3.7.2.

which may be further generalized by replacing an electron coming in by a positron going out

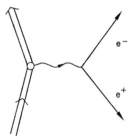

FIG. 3.7.3.

A virtual photon in the coulomb field can convert into a pair of electrons, provided that the nuclear transition energy is $> 2m_e c^2$. The best-known example is the decay of the first excited state of ^{16}O at 6 MeV, a $0^+ \to 0^+$ transition with emission of an electron pair.

131

Nuclear Physics

To calculate the transition rate for such a process we would need the phase space factor for three particles in the final state, which we shall study anyway for β-decay, and we really need quantum electrodynamics as well. This, however, is beyond the scope of this book.

We complete this section with some remarks about the virtual photons involved in $0 \to 0$ transitions. Consider first internal conversion. The electron is emitted radially, and the fluctuating electric field is radial. Thus the electric vector of the virtual photon is parallel to its momentum, in contrast to a free field photon. At threshold, the photon carries just enough energy to liberate the electron from the atom, and so carries no momentum—it is *timelike*. In general

$$E_\gamma = E_e - B_K - m_e c^2,$$
$$\mathbf{p}_\gamma = \mathbf{p}_e,$$
$$p_\gamma^2 c^2 - E_\gamma^2 = p_e^2 c^2 - (E_e - B_K - m_e c^2)^2$$
$$= (p_e^2 c^2 - E_e^2 - m_e^2 c^4) - B_K^2 + 2T_e B_K$$
$$= 2T_e B_K - B_K^2.$$

As soon as $T_e > B_K/2$ the photon becomes spacelike, as it always is for elastic scattering.

For pair production, at threshold the pair is created at rest. The photon is again longitudinal and timelike. As the energy increases

$$\mathbf{p}_\gamma = \mathbf{p}_{e^+} + \mathbf{p}_{e^-} \quad E_\gamma = E_{e^+} + E_{e^-},$$
$$p_\gamma^2 c^2 - E_\gamma^2 = (\mathbf{p}_{e^+} + \mathbf{p}_{e^-})^2 c^2 - (E_{e^+} + E_{e^-})^2$$
$$= 2\mathbf{p}_{e^+} \cdot \mathbf{p}_{e^-} c^2 - 2E_{e^+} E_{e^-}$$

which is always negative, approaching zero as the electron momenta approach ∞. Thus the virtual photon leading to pair production is always timelike.

3.8. Measurement of lifetimes in electromagnetic transitions

The technology of measurement in nuclear physics changes so rapidly that any attempt to discuss it in a textbook is doomed to obsolescence. In any case, experimental technique can only be learned in the laboratory—and only in the research laboratory in any subject which is not stultified. Nevertheless, it is worth spending a little time considering the physical nature of the problems of measurement in electromagnetic decays. There are two classes: how can we measure lifetimes, and how can we determine multipolarities?

In Section 3.5 we found that we should expect γ lifetimes ranging from $\sim 10^{-16}$ sec upwards into seconds, hours and months for isomeric transitions. It is easy enough to determine lifetimes characteristic of ordinary radioactive decay: the rate of decay of activity may be measured directly by counting the number of γ's (at the right energy) detected in a given solid angle with a known efficiency per unit time. For long lifetimes, if the number of initial isomers in a sample of material is known, a single measurement suffices. The difficulties begin when lifetimes become short compared with 1 sec. For example, suppose that a β-decay leads to an excited nuclear state which subsequently emits a γ. If the β lifetime were much shorter than the γ lifetime, then several β lifetimes after irradiation of the source it

Nuclear Decay

would be possible just to measure the decay rate of the γ radiation. If we have to sample ~ 100 times within a single mean life, we need to be able to count a statistically significant number of γ in $\tau/100$ and a statistically significant number is ~ 100: this means detector plus counters responding to individual pulses every $\tau/10^4$ sec on average, with room for fluctuations. With response times of $\sim 10^{-9}$ sec, which is easily achieved, this sets the limit for this kind of technique at around $\tau \sim 10^{-4}$ sec, even if the γ source can be suitably prepared.

But if instead a clock is started by the electron emitted, and stopped by the γ, the times measured are only $\sim \tau$ and the distribution of delay times will yield the lifetime. A 1-cm wavelength signal in a cable has frequency 3×10^{10} sec^{-1}: electronic clocks can run at this kind of frequency and so coincidence methods extend down into the 10^{-10} sec region. Anything suitable may start the clock, for example a preceding decay, as in the β–γ case, or a pulse from an accelerator making the excited states. But how long does an electron or γ take to go, say, 10 cm through a scintillator, disturbing the electrons in the material that eventually lead to a signal? The answer is $\sim 10^{-10}$ sec, and electrons take longer than this to fall back into an unexcited state in the scintillator and emit light. Thus in the region of 10^{-9}–10^{-10} sec, signals last as long as their separation, and while it is still possible to unfold the delay from overlapping signals, coincidence methods get dodgy in this region.

Time may be turned into distance via velocity. If the source is very thin on an inert backing, the excited nuclei are ejected when struck by the beam from an accelerator, with velocities $\sim 10^9$ cm/sec (a proton of ~ 1 MeV has a velocity $\sim 10^9$ cm/sec). Counting rate at 90°

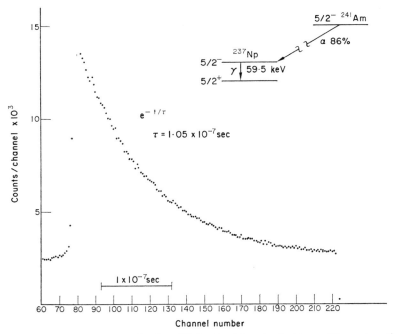

FIG. 3.8.1. An example of measurement of a short lifetime by delayed coincidence. The nucleus ^{241}Am decays by α-emission to the 5/2$^-$ excited state of ^{237}Np. The distribution of delay between the α and the subsequent 59.5 keV γ is plotted in the figure and yields a mean life of 1.05×10^{-7} sec for the 5/2$^-$ ^{237}Np state. [Nuclear Physics Teaching Laboratory, Oxford.]

Nuclear Physics

to the beam as a function of distance along it gives the lifetime. If counts can be made at ~1-mm intervals then lifetimes ~10^{-10} sec are reachable by such a method.

Finally, there are Doppler shift methods. If an excited nucleus decays in flight, a Doppler shift will be seen. If it is stopped before it decays, no Doppler shift will be seen. The Doppler shift of a γ emitted from a nucleus moving at 10^9 cm/sec is ~3% which is detectable, particularly with lithium drifted germanium detectors, which have a resolution better than 1%. So the ratio of Doppler shifted γ's to unshifted γ's as a function of the distance between the thin source and a stopper will give the lifetime. The distance can be reliably set down to ~10 microns mechanically. If the velocity is ~10^9 cm/sec this makes lifetimes ~10^{-12} sec accessible.

It is clear that the line will be smeared if the time the nucleus takes to slow down and stop after leaving the drift space between the source and stopper is comparable to the

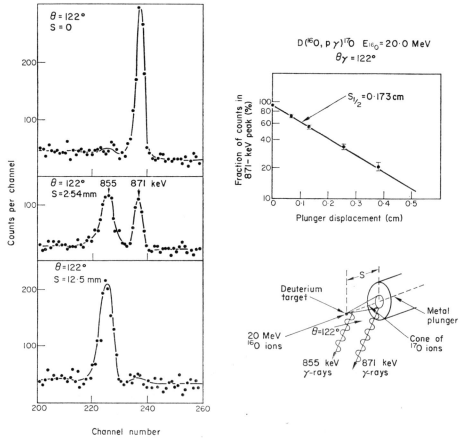

FIG. 3.8.2. An example of the measurement of short γ lifetimes by Doppler shift methods. The 871 keV level of ^{17}O was prepared through the stripping reaction D(^{16}O, p) ^{17}O. With the metal plunger against the target all ^{17}O* nuclei decay after being stopped. For a separation $s = 2.54$ mm the Doppler shifted line at 855 keV is a little more intense than the unshifted line, while for $s = 12.5$ mm almost all the ^{17}O* nuclei decay in flight. The average ^{17}O* velocity was 0.0358 c and the mean lifetime found to be $\tau = 2.33 \pm 0.26 \times 10^{-10}$ sec. [Reproduced by permission of the National Research Council of Canada from the *Canadian Journal of Physics* **43**, 1563-73 (1965).]

Nuclear Decay

lifetime. This smearing can itself be used to get at the lifetime if a thicker target is used, in which the nucleus starts slowing down as soon as it is made. If it lives much longer than the slowing-down time, no Doppler shifts appear. If the lifetime is much shorter than the slowing-down time, all γ's are Doppler shifted. When the lifetime is comparable with the slowing-down time an analysis of the line shape will yield the lifetime—and this can be checked for the same nucleus produced at different energies and in different reactions.

The slowing-down time may be computed from a knowledge of the rate of energy loss with distance which is[†]

$$\frac{dE}{dx} \approx -\frac{4\pi Z^2 e^4}{m_e v^2} N \qquad (3.8.1)$$

where N is the number of electrons per cm^3, m_e the electron mass, Ze the charge of a massive particle moving with velocity v. Rewriting we have

$$\frac{dv}{dt} \simeq -\frac{4\pi Z^2 e^4}{m_e M v^2} N \qquad (3.8.2)$$

where M is the nuclear mass whence

$$t \sim \frac{v^3 M m_e}{12\pi Z^2 e^4 N} \qquad (3.8.3)$$

for the slowing-down time t of a nucleus with charge Ze, mass M and initial velocity v. Let us evaluate this for $Z = 6$ (carbon) moving with 10^9 cm/sec in carbon.

$$M \simeq 12 \times 1.6 \times 10^{-24} \text{ g} \quad m_e \sim 9 \times 10^{-28} \text{ g},$$
$$N \sim 10^{24} \text{ electrons/cm}^3 \quad e = 4.8 \times 10^{-10} \text{ esu.}$$

Then $t \approx 10^{-12}$ sec.

With Doppler effects $\sim 1\%$ measurable, lifetime measurements can be extended into the region of $\sim 10^{-14}$ sec.

Measurement of the cross-section for photon scattering can be used to go down to $\lesssim 10^{-16}$ sec: if an excited state is formed when a γ interacts with a nucleus, the cross-section will be proportional to the width of the line, the square of the decay matrix element. The photon can be a real free photon, or a virtual longitudinal photon emitted from an energetic charged particle—this is the process of coulomb excitation. Note that a physical width of the line is not measured—a mean life of 10^{-16} seconds means a width of 6 eV and it is not possible to determine cross-section as a function of photon momentum accurately enough to find the shape of the line. Rather, the width is inferred from the cross-section for the process. We are now in the field of nuclear reactions, which will be discussed more fully in Chapter 4.

3.9. Determination of multipolarities in electromagnetic decay

Each multipole order has characteristic radiation patterns with respect to the spin axes of the transition. In a few cases it has been possible to orient nuclei at low temperatures in a magnetic field and observe this pattern directly. Determination of both the angular distribu-

[†] See, for example, E. Fermi, *Nuclear Physics*, chap. II (Chicago, 1950).

tion and the direction of the electric field at a given angle can give both the multipolarity and the parity selection rule, i.e. whether a transition is electric or magnetic.

We have already pointed out in Section 3.6 that the various internal conversion coefficients are very sensitive to both the multipolarity of a transition and the parity change. These coefficients are determined by finding the relative intensity of conversion electrons of the appropriate energy to the alternative photons—the experimental difficulties occur in determining detection efficiencies for these two kinds of radiation so as to yield a true number.

Finally, we will discuss in a little detail the properties of angular correlations in successive decays. If a photon from one decay is observed in a particular direction then the orientation of the nucleus resulting from the decay is not isotropic but reflects the angular distribution of the radiation. Decay of these aligned nuclei then results in an anisotropic distribution of radiation in a further decay following on the first, measured with respect to the first photon. This is a $\gamma\gamma$ correlation.

Information can be obtained from decay chains involving other processes than just γ-emission—for example, correlations between successive internal conversion electrons, or between a β-decay electron and a subsequent photon. In the decay of elementary particles correlations between successively emitted secondary particles can be used in order to obtain the spin and parity of the initial or intermediate states. The method is thus of some general interest: since it depends on the quantum theory of angular momentum it is highly developed but we will merely illustrate the possibilities by considering a particularly simple case. Suppose that we have two successive electric dipole transitions, between states $0^+ \to 1^- \to 0^+$ for simplicity. Our starting point is the matrix element in equation (3.3.16)

$$\int \Psi_b^* e^{-i\mathbf{k}\cdot\mathbf{r}} \boldsymbol{\epsilon}_0 \cdot \mathbf{r} \Psi_a \, dV \tag{3.9.1}$$

for an electric dipole transition this becomes

$$\int \Psi_b^* \boldsymbol{\epsilon}_0 \cdot \mathbf{r} \Psi_a \, dV \tag{3.9.2}$$

for the transition between Ψ_a and Ψ_b. The quantity $\mathbf{r}$, remember, is a vector operator. If we define a Z-axis arbitrarily in space then we have

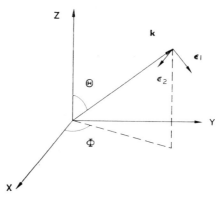

Fig. 3.9.1.

Nuclear Decay

where ϵ_1 and ϵ_2 are the two independent polarizations. The matrix element for ϵ_1 is

$$\cos\Theta\cos\Phi\langle b|x|a\rangle + \cos\Theta\sin\Phi\langle b|y|a\rangle + \sin\Theta\langle b|z|a\rangle \tag{3.9.3}$$

and for ϵ_2

$$\sin\Phi\langle b|x|a\rangle - \cos\Phi\langle b|y|a\rangle. \tag{3.9.4}$$

The two polarizations must be squared and then summed: since they are at right angles and the angular terms do not interfere on integration over Φ we can write for the function determining the angular distribution of radiation in the first decay

$$(\cos\Theta\cos\Phi + \sin\Phi)\langle b|x|a\rangle + (\cos\Theta\sin\Phi - \cos\Phi)\langle b|y|a\rangle + \sin\Theta\langle b|z|a\rangle \tag{3.9.5}$$

and if the second decay leads from Ψ_b to Ψ_c the matrix element describing the whole process is

$$\{(\cos\theta\cos\phi + \sin\phi)\langle c|x|b\rangle + (\cos\theta\sin\phi - \cos\phi)\langle c|y|b\rangle + \sin\theta\langle c|z|b\rangle\}$$
$$\times\{(\cos\Theta\cos\Phi + \sin\Phi)\langle b|x|a\rangle + (\cos\Theta\sin\Phi - \cos\Phi)\langle b|y|a\rangle + \sin\Theta\langle b|z|a\rangle\}. \tag{3.9.6}$$

The square, integrated over all orientations in space of the plane containing the two photons, gives the angular correlation between the two photons (in this case for two successive electric dipole transitions; θ, ϕ describe the second photon with propagation vector $\mathbf{k}'$).

The operators x, y can change m by ± 1 and J by ± 1, 0. The operator z can change J by ± 1 or 0 and cannot change m. The matrix elements of a vector $\mathbf{r}$ with components x, y, z are related by[†]

$$\left.\begin{array}{l}\langle J, M|x+iy|J, M-1\rangle = \langle J|r|J\rangle\sqrt{(J-M+1)(J+M)}, \\ \langle J, M-1|x-iy|J, M\rangle = \langle J|r|J\rangle\sqrt{(J-M+1)(J+M)}, \\ \langle J-1, M|x+iy|J, M-1\rangle = \langle J-1|r|J\rangle\sqrt{(J-M+1)(J-M)}, \\ \langle J, M-1|x-iy|J-1, M\rangle = \langle J|r|J-1\rangle\sqrt{(J-M+1)(J-M)}, \\ \langle J, M|x+iy|J-1, M-1\rangle = -\langle J|r|J-1\rangle\sqrt{(J+M-1)(J+M)}, \\ \langle J-1, M-1|x-iy|J, M\rangle = -\langle J-1|r|J\rangle\sqrt{(J+M-1)(J+M)},\end{array}\right\} \tag{3.9.7}$$

all other elements are zero

$$x = \frac{1}{2}\{(x+iy)+(x-iy)\},$$

$$y = \frac{i}{2}\{(x-iy)-(x+iy)\},$$

so that for our choice $0^+ \rightarrow 1^- \rightarrow 0^+$ we have

$$\langle b|x|a\rangle = \sum_m \langle 1, m|x|0, 0\rangle \rightarrow \frac{1}{\sqrt{2}}\{\langle 1, -1| - \langle 1, +1|\}r_{ab}, \tag{3.9.8a}$$

$$\langle b|y|a\rangle = \sum_m \langle 1, m|y|0, 0\rangle \rightarrow -\frac{i}{\sqrt{2}}\{\langle 1, +1| - \langle 1, -1|\}r_{ab}, \tag{3.9.8b}$$

$$\langle b|z|a\rangle = \sum_m \langle 1, m|z|0, 0\rangle \rightarrow \langle 1, 0|r_{ab}. \tag{3.9.8c}$$

[†] See, for example, L. D. Landau and E. M. Lifshitz, *Quantum Mechanics*, 2nd ed., sect. 29 (Pergamon Press, 1965).

Nuclear Physics

so the first step leads to all three magnetic substates of the 1^- nucleus, each multiplied by an angular function for the radiation:

$$\frac{\langle 1, +1|_b}{\sqrt{2}}\{-(\cos\Theta\cos\Phi+\sin\Phi)-i(\cos\Theta\sin\Phi-\cos\Phi)\}$$
$$+\frac{\langle 1, -1|_b}{\sqrt{2}}\{(\cos\Theta\cos\Phi+\sin\Phi)-i(\cos\Theta\sin\Phi-\cos\Phi)\}$$
$$+\langle 1, 0|_b \sin\Theta.$$

The operators x, y, z acting between b and c now take these terms to the final nuclear state $\langle 0, 0|_c$ and are accompanied by angular functions for emission of the second photon:

$$\langle 0, 0|x|1, +1\rangle \to -\frac{r_{bc}}{\sqrt{2}}\langle 0, 0|_c,$$

$$\langle 0, 0|x|1, -1\rangle \to \frac{r_{bc}}{\sqrt{2}}\langle 0, 0|_c,$$

$$\langle 0, 0|x|1, 0\rangle = 0$$

the x-component yields

$$r_{ab}r_{bc}(\cos\Theta\cos\Phi+\sin\Phi)(\cos\theta\cos\phi+\sin\phi).$$

Then

$$\langle 0, 0|y|1, +1\rangle \to \frac{i}{\sqrt{2}}r_{bc}\langle 0, 0|_c,$$

$$\langle 0, 0|y|1, -1\rangle \to \frac{i}{\sqrt{2}}r_{bc}\langle 0, 0|_c,$$

$$\langle 0, 0|y|1, 0\rangle = 0,$$

and the y-component yields

$$r_{ab}r_{bc}(\cos\Theta\sin\Phi-\cos\Phi)(\cos\theta\sin\phi-\cos\phi)$$

and similarly from z operating between the three magnetic substates of b, and c, we get

$$r_{ab}r_{bc}\sin\Theta\sin\theta$$

and the angular correlation between the two photons is contained in the square of

$$(\cos\Theta\cos\Phi+\sin\Phi)(\cos\theta\cos\phi+\sin\phi)+(\cos\Theta\sin\Phi-\cos\Phi)(\cos\theta\sin\phi-\cos\phi)$$
$$+\sin\Theta\sin\phi. \tag{3.9.9}$$

Since a correlation does not depend on the choice of Z-axis, it is convenient to choose the Z-axis along the direction of the first photon, and also choose ϵ_2 for the first photon to define the X-axis. θ is then the angle between the two photons and the correlation is given by

$$[\cos\theta\cos\phi+\sin\phi+\cos\theta\sin\phi-\cos\phi]^2$$

which must be integrated over ϕ to include all possible orientations of $\mathbf{k}'$ with respect to the

Nuclear Decay

electric field carried by the first photon. Terms odd in $\cos \phi$ vanish and so the correlation between the two photons is

$$1+\cos^2 \theta. \tag{3.9.10}$$

We must emphasize that not all dipole transitions give a correlation. In general, a dipole–dipole correlation contains no powers *higher* than $\cos^2 \theta$ but the two photons need not be correlated at all. For example, in $\frac{1}{2} \to \frac{1}{2}$ transitions, radiation is isotropic with respect to any system of axes, and transitions in which one stage is $\frac{1}{2} \to \frac{1}{2}$ (electric or magnetic dipole transitions, depending on the parities will show no correlation. As a simple example, work out the correlation in the two successive $E1$ transitions $\frac{1}{2}^- \to \frac{1}{2}^+ \to \frac{1}{2}^-$—there is none.

While angular correlations yield the multipolarity of any transition, they do not give information on the parity changes involved. However, it is possible also to measure a correlation between the polarisations of the two photons and complete the picture.

With these remarks we leave the subject of electromagnetic de-excitation of nuclei and go on to the subject which motivated our treatment of radiation through the Fermi Golden Rule—the theory of β-decay.

3.10. Nuclear decay through the weak interactions—β-decay. Introduction

In Chapter 1 we noted three modes of radioactive decay: α-, β- and γ-emission. The energy spectra of emitted radiation are discreet for α- and γ-decay, but the spectrum of electrons (either positive or negative) emitted in β-decay is continuous, corresponding to a three-body final state. A nucleus making a transition to a lower energy state via β-decay emits both an electron and a neutrino. The energy available in the transition is shared between the electron and the neutrino, the electron getting all when the neutrino gets none. The nucleus in the final state recoils away from the emitted *leptons*† but being very massive in comparison carries a negligible amount of energy.

The simplest example of nuclear β-decay is the decay of the neutron in free space:

$$n \to p + e^- + \bar{\nu}$$

(see Fig. 1.1.2).

The neutrino was first postulated to explain the continuous energy distribution of emitted electrons in β-decay without discarding conservation of energy: with the determination of the spins of both initial and final nuclei it became clear that it is also necessary in order to conserve angular momentum. In the decay $n \to p + e^- + \bar{\nu}$ the neutron, electron and proton all have spin $\frac{1}{2}$. If angular momentum is to be conserved, a particle of half integral spin must accompany the proton and electron. The neutrino must have a very small rest mass: the mass difference between the neutron and proton is 1.29 MeV/c², of which 0.51 MeV/c² is needed for the electron rest mass, leaving only 0.78 MeV/c²: the neutrino mass must be less than this. The decay $^3\text{H} \to {}^3\text{He} + e^- + \bar{\nu}$ in which a bound proton decays into a bound neutron, sets a much tighter limit. The energy left over after creating the electron is only

---

† Leptons are fermions (particles with half integral spin) which do not couple via the strong interactions: the only known leptons are the electron, the muon, their antiparticles and associated neutrinos.

Nuclear Physics

18 keV: the neutrino mass must be less then 18 keV/c²; substantially less since the electron gets some energy. A study of the spectrum of electrons in the β-decay of tritium (³H) leads to the conclusion that the mass of the neutrino is less than 60 eV/c².

Neutrino interactions were first observed by Reines and Cowan[†] in 1953, using neutrinos from the β-decay of fission fragments produced in the operation of a nuclear reactor. In the last 10 years a number of experiments on the interaction of high-energy neutrinos have been made at proton synchrotron laboratories, using neutrinos produced in the decay of secondary beams of π-mesons, and an attempt has been made to detect neutrinos produced in the nuclear reactions which power the sun.[‡] At the moment, however, we are concerned with nuclear β decay; the processes

$$n \to p + e^- + \bar{\nu},$$
$$p \to n + e^+ + \nu$$

(The antineutrino, $\bar{\nu}$, is different from the neutrino. In β-decay electrons are accompanied by an antineutrino and positrons by a neutrino—we have a law of conservation of leptons.) The first can occur for free neutrons, and for neutrons bound in a nucleus if the final state has a lower energy than the initial state. The second cannot occur for free protons (because the neutron is more massive than the proton) but can occur for bound protons if the energy is available. The point is illustrated neatly by the decay

$$^3H \to {}^3He + e^- + \bar{\nu}$$

in which a bound neutron decays into a bound proton. The energy released in this decay is so small because the coulomb energy of ³He is larger than the coulomb energy of ³H. If the difference exceeded the mass difference between the proton and neutron, the decay would go the other way, as in

$$^{15}O \to {}^{15}N + e^+ + \nu$$

in which the bound proton decays into a bound neutron.

An alternative to positron emission is electron capture, in which $p + e^- \to n + \nu$ (note conservation of leptons) where both nucleons are bound in a nucleus (note that this reaction cannot occur for atomic or molecular hydrogen, there is not enough energy available).

Before discussing the theory of these processes, we shall first consider briefly the energetics of β-decay, and we shall assume the rest mass of the neutrino to be zero.

Except for decay of the free neutron, we practically never observe decay of the nucleus unaccompanied by its atomic electrons. Therefore we must rather consider a transition from an initial *atom* to a final *atom* in which the nuclear charge has changed by one unit. In most cases the binding energy of the electrons may be neglected: electron capture from an atomic state is an important exception. For electron emission the charge of the nucleus increases by one: the mass of the final state is thus the mass of the final atom minus one electron, plus the mass of the emitted electron (the nucleus in general recoils so slowly that it takes its electrons along with it). Thus (if the neutrino has zero rest mass) the kinetic energy released in the decay is just

$$Q^- = (M_Z - M_{Z+1})c^2 \tag{3.10.1}$$

[†] F. Reines and C. L. Cowan, *Phys. Rev.* **92**, 830 (1953).
[‡] R. Davis, D. S. Harmer and K. C. Hoffman, *Phys. Rev. Lett.* **20**, 1205 (1968).

Nuclear Decay

(M_Z, M_{Z+1} are atomic masses) and is shared between the electron and antineutrino. This expression includes both the energy released in the nuclear decay and a small contribution due to the increase in the binding of the electrons as Z increases. Strictly we should subtract the binding energy of the outermost electron of the nucleus with charge $Z+1$ (assuming that the final atom is in its lowest singly ionized state) but this is usually ignored.

For positron emission a nucleus with charge Z goes to a nucleus with charge $Z-1$; leaving over one extra electron and with the creation of a positron. The sum of the masses in the final state are thus

$$M_{Z-1} + 2m_e$$

and the kinetic energy released is

$$Q^+ = (M_Z - M_{Z-1} - 2m_e)c^2 \tag{3.10.2}$$

(M_z, M_{z-1} are atomic masses) and is shared between the positron and neutrino. This energy is mostly provided by the nuclear mass difference, with a small subtraction to account for the decrease in binding energy of the electrons.

Finally, if as an alternative to positron emission an atomic electron is captured (usually an s-wave electron in the K electron shell) then we start with a neutral atom of mass M_Z and end up with a final state containing an atom which has mass M_{Z-1} in its ground state, but which in fact has lost an electron from the K-shell. The final state atom thus has a mass M_{Z-1}, minus the mass of one electron plus the contribution of the binding energy of a K-shell electron $B_K(Z-1)$. There is also a free electron left over, so that

$$Q^K = (M_Z - M_{Z-1})c^2 - B_K(Z-1). \tag{3.10.3}$$

The excited atom decays by emission of X-rays or Auger electrons, but too late to change the kinetic energy given to the neutrino.

There is, of course, a finite probability of the final atom being left in any state of excitation allowed by conservation of energy, but these are the important cases.

A particular β-decay process is energetically allowed if the appropriate Q value is positive. Of course, if such a condition is satisfied, this does not prevent the rate of decay, which depends on the detailed dynamics, being immeasurably long.

It is worth noting that if $Q^+ < 0$ and $Q^K > 0$ only electron capture is allowed. The condition is thus

$$B_K(Z-1) < (M_Z - M_{Z-1})c^2 < 2m_e c^2. \tag{3.10.4}$$

This condition is relaxed even further if capture from less tightly bound states is considered. Thus if $(M_Z - M_{Z-1}) \gtrsim 0$ electron capture is allowed, and if $(M_Z - M_{Z-1}) > 2m_e c^2$ both electron capture and positron emission are allowed. The decay $^7\text{Be} + e^- \to {}^7\text{Li} + \nu$ is a beautiful illustration of energetics—only electron capture is energetically possible.

Remembering the discussion of Section 1.5, only one atom of given atomic number A odd is stable against decay processes, and no more than three of given atomic number A even—but of course some may live for so long that their decay cannot be detected.

We may now set up the theory of β-decay, *our third example of the application of the Fermi Golden Rule.*

Nuclear Physics

3.11. The theory of β-decay

In the first part of this chapter we developed the theory of electromagnetic decay of nuclei using time dependent perturbation theory and the Fermi Golden Rule. We shall establish the theory of β-decay in an analogous way, but there are important differences in procedure. In discussing electromagnetic decay we had the advantage of starting from a thoroughly understood classical theory which was also relativistically invariant, while the weak interactions do not manifest themselves on a large scale at all (at least under terrestrial conditions). It is therefore necessary to guess at the form of the interaction responsible for coupling an initial nucleon to a final nucleon plus a lepton pair. In addition, while for many nuclear β-decay processes the energy is sufficiently low that the electron may be treated non-relativistically, the neutrino, having a mass <60 eV, may not be. However, we shall proceed under the assumption that all wave functions are solutions of non-relativistic equations and indicate modifications introduced by a theory which treats the leptons as relativistic as they arise.

We suppose the initial nucleus to be represented by a wave function Ψ_i and the final nucleus by Ψ_f. The leptons appear only in the final state in a β-decay process and we give them wave functions ϕ_e and $\phi_{\bar{\nu}}$, which we take to be plane waves. The matrix element connecting the initial and final states must do several things. If emission of an electron and an antineutrino is considered, then in the matrix element a neutron must be annihilated and a proton created. The nuclear part of the matrix element must also reorganize the nuclear structure to accommodate the change of a neutron into a proton. The lepton part of the matrix element must create an electron and an antineutrino and orient their spins in such a way that angular momentum is conserved in conjuction with the nuclear part. Finally, the matrix element must contain details of the energy dependence (if any) of the couplings—for example, the coupling might depend on the momentum of the leptons and if so this information will have to appear through ∇ operators in the matrix element.

In the very elementary theory of β-decay, we suppose that we can write

$$\frac{1}{\tau} = \frac{2\pi}{\hbar} |M|^2 \varrho$$

where ϱ is the available phase space for a three-particle final state and

$$M = g \int \phi_e^* \phi_{\bar{\nu}}^* \Psi_f^* \Psi_i \, dV \qquad (3.11.1)$$

where g is a constant measuring the strength of the weak interaction coupling. Once more it is worth paying attention to the form of this matrix element. A general state of a nucleus and of two leptons is

$$\phi_e(\mathbf{r}_e) \, \phi_{\bar{\nu}}(\mathbf{r}_{\bar{\nu}}) \, \Psi_f(\mathbf{r}) \qquad (3.11.2)$$

and if we define the weak interaction operator in a matrix element connecting this final state with the initial state $\Psi_i(\mathbf{r})$ then the integration must be performed over all the variables:

$$\int \phi_e^*(\mathbf{r}_e) \, \phi_{\bar{\nu}}^*(\mathbf{r}_{\bar{\nu}}) \, \Psi_f^*(\mathbf{r}) \, O \, \Psi_i(\mathbf{r}) \, d^3r \, d^3r_e \, d^3r_{\bar{\nu}}. \qquad (3.11.3)$$

We assume the interaction is local and this is expressed by writing

$$O = g\, \delta(\mathbf{r}_e - \mathbf{r})\, \delta(\mathbf{r}_{\bar{\nu}} - \mathbf{r})$$

thus suppressing the variables $\mathbf{r}_e$ and $\mathbf{r}_{\bar{\nu}}$. Again, in writing this matrix element we implicitly assume a sum over all nucleons.

Ψ_f and Ψ_i are localized wave functions normalized to unity

$$\int \Psi_f^* \Psi_f\, dV = 1; \quad \int \Psi_i^* \Psi_i\, dV = 1.$$

ϕ_e and $\phi_{\bar{\nu}}$ are plane wave functions normalized to the volume V of our imaginary box

$$\phi_e = \frac{e^{i\mathbf{p}\cdot\mathbf{r}/\hbar}}{\sqrt{V}} \quad \phi_{\bar{\nu}} = \frac{e^{i\mathbf{q}\cdot\mathbf{r}/\hbar}}{\sqrt{V}} \tag{3.11.4}$$

where $\mathbf{p}$ is the electron momentum and $\mathbf{q}$ the momentum of the antineutrino.

If $\mathbf{p}$ and $\mathbf{q}$ are both fixed, the recoil momentum of the nucleus is also fixed. The three-body phase space thus involves two factors, one for the variation in $\mathbf{p}$, the other for the variation in $\mathbf{q}$, the two independent momenta.

If $\mathbf{q}$ is fixed, the number of states between p and $p + dp$ is, using the analysis of Section 3.3,

$$\frac{p^2\, dp\, d\Omega_e V}{(2\pi\hbar)^3}$$

where $d\Omega_e$ is an element of solid angle into which the electron is emitted. Then the number of states between q and $q + dq$ is

$$\frac{q^2\, dq\, d\Omega_{\bar{\nu}} V}{(2\pi\hbar)^3}$$

and so the total number of states about values of $\mathbf{p}$ and $\mathbf{q}$ is the product:

$$\frac{p^2\, dp\, q^2\, dq\, d\Omega_e\, d\Omega_{\bar{\nu}}\, V^2}{(2\pi\hbar)^6}. \tag{3.11.5}$$

The transition rate w thus becomes

$$w = \frac{2\pi}{\hbar} \left| g \int \phi_e^* \phi_{\bar{\nu}}^* \Psi_f^* \Psi_i\, dV \right|^2 \frac{p^2\, dp\, q^2\, dq\, d\Omega_e\, d\Omega_{\bar{\nu}} V^2}{(2\pi\hbar)^6\, dE} \tag{3.11.6}$$

where E is the energy of neither the electron nor the neutrino but of the transition. As the normalization volume for the leptons is allowed to go to infinity (the leptons go out but do not return) this becomes

$$w = \frac{2\pi}{\hbar} g^2 \left| \int e^{-i(\mathbf{p}+\mathbf{q})\cdot\mathbf{r}/\hbar} \Psi_f^*(\mathbf{r}) \Psi_i(\mathbf{r})\, dV \right|^2 \frac{p^2\, dp\, q^2\, dq\, d\Omega_e\, d\Omega_{\bar{\nu}}}{(2\pi\hbar)^6\, dE} \tag{3.11.7}$$

A little attention to dimensions is worth while here. The phase space should be a number divided by an energy. pc has dimensions of energy, so $p^2 dp\, q^2 dq$ has dimensions erg^6 cm^{-6}

sec⁶. The dimensions of $\hbar$ are erg sec and so the phase space, including the normalization volume, has indeed dimensions erg⁻¹. Then if the matrix element, with the lepton wave functions normalized properly, has dimensions of ergs, as it should have since it represents an interaction energy, eq. (3.11.7) is dimensionally correct. This requires that g have dimensions erg cm³.

We now expand the lepton wave functions, writing

$$e^{-i(\mathbf{p}+\mathbf{q})\cdot\mathbf{r}/\hbar} = 1 - \frac{i(\mathbf{p}+\mathbf{q})\cdot\mathbf{r}}{\hbar} \cdots \qquad (3.11.8)$$

and suppose $(\mathbf{p}+\mathbf{q})\cdot\mathbf{r}/\hbar$ to be small over the nuclear dimensions in comparison with 1. This is our usual approximation of neglecting p-waves near the origin in comparison with s-waves. In this approximation, the probability per unit time of the nucleus decaying to yield an electron with momentum between p and $p+dp$ and a neutrino with momentum between q and $q+dq$ is given by

$$\begin{aligned} w(p,q) &= \frac{2\pi}{\hbar} g^2 \left| \int \Psi_f^* \Psi_i \, dV \right|^2 (4\pi)^2 \frac{p^2 \, dp \, q^2 \, dq}{(2\pi\hbar)^6 \, dE} \\ &= g^2 |M_{if}|^2 \frac{p^2 \, dp \, q^2 \, dq}{2\pi^3 \hbar^7} \end{aligned} \qquad (3.11.9)$$

where $M_{if} = \int \Psi_f^* \Psi_i \, dv$ is the nuclear matrix element.

In general we do not observe the neutrino and the electron energy spectrum is obtained by integrating over all allowed q for a given p. This is constrained by conservation of energy so that the effect of such an integration is to remove the factor dq/dE at fixed momentum p. If the maximum kinetic energy of the electron is E_0 and at a particular point in the spectrum the electron has momentum p and energy E_e, then the neutrino gets the difference (the nucleus, being relatively massive, can absorb the recoil momentum $(\mathbf{p}+\mathbf{q})$ but the energy it carries, $\sim (\mathbf{p}+\mathbf{q})^2/2M_{\text{nucleus}}$ is negligible). Thus we have

$$E_\nu = E_0 - E_e \qquad (3.11.10)$$

and if the neutrino has zero rest mass then

$$qc = E_0 - E_e \qquad (3.11.11)$$

at a fixed electron momentum.

E_0 is just the energy available in the transition (apart from constant terms like the electron mass) so that

$$\frac{dE_0}{dE} = 1$$

and at fixed electron momentum

$$\left.\frac{dq}{dE}\right|_{E_e} = \frac{1}{c}$$

so that

$$w(p) = g^2 |M_{if}|^2 \frac{(E_0 - E_e)^2 p^2 \, dp}{2\pi^3 \hbar^7 c^3} \qquad (3.11.12)$$

and this gives the electron spectrum under the assumption that the spectrum is determined entirely by phase space. This is the spectrum found in *allowed transitions*, namely those for which the leptons carry zero orbital angular momentum.

If the neutrino had a discernible rest mass, then

$$\sqrt{q^2c^2 + m_\nu^2 c^4} = E_0 - E_e$$

when

$$\frac{qc^2}{E_\nu}\frac{dq}{dE}\bigg|_{E_e} = 1, \quad \frac{dq}{dE}\bigg|_{E_e} = \frac{E_\nu}{qc^2}.$$

The function $w(p)$ would then be given by

$$w(p) = g^2 |M_{if}|^2 \left\{ \frac{(E_0 - E_e)\{(E_0 - E_e)^2 - m_\nu^2 c^4\}^{1/2} p^2 \, dp}{2\pi^3 \hbar^7 c^3} \right\}. \qquad (3.11.13)$$

The limit[†] $m_\nu c^2 < 60$ eV comes from the fitting of a theoretical spectrum of this kind to the experimental spectrum of $^3\text{H} \to {}^3\text{He} + e^- + \bar{\nu}$; the effect of departures from the theoretical spectrum of (3.11.11) is most easily seen by making a linear plot, the Kurie plot, in which the quantity

$$\sqrt{\frac{w(p)}{p^2 F(Z, p)}}$$

is plotted against the electron energy. $F(Z, p)$ is a correction factor which has to be introduced to take account of the distortion of the assumed plane waves by the coulomb field of the nucleus. Qualitatively, an electron wave function will be concentrated near the nucleus by the effect of the attractive coulomb field, and so the transition rate will be greater for a given p than the rate given by (3.11.11); the rate will be increased to the greatest extent for low p. Similarly, a positron wave function will be repelled from the nucleus, depressing the decay rate. The problem is no different in principle from the problem of scattering by a coulomb field and the function $F(Z, p)$ can be computed numerically. An example of a Kurie plot for ^{3}H decay is shown in Fig. 3.11.1.

Just as in electromagnetic decay, the allowed transitions are those for which the first term in the expansion

$$e^{i\mathbf{k}\cdot\mathbf{r}} = 1 + i\mathbf{k}\cdot\mathbf{r}\ldots$$

may be taken. Since no orbital angular momentum is carried off, the answer to the question "Does the nuclear parity change?" is NO. If, however, the nuclear matrix element $\int \Psi_f^* \Psi_i \, dV$ is equal to zero (because of parity or angular momentum selection rules) but the nuclear matrix element $\int \Psi_f^* \mathbf{r} \Psi_i \, dV$ is not equal to zero, then we have a *first forbidden* transition, corresponding to one unit of orbital angular momentum being carried off by the leptons. The matrix element is

$$\frac{g(\mathbf{p} + \mathbf{q})}{\hbar} \cdot \int \Psi_f^* \mathbf{r} \Psi_i \, dV \qquad (3.11.14)$$

and the decay rate for first forbidden transitions is down on that for allowed transitions

[†] K. E. Bergkvist, in *Proc. Top. Conf. on Weak Interactions*, CERN 69-7 (1969).

Nuclear Physics

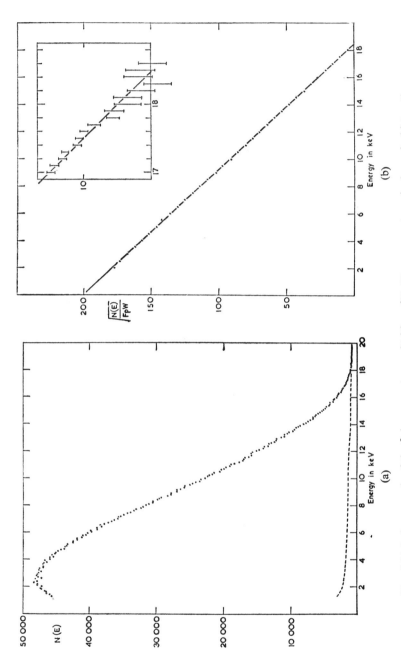

FIG. 3.11.1. An example of the β-decay spectrum of tritium, ³H. The raw spectrum is shown in (a) together with a background correction, and was obtained by implanting tritium ions in the depletion layer of a lithium-drifted silicon detector. The Kurie plot corresponding to this spectrum is shown in (b) with a fitted end point of 18.54 keV. [From V. E. Lewis, *Nucl. Phys.* A, **151**, 120 (1970).]

Nuclear Decay

by $\sim(kR)^2$ where $\mathbf{k} = (\mathbf{p}+\mathbf{q})/\hbar$ and R is the nuclear radius—this number is typically $\sim 10^{-2}$ for nuclear β-decay.

In such a first forbidden transition, there are other differences from the characteristics of an allowed transition. A term in $(\mathbf{p}+\mathbf{q})^2$ introduces an angular correlation between the directions of the electron and neutrino and also changes the spectrum. The decay rate becomes

$$g^2 \left| \frac{(\mathbf{p}+\mathbf{q})}{\hbar} \cdot \int \Psi_f^* \mathbf{r} \Psi_i \, dV \right|^2 \frac{(E-E_0)^2 \, p^2 \, dp}{2\pi^3 \hbar^7 c^3} \frac{d\Omega_e}{4\pi} \frac{d\Omega_{\bar{\nu}}}{4\pi}. \quad (3.11.15)$$

We may replace the solid angle part of the phase space by the angular interval of the vector $\mathbf{p}+\mathbf{q}$ and the element of solid angle of $\mathbf{p}$ measured relative to $\mathbf{q}$. Then if we are not observing the direction of nuclear spin change the decay rate becomes

$$g^2 \frac{(\mathbf{p}+\mathbf{q})^2}{\hbar^2} \left| \int \Psi_f^* \mathbf{r} \Psi_i \, dV \right|^2 \frac{1}{3} \frac{(E_0-E_e)^2 \, p^2 \, dp}{2\pi^3 \hbar^7 c^3} \frac{d\Omega}{4\pi} \quad (3.11.16)$$

where Ω is the solid angle of the electron. Now $(\mathbf{p}+\mathbf{q})^2 = p^2 + q^2 + 2\mathbf{p}\cdot\mathbf{q}$ which introduces the angular correlation between the electron and neutrino. Integration over $d\Omega$ averages the term in $\mathbf{p}\cdot\mathbf{q}$ to zero, leaving only

$$p^2 + \frac{(E_0-E_e)^2}{c^2} = \frac{1}{c^2}\{p^2 c^2 + (E_0-E_e)^2\}.$$

Then for a first forbidden transition

$$w(p) = g^2 \left| \int \Psi_f^* \mathbf{r} \Psi_i \, dV \right|^2 \frac{1}{3} \frac{\{p^2 c^2 + (E_0-E_e)^2\}(E_0-E_e)^2 \, p^2 \, dp}{2\pi^3 \hbar^9 c^5}. \quad (3.11.17)$$

The extra term in the electron spectrum introduces a wiggle in the Kurie plot, an example being shown in Fig. 3.11.2.

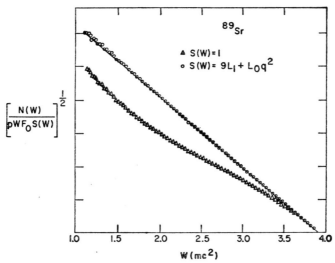

FIG. 3.11.2. An example of first forbidden β-decay. The lower curve shows a Kurie plot of the energy spectrum from the decay of ^{89}Sr and is markedly non-linear. Application of a theoretical correction for a first forbidden decay produced the linear plot shown in the upper curve. [From F. K. Wohn and W. L. Talbert, Jr., *Nucl. Phys.* A **146**, 33 (1970).]

Nuclear Physics

Second forbidden transitions correspond to the third term in the expansion of $e^{i\mathbf{k}\cdot\mathbf{r}}$, with two units of angular momentum being carried off, and so on. We shall not discuss this in any more detail, since our treatment is non-relativistic, and while this is frequently justified for the electron, it is never justified for the neutrino.[†] (Many of the possible forms of the matrix element for *allowed* β-decay would induce significant departures from the simple spectrum of equation (3.11.12): if such forms were dominant the simple non-relativistic treatment we have given would become pointless.)

A parameter frequently used in the classification of β-decay processes is the *Konopinski ft value*. If we write for any β-decay process

$$\frac{1}{\tau} = \int_0^{p_{\max}} g^2 \, |\mathcal{M}_{if}|^2 \, \frac{(E_0 - E_e)^2 \, p^2}{2\pi^3 \hbar^7 c^3} \, F(Z, p) \, dp \qquad (3.11.18)$$

where $\mathcal{M}_{if}$ is the complete matrix element (not necessarily the nuclear part only) then the half-life t is given by

$$t = \tau \ln 2 = \frac{\ln 2}{\displaystyle\int_0^{p_{\max}} g^2 \, |\mathcal{M}_{if}|^2 \, \frac{(E_0 - E_e)^2 \, p^2 F(Z, p) \, dp}{2\pi^3 \hbar^7 c^3}}$$

and the Konopinski *ft* value is defined as

$$ft = t \int_0^{p_{\max}} \frac{(E_0 - E_e)^2 \, p^2 F(Z, p) \, dp}{2\pi^3 \hbar^7 c^3}. \qquad (3.11.19)$$

Thus if the nuclear matrix element was the same for all allowed transitions, all allowed transitions would have the same *ft* value. Variations in the *ft* value for allowed transitions reflect variations in the magnitude of the nuclear matrix element. The matrix element $\mathcal{M}_{if}$ for first forbidden transitions involves the integration of

$$\frac{(\mathbf{p}+\mathbf{q})\cdot\mathbf{r}}{\hbar}$$

over the nucleus and thus suppresses the decay rate, raising the *ft* value. Roughly speaking, β-decays may be classified by:

	$\ln_{10} ft$
Allowed transitions	3–6
First forbidden	6–10
Second and higher forbidden	> 10

If the nuclear part of the matrix element is close to unity, a transition is called *favoured*; if it is small the transition is *unfavoured*. The fastest allowed decays, in which $M_{if} \sim 1$

[†] A detailed account of the full treatment may be found in E. J. Konopinski, *The Theory of β-radioactivity* (Oxford, 1966).

Nuclear Decay

are called *super allowed*: the obvious example is the decay of the neutron in which, neglecting nucleon recoil, the wave function of the proton is the same as that of the initial neutron. If the decaying neutron is bound in tritium, the decay leads to a proton bound in ^{3}He which has the same wave function as the neutron, apart from small differences induced by the coulomb interaction. In general, β-decay transitions between *mirror nuclei* are super allowed. A pair of mirror nuclei are two nuclei in which the number of protons in the first is equal to the number of neutrons in the second, and vice versa. If a β-transition can take place between a pair of mirror nuclei, then they consist of an even number of neutrons, the same number of protons and a single unpaired nucleon which makes the transition. Because of the charge symmetry of nuclear forces, the wave function of the unpaired nucleon is the same regardless of its charge, apart from coulomb corrections. Such a pair of nuclei forms an *isotopic doublet*. The nuclear matrix element for super allowed transitions between members of an isotopic doublet is only negligibly different from unity. Transitions between neighbouring members of an *isotopic triplet* are also super allowed. Consider an even–even core containing an equal number of protons and neutrons, and to this core add not one nucleon but two. Two neutrons or two protons will pair off; a neutron and a proton may occur in a relative singlet or a triplet state. The triplet state is the third member of the isotopic triplet; the two-nucleon configuration that exists only for a neutron and a proton is an isotopic singlet. The wave functions of the three members of the triplet are the same, again apart from coulomb corrections. An important example is the ^{14}C, ^{14}N, ^{14}O triplet: In the

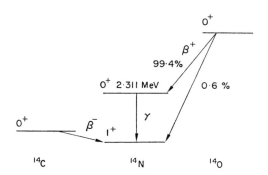

FIG. 3.11.3. The ^{14}C, ^{14}N, ^{14}O system. The three 0^+ states form an isotopic triplet and the transition between ^{14}O and the 0^+ excited state of ^{14}N is superallowed. The half-life of ^{14}O is 71 sec. In contrast, the allowed ^{14}C → ^{14}N transition is highly unfavoured, having a half-life of 5730 years (and is extensively used for dating organic remnants in archaeology).

decay ^{14}O → ^{14}N* + e^+ + ν the wave functions do not change. The square of the nuclear matrix element is 2, not 1, since either additional proton has an equal probability of decay. (Only the least tightly bound nucleons in a nucleus can make the β transition, the Pauli exclusion principle prevents low-lying nucleons from changing their nature.) The ft value for ^{14}O is 3075 sec and the coupling constant for this decay is 1.41×10^{-49} erg cm^3.

So far we have completely neglected the fact that the fundamental β-decay process involves four particles, each with spin $\frac{1}{2}$. If we restrict ourselves to allowed transitions, in which no orbital angular momentum is carried off by the leptons (parity selection rule NO) then if the electron and neutrino are created in a singlet state, no angular momentum of any kind is

Nuclear Physics

carried off, while if the electron and neutrino are formed in a triplet state, then one unit of angular momentum is carried off. Thus we should write the matrix element in such a way as to distinguish between these two cases; and if the leptons are produced in the triplet state, then the nuclear matrix element must contain information on the reorganization of the spin orientations of nucleons in the nucleus.

Transitions in which the leptons (in the non-relativistic limit) are emitted in singlet states are called Fermi Transitions, while if the leptons are emitted in a triplet spin state the transitions are Gamow–Teller. Then for allowed transitions the selection rules are

$$\text{Fermi:} \quad \Delta J = 0; \quad \text{NO}$$

$$\text{Gamow–Teller:} \quad \Delta J = \pm 1, 0 \quad \text{except} \quad 0 \to 0, \text{NO}$$

Thus the decay $^{14}\text{O} \to {}^{14}\text{N}^* + e^+ + \nu$, a transition between two nuclei with spin-parity 0^+, is an Allowed Fermi transition. The decay $^6\text{He} \to {}^6\text{Li} + e^- + \bar{\nu}$ is a transition in which the 0^+ state of ^6He goes to the 1^+ state of ^6Li, and so is an allowed Gamow–Teller transition. The decay of the neutron, $n \to p + e^- + \bar{\nu}$ is $\frac{1}{2}^+ \to \frac{1}{2}^+$ and so may be a mixture of the two.

The triplet state of an electron and neutrino with $m = +1$ is just $\psi_e(+)\psi_{\bar{\nu}}(+)$, with $m = -1$, $\psi_e(-)\psi_{\bar{\nu}}(-)$.

These two are symmetric under interchange of the spins of the leptons, the third component has the same symmetry and so the triplet state with $m = 0$ is

$$\frac{1}{\sqrt{2}} \{\psi_e(+)\psi_{\bar{\nu}}(-) + \psi_e(-)\psi_{\bar{\nu}}(+)\}.$$

The singlet state has $m = 0$ and under summation over spins must be orthogonal to the triplet state with $m = 0$. It is therefore the antisymmetric combination

$$\frac{1}{\sqrt{2}} \{\psi_e(+)\psi_{\bar{\nu}}(-) - \psi_e(-)\psi_{\bar{\nu}}(+)\}.$$

If we write

$$\phi_e = \begin{pmatrix} \psi_e(+) \\ \psi_e(-) \end{pmatrix} e^{i(\mathbf{p}\cdot\mathbf{r})/\hbar} \quad \text{and} \quad \phi_{\bar{\nu}} = \begin{pmatrix} \psi_{\bar{\nu}}(+) \\ \psi_{\bar{\nu}}(-) \end{pmatrix} e^{i(\mathbf{q}\cdot\mathbf{r})/\hbar}$$

then a matrix operator B between the electron and neutrino wave functions generates the singlet state; if

$$B = \frac{1}{\sqrt{2}} \begin{pmatrix} 0 & 1 \\ -1 & 0 \end{pmatrix}$$

then

$$(\phi_e^* B \phi_{\bar{\nu}}^*)$$

is the lepton part of the Fermi matrix element.[†] Since the nucleon spins are not rearranged at all, only the unit matrix will appear between the nuclear wave functions. The matrix element is then

$$\int (\phi_e^* B \phi_{\bar{\nu}}^*)(\Psi_f^* \Psi_i) \, dV \tag{3.11.20}$$

[†] The non-relativistic theory of β-decay is discussed in detail in J. M. Blatt and V. Weisskopf, *Theoretical Nuclear Physics*, chap. XIII (Wiley, 1952).

for Fermi transitions. (You should note that the wave functions Ψ_f and Ψ_i belonging to different nuclei, are orthogonal. We should therefore include an operator to change the nuclear charge, but we have suppressed this.)

The Gamow–Teller matrix element may be constructed with the aid of the Pauli spin matrices:

$$\sigma_x = \begin{pmatrix} 0 & 1 \\ 1 & 0 \end{pmatrix} \quad \sigma_y = \begin{pmatrix} 0 & -i \\ i & 0 \end{pmatrix} \quad \sigma_z = \begin{pmatrix} 1 & 0 \\ 0 & -1 \end{pmatrix}.$$

The combinations

$$\sigma^+ = \frac{1}{\sqrt{2}}(\sigma_x + i\sigma_y) \quad \text{and} \quad \sigma^- = \frac{1}{\sqrt{2}}(\sigma_x - i\sigma_y)$$

are the spin flip matrices and the $m = 1$ triplet state is

$$\phi_e^* \sigma^+ B \phi_{\bar{\nu}}^*$$

accompanied by the spin flip nuclear term $\Psi_f^* \sigma^+ \Psi_i$. Thus the Gamow–Teller matrix element may be written as

$$\int (\phi_e^* \boldsymbol{\sigma} B \phi_{\bar{\nu}}^*) \cdot (\Psi_f^* \boldsymbol{\sigma} \Psi_i) \, dV \tag{3.11.21}$$

the scalar product of two axial vectors. If the spatial part of the wave function is the same, on squaring and summing over all spins for allowed transitions we find that the square of the Gamow–Teller matrix element for a single nucleon is 3 rather than 1, corresponding to the three possible orientations in space of the lepton spins. Similarly, if two nucleons have an equal probability of decay, as in the decay $^6\text{He} \rightarrow {}^6\text{Li} + e^- + \bar{\nu}$, the square of the Gamow–Teller matrix element is 6 rather than 2 as for the Fermi transition $^{14}\text{O} \rightarrow {}^{14}\text{N} + e^+ + \nu$. Then if the coupling constants for the two types of transition are the same (and they are very nearly) neutron decay will be one-quarter Fermi and three-quarters Gamow–Teller. In general, though, the mixture, in those decays where the selection rules allow both Fermi and Gamow–Teller transitions, will be determined by the details of the nuclear structure. It is worth noting that pure Gamow–Teller transitions are unlikely to be superallowed because of the coupling between the spin and spatial parts of the nucleon wave functions.

The unaesthetic lack of symmetry between the lepton and nuclear brackets in the two matrix elements may be removed by the following considerations. Suppose we have a decay in which an electron and an antineutrino are emitted. Now emission of an antineutrino with positive energy and momentum is formally the same as absorption of a neutrino from the initial state with negative energy and negative momentum (and of course reversed spin). This is hardly surprising, in relativistic field theory the negative energy states of a particle are interpreted as being the antiparticle states. If we take the antineutrino wave function

$$\phi_{\bar{\nu}} = \begin{pmatrix} \psi(+) \\ \psi(-) \end{pmatrix} e^{i(\mathbf{q} \cdot \mathbf{r})/\hbar}$$

and reverse momentum and spin we get a state

$$\begin{pmatrix} \psi(-) \\ \psi(+) \end{pmatrix} e^{-i(\mathbf{q} \cdot \mathbf{r})/\hbar}$$

Nuclear Physics

which we interpret as the state of a negative energy neutrino. We now no longer need the operator B to invert the functions $\psi(+)$ and $\psi(-)$, but a relative sign is still wrong. We have at our disposal some phases which in the non-relativistic theory are arbitrary. If we write the neutrino spin wave function as

$$\begin{pmatrix} \psi_1 \\ \psi_2 \end{pmatrix}$$

and the antineutrino wave function as

$$\begin{pmatrix} \psi_1 \\ -\psi_2 \end{pmatrix}$$

then we can write the Fermi lepton bracket as $(\phi_e^* \phi_\nu)$ and the Gamow–Teller lepton brackets as $(\phi_e^* \boldsymbol{\sigma} \phi_\nu)$ provided that we take negative energy and momentum for the neutrino (if an electron and antineutrino are emitted) or negative energy and momentum for the electron (if a positron and neutrino are emitted). The full relativistic theory of fermions relates explicitly the wave functions of particle and antiparticle and justifies our writing $B\phi_{\bar{\nu}}^* \rightarrow \phi_\nu$. In the non-relativistic theory, of course, there is no advantage in writing the matrix elements in this way: we make the point because in any modern treatment of β-decay the matrix elements are written with this symmetry between the two brackets and this can prove confusing if you have previously only encountered the elementary non-relativistic treatment. The relationship is perhaps clarified by comparing the rival processes of positron emission and electron capture:

$$N(Z, A) \rightarrow N(Z-1, A) + e^+ + \nu,$$
$$N(Z, A) + e^- \rightarrow N(Z-1, A) + \nu.$$

In the first process the nucleus emits a positron (with positive energy and momentum) and a neutrino (with positive energy and momentum). We would interpret this as absorption of an electron with negative energy and momentum, and emission of a neutrino with positive energy and momentum. In the second process an electron really is absorbed (although it has positive energy) and a neutrino emitted. The matrix elements are written in precisely the same way. The diagram for electron capture is

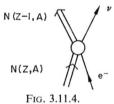

Fig. 3.11.4.

and the diagram for positron emission is reached by reversing the electron line and going to the antiparticle state

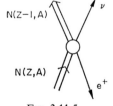

Fig. 3.11.5.

Nuclear Decay

If the weak interactions are in fact mediated by exchange of an intermediate boson, as electromagnetic interactions are mediated by exchange of a photon, then the point is made even more clearly by the diagrams:[†]

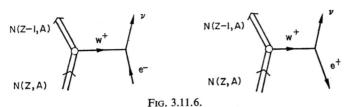

FIG. 3.11.6.

compare with

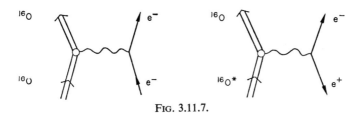

FIG. 3.11.7.

For electron capture we have a simple non-relativistic matrix element:

$$\int \phi_\nu^* \phi_e \Psi_f^* \Psi_i \, dV \tag{3.11.22}$$

and two-body phase space. If the electron is in an s state then

$$\phi_e = \frac{1}{\pi^{1/2}} \left(\frac{Z}{a}\right)^{3/2} e^{-Zr/a} \tag{3.11.23}$$

where a is the Bohr radius. Evaluating both wave functions at the origin yields

$$\frac{1}{\tau_K} = \frac{2\pi}{\hbar} \frac{1}{\pi} \left(\frac{Z}{a}\right)^3 g^2 |M_{if}|^2 \varrho \tag{3.11.24}$$

where

$$\varrho = \frac{4\pi q^2 \, dq}{(2\pi\hbar)^3 \, dE} = \frac{4\pi q^2}{(2\pi\hbar)^3 c}$$

so that

$$\frac{1}{\tau_K} = \frac{2\pi}{\hbar} g^2 |M_{if}|^2 \frac{1}{\pi} \left(\frac{Z}{a}\right)^3 \frac{E_\nu^2}{2\pi^2 \hbar^3 c^3}. \tag{3.11.25}$$

The probability of K capture increases as Z^3 while the probability of positron emission decreases as Z increases because the final state positron wave function is pushed away from the nucleus. For values of $A \gtrsim 200$ electron capture is overwhelmingly predominant.

[†] If the weak interactions are mediated by an intermediate boson, then the four-fermion interaction is no longer local, but is replaced by local interactions of the boson exchanged with two fermion currents.

Nuclear Physics

When negative muons are stopped in matter, they are captured by an atom and rapidly reach a low-lying state, from which they may either decay spontaneously

$$\mu^- \to e^- + \bar{\nu} + \nu_\mu \dagger$$

or may be captured by the nucleus: $\mu^- + N(Z, A) \to N(Z-1, A) + \nu_\mu$. The process is the same, the capture rate greater than for electron capture because of the increased phase space factor and the reduced Bohr radius.

There is little point in pursuing a non-relativistic theory of the *weak interactions* any further. In particular, it is well known that the weak interactions violate conservation of parity, and the violation of parity appears as a purely relativistic effect. We must therefore consider relativistic effects in β-decay.

3.12. Relativistic effects in β-decay

The neutrino emitted in β-decay is always relativistic and the electron frequently is relativistic. Both particles are leptons, having spin $\frac{1}{2}$ and not interacting through the strong interactions and so are properly described by the Dirac equation.

It is no part of this book to deal in detail with the properties of the Dirac equation, but a summary of the properties is desirable in order to obtain some idea of where the relativistic effects in β-decay come from.

The Schrödinger equation for a free particle is

$$\left\{ \frac{\hbar^2}{2m} \nabla^2 + i\hbar \frac{\partial}{\partial t} \right\} \psi = 0$$

and, having no internal degree of freedom, cannot describe a particle with spin. When modified by writing ψ as a two-component column matrix, the equation becomes a pair of equations, one describing the spin up solution, and one the spin down. The two solutions are not mixed unless a spin-dependent potential is included. Whether ψ has only one component or two, the equation can only be a low velocity approximation because in a relativistic equation derivatives with respect to x, y, z and ict must enter symmetrically. The Dirac equation is first order in all derivatives and may be written‡

$$(E + c\boldsymbol{\alpha} \cdot \mathbf{p} + \beta mc^2) \psi = 0 \tag{3.12.1}$$

for a free particle. The properties of the internal variables α_x, α_y, α_z and β can be inferred from the requirement that

$$(E^2 - p^2 c^2 - m^2 c^4) \psi = 0. \tag{3.12.2}$$

The three components of α and β cannot be ordinary numbers but may be written as matrices. Multiplication by β puts the equation into a more symmetric form

$$\left(\gamma_\mu \frac{\partial}{\partial x_\mu} + \frac{mc^2}{\hbar^2} \right) \psi = 0 \tag{3.12.3}$$

where $\gamma_\mu \partial/\partial x_\mu$ is summed over all four values of the index μ.

---

† The neutrinos associated with electrons and those associated with muons are distinguishable.
‡ Excellent discussions of the Dirac equation may be found in L. I. Schiff, *Quantum Mechanics*, 3rd ed., sect. 52 (McGraw-Hill, 1968); L. Mandl, *Quantum Mechanics*, 2nd ed., chap. X (Butterworth, 1957).

The operators α, β (or the equivalent operators γ_μ) must be 4×4 square matrices (operating on a four-component column matrix)—smaller matrices do not allow the condition $E^2 - p^2c^2 - m^2c^4 = 0$ to be satisfied. They obey the relations

$$\alpha_i\alpha_j + \alpha_j\alpha_i = 2\delta_{ij},$$
$$\alpha_i\beta + \beta\alpha_i = 0,$$
(3.12.4)

or equivalently

$$\gamma_\mu\gamma_\nu + \gamma_\nu\gamma_\mu = 2\delta_{\mu\nu}.$$
(3.12.5)

Thus the Dirac equation as written here is shorthand for a set of four equations which must be solved simultaneously to yield the components of ψ. However, only one of the γ-matrices may be made diagonal and so the four components of ψ are mixed in each of the four equations, even in the absence of a potential. The condition that a solution shall exist is that the determinant of the coefficient matrix shall vanish, which turns out to be that $E^2 = p^2c^2 + m^2c^4$. Thus negative energy solutions appear symmetrically with the positive energy solutions, and the relation between wave functions for positive and negative energies is specified by the equation.

Although the equation has solutions with four components, in the low-velocity limit two always go to zero and the equation reduces to the non-relativistic Schrödinger equation with a two-component wave function. If the equation is solved in a central potential, the orbital angular momentum **L** is not conserved, but the quantity **L**+**S** is conserved, where **S** has eigenvalues $\pm\hbar/2$. In a magnetic field **B** the Hamiltonian in the non-relativistic limit contains a term $e/mc\,\mathbf{S}\cdot\mathbf{B}$. Thus the Dirac equation has all the properties required of an equation describing spin $\frac{1}{2}$ particles, and even predicts the magnetic moment associated with electron spin.

The free field solutions for positive energy may be written as

$$\begin{bmatrix} -\dfrac{cp_z}{|E|+mc^2} \\ -\dfrac{c(p_x+ip_y)}{|E|+mc^2} \\ 1 \\ 0 \end{bmatrix} e^{\frac{i}{\hbar}(\mathbf{p}\cdot\mathbf{x}-Et)}$$
(3.12.6)

which *in the non-relativistic limit* corresponds to spin up (in the z-direction) and

$$\begin{bmatrix} -\dfrac{c(p_x-ip_y)}{|E|+mc^2} \\ \dfrac{cp_z}{|E|+mc^2} \\ 0 \\ 1 \end{bmatrix} e^{\frac{i}{\hbar}(\mathbf{p}\cdot\mathbf{x}-Et)}$$
(3.12.7)

for spin down (in the z-direction) in the non-relativistic limit.

Nuclear Physics

These solutions have the property that if

$$\begin{bmatrix} \psi_1 \\ \psi_2 \\ \psi_3 \\ \psi_4 \end{bmatrix} \text{ is a solution with positive energy and momentum,}$$

then

$$\begin{bmatrix} \psi_4^* \\ -\psi_3^* \\ -\psi_2^* \\ \psi_1^* \end{bmatrix} \text{ is a solution with negative energy and negative momentum.}$$

It is this property that justifies our earlier interpretation of absorption of a particle with negative energy and momentum being equivalent to emission of an antiparticle with positive energy and momentum.

It is essential to note that these solutions only correspond to spin up (or down) in the z-direction in the extreme non-relativistic limit when $|p| \to 0$ and in general are NOT eigenstates of the operator S_z unless the z direction coincides with the direction of momentum of the particle.† Thus in general any spin state of a relativistic lepton must be represented as the sum of the two *helicity states* in which the lepton must have spin component $\pm\frac{1}{2}$ with respect to its direction of motion. (A manifestation of this in atomic physics is the Thomas Precession.)

The relativistic theory of β-decay is set up by forming all possible Lorentz invariants out of two fields and the γ_μ matrices. For example, the quantity

$$\bar{\psi}_e \psi_\nu$$

remains unchanged under a Lorentz transformation and so is a 4-scalar. The quantity

$$\bar{\psi}_e \gamma_\mu \psi_\nu$$

transforms as a 4-vector (like $x_\mu = \mathbf{x}, ict$) and the scalar product of two such quantities gives a scalar of the kind needed for the interaction Hamiltonian in β-decay:

$$\sum_\mu (\bar{\psi}_e \gamma_\mu \psi_\nu)(\bar{\Psi}_f \gamma_\mu \Psi_i). \tag{3.12.8}$$

$\bar{\psi}$ means $\psi^* \gamma_4$ so $\bar{\psi}\gamma_\mu\psi \equiv \psi^* \gamma_4 \gamma_\mu \psi$. It is not obvious why $\bar{\psi}$ replaces ψ^* in relativistic theory without going too deeply into the theory for this book, but here is an illustration.

γ_μ has four indices, and so one might expect a combination like $e\,(\psi_e^* \gamma_\mu \psi_e)$ to give the four components of the electromagnetic charge and current density, j_μ, the fourth component of which gives the charge density. But $e\psi_e^* \psi_e$ is the charge density, so that the factor of γ_4 is necessary and $\bar{\psi}_e \gamma_\mu \psi_e = j_\mu$ which is a 4-vector.

---

† The operator S_z is a matrix $\begin{pmatrix} \sigma_z & 0 \\ 0 & \sigma_z \end{pmatrix}$ which is shorthand for a 4×4 matrix, the diagonal submatrices of which are the Pauli spin matrix σ_z. The result follows at once.

There are five quantities that are independent and Lorentz invariant:

$\bar{\psi}\psi$	1 component	4-scalar,
$\bar{\psi}\gamma_\mu\psi$	4 components	4-vector,
$\bar{\psi}\dfrac{\gamma_\mu\gamma_\nu - \gamma_\nu\gamma_\mu}{2i}\psi$	6 components	4-tensor,
$i\bar{\psi}\gamma_5\gamma_\mu\psi$	4 components	4-axial vector,
$\bar{\psi}\gamma_5\psi$	1 component	4-pseudoscalar,

$$\gamma_5 = \gamma_1\gamma_2\gamma_3\gamma_4.$$

Under a velocity transformation, the components of each quantity transform among themselves only: for example, the four components of the 4-vector $\bar{\psi}\gamma_\mu\psi$ mix in the same way as the four components of $(\mathbf{x}, ict)$. The axial vector and pseudoscalar have the same properties as the vector and scalar under a velocity transformation. They differ under a reflection, the axial vector not changing sign and the pseudoscalar changing sign. An axial vector has opposite parity to a vector, and a pseudoscalar opposite parity to a scalar. [A simple example of a pseudoscalar is the so-called scalar triple product $(\mathbf{A}\cdot\mathbf{B}\times\mathbf{C})$, the dot product of the vector $\mathbf{A}$ and the axial vector formed from the cross product of the vectors $\mathbf{B}$ and $\mathbf{C}$.]

A relativistic interaction Hamiltonian for β-decay may be formed from the scalars

$$(\bar{\psi}_e\psi_\nu)(\bar{\Psi}_f\Psi_i), \tag{3.12.9a}$$

$$\sum_\mu (\bar{\psi}_e\gamma_\mu\psi_\nu)(\bar{\Psi}_f\gamma_\mu\Psi_i), \tag{3.12.9b}$$

$$\sum_{\mu,\nu} \left(\bar{\psi}_e \frac{\gamma_\mu\gamma_\nu - \gamma_\nu\gamma_\mu}{2i} \psi_\nu\right)\left(\bar{\Psi}_f \frac{\gamma_\mu\gamma_\nu - \gamma_\nu\gamma_\mu}{2i} \Psi_i\right), \tag{3.12.9c}$$

$$\sum_\mu (\bar{\psi}_e i\gamma_5\gamma_\mu\psi_\nu)(\bar{\Psi}_f i\gamma_5\gamma_\mu\Psi_i), \tag{3.12.9d}$$

$$(\bar{\psi}_e\gamma_5\psi_\nu)(\bar{\Psi}_f\gamma_5\Psi_i). \tag{3.12.9e}$$

In the non-relativistic limit

$$\bar{\psi}\psi \to \psi^*\psi, \tag{3.12.10a}$$

$$\bar{\psi}\gamma_\mu\psi \to \psi^*\psi, \tag{3.12.10b}$$

$$\bar{\psi}\frac{\gamma_\mu\gamma_\nu - \gamma_\nu\gamma_\mu}{2i}\psi \to \psi^*\boldsymbol{\sigma}\psi, \tag{3.12.10c}$$

$$\bar{\psi}i\gamma_5\gamma_\mu\psi \to \psi^*\boldsymbol{\sigma}\psi, \tag{3.12.10d}$$

$$\bar{\psi}\gamma_5\psi \to 0. \tag{3.12.10e}$$

Thus both the scalar and vector couplings reduce in the limit where all four fermions are non-relativistic to the Fermi interaction. Similarly, both the tensor and axial vector couplings reduce to the Gamow–Teller interaction, while the pseudoscalar would not contribute in the extreme non-relativistic limit, even if present. One of the problems in the elucidation of the β-decay coupling was to sort out the relative importance of the two Fermi terms and of the two Gamow–Teller terms.

Nuclear Physics

If the electron momentum and energy are **p**, E and the neutrino momentum is **q**, then for allowed transitions, summing over all spins,

$$\left| \int (\bar{\psi}_e \psi_\nu)(\Psi_f \Psi_i) \, dV \right|^2 = \left\{ 1 - \frac{\mathbf{p} \cdot \mathbf{q} c}{|E||q|} \right\} \left| \int \Psi_f^* \Psi_i \, dV \right|^2 \qquad (3.12.11)$$

and

$$\left| \sum_\mu \int (\bar{\psi}_e \gamma_\mu \psi_\nu)(\Psi_f \gamma_\mu \Psi_i) \, dV \right|^2 = \left\{ 1 + \frac{\mathbf{p} \cdot \mathbf{q} c}{|E||q|} \right\} \left| \int \Psi_f^* \Psi_i \, dV \right|^2 \qquad (3.12.12)$$

if the leptons are treated as relativistic and the nucleons as non-relativistic. If θ is the angle between the two leptons then the scalar interaction gives a matrix element squared of

$$\left\{ 1 - \frac{v}{c} \cos \theta \right\} \left| \int \Psi_f^* \Psi_i \, dV \right|^2 \qquad (3.12.13)$$

and the vector

$$\left\{ 1 + \frac{v}{c} \cos \theta \right\} \left| \int \Psi_f^* \Psi_i \, dV \right|^2 \qquad (3.12.14)$$

where v is the electron velocity (the neutrino velocity is taken as c).

For the Gamow–Teller transitions, if we observe neither the lepton spins nor the change in direction of the nuclear spin, we have the result

$$\left\{ 1 - \frac{1}{3} \frac{v}{c} \cos \theta \right\} \left| \int \Psi_f^* \boldsymbol{\sigma} \Psi_i \, dV \right|^2 \qquad (3.12.15)$$

for the axial vector coupling and

$$\left\{ 1 + \frac{1}{3} \frac{v}{c} \cos \theta \right\} \left| \int \Psi_f^* \boldsymbol{\sigma} \Psi_i \, dV \right|^2 \qquad (3.12.16)$$

for the tensor coupling. The additional factor of $\frac{1}{3}$ appears essentially because averaging over the three components of the lepton spin vector washes out the electron–neutrino angular correlation. The terms in **p**·**q** are not present in the non-relativistic limit, being contributed by the "small" components of the spin functions, which are just those responsible for the time dependence of σ_z. The electron–neutrino angular correlation may be regarded as a consequence of the spin–spin correlations (leading to the singlet non-relativistic states in the Fermi transitions and the triplets in the Gamow–Teller transitions) and the increasing importance of the momentum as the axis of quantization as the momentum increases.

For a given interaction, integration over the angles of emission of the leptons averages the electron–neutrino correlation to zero, so that the electron spectrum is not affected and is as given by the non-relativistic theory. However, if either the two Fermi terms or the two Gamow–Teller terms are both present and interfere, the interference term is of the form

$$\pm \frac{m_e c^2}{|E|}$$

(opposite signs for electron and positron) and such terms would drastically affect the electron

Nuclear Decay

spectrum. These terms are called Fierz interference terms, and have never been observed. This tells us that the Fermi interaction is either scalar (S) or vector (V) but probably not both, while the Gamow–Teller interaction is either axial vector (A) or tensor (T) but probably not both. However, the absence of Fierz interference terms cannot entirely rule out two types of coupling in either the Fermi or the Gamow–Teller transitions—the couplings could be out of phase and hence yield no interference term.

The proper form of the weak interaction Hamiltonian was not determined until after the discovery of the non-conservation of parity in weak interactions, and indeed was postulated by Feynman and Gell-Mann in the teeth of contrary experimental evidence. It is this subject we must discuss next, but before doing so we will take events out of their chronological order and examine the information on the weak interaction coupling available from the electron–neutrino correlation. This cannot be studied directly, but reflects itself in the energy spectrum of the recoiling nucleus (and in the electron spectrum when measured at a fixed angle to the recoiling nucleus). The effect of the relativistic couplings on the energy spectrum of the recoiling nucleus is shown in Fig. 3.12.1.

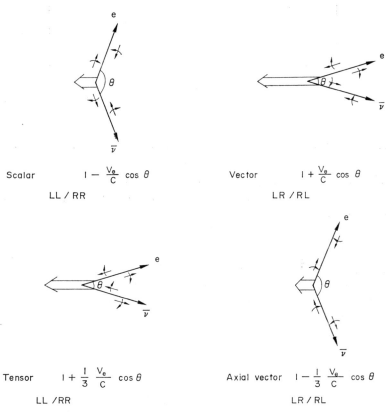

Scalar $1 - \frac{v_e}{c} \cos\theta$ Vector $1 + \frac{v_e}{c} \cos\theta$
LL / RR   LR / RL

Tensor $1 + \frac{1}{3} \frac{v_e}{c} \cos\theta$ Axial vector $1 - \frac{1}{3} \frac{v_e}{c} \cos\theta$
LL / RR   LR / RL

FIG. 3.12.1. Angular correlations between electron and neutrino are shown for the four couplings which survive in the non-relativistic limit. For a scalar interaction large angles and consequently low-energy recoils are favoured, and the leptons have the same handedness. For a vector interaction small angles and higher energy recoils are favoured, and the leptons have opposite handedness. A tensor interaction would favour small angles and the same lepton handedness, while the axial vector interaction favours large angles and leptons with opposite handedness.

Nuclear Physics

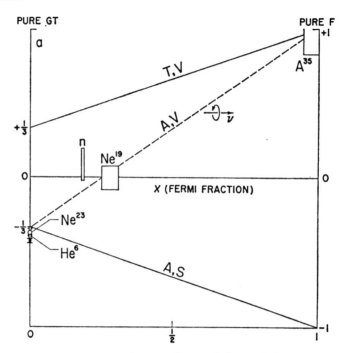

FIG. 3.12.2. Results of some measurements of the angular correlation coefficient a as a function of the Fermi fraction x. [From E. J. Konopinski, The experimental clarification of the laws of β-decay, Ann. Rev. Nucl. Sci. **9**, 99 (1959).]

Thus measurements of the recoil spectra could in principle resolve S from V and T from A. Some of the early results are results shown in Fig. 3.12.2.

The coefficient a of $\mathbf{p}\cdot\mathbf{q}/|E||q|$, extracted from measurements of the recoil spectrum, is shown against the Fermi Fraction x. The Fermi Fraction is 1 for a pure Fermi transition, 0 for a pure Gamow–Teller transition, and must be computed from the nuclear structure for a mixed transition. For the neutron it is $\frac{1}{4}$, because of the close equality of the Fermi and Gamow–Teller couplings and the three orientations in space of the lepton triplet. ^{19}Ne is a $\frac{1}{2}^+ \rightarrow \frac{1}{2}^+$ transition and is $\frac{1}{4}$ Fermi and $\frac{3}{4}$ Gamow–Teller. It has ten protons (two over a closed shell of eight) and nine neutrons (one over a closed shell of eight) and one of the protons converts to a neutron. It should be remembered, however, that with several nucleons outside a closed shell results from the Fermi fraction obtained using the single particle shell model must be viewed with caution. ^{35}A $\rightarrow$ ^{35}Cl is a $\frac{3}{2}^+ \rightarrow \frac{3}{2}^+$ transition, and is almost pure Fermi (obtained from the lifetimes) and this is a result to be explained in terms of detailed nuclear structure rather than a result which may be trivially calculated.

The difficulty of these experiments may be brought home by noting that prior to the discovery of the non-conservation of parity, the weak interaction was thought to be S–T. This came from the observations on neutron decay ($a \sim 0$, $x = \frac{1}{4}$) and on ^{6}He $\rightarrow$ ^{6}Li. The latter is pure Gamow–Teller $0^+ \rightarrow 1^+$ and the early experiments yielded a value of a in the range 0.32–0.34.[†] It was only after Feynman and Gell-Mann produced the appealing

[†] B. M. Rustad and S. L. Ruby, Phys. Rev. **97**, 991 (1955).

V–A theory that the experiment was redone[†] and the result $a = -0.39 \pm 0.05$ obtained, giving V–A rather than S–T. A more recent value[‡] is $a = -0.334 \pm 0.003$.

The non-conservation of parity gives point to a set of much easier measurements on electron longitudinal polarization which reflect the nature of the coupling, but which would in a world where parity was conserved yield nothing. We must therefore go on to consider the effect of non-conservation of parity.

3.13. Non-conservation of parity in the weak interactions

Until 1956 it was taken for granted that no interactions existed that violated conservation of parity; that is, that given an initial state of well-defined parity, no interaction would connect that state with a final state of opposite parity. This was not seriously questioned until it became clear in the mid-1950s that the so-called τ and θ mesons were in reality the same particle. The τ-meson has a decay mode $\tau^+ \to \pi^+\pi^-\pi^+$ and the θ-meson $\theta^+ \to \pi^+\pi^0$. Both have the same mass and the same lifetime and are known to be two different decay modes of the K-meson.

$$K^+ \to \pi^+\pi^-\pi^+,$$

$$K^+ \to \pi^+\pi^0.$$

The π has negative intrinsic parity and the final state pions are all in S-states—therefore the K with (presumably) well-defined parity decays via the weak interaction (the lifetime is 1.23×10^{-8} sec) into two states of *opposite* parity.

Many attempts were made to evade the conclusion that parity was violated in the weak interactions; for example, the hypothesis that there just happen to be two mesons, τ and θ, with opposite parity and the *same* mass and lifetime. Lee and Yang[§] examined the hypothesis that parity was not conserved. They concluded that all experiments on nuclear β-decay that had been done would not reveal a lack of conservation of parity in the weak interactions, even if it were present, and suggested experiments that would reveal non-conservation of parity in the weak interactions. The experiments were done and Lee and Yang proved right. To go further we need to look at the kind of effects that parity violation could produce.

If we wish to observe the effects of a parity-violating part to the weak interactions, it is no good looking at the nuclei alone. The wave functions of nucleons in the nucleus are overwhelmingly determined by the strong interactions, with corrections from electromagnetism and both of these interactions do not violate conservation of parity. Both the initial and final nuclei are thus in states of well-defined parity and so the nucleus alone will not show any effects. The wave functions of the leptons, however, are determined by the weak interactions that create them (with coulomb corrections in the case of electrons) and so it is among the properties of the emitted leptons that we should search. If a lepton is produced in a state

[†] J. S. Allen et al., *Phys. Rev.* **116**, 134 (1959).
[‡] C. Johnson et al., *Phys. Rev.* **132**, 1149 (1963).
[§] T. D. Lee and C. N. Yang, *Phys. Rev.* **104**, 254 (1956).

Nuclear Physics

of mixed parity, its wave function will be written as $\psi = \psi^+ + \psi^-$ where $P\psi^+ = \psi^+$, $P\psi^- = -\psi^-$. Then the square is

$$|\psi^+|^2 + |\psi^-|^2 + 2Re\psi^{+*}\psi^- \qquad (3.13.1)$$

while the square of $P\psi$ is

$$|\psi^+|^2 + |\psi^-|^2 - 2Re\psi^{+*}\psi^-. \qquad (3.13.2)$$

The effect of parity violation will thus show up in some observable property of the leptons that has both even and odd parts. The leptons are characterized by their momenta and spins, and the nuclear spin is also available as a direction to which these quantities may be referred.

Quantities such as $\mathbf{p}\cdot\mathbf{q}$, $\sigma_e\cdot\sigma_\nu$, or $\sigma_e\cdot(\mathbf{p}\times\mathbf{q})$ are scalars and do not change sign under reflections. Quantities such as $\mathbf{p}\cdot\sigma_e$ are the scalar product of a vector and an axial vector, pseudoscalars, and do change sign under reflections. Thus if the electron wave function contains a term $1 + a\,\mathbf{p}\cdot\sigma_e$, this will represent an electron state of undefined parity—an electron polarized longitudinally with respect to its direction of motion. If a is positive, this means the probability of finding an electron spinning right-handedly with respect to its direction of motion is greater than the probability of finding an electron which is left-handed. Under a reflection the spin does not reverse but the momentum does and right-handedness becomes left-handedness. Thus the observation of a longitudinal polarization in decay from an initial state of well-defined parity implies that the interaction responsible for the decay is parity violating. As an extreme example consider a neutrino which is completely longitudinally polarized with respect to its direction of motion. A general state is

$$a\psi_L + b\psi_R \qquad \psi_L \text{ is left-handed,}$$
$$\psi_R \text{ is right-handed.}$$

The operator $\frac{1}{2}\{1 + (\mathbf{q}\cdot\sigma)/|q|\}$ is a projection operator with eigenvalue $+1$ acting on ψ_R and zero on ψ_L. $\frac{1}{2}\{1 - (\mathbf{p}\cdot\sigma)/|q|\}$ has eigenvalues $+1$ on ψ_L and zero on ψ_R.

Under reflection $\mathbf{q}\cdot\sigma \to -\mathbf{q}\cdot\sigma$. Thus if the operator $\frac{1}{2}\{1 - (\mathbf{q}\cdot\sigma)/|q|\}$ enters the lepton bracket, neutrinos will always be left-handed, corresponding to maximum violation of parity—and this is what happens. The effects of parity violation in the weak interactions were searched for in nuclear β-decay after Lee and Yang made their original suggestions. The race was won by Madame Wu and collaborators[†] at Columbia who measured the angular distribution of electrons emitted in the decay of cobalt-60. ^{60}Co is an electron emitter going to an excited state of ^{60}Ni. The transition is between states of angular momentum and parity $5^+ \to 4^+$ and so is an allowed Gamow–Teller transition. The decay scheme is shown in Fig. 3.13.1.

This nucleus was chosen because its decay scheme had been thoroughly studied, and it was known that it could be lined up in a magnetic field at low temperatures. Furthermore, the alignment of the ^{60}Co spin could be monitored through the degree of anisotropy (not asymmetry) of the subsequent γ-radiation from the excited states of ^{60}Co. The experiment, one of those suggested by Lee and Yang, looked for an asymmetry in the distribution of electrons with respect to the aligned spin of the ^{60}Co nuclei. The nuclei were lined up in

---

[†] C. S. Wu et al., Phys. Rev. **105**, 1413 (1957).

Nuclear Decay

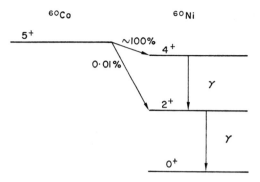

FIG. 3.13.1. The ^{60}Co–^{60}Ni decay scheme. The 5$^+$ ground state of ^{60}Co decays to the 4$^+$ excited state of ^{60}Ni, an allowed Gamow–Teller transition. The 4$^+$ state of ^{60}Ni decays to the ground state with the emission of two E2 photons.

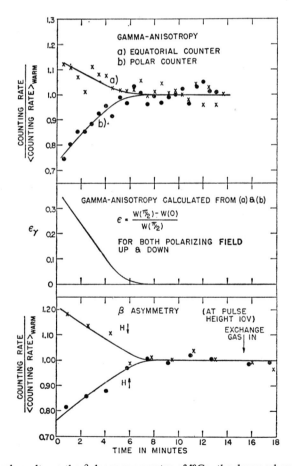

FIG. 3.13.2. The original results on the β-decay asymmetry of ^{60}Co, the decay scheme of which is shown in Fig. 3.13.1. The β-asymmetry decays away with time as the sample warms up and demagnetizes, and this is reflected in the decaying anisotropy of the γ's emitted from ^{60}Ni, which is left aligned in the decay of aligned ^{60}Co. [From C. S. Wu et al., Phys. Rev. **105**, 1413 (1957).]

Nuclear Physics

a magnetic field at 0.01°K and the equatorial and polar electron intensities measured as a function of time with frequent reversals of the magnetic field. A substantial asymmetry was observed, which decayed away with a time constant of some 5 min, as the sample containing the ^{60}Co warmed up and the alignment was destroyed. The associated γ-radiation became isotropic with the same time scale: the results are reproduced in Fig. 3.13.2 and were found to be consistent with form

$$1 - \frac{1}{3} \frac{\langle \mathbf{J} \rangle}{J} \cdot \frac{\mathbf{p}c}{|E|}$$

appropriate to maximal parity violation in an allowed Gamow–Teller transition. The experiment and its mirror image are shown in Fig. 3.13.3. Such an experiment provides a way of defining handedness. The recipe goes: take ^{60}Co and line it up in a magnetic field. Find which direction the most electrons are emitted in. The nuclear spin is left-handed with respect to this direction (and the direction of nuclear spin is defined by the magnetic field).

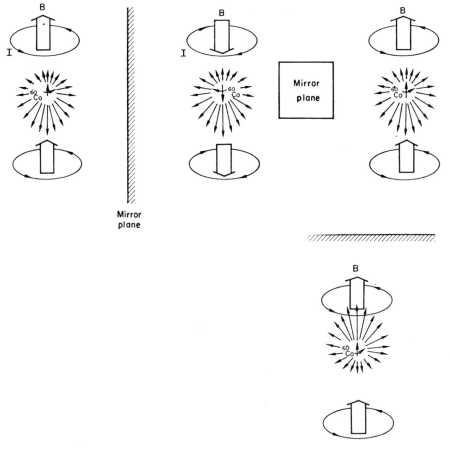

FIG. 3.13.3. The images of the decay of ^{60}Co formed by the successive reflections which together make up the parity operation. In the real world more electrons are emitted in a direction opposite to the aligning field than along it. In the mirror world reached by the physically unrealizable parity operation the situation is reversed.

Nuclear Decay

It will also be clear from the diagram that the electrons emitted in the decay of ^{60}Co will be longitudinally polarized in the left-handed sense. This is required by the longitudinal alignment and the anisotropy of the electrons emitted. Now a correlation between nuclear spin and electron momentum can only occur, even if parity is violated, for Gamow–Teller transitions or a mixture of Fermi and Gamow–Teller transitions. For a pure Fermi transition, even if the nuclear spins are not zero, the operator σ which changes nuclear spin does not come into the interaction and so the direction of the nuclear spin cannot be correlated with any other variables. Longitudinal electron polarization can occur in Fermi transitions, however. The study of longitudinal electron polarization is much easier than experiments involving alignment of nuclei at low temperature and after the initial observations of parity violation became the most important method of sorting out the β interactions.

An experiment made almost simultaneously with the ^{60}Co work depended on the observation of both these types of parity violating effect. Again, the experiment was suggested in Lee and Yang's original paper. The π-meson is a particle of spin 0 and negative intrinsic parity and decays into a muon and a neutrino with a lifetime of 2×10^{-8} sec—a weak interaction, but not of the standard four-fermion kind. Non conservation of parity in this decay could give a longitudinal polarization of the muon. The muon decays into an electron and two neutrinos, a standard four-fermion β-decay. Non-conservation of parity in muon decay could lead to an asymmetry in the angular distribution of the electrons relative to the muon spin, and hence relative to the direction of emission of the muon in π-decay if this resulted in a longitudinally polarized muon. The experiment was made by Garwin, Lederman and Weinrich and is reported[†] in the paper following that on the ^{60}Co experiment. Garwin and his collaborators took muons from a small part of the solid angle into which they were emitted by pion decay, stopped them and let them precess in a magnetic field. The rate of electron emission in a fixed direction in space was found to vary sinusoidally at the precession frequency of the muons, thus establishing parity violation in both processes of the π-meson decay chain.

Subsequent experiments performed in many laboratories rapidly established that electrons emitted in nuclear β-decay are longitudinally polarized to an extent v/c and are left-handed, while positrons are right-handed to the same degree. This again corresponds to maximal parity violation in β-decay, the parity conserving and violating parts of the lepton bracket being of equal strength. (There are various ways of measuring such a longitudinal polarization. For example, the polarization can be transformed into a transverse polarization by electric and magnetic deflection (or by scattering off a light atom) and then analysed by determining the left–right asymmetry in scattering off a heavy atom such as gold in which the spin–orbit coupling is strong, or the circular polarization of γ-rays emitted in a bremsstrahlung process or positron annihilation may be measured.[‡])

Reference to Fig. 3.12.1 indicates that a left-handed electron will be accompanied by a left-handed antineutrino in S and T transitions, but by a right-handed antineutrino in V and A transitions. Thus if the interaction is vector and axial vector the results on electrons and positrons imply that neutrinos are left-handed and antineutrinos right-handed. This gives

---

[†] R. Garwin et al., Phys. Rev. **105**, 1415 (1957).
[‡] Detailed discussions are given in K. Siegbahn (Ed.), *α-, β- and γ-Ray Spectroscopy*, Vol. 2, chap. XXIV (North Holland, 1965).

Nuclear Physics

a natural way of distinguishing between neutrinos and antineutrinos and embodies the two-component neutrino theory. (In general a neutrino has four components—two spins and both particle and antiparticle states. Two of the spin states never appear in the parity violating two-component theory.) The elegance of this hypothesis led Feynman and Gell-Mann to suggest that the weak interaction is vector (the Fermi part) and axial vector (the Gamow–Teller part) and maximally parity violating.

Repetition of the ^{6}He-decay experiments provided confirmatory evidence of this, and the prediction that leptons are left-handed was confirmed by an enormously ingenious experiment in which the helicity (longitudinal polarization with respect to the direction of motion) of the neutrino was measured and found to be -1, left-handed polarization. This experiment was made by Goldhaber, Grodzins and Sunyar in 1958.[†] The neutrino helicity was turned into circular polarization of a γ-ray through the following sequence of events. A sample of Europium (10 mg of Eu_2O_3) was irradiated in the Brookhaven reactor to produce ^{152}Eu in an isomeric state, denoted ^{152}Eu^m. This state has spin and parity 0^- and can decay by electron capture, principally from the K-shell, to the 1^- excited state of Samarium, denoted ^{152}Sm*:

$$^{152}\text{Eu}^m + e^- \rightarrow \nu + {}^{152}\text{Sm}^*.$$

The initial angular momentum is $\frac{1}{2}$. If the neutrino is longitudinally polarized, then the excited state of Samarium is also longitudinally polarized, by conservation of angular momentum

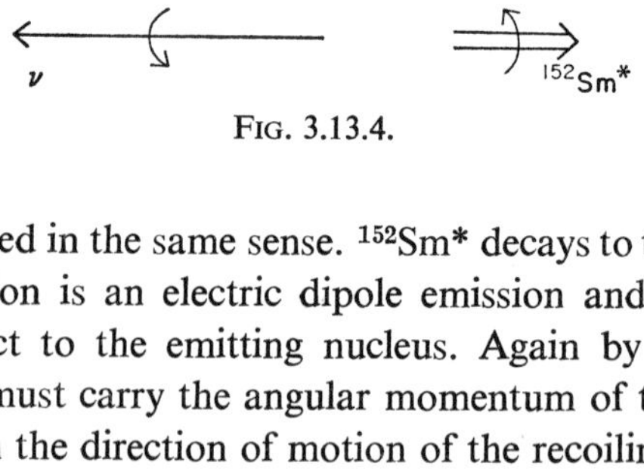

FIG. 3.13.4.

and furthermore is polarized in the same sense. ^{152}Sm* decays to the 0^+ ground state radiatively. A $1^- \rightarrow 0^+$ transition is an electric dipole emission and the photon has angular momentum 1 with respect to the emitting nucleus. Again by conservation of angular momentum, this photon must carry the angular momentum of the emitting ^{152}Sm*. Thus if the photon is emitted in the direction of motion of the recoiling ^{152}Sm*, it will have the same longitudinal polarization

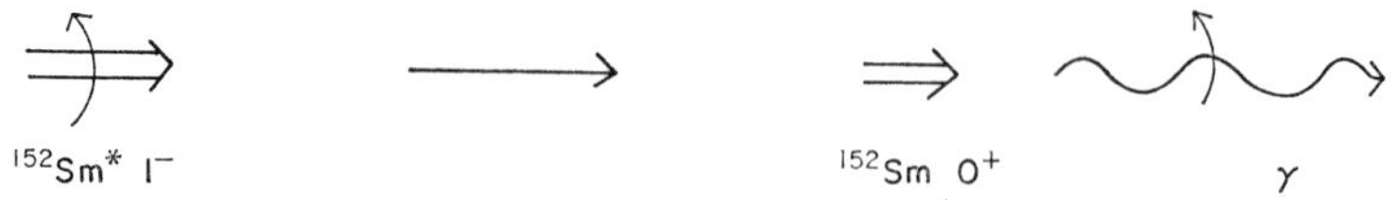

FIG. 3.13.5.

and hence will be polarized in the same sense as the neutrino, while if it is emitted opposite to the direction of motion, it will have the opposite polarization. The circular polarization of γ-rays may readily be measured from the transmission through magnetized iron. (Iron is used as a convenient source of aligned electrons and is magnetized so that the electrons

[†] M. Goldhaber, L. Grodzins and A. W. Sunyar, *Phys. Rev.* **109**, 1015 (1958).

are pointing either in or against the direction of motion of the emitted photons. If the photon spin and the electron spin are antiparallel, electron spin flip in the scattering increases the scattering cross-section over the parallel case.)

The crucial point of the experiment is the selection of those γ's emitted forwards— clearly if γ's emitted at all angles are detected the net circular polarization will be zero even for completely aligned ^{152}Sm*. Now the excited state ^{152}Sm* has a lifetime $\sim 3 \times 10^{-14}$ sec, less than the slowing-down time of the excited nucleus as it recoils through the source material. Thus the majority of decays take place in flight, and γ's emitted forward are Doppler shifted up in frequency and those emitted backward, down. The resulting energy difference makes possible selection of the forward γ's. Consider a nucleus in the excited state ^{152}Sm* at rest in the laboratory. When it decays, the energy difference between the excited state and the ground state is shared between the γ and the recoiling nucleus. The nucleus does not get much, because it is massive, but it gets some and so the γ has insufficient energy to induce the inverse process $\gamma + {}^{152}$Sm* $\rightarrow {}^{152}$Sm when incident on ^{152}Sm nucleus also at

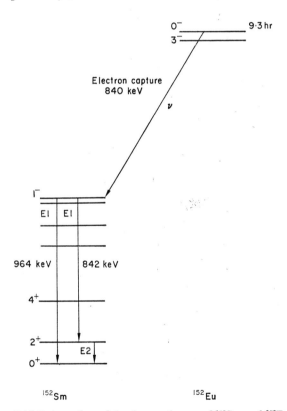

FIG. 3.13.6. A portion of the decay schemes of ^{152}Sm and ^{152}Eu.

rest in the laboratory. But the Doppler shift of the energy when the ^{152}Sm* decays while recoiling against a neutrino is sufficient to allow the process for γ's emitted forward. Thus these γ's can be resonance scattered by ^{152}Sm nuclei. The decay scheme of ^{152}Eu and ^{152}Sm is shown in Fig. 3.13.6 and the experimental arrangement in Fig. 3.13.7.

Nuclear Physics

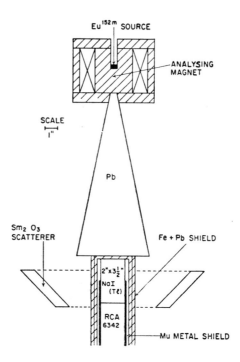

FIG. 3.13.7. The experimental arrangement used by Goldhaber, Grodzins and Sunyar to measure the neutrino helicity. [From M. Goldhaber, L. Grodzins and A. W. Sunyar, *Phys. Rev.* **109**, 1015 (1958).]

Goldhaber, Grodzins and Sunyar found a fractional difference in the counting rates for the two magnetic field orientations, N_+ and N_-, of

$$\frac{N_- - N_+}{\frac{1}{2}(N_+ + N_-)} = 0.017 \pm 0.003$$

to be compared with an expected value of 0.025 for 100% circularly polarized (left-handed) photons. The finite solid angle and depolarizing effects could account for the difference and they concluded that neutrinos were left-handed and that the measurement was consistent with 100% left-handed polarization.

The vector and axial vector interactions were thus confirmed as being responsible for β-decay. These interactions, with maximal parity violation, are noteworthy in that they combine a two-component neutrino theory with conservation of leptons. If the Gamow–Teller interaction was tensor, for example, left-handed electrons would be accompanied by left-handed neutrinos, thus keeping a two-component theory but maximally violating conservation of lepton number, or by left-handed antineutrinos, thus violating a two-component theory.

The final structure of the famous V minus A interaction proposed by Feynman and Gell-Mann[†] is at first sight complicated unless approached directly from the two-component

[†] R. P. Feynman and M. Gell-Mann, *Phys. Rev.* **109**, 193 (1958).

neutrino theory. Suppose we consider first a pure Fermi transition, and take the vector interaction. A parity-conserving interaction will be given by the scalar product of two vectors

$$\sum_\mu (\bar\psi_e \gamma_\mu \psi_\nu)(\bar\Psi_f \gamma_\mu \Psi_i). \tag{3.13.3}$$

Maximal parity violation means a pseudoscalar term of the same size must be present. The two nuclei have well-defined spin and parity and so the nuclear bracket must remain the same. A pseudoscalar must be formed by adding a term proportional to

$$\bar\psi_e \gamma_5 \gamma_\mu \psi_\nu$$

to the lepton bracket so as to form the scalar product of a vector and axial vector.

The expression

$$\{\bar\psi_e(1-\gamma_5)\gamma_\mu\psi_\nu\}\{\bar\Psi_f \gamma_\mu \Psi_i\} \tag{3.13.4}$$

violates parity maximally. The lepton bracket is of the V minus A form (rather than $V+A$): $(1-\gamma_5)$ acting to the left projects out the left-handed part of an electron spin or the right-handed part of a positron spin, to degree v/c, and operating to the right on $\gamma_\mu \psi_\nu$ it projects out a left-handed neutrino or a right-handed antineutrino. The equality of the two terms is required by the extreme longitudinal polarization of leptons, and the relative sign by the left-handedness of leptons. We can, of course, never be sure that neutrinos are completely polarized longitudinally: this form is adopted for aesthetic reasons, and is consistent with experiment.

Similarly we may set the axial vector Gamow–Teller term equal to

$$\{\bar\psi_e(1-\gamma_5)\gamma_\mu\psi_\nu\}\{\bar\Psi_f \gamma_5 \gamma_\mu \Psi_i\} \tag{3.13.5}$$

and the β-interaction is written as

$$\{\bar\psi_e(1-\gamma_5)\gamma_\mu\psi_\nu\}\{\bar\Psi_f(g_V - g_A\gamma_5)\gamma_\mu\Psi_i\} \tag{3.13.6}$$

where g_V and g_A must be determined from experiment. g_V is determined from ^{14}O-decay, a $0^+ \to 0^+$ allowed pure Fermi transition, and g_A determined relative to g_V from the mixed transition $n \to p + e^- + \bar\nu$: the pure Gamow–Teller transitions involve the uncertainties of nuclear structure to too great an extent to be reliable. Neutron decay has a Fermi fraction of $\frac{1}{4}$ and the value of g_A may be directly obtained from the neutron lifetime, and also from correlation studies in the decay of polarized neutrons. If the electron and proton momenta are observed the neutrino momentum may be constructed and the matrix element for polarized neutron decay takes the form

$$|M|^2 = \xi \left[\left\{ 1 + a\frac{\mathbf{p}\cdot\mathbf{q}c}{|E||q|} + b\frac{mc^2}{|E|} \right\} + \langle 1\rangle\langle\sigma\rangle \cdot \left\{ A\frac{\mathbf{p}c}{|E|} + B\frac{\mathbf{q}}{|q|} + D\frac{\mathbf{p}\times\mathbf{q}c}{|E||q|} \right\} \right]. \tag{3.13.7}$$

Integration over the angular correlations leaves the neutron lifetime determined by a factor

$$1 + b\frac{mc^2}{E}.$$

Nuclear Physics

b is the coefficient of the Fierz interference term and is consistent with zero.

$$\xi = |g_V|^2 + 3|g_A|^2$$

and so a knowledge of the neutron lifetime and the value of $|g_V|^2$ from ^{14}O decay gives $|g_A|$.

The coefficient of electron–neutrino angular correlations, $a\xi = |g_V|^2 - |g_A|^2$ can also give a value for $|g_A|$ but since $g_A \sim g_V$ the correlation is small and difficult to measure. However, the parity violating terms with coefficients $A\xi$ and $B\xi$ can give independent information. $A\xi$ and $B\xi$ contain the real part of the interference between the vector and axial vector parts of the nucleon bracket. $D\xi$, the coefficient of the term in

$$\langle \sigma \rangle \cdot \frac{\mathbf{p} \times \mathbf{q}}{|E||q|},$$

contains the imaginary part, and is consistent with zero. It should be precisely zero if the weak interactions do not violate *time reversal invariance*: a term of this form reverses sign if the direction of time is reversed.

From the neutron lifetime $T_{1/2}$ (10.8 min) a value $g_A/g_V = 1.24 \pm 0.01$ has been obtained, and from the coefficients A and B a value of 1.26 ± 0.02.[†] Thus the weak interaction matrix element may be written

$$\sum_\mu \{\bar{\psi}_e(1-\gamma_5)\gamma_\mu\psi_\nu\}\left\{\bar{\Psi}_f\left(1-\frac{g_A}{g_V}\gamma_5\right)\gamma_\mu\Psi_i\right\}\frac{g_V}{\sqrt{2}} \quad (3.13.8)$$

(the $1/\sqrt{2}$ is a convention introduced to give a definition of g_V consistent with that used earlier in non-parity violating theories) and the value of $|g_V|$ obtained from ^{14}O decay is 1.415×10^{-49} erg cm^3.

A comparison with the decay of the muon is instructive. The theory of muon decay is more complicated because all three particles in the final state are relativistic:

$$\mu^+ \to e^+ \, \nu \, \bar{\nu}_\mu,$$
$$\mu^- \to e^- \, \bar{\nu} \, \nu_\mu.$$

Electron polarization, angular correlations in the decay of polarized muons, and the electron energy spectrum (which is sensitive to the interaction when all particles are relativistic) are all consistent with an interaction

$$\frac{g_\mu}{\sqrt{2}}\bar{\psi}_e\{(1-\gamma_5)\gamma_\mu\psi_\nu\}\{\bar{\psi}_\mu(1-\gamma_5)\gamma_\mu\psi_{\nu_\mu}\}. \quad (3.13.9)$$

That is, in muon decay the vector and axial vector parts are of the same strength. Furthermore, from the lifetime of the muon the constant g_μ is found to be very nearly equal to g_V but 2% greater even after all necessary corrections to the simple theory have been calculated. The striking equality of g_μ and g_V suggest that the strength of the fundamental β coupling is unaffected by the presence of the strong interactions in the vector part, but changed by their presence in the axial vector part of a bracket embracing strongly interacting fermions.

[†] See, for example, Christensen *et al.*, *Phys. Lett.* **28B**, 411 (1969), and papers cited therein. This value is more recent than the often quoted value of 1.18 ± 0.02.

Nuclear Decay

The electric current $e\bar{\psi}\gamma_\mu\psi$ has this property: the charge on a proton is the same as the charge on a positron, the charge on a neutron is zero—they do not have charges 1.25 and 0.3, for example. This is the basis of the conserved vector current hypothesis which predicts that the vector part of the weak interaction is proportional to the electromagnetic interaction of a nucleon. In particular, among the transitions which correspond to the second term in an expansion of $e^{i\mathbf{k}\cdot\mathbf{r}}$ a weak analogue of magnetic transitions is predicted, with a strength proportional to the whole nucleon magnetic moment rather than just the Dirac part. This prediction has been borne out by experiment.[†]

The anomalous magnetic moments of the nucleons are presumed to be due to the virtual meson clouds surrounding them:

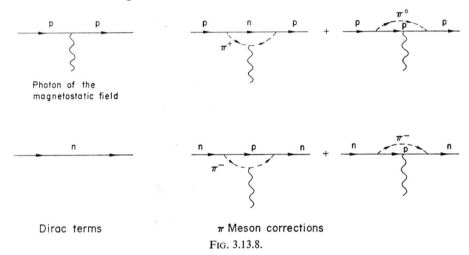

FIG. 3.13.8.

Thus in neutron decay a pure fermion term and a correction contribute:

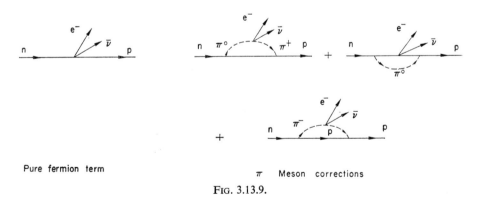

FIG. 3.13.9.

This hypothesis leads to the prediction of a β-decay mode $\pi^+ \to \pi^0 e^+ \nu$ for the π-meson—very different from the original four-fermion interaction, and furthermore predicts that the ft value should be identical with that for ^{14}O decay (both are allowed Fermi transitions,

---

[†] See C. S. Wu, *Rev. Mod. Phys.* **36**, 618 (1964).

the π has spin 0). Within experimental error, it is, but the accuracy is only ~10%, the predicted decay rate leads to a branching ratio for $\pi^+ \to \pi^0 e^+ \nu$ of ~10^{-8} against the other decay modes (mostly $\pi^+ \to \mu^+ \nu_\mu$).

The corrections to the axial part are believed to be due to diagrams like

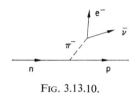

FIG. 3.13.10.

where the virtual pion does not return to its parent nucleon.

The β-decay interaction is really a vector interaction involving only left-handed fermions and right-handed antifermions. It thus resembles the original form proposed by Fermi, differing only in the inclusion of the helicity projection operators which have the properties

$$\tfrac{1}{2}(1+\gamma_5)\psi_L = \psi_L; \quad \tfrac{1}{2}(1+\gamma_5)\psi_R = 0$$

where ψ_L represents a lepton with left-handed longitudinal polarization (helicity) and ψ_R a right-handed lepton. When strongly interacting fermions (baryons) are included, the vector term in the baryon bracket is reduced by ~2% and the axial vector term is increased by 25% if both baryons are nucleons. (In the β-decay of strange particles the changes are more drastic—the β-decay rate of the Λ hyperon $\Lambda \to p\, e^- \bar{\nu}$ is down by a factor of ~10 on the value expected if the couplings are the same as in neutron decay.)

To conclude this section, we write down the V–A β-interaction and compare its form with that for a non-parity violating V and A interaction:

V–A:

V and A:

$$\{\bar{\psi}_e(1-\gamma_5)\gamma_\mu \psi_\nu\}\{\bar{\Psi}_f(g_V - g_A\gamma_5)\gamma_\mu \Psi_i\}, \qquad (3.13.10)$$

$$\{\bar{\psi}_e\gamma_\mu\psi_\nu\}\{\bar{\Psi}_f g_V \gamma_\mu \Psi_i\} + \{\bar{\psi}_e\gamma_5\gamma_\mu\psi_\nu\}\{\bar{\Psi}_f g_A\gamma_5\gamma_\mu \Psi_i\}. \qquad (3.13.11)$$

3.14. The weak interactions and parity, charge conjugation and time reversal

We have seen that the weak interactions violate parity maximally—the scalar and pseudo-scalar terms in the β-decay matrix element have the same strength (and are relatively real). A law of lepton conservation holds, and while leptons are left-handed, antileptons are right-handed. You can now easily see that charge conjugation invariance is also maximally violated by the weak interactions. Until 1956 this was another sacred cow of physics, which may be summarized by saying that *until* the discovery of parity violation, no one could see why a system of antiparticles should behave at all differently from a system of particles. If every particle is changed with its antiparticle—the operation of charge conjugation—then a perfectly possible physical system should result. The breakdown of this symmetry between particle and antiparticle is implied by the two-component neutrino theory. The operation of charge conjugation (C) applied to a left-handed neutrino gives a left-handed antineutrino

—and left-handed antineutrinos do not exist, the antineutrino is right-handed. Consider the process $\pi^+ \to \mu^+ \nu_\mu$ in the rest system of the π:

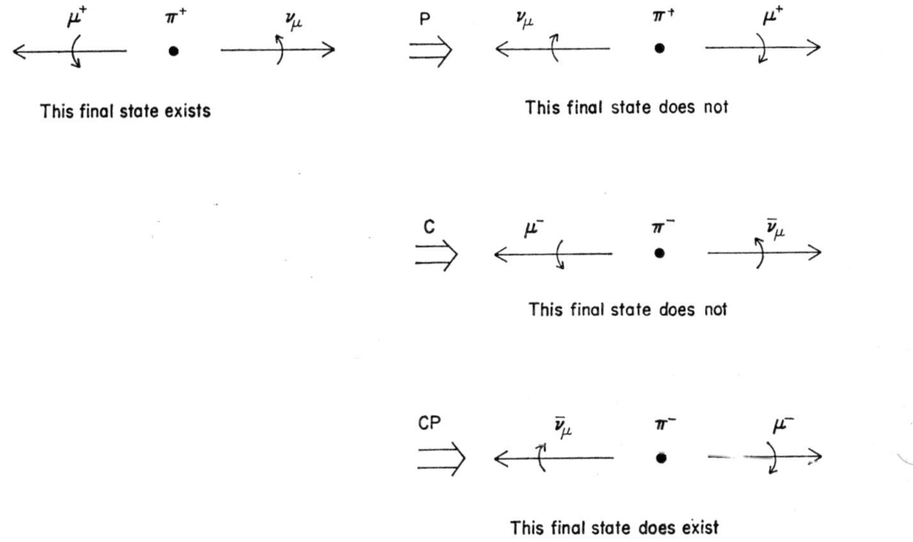

FIG. 3.14.1.

The weak interactions are thus said to be *CP* invariant. This statement means that if you wish to use the π-meson decay in order to provide a universal definition of left-handedness, it is necessary first to specify whether the laboratory is constructed of matter or antimatter. A π^- in a laboratory constructed of antimatter will behave precisely like a π^+ in a laboratory constructed of matter—except that the neutrino emitted in its decay will be right-handed. It would seem, then, that if the antiphysicist exists (the famed Dr. Edward Anti-Teller of a well-known piece of doggerel) it is not possible to provide all physicists everywhere with a universal prescription for labelling their left hands.

Consider the operation of time reversal (T) applied to our decaying π^+. Under time reversal (in picturesque terms, the result of running a film backwards) the momentum vectors are reversed and spins go around in the opposite direction. The final state $\mu\nu$ is affected as shown below:

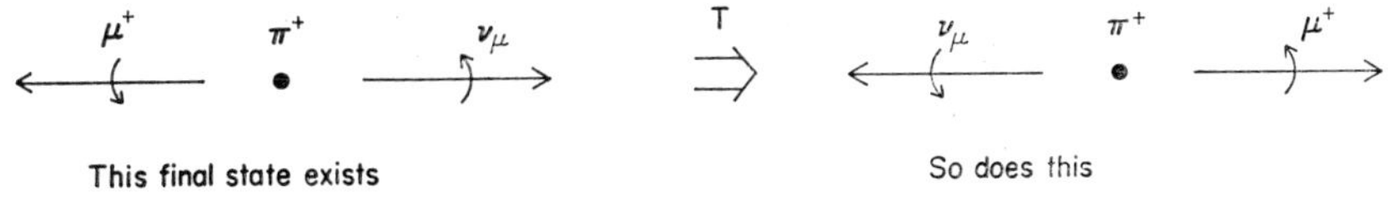

FIG. 3.14.2.

Applied to the full process rather than just the final state we have ν_μ and μ^+ with the proper handedness interacting to form a pion.

The *V–A* interaction is thus time reversal invariant. (There exists a very general theorem that says that all systems must be *PCT* invariant—thus an interaction which is *PC* invariant must be *T* invariant.)

Nuclear Physics

Since 1964 it has been known that the interaction responsible for the decay of K-mesons is *not CP* invariant and so presumably not *T* invariant. K-meson decay is peculiarly sensitive to such effects, and they have not been seen anywhere else. It is not yet clear whether a small violation of *CP* occurs in the ordinary weak interactions, or whether there is a superweak interaction which is maximally *CP* violating. In K^0-meson decay it was observed that a K^0-meson state of well-defined *CP* decayed into two final states with opposite *CP* eigenvalues. In ordinary β-decay correlations of the kind **σ**·(**p**×**q**) have been searched for, unsuccessfully. Other attempts to observe effects of time reversal violation have been experiments to look for electric dipole moments of the proton, neutron and electron. Consider a neutron and suppose it to have an intrinsic electric dipole moment, which can only be parallel or antiparallel to the spin.

Under *T* the electric dipole moment remains the same but the spin is reversed and consequently an elementary particle with an electric dipole moment is not an eigenstate of *T*, implying that there exist *T*-violating interactions. At the time of writing no such dipole moments have been observed.

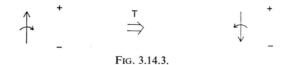

Fig. 3.14.3.

3.15. Odd remarks on β-decay

We conclude our discussion of β-decay with some general remarks. The decay process is surprisingly analogous to electromagnetic decay and may be used similarly as a probe of nuclear structure. While we have concentrated on allowed β-decay, the forbidden processes may occur either alone (when allowed decay is impossible) or with allowed decay, when

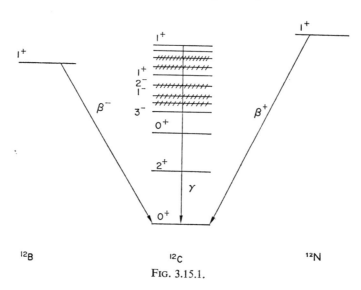

Fig. 3.15.1.

Nuclear Decay

interference between the two processes may take place, as between M1 and E2 electromagnetic processes. The observation of a weak anomalous magnetic moment[†] depended on the observation of such interference in the decay schemes of ^{12}B, ^{12}C, ^{12}N. In contrast, β^+-, β^-- and γ-decay from such an isotopic multiplet may be used to sort out nuclear structure. As an example, attempts to explain the surprisingly long lifetime of ^{14}C in terms of nuclear structure have used as input data from all these decay processes from the isotopic triplet ^{14}C, ^{14}N, ^{14}O.

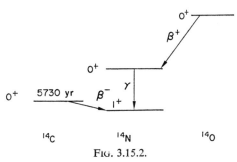

FIG. 3.15.2.

Just as γ-γ correlations in a decay chain may be used to find the spin and parity of nuclear levels, successive β-β correlations may be used in the same way, as may successive β-γ correlations. The details are for the professionals, and at this point we terminate our discussion of β-decay.

3.16. α-decay

We conclude our discussion of nuclear decay by considering a process very unlike the weak and electromagnetic de-excitations of the previous sections: α-emission. An α-emitter decays not by emission of elementary particles created from the available energy but by emitting an object which is a nucleus in its own right, assembled from two protons and two neutrons of the parent nucleus and ejected through an enormous coulomb barrier. The process of α-emission is perhaps best regarded as a highly asymmetric example of spontaneous fission. While β- and γ-decay can be treated with some degree of success in the shell model, α-emission is a many-body process which cannot by any stretch of the imagination be discussed in terms of individual nucleons moving in a time-independent potential.

The energy condition for an α-decay

$$^A_Z N \rightarrow ^{A-4}_{Z-2}N + \alpha$$

(which leaves a neutron rich daughter) to occur is

$$M(Z, A) > M(Z-2, A-4) + M_\alpha,$$

or in terms of the binding energies

$$B(Z, A) < B(Z-2, A-4) + B_\alpha.$$

[†] For a review see C. S. Wu, *Rev. Mod. Phys.* **36**, 618 (1964).

Nuclear Physics

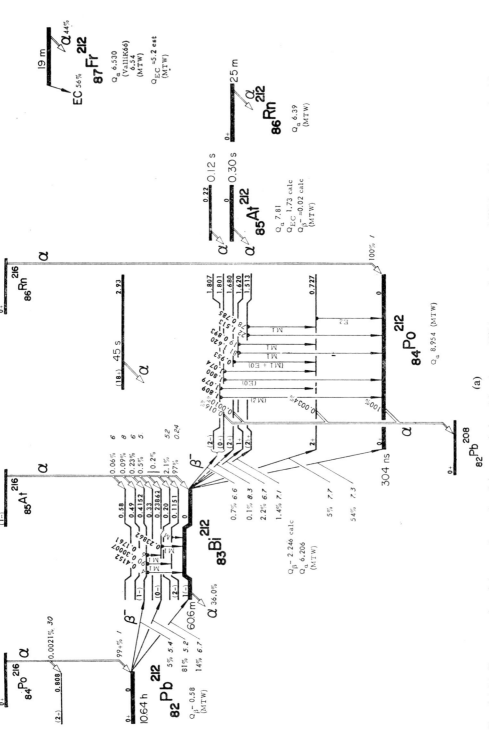

FIG. 3.16.1. Two examples of decay schemes involving α-emission. (a) α-decay of ²¹⁶At to excited states of ²¹²Bi provides an example of fine structure in α-emission, while α-decay of excited states of Po²¹² provides an example of long-range α's. (b) The α-decay of ²⁴¹Am provides an example of α-decay preferentially to an excited state. The ground state of ²⁴¹Am has $J^P = 5/2-$ and α-decay goes to the 59.54 keV excited state of ²³⁷Np with $J^P = 5/2-$ rather than to the ground state with $J^P = 5/2+$. [From Lederer, Hollander and Perlman, *Table of Isotopes* (Wiley, 1967).]

Nuclear Decay

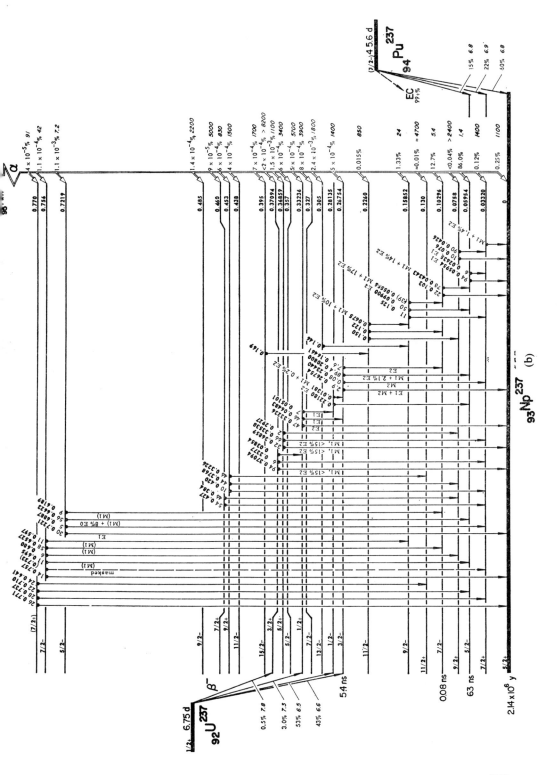

Nuclear Physics

The binding energy of ⁴He is 28.3 MeV. The limit of instability is then given by

$$B(Z, A) - B(Z-2, A-4) = 28.3 \text{ MeV}.$$

Using the semi-empirical mass formula of Chapter 1 for the binding energies shows that nuclei with $A \sim 150$ are becoming unstable against α-emission. This result does not apply to individual nuclei, however. In obtaining it the pairing energy which causes the odd–even effect is neglected, as are the effects of closed shells of nucleons. Atomic masses of 150 are reached in the middle of the Lanthanide series of rare earths, but the naturally occurring radioactive series terminate at lead, $A \simeq 208$ (with the exception of the Neptunium Series which terminates at $^{209}_{83}$Bi).

The lifetime for α-emission varies very rapidly with the available energy: those nuclei classified as α stable but energetically capable of α-decay simply have very long lifetimes, even when measured on an astronomical time scale.

α-decay between two nuclear states leads to two assemblies of nucleons recoiling from each other: the α-particle and the daughter nucleus. The spectrum for such a transition is thus a single line. However, the α-spectrum of a transition between two nuclei may contain a number of lines. The fine structure of α-spectra is due to decay from the ground state of the parent nucleus to both the ground and excited states of the daughter. (This can be checked by adding up the energies of α's and the coincident γ's which accompany α-decay to an excited state of the daughter.) Because of the rapid variation of decay rate with α-energy the most energetic gives the most intense line. This corresponds to transitions to the ground state of the daughter (unless the ground state has a very different nuclear structure in which case the decay goes to the lowest-lying excited state with a good overlap integral—usually the same spin).

Long-range α's are a related phenomenon. These are α's with a higher energy than the ground → ground transition, but are rare. (The name dates from the time when the energy of α-particles was determined by measuring their range in air.) The long range α's come from excited states of the parent nucleus. Their intensity depends on the population of the excited states when the parent is formed (for example, by a previous radioactive decay) and also on the branching ratio of α- against γ-decay for each particular excited state. Examples of both fine structure and long-range α's are given in Fig. 3.16.1.

Before we consider the physics of the α-decay process, we note a very striking relation between the lifetime and the energy of the emitted α. This is the Geiger–Nuttall Rule, which has been known since $\sim$ 1911.

$$\lambda(E_\alpha) = CE_\alpha^x$$

where the exponent x is ~ 80. This means a factor of 2 in the energy released gives a factor of 10^{24} in the decay rate. (The original form this rule took was the relation $\lambda(R) \sim R^{57.5}$, where R is the range of the emitted α-particle in air, which varies roughly as $E^{1.5}$.) The enormous value of the exponent suggests that the power law is only an approximation to a portion of a rapidly varying function: we shall see that the decay rate has a variation proportional to $\exp(-cE_\alpha^{-1/2})$.

We approach the problem of the theory of α-decay by considering the variation of the potential energy of the α-particle and daughter nucleus. Let us time reverse the process of

Nuclear Decay

α-decay and consider the α-particle approaching the daughter nucleus from infinity with the right energy to form the parent. The total energy of this system is positive. At large distances from the daughter, the potential energy of the system is just the long-range coulomb energy, varying inversely with the separation between the centres of the α and the daughter. Then at some distance comparable to the nuclear radius, the strong forces start acting and the enormous repulsive coulomb potential is wiped out by the strong attractive potential due to the nuclear forces, which will distort the daughter and the α-particle. A little closer in and the α-particle merges with the daughter nucleus and loses its identity. The theory of α-decay depends on the assumption that the energy of the system can be represented as the potential energy of the α particle–daughter system down to some separation at which the attractive short-range nuclear forces predominate, but at which the α-particle still retains its identity, as shown in Fig. 3.16.2.

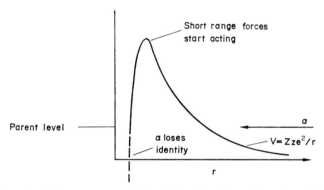

FIG. 3.16.2. The interaction between an α-particle and the daughter nucleus formed in α-decay. At large distances the interaction is just the coulomb repulsion. Near the nuclear surface the short-range attractive interactions dominate and a little closer in the nucleus an α-particle becomes distorted and then the α loses its identity in entering the nucleus and a parametrization in terms of r breaks down.

Thus in the decay process the α-particle pops out of the nucleus and is promptly confronted by an enormous coulomb barrier which most of the time reflects it straight back into the nucleus again. The decay rate will thus be governed by two factors. The first is the frequency with which a fully formed α-particle emerges from the nucleus to assault the coulomb barrier, and the second is the probability of any one assault succeeding. The first factor cannot be calculated at present (although many estimates of a varying degree of crudity have been made) while the second may be calculated fairly straightforwardly. We shall consider the general problem of barrier penetration in some detail in Section 4.13.

We first assume that the α-particle approaches the coulomb barrier in a constant potential $-V_0$, representing the effect of the nuclear forces, and then at a radius R the potential changes discontinuously to the coulomb potential $z(Ze^2/r)$.

You should note that this involves no assumption about the existence of the α-particle in the nucleus, but merely the very reasonable assumption that the short-range nuclear forces are still acting after the α-particle has assumed its identity. The assumed discontinuous change in potential is merely an expression of our ignorance of the true variation of the potential near the nuclear surface.

Nuclear Physics

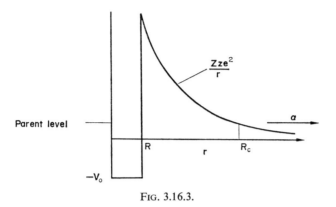

FIG. 3.16.3.

We now solve the time-independent Schrödinger equation

$$\nabla^2 \psi + \frac{2m}{\hbar^2}(E-V)\psi = 0 \qquad (3.16.1)$$

for this potential

$$\begin{aligned} V &= -V_0 & r < R, \\ V &= \frac{Zze^2}{r} & r > R. \end{aligned} \qquad (3.16.2)$$

The radial part of the Schrödinger equation is

$$\frac{1}{r^2}\frac{d}{dr}\left(r^2 \frac{d\mathcal{R}}{dr}\right) + \frac{2m}{\hbar^2}(E-V)\mathcal{R} - \frac{l(l+1)}{r^2}\mathcal{R} = 0 \qquad (3.16.3)$$

if we take a spherically symmetric potential and separate the equation. The last term contains the dependence of the radial function $\mathcal{R}$ on the orbital angular momentum and behaves like a repulsive potential, pushing wave functions away from the origin. In analogy with the coulomb barrier, terms of this kind are known as centrifugal barrier terms. Let us compare the importance of the coulomb barrier with the centrifugal barrier at the radius R:

$$\begin{array}{cc} \text{coulomb barrier} & \text{centrifugal barrier} \\ C(R) = \dfrac{Zze^2}{R} & G(R) = \dfrac{\hbar^2}{2m}\dfrac{l(l+1)}{R^2}, \end{array}$$

$z = 2$, $e = 4.8 \times 10^{-10}$ esu, $\hbar = 10^{-27}$ erg sec and $m \simeq 6.4 \times 10^{-24}$ gm.

Take $Z = 90$, $A = 230$, whence a reasonable value for R is $\sim 8 \times 10^{-13}$ cm.

$$\frac{G(R)}{C(R)} = \frac{\hbar^2}{2m}\frac{l(l+1)}{Zze^2}\frac{1}{R} \approx \frac{l(l+1)}{500}.$$

Angular momentum effects, then, may be expected to contribute to the problem of formation and ejection of the α-particle, but it seems reasonable to relegate them to this region of

Nuclear Decay

ignorance and drop the centrifugal barrier term in our evaluation of barrier penetrability, in α-decay.

With the dropping of the centrifugal barrier term the radial wave equation is reduced to a simple one-dimensional form by the substitution $\mathcal{R} = \phi/r$ whence

$$\frac{d^2\phi}{dr^2} + \frac{2m}{\hbar^2}(E-V)\phi = 0 \qquad (3.16.4)$$

and it is this equation we must solve for

$$V = -V_0 \qquad r < R,$$
$$V = \frac{Zze^2}{r} \qquad r > R$$

subject to the boundary condition that there exists only an outward going wave at large values of r. At such large values, characterized by $Zze^2/r \ll E$, we have

$$\frac{d^2\phi}{dr^2} + \frac{2mE}{\hbar^2}\phi = 0 \quad \text{or} \quad \phi \sim e^{\pm ikr} \quad k^2 = \frac{2mE}{\hbar^2}. \qquad (3.16.5)$$

For $r < R$

$$\frac{d^2\phi}{dr^2} + \frac{2m(E+V_0)}{\hbar^2}\phi = 0 \quad \phi \sim e^{\pm ik'r} \quad k'^2 = \frac{2m(E+V_0)}{\hbar^2}. \qquad (3.16.6)$$

Thus we have a wavelength $\sim h/\sqrt{2mE}$ at great distances. For intermediate values of r, the wavelength gets longer and longer as r decreases, becoming infinite at $r = R_c$, $R_c = Zze^2/E$. Thereafter the propagation factor k is imaginary until $r = R$, and for $r < R$ the wavefunction is oscillatory again, with a wavelength shorter than in free space. It is the intermediate region $R < r \lesssim R_c$ which presents a problem of calculation, because of the rapidly varying coulomb potential. Even for the simple form of coulomb barrier we have used, it is necessary to calculate the wave function numerically in order to get an accurate answer for the transition probability. However, we can get a good feeling for what goes on by using the W.K.B. approximation. The details are discussed in Section 4.13, rather than here, because of the importance of the coulomb barrier in nuclear reactions induced by charged particles. The result is that we can write

$$\phi_1 = \frac{c_1}{k_1^{1/2}} e^{\pm ik_1 r}, \quad k_1^2 = \frac{2m(E+V_0)}{\hbar^2} \qquad r < R \qquad (3.16.7)$$

which is an exact solution for k_1 constant

$$\phi_2 = \frac{c_2}{k_2^{1/2}} e^{\pm \int_R^r k_2 \, dr}, \quad k_2^2 = \frac{2m}{\hbar^2}\left(\frac{Zze^2}{r} - E\right) \qquad R < r < R_c, \qquad (3.16.8)$$

$$\phi_3 = \frac{c_3}{k_3^{1/2}} e^{\pm i \int_{R_c}^r k_3 \, dr}, \quad k_3^2 = \frac{2m}{\hbar^2}\left(E - \frac{Zze^2}{r}\right) \qquad R_c < r, \qquad (3.16.9)$$

which are approximate solutions valid provided k_2, k_3 are not too small.

Nuclear Physics

Now we make some physical assumptions. First, since the nucleus is decaying into a vacuum and beyond the coulomb barrier there is nothing to reflect the α-particle back in, we only want an outgoing wave in the region $r > R_c$.

Since the variation of the potential in the region $r \sim R_c$ is very much slower than the variation in the region $r \sim R$, reflections occurring at $r \sim R_c$ may be neglected in comparison with reflections at the discontinuity $r = R$. Thus in the region $R < r < R_c$ the wave function is like an exponential decaying with increasing r. Finally in the region $r < R$ we have sinusoidal oscillations propagating in both directions.

In treatments of α-decay it is frequently assumed that the α-particle wave function inside the coulomb barrier must be of the form

$$\phi_1 \sim \sin k_1 r = \frac{e^{ik_1 r} - e^{-ik_1 r}}{2i}.$$

This assumption is based on a requirement that the solution $\mathcal{R}$ of the radial part of the Schrödinger equation must be regular at $r \to 0$. This is only true if the α-particle exists permanently inside the nucleus and moves in a time independent potential—an independent α-particle model. In our treatment we have ignored the problems of α-particle formation and existence within the nucleus and merely assume that a fully formed α approaches the coulomb barrier. Thus we do not require the combination of $e^{ik_1 r}$ and $e^{-ik_1 r}$ to be regular at the origin, for we do not consider this region at all. While we may expect the intensities of the incoming and outgoing waves to be very nearly equal because of the efficiency of the coulomb barrier as a reflector, we may not assume any phase relation between them and write

$$\phi_1 = \frac{c_1}{k_1^{1/2}} \sin(k_1 r + \eta) \qquad r < R \tag{3.16.10}$$

where the phase between the two components is set by the detailed nuclear structure. Then

$$\phi_2 = \frac{c_2}{k_2^{1/2}} e^{-\int_R^r k_2 \, dr}, \qquad R < r < R_c, \tag{3.16.11}$$

$$\phi_3 = \frac{c_3}{k_3^{1/2}} e^{i \int_{R_c}^r k_3 \, dr}, \qquad R_c < r \tag{3.16.12}$$

and in Section 4.13 it is shown that on matching these functions across the boundaries at R, R_c we obtain

$$c_3 \sim c_1 \left[\frac{k_2(R)}{k_1}\right]^{1/2} \sin(k_1 R + \eta) \, e^{-\int_R^{R_c} k_2 \, dr} \tag{3.16.13}$$

and

$$\tan(k_1 R + \eta) = -\frac{k_1}{k_2(R)},$$

$$\phi_3 \sim -\frac{c_1}{k_1^{1/2}} \frac{k_1}{k_2(R)} \left[\frac{k_2(R)}{k_3(R)}\right]^{1/2} e^{-\int_R^{R_c} k_2 \, dr} \, e^{ik_3 r}. \tag{3.16.14}$$

Nuclear Decay

The quantity $c_1/k_1^{1/2}$ is the amplitude of the sinusoidal wave inside the barrier. At a time $t \sim 0$ all the wave function is inside and so we normalize

$$\int_{\text{nucleus}} \left|\frac{\phi_1}{r}\right|^2 dV \leq 1 \tag{3.16.15}$$

where equality means that the α-particle exists as an α all the time within the nucleus.

With

$$\phi_1 = A \sin(k_1 r + \eta),$$
$$A \sim \frac{1}{\sqrt{R}}. \tag{3.16.16}$$

The probability per unit time of an α-particle passing through a sphere of radius $r \gg R_c$ is

$$4\pi v_\alpha |\phi_3(\infty)|^2$$

where

$$v_\alpha = \sqrt{\frac{2E}{m}} = \frac{\hbar k_3(\infty)}{m}$$

and so the decay rate is given by

$$T \approx \frac{4\pi \hbar}{m} \frac{1}{R} \frac{k_1^2}{k_2(R)} e^{-2\int_R^{R_c} k_2\, dr} = F(E, R, Z, A)\, e^{-G} \tag{3.16.17}$$

where $F(E, R, Z, A)$ is a function which embodies an ignorance of the nuclear physics part of the problem, as opposed to the straightforward quantum mechanics of coulomb barrier penetration. It is to be expected that F is slowly varying in comparison with the exponential. The integral in the Gamow factor e^{-G}

$$G = 2\frac{\sqrt{2m}}{\hbar} \int_R^{R_c} \sqrt{\frac{Zze^2}{r} - E}\, dr \tag{3.16.18}$$

is evaluated in Section 4.13 and at energies $E \ll Zze^2/R$ is given by

$$G \sim \frac{2\pi Zze^2}{\hbar v}$$

where v is the velocity of the α-particle at $r \to \infty$, and is independent of R.

The α-emitters in the naturally occurring radioactive series (and among the transuranic elements) have α energies $E \sim 5\text{--}10$ MeV and the barrier height is ~ 30 MeV. The approximation

$$G \sim \frac{2\pi Zze^2}{\hbar v}$$

is not very good and the decay rate is rather sensitive to the value of the parameter R which appears in an exact expression for G (see Section 4.13). The occurrence of terms dependent

Nuclear Physics

on R in the argument of the exponential explains why values of R obtained from α-decay are in reasonable accord with the nuclear radii obtained by other methods. However, in view of the enormous area of ignorance contained in the factor $F(E, R, Z, A)$ it is clear that the values of R_0 obtained by setting $R = R_0 A^{1/3}$ must not be taken too seriously. Indeed, they are $\sim 30\%$ larger than the best values obtained by other methods.

We should finally consider briefly the order of magnitude of the factor $F(E, R, Z, A)$

$$F(E, R, Z, A) \approx \frac{4\pi\hbar}{m} \frac{1}{R} \frac{k_1}{k_2(R)} k_1. \qquad (3.16.19)$$

The dimensions of F are sec^{-1}, thus providing the correct dimensions for T. Now

$$\frac{\hbar k_1}{m} = v_1,$$

the velocity of the α inside the nucleus, and so

$$F \approx \frac{v_1}{R} \approx 10^{-20}\text{--}10^{-21} \text{ sec}^{-1}. \qquad (3.16.20)$$

The decay rate T may thus be visualized as made up of a term representing the frequency with which the α assaults the coulomb barrier and the Gamow factor e^{-G} which gives the probability of the assault succeeding. The use of a factor F of this order of magnitude in conjunction with the Gamow penetration factor gives values of $R_0 \sim 1.5 \times 10^{-13}$ cm. This value comes out about right because of the enormous variation in T with a small variation of R: for radium the calculated value of T varies by 10^{12} for a variation in R of a factor 2.[†]

3.17. α-decay and the Fermi Golden Rule

Our treatment of α-decay has been conceptually quite different from our treatments of β- and γ-decay. In the latter cases particles which in no sense exist on a semipermanent basis within the nucleus are created from the available energy, and we had no other way of proceeding than by making a direct coupling between the nuclear states and the appropriate fields. In the case of α-decay a group of four nucleons which already exist within the nucleus get together and make a united assault on the coulomb barrier, an assault which is occasionally successful. Thus for α-decay we have in non-relativistic quantum mechanics (and the most important restriction of non-relativistic mechanics is that particles are neither created or destroyed) a dynamical model of α-decay which can be calculated directly. However, it is interesting to apply the Fermi Golden Rule to the α-decay process.

This is made difficult by the fact that because of the coulomb potential we can no longer take our final state as a plane wave. We approach the problem as follows:

The initial nuclear state is composed of i nucleons and described by

$$\Psi_i(r_1 \ldots, r_i) \qquad \int \Psi_i^* \Psi_i \, d^3 r_1, \ldots, d^3 r_i = 1. \qquad (3.17.1)$$

[†] See R. D. Evans, *The Atomic Nucleus*, p. 77 (McGraw-Hill, 1955).

Nuclear Decay

The final state may be factored into a wave function describing the daughter nucleus, a wave function describing the α and a wave function describing the relative motions of the daughter and of the α-particle. The wave functions describing the internal motions of the constituent nucleons in the decay products, Ψ_d and Ψ_α, are normalized to unity, and we take the relative motion to be given by a plane wave normalized to 1 per unit volume, in the absence of a coulomb potential:

$$e^{i\mathbf{k}\cdot\mathbf{r}}$$

where $\mathbf{r}$ is the relative coordinate. Let us suppose the interaction takes place through an s-wave. The s-wave part is

$$\frac{e^{ikr} - e^{-ikr}}{2ikr}$$

in the absence of a coulomb potential, and we replace this in the region, $r > R_c$ by the form

$$u(r) = \frac{e^{i\left[\int_{R_c}^{r} k_3\, dr + \delta\right]} - e^{-i\left[\int_{R_c}^{r} k_3\, dr + \delta\right]}}{2ik_3^{1/2}k^{1/2}r} \tag{3.17.2}$$

where k_3 is a function of r, $k_3 \to k$ as $r \to \infty$ and δ a constant factor. As $r \to \infty$ this gets more and more like the s-wave part of a plane wave and it also agrees with our W.K.B. approximation on writing

$$u(r) = \phi_3(r)/r.$$

Thus

$$\phi_3(r) = \frac{\sin\left[\int_{R_c}^{r} k_3\, dr + \delta\right]}{k_3^{1/2}k^{1/2}} \tag{3.17.3}$$

may be matched to an interior solution. Since we are coupling a wave function determined without the nuclear interaction to an initial nuclear state, the interior function will be of the form

$$\phi_1(r) = A_1 \frac{\sinh\{k_2(R)r\}}{k_2(R)} \tag{3.17.4}$$

so that the unperturbed wave function ϕ_1/r is regular at the origin, where we take as the approximate form of the potential in the absence of strong interactions

$$V = \frac{Zze^2}{r} \quad r > R,$$

$$V = \frac{Zze^2}{R} \quad r < R.$$

It is this potential that would be felt by a positron. Matching ϕ_1 to ϕ_3 through the coulomb barrier gives us the complete wave function for the s-wave part of the final state in the presence of a coulomb barrier but before we switch on the nuclear interaction. The details of

Nuclear Physics

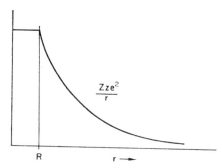

FIG. 3.17.1. Approximate form for the potential between an α-particle and the daughter nucleus in the absence of strong interactions. For simplicity we have taken a constant potential for $r < R$, corresponding to all the charge being concentrated at the nuclear surface.

such matching are discussed in Section 4.13 and we just quote the result here:

$$\frac{A_1}{k_2(R)} \sinh\left(k_2(R) R\right)$$

$$= \frac{1}{k_2(R)^{1/2} k^{1/2}} \left\{ (\sin \delta + \cos \delta) e^{-\int_R^{R_c} k_2\, dr} + (\sin \delta - \cos \delta) e^{\int_R^{R_c} k_2\, dr} \right\}, \qquad (3.17.5)$$

$$A_1 \cosh\left(k_2(R) R\right) =$$

$$= \left(\frac{k_2(R)}{k}\right)^{1/2} \left\{ -(\sin \delta + \cos \delta) e^{-\int_R^{R_c} k_2\, dr} + (\sin \delta - \cos \delta) e^{\int_R^{R_c} k_2\, dr} \right\} \qquad (3.17.6)$$

and

$$(\sin \delta - \cos \delta) e^{\int_R^{R_c} k_2\, dr} \ll (\sin \delta + \cos \delta) e^{-\int_R^{R_c} k_2\, dr} \qquad (3.17.7)$$

unless

$$\frac{k_2(R)}{k} \tanh\left(k_2(R) R\right) \simeq -1$$

which it never does, for k_1, $k_2(R)$ and R are all positive quantities. Then

$$A_1 \sim \left[\frac{k_2(R)}{k}\right]^{1/2} e^{-\int_R^{R_c} k_2\, dr} \approx e^{-\int_R^{R_c} k_2\, dr}. \qquad (3.17.8)$$

The matrix element to be inserted in the Golden Rule is thus

$$\int^R \left[\Psi_d^* \Psi_\alpha^* \frac{A_1 \sinh\left(k_2(R) r\right)}{k_2(R) r} V_s \Psi_i \right] d^3 r_1, \ldots, d^3 r_{i-1}\, d^3 r \qquad (3.17.9)$$

where V_s is responsible for the coupling of the initial to the final state and is a function of all the coordinate variables.

Nuclear Decay

Now the relative wave function $u(r)$ has been normalized to 1 per unit volume, while Ψ, Ψ_d and Ψ_α are all normalized to unity. Ψ_i contains one more coordinate variable than Ψ and Ψ_α because one of these variables is accounted for as r in the final state function. Then integration over everything except r yields a term

$$\approx \int^R \frac{1}{R^{3/2}} \frac{A_1 \sinh(k_2(R)r)}{k_2(R)r} \bar{V}_s \, d^3r \tag{3.17.10}$$

where $\bar{V}_s$ is the appropriate average of V_s over all variables except r. The final integration gives a matrix element M

$$M \approx R^{3/2} A_1 \bar{V}_s \approx R^{3/2} e^{-\int_R^{R_c} k_2 \, dr} \bar{\bar{V}}_s \tag{3.17.11}$$

where $\bar{\bar{V}}_s$ is the overall average interaction responsible for formation of the α-particle. Inserting this in the Golden Rule

$$T = \frac{2\pi}{\hbar} |M|^2 \varrho$$

we obtain

$$T = \frac{2\pi}{\hbar} [\bar{\bar{V}}_s^2 R^3 e^{-G}] \varrho \tag{3.17.12}$$

where

$$\varrho = \frac{4\pi p^2 \, dp}{(2\pi\hbar)^3 dE}$$

where p is the α particle momentum as $r \to \infty$.

Now because the daughter is massive, $dE \approx dE_\alpha$ and so

$$\varrho = \frac{4\pi}{(2\pi\hbar)^3} m^2 v_\alpha = \frac{mk}{2\pi^2 \hbar^2}$$

so

$$T \approx \frac{mk}{\pi\hbar^3} \bar{\bar{V}}_s^2 R^3 e^{-G}$$

and

$$F(E, R, Z, A) \approx \frac{mk \bar{\bar{V}}_s^2 R^3}{\pi\hbar^3} \tag{3.17.13}$$

which has dimensions sec^{-1}, the value depending on $\bar{\bar{V}}_s$. For a 10 MeV α we find on putting in numbers

$$F \approx 10^{23} \bar{\bar{V}}_s^2 \text{ sec}^{-1} \tag{3.17.14}$$

for $R = 10^{-12}$ cm and $\bar{\bar{V}}_s$ in MeV. Thus if $\bar{\bar{V}}_s$ is 0.1 MeV, 10^{-2} of the particle energy, F is indeed $\sim 10^{21}$ sec^{-1}.

With this treatment of α-decay, our fourth application of the Fermi Golden Rule, we terminate this discussion of nuclear decay. However, a nuclear decay can only take place after the state has been formed, and so in this chapter we have really chopped nuclear reaction processes in half. In the next chapter we consider the complete chains.

CHAPTER 4

Nuclear Reactions

4.1. Introduction

A nuclear reaction takes place when an initial state involving nucleons is converted into a different final state involving nucleons. There is thus no cut and dried distinction between nuclear and atomic reactions. Suppose a high-energy neutron is being scattered by a nucleus under ordinary laboratory conditions. At a very small scattering angle the whole atom will recoil, while at large scattering angles the impulse to the nucleus can be large enough to disrupt the atom, the nucleus recoiling and some of the less strongly bound atomic electrons staying where they are, because of their inertia, for a time sufficient to break the bonds. Whether a process of this kind is called a nuclear reaction or an atomic reaction depends only on what is being studied. We talk about nuclear reactions when we do not care what happens to the atomic electrons. Thus most nuclear reactions involve energies sufficient to ionize atoms—a notable exception being reactions involving thermal neutrons, which are exceptions only because of the very low interaction of neutrons and electrons. At the other end of the energy scale nuclear reactions merge into particle physics. A convenient marker which may be used to arbitrarily separate nuclear reactions and particle physics is the kinetic energy needed to create a single pion, of mass ~ 140 MeV/c^2. In a reaction with a very massive nucleus this kinetic energy is 140 MeV, but in an interaction between two nucleons the single pion threshold is ~ 300 MeV. Loosely we may take energies below 100 MeV as the domain of nuclear reactions, energies between 100 and 1000 MeV as the domain of intermediate energy physics and energies greater than 1000 MeV (1 GeV) as the domain of high energy, or particle, physics.

Thus in this chapter we shall discuss processes in which a projectile interacts with a target and the total kinetic energy in the centre of mass is $\lesssim 100$ MeV. The targets are nuclei—either a single proton or a complex nucleus. The projectiles may be nucleons, complex nuclei, photons, electrons, neutrinos or any particle with a lifetime $\gtrsim 10^{-8}$ sec. This again is not a clear-cut line—a highly unstable particle can be produced in one interaction and interact before decaying in either the same nucleus or a neighbouring nucleus. If we restrict ourselves, however, to reactions in which the projectiles can be produced in a well-collimated beam with good momentum resolution then the lifetime restriction applies.

While we can have any number of particles in the final state (subject to the constraints of

Nuclear Reactions

conservation of momentum and energy) we will only be considering initial states consisting of two particles (or nuclei). Under laboratory conditions it is not possible to study nuclear reactions involving a three-particle initial state, and such reactions could only be of comparable importance with two-particle reactions under extreme conditions. For example, in the cores of highly evolved stars helium burning proceeds through

$$^4He + {}^4He + {}^4He \rightarrow {}^{12}C + photons$$

and even this can be represented as a two-stage process involving the (unstable) ground state of ^8Be:

$$^4He + {}^4He \rightleftharpoons {}^8Be; \quad {}^4He + {}^8Be \rightarrow {}^{12}C + photons.$$

4.2. Qualitative features of nuclear reactions

It is convenient to distinguish three categories of reaction.

1. *Elastic scattering*

The initial and final state particles (or nuclei) are the same before and after the interaction, and in the centre of mass emerge from the interaction with the same energies (but different directions) that they had before the interaction took place, for example

$$e^- + {}^{12}C \rightarrow e^- + {}^{12}C$$

or

$$n + {}^{238}U \rightarrow n + {}^{238}U.$$

You should note, however, that charge exchange reactions are sometimes referred to as elastic scattering

$$\pi^- + p \rightarrow \pi^0 + n$$

and even

$$\nu + {}^A_Z N \rightarrow e^- + {}^A_{Z+1} N.$$

2. *Inelastic scattering*

In this case one of the initial particles retains its identity and the recoiling nucleus is raised to an excited state. The kinetic energy in the centre of mass is lower after the interaction than before

$$p + {}^{14}N \rightarrow {}^{14}N^* + p; \quad {}^{14}N^* \rightarrow {}^{14}N + \gamma$$

and an example from high-energy physics

$$\pi^+ + p \rightarrow N^*(1236) + \pi^+; \quad N^*(1236) \rightarrow p + \pi^0$$

where the highly unstable state $N^*(1236)$ with a mass of 1236 MeV/c^2 is frequently called a nucleon isobar.

It is clear that this category tends to overlap with the next:

189

Nuclear Physics

3. *Inelastic reactions*

This category may be held to embrace everything not in the previous two, for example

$$n + {}^{113}\text{Cd} \to {}^{114}\text{Cd} + \gamma,$$
$$p + {}^{14}\text{N} \to p + p + {}^{13}\text{C}$$

and so on.

The second example serves to indicate the lack of distinction between inelastic scattering and inelastic reactions. Under the heading of inelastic scattering we considered formation of an excited state of ^{14}N stable against nucleon emission. Some part of the reaction above will go via formation of excited states of ^{14}N unstable against proton emission, and the distinction between this inelastic scattering process and the reaction background is lost unless the excited state has a narrow width. For high excitations the inelastic scattering process merges into the background inelastic reaction.

A further approximate classification may be made on the basis of energy and wavelength. If the wavelength of the incident particle is large in comparison with the size of the nucleus, then the scattering will be insensitive to the detailed structure of the nucleus and the nucleus as a whole will act coherently. At higher energies, as the wavelength of the incident particle becomes shorter the interaction takes place with a local cluster of nucleons, and at energies such that the wavelength is about the size of a nucleon the primary interaction will take place with an individual nucleon in the nucleus, although secondary interactions with other nucleons can leave the residual nucleus in a highly excited state which can evaporate nucleons and eventually decay to a ground state electromagnetically. At several hundred MeV and greater energies meson production becomes important and there may be enough energy left over to disrupt the nucleus more or less completely.

The wavelength of an incident nucleon of kinetic energy T and velocity v is

$$\lambda = \frac{h}{p} = \frac{h}{mv} = \frac{h}{\sqrt{2mT}} = \frac{hc}{\sqrt{2mc^2 T}},$$
$$T = \frac{h^2}{2m\lambda^2}. \tag{4.2.1}$$

If $\lambda \simeq 10^{-12}$ cm, $T \simeq 7$ MeV. Thus from wavelength considerations alone we expect the nucleus to act as a whole until energies $\gtrsim 10$ MeV are reached. This is hardly surprising —the mean binding energy is ~ 8 MeV/nucleon through most of the periodic state, and the nucleon separation energy (which, of course, is not necessarily the same) is also ~ 5–8 MeV for medium and heavy nuclei.

A wavelength of 10^{-13} cm implies a kinetic energy up by a factor of 100 on the energy for a wavelength of 10^{-12} cm, and so the non-relativistic formula is no longer valid. We must write instead

$$E = T + mc^2; \quad p = \sqrt{E^2/c^2 - m^2 c^2} \tag{4.2.2}$$

whence $T \sim 500$ MeV, and this is in the range where meson production has already become important.

Nuclear Reactions

We may summarize these points as follows:

1. Below the nucleon separation energy ($\sim$ 5–8 MeV) the nucleus will act as a whole.
2. Interactions in which the nucleus acts coherently may be expected to predominate over incoherent interactions in an energy range reaching into a few tens of MeV.
3. In the high tens and low hundreds of MeV interactions with local clusters of nucleons may be expected to be most important, dominating both coherent interactions and interactions with a single nucleon.
4. As the bombarding energy increases into the region of several hundred MeV, meson production dominates the inelastic processes and the structure of the nucleus is of little importance, merely complicating the study of the elementary particle physics—most experiments in high-energy physics are made with protons as the target particles.

The angular momentum states likely to be important may also be estimated in a trivial way. We have used before the result that the partial wave components of an incident plane wave behave as $\sim [kr]^l$ for $kr \ll 1$, where l is the orbital angular momentum of a given partial wave. If R is $\sim$ the nucleon radius then for $kR \ll 1$ we may expect only the s-wave components of the incident plane waves to be important, or more precisely we expect cross-sections varying roughly as $[kR]^{2l}$, the lowest partial waves predominating.

If $R = 5 \times 10^{-13}$ cm, a nucleon of 1 MeV corresponds to $kR \approx 1$ to be compared with photon or electron momenta of $\sim$ 40 MeV. Thus we expect nuclear interactions at energies of $\lesssim 1$ MeV to involve s-waves in most cases, higher partial waves coming in with comparable frequencies only above an energy of ~ 1 MeV for nucleons, and higher for lighter particles.

The time scale for a nuclear reaction is very dependent on the kind of reaction. If we consider energies sufficiently high that the projectile is not absorbed so as to yield a long-lived excited state, then the time for a reaction must be very approximately the time taken to cross the nucleus, for those reactions with a cross-section close to geometric, πR^2. That is, in a semiclassical picture, if the projectile entering the nucleus at all is followed by the reaction, the cross-section is πR^2 and the time taken must be $\sim R/v$ where v is the velocity of the projectile. For a nucleus $\pi R^2 \sim 10^{-26}$–10^{-24} cm^2. The quantity 10^{-24} cm^2 is the unit of cross-section used in particle and nuclear physics, and is called 1 barn. Velocities are typically $\sim 10^9$–10^{10} cm/sec (a nucleon of energy only 1 eV has a velocity $\sim 10^6$ cm/sec and at 1 MeV a velocity of $\sim 10^9$ cm/sec). We may therefore expect, knowing that the cross-section for nuclear reactions is typically 1 barn, that the time scale for direct reactions will be $\sim 10^{-20}$ sec. A coherent reaction which proceeds via formation of an intermediate state, living perhaps 10^{-16} sec, can be broken into three stages: the formation process which must take place in the time it takes for the projectile to cross the nucleus, that is $\sim 10^{-20}$ sec, the time during which the intermediate state propagates through space, and finally the time during which the state is breaking up, again $\sim 10^{-20}$ sec. Thus we may regard 10^{-20} sec as a typical nuclear time scale, regardless of the fact that low-energy resonant reactions may take many orders of magnitude longer for completion even though they have a large cross-section. Indeed, the conjunction of a large cross-section with a long-lived intermediate state, corresponding to a narrow width in energy, is the signature of compound nucleus formation in a nuclear reaction.

Nuclear Physics

4.3. The concept of cross-section

Throughout this book so far we have used the term cross-section without ever discussing the concept, in the hope that you have an intuitive understanding of this idea. In fact, on the quantum mechanical level the concept of cross-section exhibits such differences from a classical picture that it is necessary to discuss it in some considerable detail.

We will begin with a discussion of the classical definition. Suppose that we are considering the scattering of two billiard balls (in space, so as to have motion in three dimensions). Drop a perpendicular from the centre of the target back onto the initial path of the projectile ball: this is the *impact parameter b*

FIG. 4.3.1.

If $b > 2R$, where R is the radius of the balls, then there is no scattering. If $b < 2R$ then scattering takes place. We associate with the target an imaginary disc, of such area that if the line of motion of the centre of the projectile intersects this disc, then a scattering takes place. The area of this disc is the *total scattering cross-section*, and for this case is clearly $4\pi R^2$. The differential cross-section may be defined similarly. Suppose that for an impact parameter b, the scattering angle is θ.

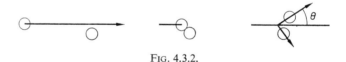

FIG. 4.3.2.

Then we associate with the target an imaginary annulus, such that if the line of motion of the centre of the projectile intersects this annulus, then the scattering angle lies between θ and $\theta + d\theta$.

Then

$$d\sigma = 2\pi b(\theta)\, db(\theta) = 2\pi b(\theta)\, \frac{db(\theta)}{d\theta}\, d\theta \qquad (4.3.1)$$

and we can define a differential scattering cross-section

$$\frac{d\sigma}{d\theta} = 2\pi b(\theta)\, \frac{db(\theta)}{d\theta} \qquad (4.3.2)$$

where $b(\theta)$ can be expressed in terms of θ (and the bombarding energy) from the dynamics of the collision of perfect billiard balls: the impulse acts along the line joining the centres of the two and this gives the change in momentum. It is usually more convenient to define

Nuclear Reactions

a differential cross-section in terms of solid angle, and write

$$\frac{d\sigma}{d\Omega} = \frac{1}{2\pi \sin\theta} \frac{d\sigma}{d\theta} = \frac{b(\theta)}{\sin\theta} \frac{db(\theta)}{d\theta} \tag{4.3.3}$$

which gives the probability of scattering into unit solid angle at an angle θ, when b is randomly distributed.

If instead of billiard balls we want to work with a long-range interaction, consider scattering of a particle in an inverse square force field generated by an infinitely massive object. The orbits of a particle in such a field are conic sections, and if the system has positive energy they are hyperbolae with the massive source at one focus

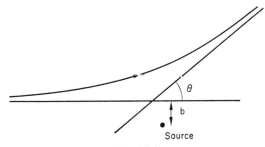

Fig. 4.3.3.

This is the scattering of a glob of matter unattached to the sun under the influence of solar gravitational field—if any of the comets do not follow closed orbits then their motion is described by this process. It is also the scattering of an electron by a charged sphere, and of course gives the famous formula for Rutherford scattering of α-particles by nuclei, which was first worked out with classical assumptions. In almost any elementary book on atomic physics it is shown that the relation between the impact parameter b and the scattering angle θ for an inverse square field is

$$b = \frac{zZe^2}{mv^2} \cot\frac{\theta}{2} \tag{4.3.4}$$

where ze is the charge on the projectile, Ze the charge on the scattering centre, m the mass of the projectile and v its asymptotic speed. For a gravitational field, Zze^2 is replaced by GmM where M is the mass of the source of the field.

Then

$$\frac{d\sigma}{d\theta} = 2\pi \left\{ \frac{zZe^2}{mv^2} \cot\frac{\theta}{2} \right\} \left\{ \frac{zZe^2}{mv^2} \frac{1}{2} \operatorname{cosec}^2\frac{\theta}{2} \right\} \tag{4.3.5}$$

whence the Rutherford formula

$$\frac{d\sigma}{d\Omega} = \frac{1}{4} \left(\frac{zZe^2}{mv^2} \right)^2 \operatorname{cosec}^4\frac{\theta}{2}. \tag{4.3.6}$$

Of course, if we know the impact parameter for a given case, the concept of cross-section has no use. It is important, however, if we throw a projectile at a target, or an array of targets, and we want to know the *probability* of a particular scattering taking place. Note

Nuclear Physics

that for the classical limit it is not a question of our being unable in principle to predict exactly what happens, it is merely that we do not have all the information beforehand. So suppose we have a single target, and hurl at it a very large number of projectiles, all with the same initial direction and speed, but located at random in a plane normal to this initial direction. On average, how many will be scattered into unit solid angle at an angle θ by the target? If there are N projectiles spread over a large area A, then the probability of any one passing through an annulus of radius between b and $b+db$ is given by

$$\frac{2\pi b\, db}{A} \tag{4.3.7}$$

and the number passing through this annulus is

$$\frac{2\pi b\, db}{A} N. \tag{4.3.8}$$

The number scattered through an angle between θ and $\theta+d\theta$ is thus

$$2\pi b(\theta)\, db(\theta)\, \frac{N}{A} \tag{4.3.9}$$

and the number scattered into a unit of solid angle between θ and $\theta+d\theta$

$$\frac{b(\theta)\, db(\theta)}{\sin\theta\, d\theta}\, \frac{N}{A} = \frac{N}{A}\, \frac{d\sigma}{d\Omega}. \tag{4.3.10}$$

In any real situation we produce a beam of projectiles spread over a certain area and with a well-defined velocity. The number each second passing through a plane of area A normal to the beam direction is nvA where there are n projectiles per unit volume, of velocity v. The number scattered per second into a unit of solid angle at θ is thus on average

$$n(\theta) = \frac{nvA}{A}\, \frac{d\sigma}{d\Omega}. \tag{4.3.11}$$

The number passing through unit area each second is the flux f, given by $f = nv$. Thus the number scattered each second into unit solid angle at θ is given by

$$n(\theta) = f\, \frac{d\sigma}{d\Omega} \tag{4.3.12}$$

and the differential scattering cross-section is thus equivalent to the number of particles scattered through an angle θ into unit solid angle each second when a single target is immersed in an (infinite) beam of unit flux. Integration shows at once that the total scattering cross-section is equal to the number of particles scattered *at all angles* each second when a single target is immersed in a beam of unit flux. The Rutherford total cross-section is infinite—which simply means no particles emerge undeviated.

Let us apply the concept of cross-section to elastic scattering in a statistical situation, still sticking to classical physics. Suppose we direct a beam of particles onto a thick target and we want to know how many emerge undeviated—a single scattering alone being sufficient to remove a particle from the beam.

We first consider an initial slice of target material so thin that the imaginary disc associated with each individual target overlaps no other. Let there be N targets per unit volume of target material. Then the scattering cross-section of each is σ and in a thin slice of thickness dt there are $N\,dt$ per unit area. The cross-sectional area for scattering presented to the beam is thus $N\,dt\,\sigma$ per unit area of target. The probability of scattering of a beam particle is thus $N\sigma\,dt$ and so if the flux is f then

$$df = -fN\sigma\,dt \tag{4.3.13}$$

and

$$f = f_0 e^{-N\sigma t} \tag{4.3.14}$$

(note dimensions are correct) or

$$f = f_0 e^{-t/\lambda} \quad \text{where} \quad \lambda = \frac{1}{N\sigma} \tag{4.3.15}$$

and is the mean distance travelled without deflection. If particles are absorbed as well as scattered, then the flux emerging undeflected after traversing a thickness of target material t is

$$f = f_0 e^{-N(\sigma_{\text{sc}}+\sigma_{\text{ab}})t} \tag{4.3.16}$$

and the number of reactions taking place per second in a thin slice is

$$fN\sigma_{\text{abs}}\,dt.$$

Thus cross-section is essentially a statistical concept even classically, but can be readily measured experimentally and is intimately linked to the dynamics of individual scattering processes. If these are understood, the cross-section can be computed from the relationship between impact parameter and the quantity of interest, for example θ. But classically the impact parameter is a vital quantity in linking the dynamics of individual processes with the operational definition of the cross-section, and herein lies the difficulty in making the transition to quantum processes: the uncertainty principle knocks on the head the idea of a well-defined impact parameter.

A quantum mechanical treatment of any process must reduce to the classical treatment in the macroscopic limit: one way of moving to this limit is to let $\hbar \to 0$. Classically we may consider a particle localized in space, with both its position and momentum as well-defined as may be desired. Microscopically the uncertainty principle sets a limit: $\Delta x \Delta p_x \sim \hbar$. If the impact parameter b is to be a meaningful concept, then this requires $\Delta p \sim \hbar/b$ and $b \sim 10^{-12}$ cm for nuclear processes. Then $\Delta p \sim 10^{-15}$ cgs units ~ 20 MeV/c which goes up to ~ 200 MeV/c, for $b \sim 10^{-13}$ cm. To invert the process, consider how a beam of particles is defined. The energy or momentum may be defined either in the acceleration process, or by passing through a momentum or velocity analyser. The line of flight may be defined by a series of collimators, or by counters. For a beam well defined in position, such collimators or counters may have a side ~ 1 mm. The uncertainty in momentum induced thereby is $\sim 10^{-26}$ cgs units, which is completely negligible. Without using nuclear processes themselves to define where a particular beam particle went on its way to the target, the *best* definition we can imagine is $\sim 10^{-8}$ cm, the diameter of an atom. If a particular atom, which we suppose we can label, has an electron ejected from it, then the beam particle went through that atom. Localization of the beam to $\sim 10^{-8}$ cm only introduces an uncertainty in momentum of $\Delta p \sim 10^{-19}$ cgs units. Thus without introducing enormous and undesired uncertainties in momentum through the use of nuclear scattering, positions in the beam may be defined down to $\sim 10^{-8}$ cm and momentum may in principal be defined down to $\sim 10^{-19}$ cgs units, $\sim 10^{-3}$ MeV/c if full use is made of this ideal accuracy. In any real set-up, a beam carries a spread in energy or momentum that is perhaps ~ 0.1–1%.

Thus in any real experiment with a reasonably well-defined momentum (better than 1%) the bombarding particles are never experimentally localized to better than $\sim 10^{-8}$ cm which is $\sim 10^5$ nucleon diameters.

The wave mechanical representation of all this depends on the concept of a wave packet, already familiar from other wave phenomena. The Schrödinger equation may be written

$$\nabla^2 \psi + k^2 \psi = 0$$

where k is independent of position for the beam, and is instantly solved to give a solution

$$\psi = A e^{i\mathbf{k} \cdot \mathbf{r}}$$

for a given value of k. If more than one value of k is allowed, then

$$\psi = \int_{-\infty}^{\infty} A(k) e^{i\mathbf{k} \cdot \mathbf{r}} \, dk \tag{4.3.17}$$

is a general solution, the form of $A(k)$ being determined by the boundary conditions. It is convenient to consider the usual example in which $A(k)$ is gaussian, and for simplicity we consider one dimension alone

$$\psi(x) = \int_{-\infty}^{\infty} A(k_x) e^{i k_x x} \, dk_x. \tag{4.3.18}$$

Set $A(k_x) = A_0 e^{-(k_x - k_0)^2/2\sigma^2}$ whence a little algebra rapidly yields

$$\psi(x) \propto e^{-\sigma^2 x^2 / 2} e^{i k_0 x}. \tag{4.3.19}$$

Putting in time dependence explicitly

$$\psi(x, t) = \int_{-\infty}^{\infty} A(k_x) e^{i(k_x x - \omega t)} \tag{4.3.20}$$

where ω is a function of k_x alone for a one-dimensional problem. For a photon $\omega = kc$; empty space is not a dispersive medium for a photon. For a massive particle ω is not a linear function of k, and empty space is a dispersive medium. (This is clear from the difference between phase velocity and particle velocity. The phase velocity v_p is given by $px - Et$ = constant, or $v_p = E/p$ while the particle velocity is equal to pc^2/E, where E is the relativistic energy. The two are only the same for massless particles with $E = pc$.) So as not to introduce the complications of dispersion into the problem, we suppose ω to be a linear function of k, $\omega = v_p k$

$$\psi(x, t) \propto e^{-(\sigma^2/2)(x-v_p t)^2} e^{i(k_0 x - \omega_0 t)} \tag{4.3.21}$$

which is a gaussian in x (this is why we chose a gaussian in k_x of course—the Fourier transform of a gaussian is another gaussian) which propagates with velocity v_p and unchanged in shape; that is, no dispersion.[†] The width of this pulse in space, Δx, is related to the width in momentum Δp by setting

$$\frac{\sigma \Delta x}{\sqrt{2}} = 1 \quad \text{and} \quad \frac{\Delta k_x}{\sqrt{2}\sigma} = 1$$

whence

$$\Delta x \Delta k = 2 \quad \text{or} \quad \Delta x \Delta p_x = 2\hbar$$

and this is the way the uncertainty relation emerges from Schrödinger's equation. (The factor 2 is unimportant and depends merely on the definition of Δx and Δk_x.) We rewrite $\psi(x, t)$ as

$$\psi(x, t) \propto e^{i(k_0 x - \omega_0 t)} e^{-(\sigma^2/2 k_0^2)(k_0 x - \omega_0 t)^2}$$

and it is at once clear that as $\sigma \to 0$, that is, as the momentum becomes better and better defined, then $\psi(x, t)$ tends more and more to the form of a plane wave. Even if we define position and momentum by looking at electrons ejected from specific atoms, on a nuclear scale the wave function of a beam particle is admirably represented as a plane wave

$$\psi(x, t) \propto e^{i(kx - \omega t)}.$$

The square modulus of this is independent of x and t, implying that the particle has an equal probability of being anywhere in space. This is not, of course, really true. For an exact treatment we ought to use wave packets, but it is adequate and very convenient to represent the wave functions of beam particles by plane waves, remembering that they are really portions of wave packets of rather well defined momentum and do not really stretch to infinity. This is, of course, exactly the same problem as is encountered in optics, when we treat a beam of light falling on a diffracting screen as a plane wave. It is not, because the apertures of the optical components producing our "parallel" beam have themselves diffracted it. However, over a sufficiently small range of distances it may be represented as a plane wave, and it is only necessary to take account of the real form in ensuring that integrals such as

$$\int_x^\infty e^{iky} dy$$

converge, which is conveniently done by introducing some extra factor like $e^{-\varepsilon y}$ to remove the contribution from the infinite upper limit. But physically what this means is that there is no such thing as a plane wave.

Having seen that in a beam having momentum sufficiently well defined to do experiments with, the particles are well represented on the nuclear level by plane waves and consequently

---

[†] For a discussion of what happens with dispersion, we refer you to, for example, J. D. Jackson, *Classical Electrodynamics*, chap. 7.3 (Wiley, 1962), or H. J. J. Braddick, *Vibrations, Waves and Diffraction*, chap. 5 (McGraw-Hill, 1965).

Nuclear Physics

the concept of impact parameter has no meaning at this level, how do we connect the operational definition of cross-section with the dynamics of the system being studied?

In a reaction we have an initial state which we can represent by a plane wave and a target which is transformed into a different final state by the interaction between the projectile and the target. The final state may differ only in the direction of the relative momentum vector of the projectile, as for elastic scattering, or it may contain the debris of a nuclear breakup, but in all cases it is a different final state, and the nuclear reaction may be discussed in the same way as any other kind of nuclear transition. Indeed, we should note that the whole of the preceding chapter on nuclear decay really dealt with nuclear reactions, for it is necessary to make an unstable state before you can examine its decay properties. In the previous chapter we chopped the complete process in two and concentrated only on the decay stage, which was adequately isolated from the formation process by the relatively long lifetimes of the unstable states considered.

4.4. Nuclear reactions and the Fermi Golden Rule

It is convenient, then, to begin a quantum mechanical treatment of nuclear reactions by working from the Fermi Golden Rule

$$T = \frac{2\pi}{\hbar} |M|^2 \varrho$$

where T is the transition rate between the initial and final states, ϱ an element of phase space for the final state and M is the matrix element connecting the initial and final states. The matrix element M may be relatively simple and involve no intermediate states, as for *direct* nuclear reactions, or it may be complicated, involving the formation and subsequent decay of an intermediate compound nuclear state. In either case, however, the general form remains.

In Section 3.2 we derived the Fermi Golden Rule for an initial nuclear (or atomic) state. In a reaction the initial state consists not of an excited nucleus or atom which has already been prepared, but rather of a product of wave functions describing the internal structures of the projectile and the target, and a plane wave describing their relative motion. We may visualize this set up as an initial two-particle system confined within a box and being steadily turned into a final two (or more) particle system confined within the same box and in this case it is obvious that the same result obtains for the transition rate.

However, in calculating the decay rate of a state we noted that because of the decay of the quantum mechanical amplitude describing this state, it does not have a well-defined energy. In a nuclear reaction we are usually interested in a steady state situation, where the initial state has a well-defined energy and is maintained at constant amplitude by a source external to the normalization volume. The final state is continuously fed from this initial state, but does not build up to an infinite amplitude once the transient effects have died out because the wave function continuously escapes through the boundaries of the normalization volume. The problem is then analogous to determining the response of a pair of coupled oscillators driven at the natural frequency of the first, rather than of studying the decay of current in a damped oscillatory circuit. We can therefore write the coupled equations

$$i\hbar \frac{\partial A_1}{\partial t} = \langle 1|H|1\rangle A_1 + \langle 1|H|2\rangle A_2 + Ae^{\lambda t}, \\ i\hbar \frac{\partial A_2}{\partial t} = \langle 2|H|1\rangle A_1 + \langle 2|H|2\rangle A_2 - i\alpha A_2, \quad (4.4.1)$$

where $i\hbar\lambda = E_1$, the energy of the initial state, and A is the driving term. Because of the term $Ae^{\lambda t}$ the amplitude A settles down to a steady state

$$A_1(t) = A_1^0 e^{\lambda t}$$

and the amplitude A_2 settles down in the steady state to

$$A_2(t) = \frac{\langle 2|H|1\rangle A_1^0 e^{\lambda t}}{i\hbar\lambda - \langle 2|H|2\rangle + i\alpha} \quad (4.4.2)$$

where α gives the leakage rate for the final state. The rate at which A_1 feeds A_2 is given by

$$\frac{\partial A_{12}}{\partial t} = \frac{1}{i\hbar}\langle 2|H|1\rangle A_1 = \frac{1}{i\hbar}\langle 2|H|1\rangle A_1^0 e^{\lambda t}. \quad (4.4.3)$$

We may now calculate the transition rate T_{12} from state 1 into state 2. In the steady state $\partial |A_2(t)|^2/\partial t = 0$ but taking the appropriate piece of the time derivative

$$T_{12} = \left(\frac{\partial A_{12}}{\partial t}\right)^* A_{12} + \frac{\partial A_{12}}{\partial t} A_{12}^* \\ = |A_1^0|^2 |\langle 2|H|1\rangle|^2 \left\{\frac{i}{\hbar}\frac{1}{E_1 - E_2 + i\alpha} + \frac{1}{i\hbar}\frac{1}{E_1 - E_2 - i\alpha}\right\} \\ = \frac{2\alpha}{\hbar}\frac{|A_1^0|^2 |\langle 2|H|1\rangle|^2}{(E_2 - E_1)^2 + \alpha^2}. \quad (4.4.4)$$

The transition rate into the continuum states surrounding E_1 (of width $\sim\alpha$) is given by

$$T_{1f} = |A_1^0|^2 |\langle 2|H|1\rangle|^2 \frac{dn_f}{dE} \int_0^\infty \frac{2\alpha dE}{(E-E_1)^2 + \alpha^2} \\ = \frac{2\pi}{\hbar} |A_1^0|^2 |\langle 2|H|1\rangle|^2 \frac{dn_f}{dE} \quad (4.4.5)$$

which is independent of α. For an infinitely large normalization volume, the amplitude in the secondary state never escapes, and so we have transitions from an initial continuum state with well-defined energy E_1 into a final continuum state at the same energy. We can easily choose the driving term A so as to preserve the normalization of the initial state, and setting $|A_1^0|^2 = 1$ we have

$$T = \frac{2\pi}{\hbar} |M|^2 \varrho$$

once more. This treatment is easily extended to the case where an intermediate state is involved: we defer this topic until Section 4.12.

Nuclear Physics

In Chapter 3 we had plenty of experience of dealing with the particles in the final state: let us now concentrate on the description within the plane wave framework of the initial state, which we will always consider as a two-particle, or more generally two-nucleus, system. So as to suppress the details of internal nuclear motion, we suppose that the initial state consists of two particles, each described by a single coordinate vector. To make the problem even simpler, let the target particle be very massive so that the centre of mass system coincides with the laboratory system. The target is at rest in the laboratory, and therefore is not localized and must be represented by its own plane wave—which simply means that the target has equal probability of being anywhere.

The target wave function is $e^{i\mathbf{k}_T \cdot \mathbf{r}_T}$ and the particle wave function is $e^{i\mathbf{k}_p \cdot \mathbf{r}_p}$. For a massive target, the target velocity $\to 0$ as the mass $\to \infty$ but we have $\mathbf{k}_T = -\mathbf{k}_p$ if we are working in the centre of mass system, which in the limit of a massive target coincides with the laboratory system. Thus the initial wave function is

$$e^{i\mathbf{k}_p \cdot (\mathbf{r}_p - \mathbf{r}_T)}$$

and only the relative coordinates enter. The interaction will be a function of this relative coordinate. In order to normalize these wave functions, we use the same device as was used for the final state wave functions in Chapter 3 and suppose the whole system to be contained in a box, of volume V. The normalized initial wave function is thus $e^{i\mathbf{k} \cdot \mathbf{r}}/V$ where $\mathbf{k}$ is the relative momentum vector and $\mathbf{r}$ the relative coordinate vector of the two particles. The final state particles, at a great distance from the interaction, are also represented by plane waves, and one of the normalization factors of $1/\sqrt{V}$ which enters for each of these wave functions is absorbed as before into the phase space term. This factor ϱ thus contains no volume terms. Then

$$T = \frac{2\pi}{\hbar} \frac{1}{V} \left| \frac{M}{V} \right|^2 \varrho \qquad (4.4.6)$$

where the plane wave functions in M are normalized to unit volume and the phase space contains no factors of V. The factor $1/V$ remaining multiplying M arises from one factor from the initial state and one from the final state. There is only one factor of V in a two-particle phase space. Equation (4.4.6) gives the transition rate, either partial or differential, for one particle incident on a single target. The transition rate when N particles are incident on a given target is just

$$NT = \frac{2\pi}{\hbar} \frac{N}{V} \left| \frac{M}{V} \right|^2 \varrho. \qquad (4.4.7)$$

But N/V is just the density n of projectiles in space and as we let $V \to \infty$ this is clearly the relevant quantity. The flux f in the beam is just $f = nv$ where v is the velocity of the projectile (remember that for a massive target the target velocity in the centre of mass, defined by $\mathbf{k}_p = -\mathbf{k}_T$, is zero).

Then the transition rate T_f from the initial state to the final state when a single target is exposed to a flux f is

$$T_f = \frac{2\pi}{\hbar} \frac{f}{v} \left| \frac{M}{V} \right|^2 \varrho. \qquad (4.4.8)$$

Nuclear Reactions

But the operational definition of cross-section is the number of processes per second when one target is exposed to a beam of unit flux. Thus

$$\sigma = \frac{1}{v} \frac{2\pi}{\hbar} \left|\frac{M}{V}\right|^2 \varrho \qquad (4.4.9)$$

(where σ may be either a total cross-section on a differential cross-section depending on the degree of integration included in ϱ) and this may be written as

$$\sigma = \frac{\mathcal{T}}{v} \qquad (4.4.10)$$

where, defining $\mathcal{M} = M/V$,

$$\mathcal{T} = \frac{2\pi}{\hbar} |\mathcal{M}|^2 \varrho \qquad (4.4.11)$$

and has the dimensions of cm³ sec⁻¹. Equation (4.4.11) is of considerable importance and we shall frequently return to it: it is our *fifth application of the Fermi Golden Rule*.

To illustrate the content of this formula, we will return to the old problem of coulomb scattering of an electron (any charged particle would do) by a nucleus. The nucleus is described by the coordinates of all the individual particles within it, but in this problem the nuclear structure is the same before and after the scattering, so that the internal nuclear coordinates do not enter the problem and may be forgotten. The relevant coordinates are the nuclear coordinate, that is the coordinate of the centre of mass of the nucleus, and the relative coordinate of the nucleus and the electron. We consider a situation in which it is adequate to describe the initial and final wave functions as plane waves, even very close to the nucleus. Then the potential acting is

$$\frac{Ze^2}{r}$$

where r is the relative coordinate, and the matrix element is written as

$$\iint e^{-i\mathbf{k}'\cdot\mathbf{r}} \frac{Ze^2}{r} e^{i\mathbf{k}\cdot\mathbf{r}} d^3r \cdot \frac{d^3r_T}{V} \qquad (4.4.12)$$

where $\mathbf{k}'$ is the electron propagation vector after the interaction and $\mathbf{k}$ the electron propagation vector before the collision. In the centre of mass, of course, $|\mathbf{k}'| = |\mathbf{k}|$.

There is no explicit dependence of potential or wave function on r_T and integration over this variable merely serves to remove the extra factor $1/V$ left in the matrix element after cancelling the volume term in the phase space and after extraction of one factor of $1/V$ in the definition of the cross-section. Thus all volume terms have disappeared and the cross-section now depends only on an integration over the relative coordinate $\mathbf{r}$. Setting $\mathbf{q}/\hbar = \mathbf{k}' - \mathbf{k}$ yields

$$Ze^2 \int_0^\infty \frac{e^{-i\mathbf{q}\cdot\mathbf{r}/\hbar}}{r} d^3r. \qquad (4.4.13)$$

Nuclear Physics

This integral was evaluated in Section 1.4 and is

$$4\pi Ze^2 \left(\frac{\hbar}{q}\right)^2 \int \sin x \, dx. \tag{4.4.14}$$

Remembering that there is no such thing as a plane wave and hence that the contribution from the infinite limit will be zero, this becomes $4\pi Ze^2(\hbar/q)^2$. The quantity q is the momentum change in the scattering process, given by

$$q^2 = 2p^2(1-\cos\theta) = 4p^2 \sin^2\frac{\theta}{2},$$

where p is centre of mass momentum and θ the scattering angle. In Chapter 1 it was also demonstrated that if the charge on the nucleus is distributed then this results in the introduction of an extra factor, $F(q^2)$, which is the form factor of the nucleus, the Fourier transform of the charge distribution with respect to the variable q. Now in Chapter 1 we were concerned merely with the effect on the scattering of a finite charge distribution, and did not pursue the calculation to the evaluation of a cross-section. We now complete this process by the inclusion of the phase space factor ϱ and the flux factor $1/v$:

$$d\sigma = \frac{1}{v}\frac{2\pi}{\hbar}|\mathcal{M}|^2 \, d\varrho.$$

For a two-body final state, the differential phase space is

$$d\varrho = \frac{p^2 \, dp \, d\Omega}{(2\pi\hbar)^3 \, dE}$$

where E is the total energy and p the centre of mass momentum of one of the two final particles. The massive nucleus carries no kinetic energy, and so dp/dE may be evaluated solely with respect to the electron. If the total energy of the electron is E and the kinetic energy T, $E = T + mc^2$ and $dE = dT$

$$pc = \sqrt{E^2 - m^2c^4},$$
$$\frac{dp}{dE} = \frac{1}{c}\frac{E}{pc} = \frac{1}{v}.$$

(Of course this formula, being derived from the relativistic relation between momentum and energy, also holds non-relativistically:

$$\frac{dp}{dT} = \frac{d}{dT}(2mT)^{1/2} = \frac{1}{2}\left(\frac{2m}{T}\right)^{1/2} = \frac{1}{v}.)$$

Then

$$d\sigma = \frac{1}{v}\frac{2\pi}{\hbar}\left|4\pi Ze^2\left(\frac{\hbar}{q}\right)^2\right|^2 \frac{p^2 \, d\Omega}{(2\pi\hbar)^3 \, v}$$

$$= \frac{4p^2}{v^2}\frac{(Ze^2)^2}{q^4} d\Omega, \tag{4.4.15}$$

$$\frac{d\sigma}{d\Omega} = \frac{1}{4}(Ze^2)^2 \frac{1}{p^2v^2} \operatorname{cosec}^4\frac{\theta}{2}$$

and on going to the non-relativistic limit $pv = mv^2$ we get

$$\frac{d\sigma}{d\Omega} = \frac{1}{4}\left(\frac{Ze^2}{mv^2}\right)^2 \operatorname{cosec}^4 \frac{\theta}{2} \qquad (4.4.16)$$

which is the familiar Rutherford formula for the differential scattering cross-section in coulomb scattering, which we obtained classically earlier in this chapter. It is perhaps rather surprising that we obtain exactly the same answer from this wave mechanical calculation. That the formula survives unchanged in the classical limit as $\hbar \to 0$ is obvious. It is a peculiarity of the coulomb field that all factors of $\hbar$ cancel, and this is true only for a coulomb field. In general the wave mechanical result does not survive unchanged in the classical limit. There is a second freak result, however. We evaluated the matrix element in first-order perturbation theory only, considering the potential to act only once so that the only distortion the incident phase wave suffered was its conversion into the outgoing wave, and this approximation gives the classical result, which also coincides with the result obtained with wave mechanics when no approximations are made. The process may be graphically represented by a diagram

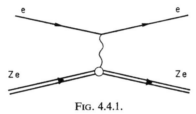

FIG. 4.4.1.

that is, single photon exchange, while the higher-order corrections give diagrams like

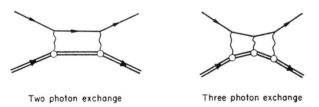

Two photon exchange   Three photon exchange

FIG. 4.4.2.

etc. Thus our approximation, which happens to give the right answer even in the low-velocity limit, is equivalent classically to calculating the impulse to the electron supposing its path always to be at a fixed perpendicular distance from the nucleus rather than evaluating the force along the real orbit.[†] This classical treatment gives the wrong relation between b and θ, but the right expression for the product $b\,db$.

In general, of course, it is not adequate to suppose that the centre of mass system and the laboratory system coincide: we could very well be interested in the scattering of two particles of equal mass. Our original definition specified a single target in a flux of one particle per

† See E. Fermi, *Nuclear Physics*, chap. II (Chicago, 1950).

Nuclear Physics

unit area per unit time, and we must investigate how this changes as we go to a centre of mass system in which both protagonists are moving. Part of the phase space term remains the same: we already defined p to be the momentum of one particle in the centre of mass. The phase space is

$$\frac{p^2 \, dp \, d\Omega}{(2\pi\hbar)^3 \, dE}$$

and we now have $E = \sqrt{p^2c^2+m_3^2c^4}+\sqrt{p^2c^2+m_4^2c^4}$ where the masses involved are m_3 and m_4. Then

$$\begin{aligned}\frac{dE}{dp} &= \frac{pc^2}{\sqrt{p^2c^2+m_3^2c^4}} + \frac{pc^2}{\sqrt{p^2c^2+m_4^2c^4}} \\ &= pc^2\left\{\frac{1}{E_3}+\frac{1}{E_4}\right\}.\end{aligned} \quad (4.4.17)$$

The matrix element only involved the change of momentum q defined already in the centre of mass, so this does not change when we allow both particles to be moving in the centre of mass. The other term that is affected is the flux factor f. The transition rate is still

$$T = \frac{2\pi}{\hbar}\frac{1}{V}|\mathcal{M}|^2 \varrho$$

when one incident particle and one target are being considered. In the centre of mass system, of course, you do not know which is the incident particle and which is the target—the distinction is always artificial. If we define the flux in a symmetric way, it is equal to the number of projectiles passing the target per unit area per second, or if we reverse the situation we get the number of targets passing the projectile per unit area per second. With this definition of flux, the cross-section becomes

$$\sigma = \frac{2\pi}{\hbar}\frac{1}{v_{12}}|\mathcal{M}|^2 \varrho \quad (4.4.18)$$

where v_{12} is the relative velocity of approach of the two particles. Non-relativistically this introduces no problems, $v_{12}=v_1+v_2$ is the relative velocity of the two particles in the centre of mass, or in the rest frame of either particle. If either or both velocities v_1 and v_2 in the centre of mass approach the velocity of light, we have to be a bit more careful about our definitions. If the relative velocity of two particles is defined as the velocity of one in the rest frame of the other, then since no such velocity can exceed the velocity of light, c, it is clear that the relative velocity v_1+v_2 is not the relative velocity in the rest frame of either of the two particles: the relative velocity is not a relativistic invariant. The resolution of this problem is straightforward. The wave functions have been normalized in the centre of mass, the matrix element is being evaluated in the centre of mass and we are interested in the number of scatters per second which occur, as measured in the centre of mass, relative to the rate, measured in the centre of mass, at which particles of one kind pass a particle of the other kind. From the point of view of an observer at rest in the centre of mass, the rate per unit area normal to the beam is $n(v_1+v_2)$ where n, v_1 and v_2 are all centre of mass quantities.

Nuclear Reactions

Thus when one particle is not infinitely massive

$$d\sigma = \frac{2\pi}{\hbar} \frac{1}{(v_1+v_2)} |\mathcal{M}|^2 \frac{p^2 \, dp \, d\Omega}{(2\pi\hbar)^3 \, dE}, \tag{4.4.19}$$

$$\frac{d\sigma}{d\Omega} = \frac{2\pi}{\hbar} \frac{1}{(v_1+v_2)} |\mathcal{M}|^2 \frac{p^2}{(2\pi\hbar)^3} \frac{E_3 E_4}{pc^2(E_3+E_4)}. \tag{4.4.20}$$

As $M_3 \to \infty$, $v_3 \to 0$ and $\dfrac{E_3 E_4}{c^2(E_3+E_4)} \to \dfrac{E_4}{c^2}$

thus giving back the form derived, for a massive target. Now

$$\frac{pc^2(E_3+E_4)}{E_3 E_4} = pc^2 \left\{ \frac{1}{E_3} + \frac{1}{E_4} \right\} = v_3 + v_4,$$

so for *elastic scattering*

$$\frac{d\sigma}{d\Omega} = \frac{2\pi}{\hbar} \frac{1}{(v_1+v_2)^2} |\mathcal{M}|^2 \frac{p^2}{(2\pi\hbar)^3} \tag{4.4.21}$$

and the expression for the cross-section in the centre of mass when neither particle is massive differs from that when one particle is massive simply in the replacement of $1/v_1^2$ by $1/(v_1+v_2)^2$. When one particle is massive, in the non-relativistic limit $p/v \to m$, the mass of the light projectile. When neither is massive, $p = m_1 v_1 = m_2 v_2$ and so m is replaced in the product of phase space and flux factor by $m_1 m_2/(m_1+m_2)$: the familiar reduced mass.

Any cross-section integrated over all angles has the same value when either measured or evaluated in any system—for example, the centre of mass system or the laboratory system.

We conclude this section with an elementary discussion of the behaviour of cross-sections for various types of reactions that can be inferred from the Golden Rule.

$$\sigma \propto \frac{1}{v_{\text{in}}} |\mathcal{M}|^2 \frac{p_{\text{out}}^2}{v_{\text{out}}}$$

(where all quantities are measured in the centre of mass). Following Fermi,[†] we consider the behaviour of the factor

$$\frac{p_{\text{out}}^2}{v_{\text{in}} v_{\text{out}}}. \tag{4.4.22}$$

For *elastic scattering* $v_{\text{in}} = v_{\text{out}}$ and so this factor is independent of the energy, and all the energy dependence of the cross-section arises from the energy dependence of the matrix element M. If M is constant, then the scattering cross-section is also constant. For *exothermic reactions* with a positive Q value, $p_{\text{out}}^2/v_{\text{out}}$ is determined near threshold by the value of Q alone and so is constant (this also applies, of course, if there are more than two particles in the final state). If the matrix element M is constant near threshold, then

$$\sigma \propto \frac{1}{v_{\text{in}}}.$$

[†] E. Fermi, *Nuclear Physics* (Chicago, 1950).

Nuclear Physics

This is the famous $1/v$ law which applies to neutron capture near threshold (see, for example, Fig. 5.5.1). For *endothermic reactions* which have a negative Q value, the situation is reversed. At threshold the phase space factor varies rapidly with only small variations of v_{in} and so if the matrix element is constant, the cross-section (for a two-body final state) varies as p_{out}. Even if the nuclear part of the matrix element is independent of energy near threshold, if either the particle incident on the initial nucleus or the particle emitted with the final nucleus is charged, then coulomb barrier factors must be applied and the matrix element becomes

$$\approx e^{-G_{in}/2} \mathcal{M} e^{-G_{out}/2} \tag{4.4.23}$$

where $\mathcal{M}$ is the nuclear part of the matrix element. (It is essentially these factors that determine the form of the Rutherford scattering cross-section which we already computed.) For an *exothermic reaction* the important factor is $e^{-G_{in}}$, for $e^{-G_{out}}$ is independent of the incident energy near threshold. Thus $n+p \to d+\gamma$ obeys a $1/v$ law near threshold and $n+{}^{235}U \to$ fission fragments, which has an enormous coulomb barrier, also obeys a $1/v$ law near threshold. In contrast, the reaction $p+p \to d+e^+ +\nu$ which is exothermic, is completely suppressed as the energy decreases by the coulomb repulsion of the two protons. For endothermic reactions, in contrast, it is the factor $e^{-G_{out}}$ that dominates coulomb modifications to the behaviour of the cross-section near threshold.

4.5. The partial wave analysis

The use of the expression

$$d\sigma = \frac{1}{v} \frac{2\pi}{\hbar} |\mathcal{M}|^2 d\varrho$$

(where $1/v$ is the flux factor appropriate to the problem, and $\mathcal{M}$ is the transition amplitude, evaluated either exactly or in some appropriate approximation) is at first sight totally at variance with another method which is used for evaluating cross-sections in those cases where the interaction is understood, and as a framework within which to discuss the reaction when the interaction is not understood. In the centre of mass of two particles we may write the Schrödinger equation as

$$\nabla^2 \psi(\mathbf{r}, t) + \frac{2m}{\hbar^2} \left\{ \frac{\hbar}{i} \frac{\partial}{\partial t} - V(\mathbf{r}) \right\} \psi(\mathbf{r}, t) = 0 \tag{4.5.1}$$

where $\mathbf{r}$ is the relative coordinate and m is the reduced mass. If we have a system of infinitely well-defined energy, this equation may be solved for the wave function $\psi(\mathbf{r})$ corresponding to a given energy E, and the equation becomes

$$\nabla^2 \psi(\mathbf{r}) + \frac{2m}{\hbar^2} (E - V(\mathbf{r})) \psi(\mathbf{r}) = 0. \tag{4.5.2}$$

If E is negative, solutions only exist for certain eigenvalues of the energy, but if E is positive solutions exist for all values of the energy. When $V(\mathbf{r}) = 0$ the solutions are plane waves with all directions of the propagation vector allowed. The equation may in principle be solved for that solution with energy E which has the boundary conditions appropriate to the scattering problem. If the potential $V(\mathbf{r}) \equiv 0$ then $\psi(\mathbf{r}) = e^{i\mathbf{k}\cdot\mathbf{r}}$ for a plane wave with propagation vector $\mathbf{k}$. In the presence of a potential this solution is no longer a solution of the whole equation, but the sum of this plane wave plus a spherical wave $(e^{ikr}/r)f(\theta, \phi)$ is still a solution in the distant regions where $V(\mathbf{r}) \simeq 0$. This outgoing spherical wave is to be identified with the scattered wave. At very large values of the relative coordinate $\mathbf{r}$, the scattered flux is negligible in comparison with the flux due to the plane wave term, and it is in this way that it is possible to define the boundary conditions appropriate to the problem. For simplicity we suppose the potential $V(\mathbf{r})$ to be spherically symmetric, that is, neither particle has spin. (The method is easily generalized to those cases where one or both particles have spin.) In this case

$$\psi(\mathbf{r}) = \sum_{l,m} R_l(r) P_l^m(\cos\theta) e^{im\phi}$$

is a general solution. The scattering problem has symmetry about the axis of the incoming beam, and if this is chosen as the z-axis so that the polar angle θ corresponds to the scattering angle, then this symmetry is ensured by setting $m = 0$. We now proceed with the partial wave expansion of an incident plane wave, when $V(\mathbf{r}) \equiv 0$. We set

$$e^{i\mathbf{k}\cdot\mathbf{r}} \equiv \sum_l R_l(r) P_l(\cos\theta) \tag{4.5.3}$$

and evaluate R_l by using the orthogonality properties of the Legendre Polynomials $P_l(\cos\theta)$. Consider that part with zero orbital angular momentum, $R_0(r) P_0(\cos\theta) = R_0(r)$. Multiply both sides by $P_0(\cos\theta)$ and integrate with respect to $\cos\theta$ from -1 to $+1$. The integral

$$\int_{-1}^{+1} P_l(\cos\theta) P_m(\cos\theta) \, d\cos\theta$$

vanishes unless $l = m$ so that

$$\int_{-1}^{+1} e^{i\mathbf{k}\cdot\mathbf{r}} \, d\cos\theta = R_0(r) \int_{-1}^{+1} d\cos\theta$$

or

$$R_0(r) = \frac{\sin kr}{kr} \simeq 1 \quad \text{for} \quad kr \ll 1. \tag{4.5.4}$$

Similarly, for the p-wave portion of a plane wave

$$\left.\begin{aligned}
\int_{-1}^{+1} e^{i\mathbf{k}\cdot\mathbf{r}} \cos\theta \, d\cos\theta &= \int_{-1}^{+1} R_1(r) \cos^2\theta \, d\cos\theta, \\
\int_{-1}^{+1} e^{ikr\cos\theta} \cos\theta \, d\cos\theta &= \tfrac{2}{3} R_1(r), \\
R_1(r) &= \frac{3i}{(kr)^2} \{\sin kr - kr \cos kr\} \\
&\sim kr \quad \text{when} \quad kr \ll 1.
\end{aligned}\right\} \tag{4.5.5}$$

Nuclear Physics

For $kr \gg 1$ the second term dominates and we have

$$R_1(r) \xrightarrow[kr \to \infty]{} -\frac{3i}{kr}\cos kr = \frac{3i}{kr}\sin\left(kr - \frac{\pi}{2}\right).$$

In general, we have $R_l(r) \sim (kr)^l$ for $kr \ll 1$ and

$$R_l(r) \xrightarrow[kr \to \infty]{} \frac{(2l+1)}{kr} i^l \sin\left(kr - \frac{l\pi}{2}\right) \tag{4.5.6}$$

(over the whole range,

$$R_l(r) = (2l+1) i^l j_l(kr)$$

where $j_l(kr)$ are the spherical Bessel functions). You will note that all the solutions are well behaved as $r \to 0$, as we would have expected because the function we are expanding is everywhere finite.

The partial wave expansion is discussed more generally in most books on quantum mechanics.[†] There are also a series of solutions to the wave equation in spherical coordinates which are not well behaved but which blow up at the origin—for $l = 0$, $\cos kr/kr$, and so on. A general solution in a region of space where $V(r) = 0$ will be a sum of the regular and irregular solutions, so long as the solutions in the region of space where $V(r) \neq 0$ match smoothly on to this general solution and are well behaved as $r \to 0$. Well beyond the range of the potential, we may write

$$R_l(r) \simeq \frac{A_l}{kr}\sin\left(kr - \frac{l\pi}{2} + \delta_l\right) \tag{4.5.7}$$

where δ_l is the phase shift in the partial wave of angular momentum l.

We may get at the physical content of this equation in the following way. Each of the partial waves making up an incident plane wave is composed of both an ingoing and outgoing part—for example, the s-wave term is

$$\frac{\sin kr}{kr} = \frac{1}{2ikr}\{e^{ikr} - e^{-ikr}\} \tag{4.5.8}$$

and the relative phases and magnitudes of these components are adjusted as to add up to a plane wave. This should not cause any surprise. If you are sitting in the centre of mass, looking in one direction you see a particular wave crest approaching you, while turning round you see another receding. We now appeal to microscopic causality. If at a large distance the boundary conditions correspond to a plane wave $e^{i\mathbf{k}\cdot\mathbf{r}}$ then the presence of a potential in a region close to $r = 0$ cannot alter any of the properties of the incoming spherical wave components at a large distance. On the other hand, the outgoing components have been through the potential, and indeed were generated near the origin by the incoming component coming in through the potential. Their characteristics can be different from

---

[†] See, for example, L. I. Schiff, *Quantum Mechanics*, 3rd ed. (McGraw-Hill, 1968), or R. H. Dicke and J. P. Wittke, *Introduction to Quantum Mechanics* (Addison-Wesley, 1960).

those of plane wave components—for example, if absorption takes place there will be less amplitude in the outgoing spherical waves than in the incoming ones. (It should be clear that an implicit appeal to a wave packet picture is involved in making these arguments: our plane wave scattering solutions must be capable of being combined in a Fourier integral to give the right behaviour for the scattering of a localized wave packet). Now the presence of a scattering centre can only do two things to an outgoing wave. It can reduce its amplitude by some absorption process, and it can shift the phase of the outgoing component, thereby introducing some of the irregular solution at long range. These properties are most easily demonstrated for the case of the s-wave for which the one-dimensional solution in r is $kr\,R_l(r) = \sin kr$ in the absence of a potential. The presence of a potential introduces some $\cos kr$ component, and this is done by upsetting the amplitude of the outgoing spherical component e^{ikr} so that it no longer adds with e^{-ikr} to make $\sin kr$. For simplicity consider a potential

$$V(r) = -U \quad r < R,$$
$$V(r) = 0 \quad r > R.$$

For $r < R$ the solution must be $\sin k'r$ where $k'^2 = 2m(E+U)/\hbar^2$, and so the wavelength is shorter inside the attractive potential than outside. Outside, the solution must be a linear combination of e^{ikr} and e^{-ikr} with the relative amplitudes determined by the requirement that this solution matches to the completely determined interior function at $r = R$. In general this combination is NOT equivalent to $\sin kr$ and we set

$$\psi_{\text{inside}} = \frac{\sin k'r}{k'r} \quad \psi_{\text{outside}} = \frac{e^{2i\delta}e^{ikr} - e^{-ikr}}{2ikr} \tag{4.5.9}$$

(if δ is allowed to be complex this takes into account absorption as well). The quantity δ is in this example entirely determined by the requirement that ψ_{in} and ψ_{out} are equal at $r = R$, the requirement that their first derivatives are also equal at $r = R$ and by the poten-

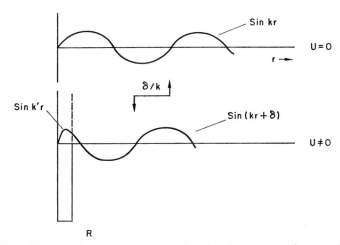

FIG. 4.5.1. This figure illustrates how in a steady state situation the presence of a potential shifts the wave crests by an amount δ/k, a phase shift of δ. It should also be clear that δ is positive for an attractive potential, negative for a repulsive potential.

Nuclear Physics

tial parameters U, R and the energy of the system E. The phase shift δ is thus a function of the energy. Then $\psi_{\text{outside}} = e^{i\delta} \sin(kr+\delta)/kr$. The total s-wave function ψ outside is shifted in phase by δ from the value obtaining when there is no potential present. This is illustrated in Fig. 4.5.1. It must clearly be understood that the outgoing wave comprises both the scattered wave (which is only outgoing) and a portion of the incident plane wave (which is both incoming and outgoing). The scattered portion is extracted by writing

$$\psi(r) = \psi_{\text{incident}}(r) + \psi_{\text{scattered}}(r) \tag{4.5.10}$$

and outside the potential this can be written as

$$e^{i\delta}\left\{\frac{e^{i(kr+\delta)} - e^{-i(kr+\delta)}}{2ikr}\right\} = \left\{\frac{e^{ikr} - e^{-ikr}}{2ikr}\right\} + \psi_{\text{scattered}}(r) \tag{4.5.11}$$

for s-waves or

$$\psi_{\text{scattered}}(r) = \frac{e^{ikr}}{2ikr}\{e^{2i\delta} - 1\} = \frac{e^{ikr}}{kr} e^{i\delta} \sin\delta. \tag{4.5.12}$$

We may now evaluate the cross-section for scattering into an s-wave state from the incident beam. With the normalization we have employed, $\psi(r)_{\text{incident}} = e^{i\mathbf{k}\cdot\mathbf{r}}$, so the flux, $v|\psi_{\text{incident}}|^2$ is just equal to v.

The scattered flux at distance r is just

$$v|\psi_{\text{scattered}}|^2 = \frac{v \sin^2\delta}{k^2 r^2}.$$

This scattered flux when integrated over the surface of a sphere of radius r gives the number of particles scattered out of the beam and passing through this sphere each second—this is the number of particles scattered per second. Dividing this by the incident flux this gives

$$\sigma = \frac{4\pi}{k^2} \sin^2\delta \tag{4.5.13}$$

for s-wave scattering, and if we do not integrate over the sphere we get instead

$$\frac{d\sigma}{d\Omega} = \frac{\sin^2\delta}{k^2}.$$

It is now straightforward to generalize this treatment which we have applied to s-waves only and get the general result for scattering by a short-range spherically symmetric potential, which is

$$\frac{d\sigma}{d\Omega} = \frac{1}{k^2} \left| \sum_{l=0}^{\infty} e^{i\delta_l}(2l+1) \sin\delta_l \, P_l(\cos\theta) \right|^2. \tag{4.5.14}$$

The different spherical components interfere in the differential cross-section. On integrating over solid angle, the interference terms drop out and we have

$$\sigma = \frac{4\pi}{k^2} \sum_{l=0}^{\infty} (2l+1) \sin^2\delta_l$$

Nuclear Reactions

where an extra factor of $1/(2l+1)$ has been picked up from integration over the angular distribution.

For the case of the square well potential and an s-wave, the phase shift is given by

$$\tan k'R = \frac{k'}{k} \tan(kR+\delta); \quad \tan\delta = \frac{k \tan k'R - k' \tan kR}{k + k' \tan k'R \tan kR} \qquad (4.5.15)$$

which must be solved numerically to give the phase shift δ as a function of energy. A square well potential is not very realistic and for a real potential the wavelength, and so the propagation constant, varies continuously as r increases, tending asymptotically to the values appropriate to an incident plane wave, λ and k. The argument is in no way affected. If the displacement of wave crests from the position they would occupy without a potential present (and this position is fixed by the requirement of a regular solution) is measured, this displacement changes with r and in the asymptotic limit becomes a constant, δ_l/k.

However, there exist potentials for which the displacement does not tend to a constant asymptotic value. These are not short-range potentials, but are those potentials varying as slowly as or more slowly than $1/r$. The only important example is the coulomb potential. The solution can none the less be obtained algebraically, and constitutes the exact derivation in wave mechanics of the Rutherford formula, eq. (4.4.16), which we derived only under an approximation.[†]

The effect of absorption of the incident wave may, as we remarked earlier, be included by letting δ_l be complex. It is perhaps a little clearer to take account of the effect explicitly. Absorption is caused by any inelastic process—and from the point of view of scattering the characteristic of an inelastic process is that it produces a final state orthogonal to the scattered states. In the limit of plane waves, this is even true of inelastic scattering, although in principle with an ill-defined wave packet some interference could occur. Supposing, then, that the scattered waves can be picked out from all other processes, we return to eq. (4.5.9) for s-waves and write it in a modified form

$$\psi_{\text{inside}} = \frac{\sin k'r}{k'r} \quad \psi_{\text{outside}} = \frac{\eta e^{2i\delta} e^{ikr} - e^{-ikr}}{2ikr} \qquad (4.5.16)$$

where δ and η are both real parameters.

If η is zero, there is no outgoing s-wave, and since it is impossible for more amplitude to go out than comes in, the bounds on η are $0 \leqslant \eta \leqslant 1$.

Then

$$\psi_{\text{scattered}} = \frac{e^{ikr}}{2ikr} \{\eta e^{2i\delta} - 1\}.$$

So that in general we can write

$$\frac{d\sigma}{d\Omega} = \frac{1}{4k^2} \left| \sum_{l=0}^{\infty} (2l+1) P_l(\cos\theta)\{\eta_l e^{2i\delta_l} - 1\} \right|^2. \qquad (4.5.17)$$

If $\eta_l = 0$ the incident l-wave is completely absorbed, but regardless of the value of δ_l there is still scattering. This is, of course, just diffraction or shadow scattering, precisely analogous to the scattering of light by a black sphere.

[†] See, for example, L. I. Schiff, *Quantum Mechanics*, 3rd ed., sect. 21 (McGraw-Hill, 1968), or N. F. Mott and H. S. W. Massey, *Theory of Atomic Collisions*, 3rd ed., chap. 3 (Oxford, 1965).

Nuclear Physics

In most problems the form of the potential is unknown. The phase shifts δ_l may be obtained from a fit to the data, and then constitute a convenient set of parameters that any model potential must be capable of generating. But the parametrization is more general than this. We have so far supposed a system of two (spinless) particles, with the potential energy of the system expressed as a function of a single parameter, the relative coordinate r, and independent of time. As soon as a particle enters a complex nucleus, this picture breaks down and the potential felt fluctuates as the individual nucleons are encountered. However, the solution outside the nucleus still has the same general form and the intranuclear conditions still determine δ_l through the logarithmic derivative of the exterior wave function at the nuclear boundary. Thus while in general we have no hope of calculating δ_l under these circumstances, the parametrization stands.

Integration of eq. (4.5.17) over solid angle gives the total scattering cross-section

$$\sigma_{el} = \frac{\pi}{k^2} \sum_{l=0}^{\infty} (2l+1) |\eta_l e^{2i\delta_l} - 1|^2. \tag{4.5.18}$$

The reaction cross-section, which includes everything except elastic scattering, is determined just by the depletion in the outgoing wave, and so is given by

$$\sigma_r = \frac{\pi}{k^2} \sum_{l=0}^{\infty} (2l+1) \{|1 - |\eta_l|^2\}. \tag{4.5.19}$$

η_l can be calculated from a potential just as δ_l can, provided that this potential is allotted an imaginary part. However, absorption is a process which involves nuclear dynamics in detail, and so such a representation can never be more than an approximation.

The total cross-section is given by the sum of the two

$$\sigma_T = \sigma_{el} + \sigma_r = \frac{2\pi}{k^2} \sum_{l=0}^{\infty} (2l+1) \{1 - \eta_l \cos 2\delta_l\}. \tag{4.5.20}$$

While we have applied the partial wave expansion to the problem of elastic scattering, the expansion of a plane wave into spherical waves of well-defined angular momentum is useful in any low-energy problem where only a few angular momentum states contribute. But it is only in the case of low-energy elastic scattering that the phase shifts can be calculated from a single time independent potential—and it cannot always be done even in this simple case.

Some further insight may be gained by a comparison of the expression

$$d\sigma = \frac{1}{v} \frac{2\pi}{\hbar} |\mathcal{M}|^2 d\varrho$$

for elastic scattering with the expression (4.5.14), that is

$$\frac{d\sigma}{d\Omega} = \frac{2\pi}{\hbar} |\mathcal{M}|^2 \frac{m^2}{(2\pi\hbar)^3}$$

(where m is the reduced mass) and

$$\frac{d\sigma}{d\Omega} = \frac{1}{k^2} \left| \sum_{l=0}^{\infty} e^{i\delta_l}(2l+1) \sin \delta_l P_l (\cos \theta) \right|^2$$

which implies that

$$|\mathcal{M}_l|^2 = \left[\frac{2\pi\hbar^2}{mk}\right]^2 \left| \sum_{l=0}^{\infty} e^{i\delta_l}(2l+1) \sin \delta_l P_l (\cos \theta) \right|^2$$

so that we can write

$$\mathcal{M}_l = \frac{2\pi\hbar^2}{mk} e^{i\delta_l} (2l+1) \sin \delta_l P_l(\cos \theta). \tag{4.5.21}$$

Thus, for example, if we were studying an elastic process at high enough energy that our approximate matrix element represented coulomb scattering adequately, and at the same time a strong interaction was causing scattering mostly in the s-wave, we could write

$$\frac{d\sigma}{d\Omega} = \frac{1}{k^2} \left| \underbrace{e^{i\delta_0} \sin \delta_0}_{s\text{-wave only}} + \frac{mk}{2\pi\hbar^2} \underbrace{4\pi Z e^2 \left(\frac{\hbar}{q}\right)^2}_{\text{all waves}} \right|^2.$$

4.6. Cross-sections and spin

So far we have assumed that all particles involved in reactions are spinless, or at least that the interactions do not depend on the spins of the particles. In general all particles participating in a nuclear reaction may have spin, and as a result there will be spin–orbit and spin–spin interactions present in addition to the central potential terms. Neither spin nor orbital angular momenta will be conserved in the interaction, only the total angular momentum J and the third component J_z. For example, suppose that an initial state consists of a spin zero nucleus and an s-wave proton, with $L = 0$, $m_L = 0$ and $m_S = +\frac{1}{2}$ where L is the orbital angular momentum, m_L its third component and m_S the third component of the proton intrinsic spin. Then $J = |\mathbf{L}+\mathbf{S}| = \frac{1}{2}$ and $J_z = m_L + m_S = +\frac{1}{2}$, $L = 0$.

If the process is elastic scattering, then the final state also has $m_L = 0$, $m_S = +\frac{1}{2}$ and $J = \frac{1}{2}$. Conservation of angular momentum and parity ensure this. But now suppose that a p-wave proton is in the initial state. Then let $L = 1$, $m_L = 0$ and $m_S = +\frac{1}{2}$. The total angular momentum $J = \frac{3}{2}$ or $\frac{1}{2}$ and $J_z = +\frac{1}{2}$. Only if the scattering amplitude is identical for the two states $J = \frac{3}{2}$ and $J = \frac{1}{2}$ will the outgoing state also have $L = 1$, $m_L = 0$, $m_S = +\frac{1}{2}$. Suppose, for example, that there is no scattering in the $J = \frac{1}{2}$ state and it all takes place through the $J = \frac{3}{2}$ state. This state may be broken up into a mixture of

$$L = 1, \quad m_L = 0, \quad m_S = +\frac{1}{2}$$

and

$$L = 1, \quad m_L = +1, \quad m_S = -\frac{1}{2}.$$

In this case conservation of parity ensures that L remains equal to 1, but states of different m_L and m_S are mixed together by the interaction, subject only to the constraint $m_L + m_S = +\frac{1}{2}$,

Nuclear Physics

If instead of a proton, with spin $\frac{1}{2}$, a deuteron with spin 1 was incident on a nucleus with spin zero, even L would not have to be the same before and after the scattering process.

The scattering amplitudes are specified by the total angular momentum J (and since different values of J are not mixed together, do not depend on J_z) and it is also necessary to specify the parity. In strong and electromagnetic interactions states of different parity are not mixed together. Of course, if L is specified, this automatically fixes the parity. The matrix element for transition between an initial and a final state is given by those amplitudes taken between specified states. If we consider a process with two particles in the initial state and two in the final state, not necessarily elastic scattering, then any particular initial and final state is specified by the intrinsic quantum numbers of the particles, by the total angular momentum and its third component, and if we are interested in greater detail, by the relative orbital angular momentum and the projections of the separate angular momenta. Thus in a reaction

$$a+b \to c+d$$

the initial state can be written

$$|S_a^{P_a}, S_b^{P_b}, L, m_a, m_b, m_L, J^P, J_z\rangle. \qquad (4.6.1)$$

Some of these terms are redundant: P, the parity of the initial state, is given by

$$P = P_a P_b (-1)^L$$

and

$$J_z = m_L + m_a + m_b.$$

It is therefore adequate to write

$$|S_a^{P_a}, S_b^{P_b}, L, m_a, m_b, m_L, J\rangle \qquad (4.6.2)$$

and the matrix element of the interaction I is given by

$$\langle J', m_{L'}, m_d, m_c, L', S_d^{P_d}, S_c^{P_c}, |I| S_a^{P_a}, S_b^{P_b}, L, m_a, m_b, m_L, J\rangle \qquad (4.6.3)$$

and vanishes unless

$$\begin{rcases} (1) \quad P_a P_b (-1)^L = P_c P_d (-1)^{L'}, \\ (2) \quad m_a + m_b + m_L = m_c + m_d + m_{L'}, \\ (3) \quad J = J'. \end{rcases} \qquad (4.6.4)$$

The cross-section is then given by

$$d\sigma = \frac{1}{v} \frac{2\pi}{\hbar} |\langle J', m_{L'}, m_d, m_c, L', S_d^{P_d}, S_c^{P_c} |I| S_a^{P_a}, S_b^{P_b}, L, m_a, m_b, m_L, J\rangle|^2 \, d\varrho \qquad (4.6.5)$$

and is a differential cross-section which may be converted into a total cross-section by integration over the phase space ϱ.

Now suppose that we are working with an unpolarized beam and an unpolarized target, and all we are interested in is the angular distribution of c and d. For a specified initial state, we must sum over all variables in the final state, namely L', m_c, m_d, $m_{L'}$. We must then sum over all initial states, namely over the variables J, L, m_a, m_b, m_L. But while the L composition of the initial plane wave is given by the partial wave expansion, we are working with an initial plane wave state normalized to unity, composed of a mixture of many different states specified by the quantum numbers J, L, m_a, m_b, m_L.

An unpolarized beam means that all magnetic substates specified by m_a are equally likely. A particular value of m_a therefore has a probability of $1/(2S_a+1)$. Similarly a particular value of m_b has a probability of $1/(2S_b+1)$. The weighting function that multiplies the transition rate when we have an unpolarized beam and target is then just

$$\frac{1}{(2S_a+1)(2S_b+1)}.$$

Thus by summing over all initial states and dividing by $(2S_a+1)(2S_b+1)$ we are making an average over all initial spin states. It is not necessary to make an average over the different values of m_L—the direction of the incoming beam and the partial wave analysis specify the orbital angular momentum properties —m_L is equal to zero with respect to the direction of the beam.

Thus the recipe for working with spins in the problem is as follows: break the plane wave states into the component parts specified by the internal quantum numbers of the particles involved, by the orbital angular moment, by the orientations of the component angular momenta and by the total angular momentum. These states are the states we have written as

$$|S_a^{P_a}, S_b^{P_b}, L, m_a, m_b, m_L, J\rangle$$

for two particles. The breakup of the plane wave is done by using the partial wave expansion and the angular momentum coupling coefficients. Then the cross-section is given by

$$d\sigma = \frac{1}{(2S_a+1)(2S_b+1)} \frac{2\pi}{\hbar v} \Bigg| \sum_{\substack{J', m_{L'} \\ m_d, m_c \\ L'}} \sum_{\substack{J, m_L \\ m_a, m_b \\ L}} \langle J', m_{L'}, m_d, m_c, L', S_d^{P_d}, S_c^{P_c},$$

$$|I|S_a^{P_a}, S_b^{P_b}, L, m_a, m_b, m_L, J\rangle \Bigg|^2 d\varrho \tag{4.6.6}$$

where the summation is taken over all the quantum numbers specifying the components of the initial plane wave state and over all the quantum numbers specifying the components of the final states. The matrix elements of the interaction I vanish between any states which do not have the same J, J_z and (for strong and electromagnetic interactions) the same parity. If the target b is completely polarized, then summation over S_b is dropped, together with the weighting factor. If there are three particles in the final state, then the summation over the final states must be extended to include another projected spin, another relative orbital angular momentum.

Let us now apply eq. (4.6.6) to some simple situations. Suppose first that we have elastic scattering and that a, b, c and d are all spinless. Then the initial state is labelled by L and m_L and conservation of angular momentum requires $L' = L$ and $m_{L'} = m_L$. The cross-section is then just

$$d\sigma = \frac{1}{v} \frac{2\pi}{\hbar} \Bigg| \sum_L \langle L, m_L = 0 | I | m_L = 0, L \rangle \Bigg|^2 d\varrho$$

which is exactly what we had before introducing spin.

Nuclear Physics

Now suppose $S_b = 0$ but $S_a = \frac{1}{2}$. The total angular momentum is restricted to $L \pm \frac{1}{2}$ and the third component to $\pm \frac{1}{2}$, since m_L is equal to zero along the beam direction. There will be an amplitude for scattering in the $J = L + \frac{1}{2}$ state and another in the $J = L - \frac{1}{2}$ state, and these will not be the same if the forces are spin dependent. The values of $m_{L'}$ are restricted to the range $-1, 0, +1$ since $m_{L'} + m_a = \pm \frac{1}{2}$ and $m_a = \pm \frac{1}{2}$.

There are two initial states to consider as far as spin is concerned. Since $J_z = m_L + m_a$ and $m_L = 0$

$$J_z = +\tfrac{1}{2} \quad \text{corresponds to} \quad m_a = +\tfrac{1}{2},$$
$$J_z = -\tfrac{1}{2} \quad \text{corresponds to} \quad m_a = -\tfrac{1}{2}.$$

The factor $1/(2S_a + 1) = \frac{1}{2}$. This expresses the fact that since we have an unpolarized beam, a state of given L will be composed half of $J_z = +\frac{1}{2}$ and half of $J_z = -\frac{1}{2}$ and so the cross-section is contributed half by $J_z = +\frac{1}{2}$ states and half by $J_z = -\frac{1}{2}$ states. $J_z = +\frac{1}{2}$ is partly $(J = L + \frac{1}{2}, J_z = +\frac{1}{2})$ and partly $(J = L - \frac{1}{2}, J_z = +\frac{1}{2})$. We write for the initial states specified by L, m_L, m_a

$$|L, 0, \pm\tfrac{1}{2}\rangle = C\begin{smallmatrix} \frac{1}{2} & L & J \\ \pm\frac{1}{2} & 0 & J_z \end{smallmatrix} |J = L \pm \tfrac{1}{2}, J_z = \pm\tfrac{1}{2}\rangle$$

$$+ C\begin{smallmatrix} \frac{1}{2} & L & J \\ \pm\frac{1}{2} & 0 & J_z \end{smallmatrix} |J = L \mp \tfrac{1}{2}, J_z = \pm\tfrac{1}{2}\rangle \qquad (4.6.7)$$

where the quantities $C\begin{smallmatrix} S_a & L & J \\ m_a & m_L & J_z \end{smallmatrix}$ are the Clebsch–Gordan coefficients obtained from the theory of angular momentum. For example, for $L = 1$, $m_a = \pm\frac{1}{2}$

$$C\begin{smallmatrix} \frac{1}{2} & 1 & \frac{3}{2} \\ +\frac{1}{2} & 0 & +\frac{1}{2} \end{smallmatrix} = \sqrt{\tfrac{2}{3}}, \quad C\begin{smallmatrix} \frac{1}{2} & 1 & \frac{3}{2} \\ -\frac{1}{2} & 0 & -\frac{1}{2} \end{smallmatrix} = \sqrt{\tfrac{2}{3}},$$

$$C\begin{smallmatrix} \frac{1}{2} & 1 & \frac{1}{2} \\ +\frac{1}{2} & 0 & +\frac{1}{2} \end{smallmatrix} = -\sqrt{\tfrac{1}{3}}, \quad C\begin{smallmatrix} \frac{1}{2} & 1 & \frac{1}{2} \\ -\frac{1}{2} & 0 & -\frac{1}{2} \end{smallmatrix} = \sqrt{\tfrac{1}{3}}.$$

The states with specified J, J_z are scattered and then break up into their allowed configurations.[†] Keeping to the example of $L = 1$

$$|J = \tfrac{3}{2}, J_z = +\tfrac{1}{2}\rangle \to \sqrt{\tfrac{1}{3}} |L' = 1, m_{L'} = +1, m_c = -\tfrac{1}{2}\rangle$$
$$+ \sqrt{\tfrac{2}{3}} |L' = 1, m_{L'} = 0, m_c = +\tfrac{1}{2}\rangle, \qquad (4.6.8a)$$

$$|J = \tfrac{3}{2}, J_z = -\tfrac{1}{2}\rangle \to \sqrt{\tfrac{2}{3}} |L' = 1, m_{L'} = 0, m_c = -\tfrac{1}{2}\rangle$$
$$+ \sqrt{\tfrac{1}{3}} |L' = 1, m_{L'} = -1, m_c = +\tfrac{1}{2}\rangle, \qquad (4.6.8b)$$

$$|J = \tfrac{1}{2}, J_z = +\tfrac{1}{2}\rangle \to \sqrt{\tfrac{2}{3}} |L' = 1, m_{L'} = +1, m_c = -\tfrac{1}{2}\rangle$$
$$- \sqrt{\tfrac{1}{3}} |L' = 1, m_{L'} = -1, m_c = +\tfrac{1}{2}\rangle, \qquad (4.6.8c)$$

$$|J = \tfrac{1}{2}, J_z = -\tfrac{1}{2}\rangle \to \sqrt{\tfrac{1}{3}} |L' = 1, m_{L'} = 0, m_c = -\tfrac{1}{2}\rangle$$
$$- \sqrt{\tfrac{2}{3}} |L' = 1, m_{L'} = -1, m_c = +\tfrac{1}{2}\rangle. \qquad (4.6.8d)$$

[†] See Appendix 2.

Nuclear Reactions

Then for $L = 1$ only, summation over the final states for $m_a = +\frac{1}{2}$ gives

$$\sum_{J, m_{L'}, m_c} \langle J, +\tfrac{1}{2} | I | m_{L'}, m_c, J \rangle$$
$$= \langle \tfrac{3}{2}, +\tfrac{1}{2} | \{ \sqrt{\tfrac{2}{3}} A_{\tfrac{3}{2}} | +1, -\tfrac{1}{2} \rangle + \tfrac{2}{3} A_{\tfrac{3}{2}} | 0, +\tfrac{1}{2} \rangle \}$$
$$+ \langle \tfrac{1}{2}, +\tfrac{1}{2} | \{ -\sqrt{\tfrac{2}{3}} A_{\tfrac{1}{2}} | 0, -\tfrac{1}{2} \rangle + \tfrac{1}{3} A_{\tfrac{1}{2}} | -1, +\tfrac{1}{2} \rangle \} \quad (4.6.9)$$

where $A_{\tfrac{3}{2}}$ is the amplitude for $J = \tfrac{3}{2}$ scattering, $A_{\tfrac{1}{2}}$ is the amplitude for $J = \tfrac{1}{2}$ scattering.

The states with different m_c do not interfere. The states with the same m_c but different $m_{L'}$ interfere in the differential cross-section, but not in the total cross-section for scattering. On squaring and integrating over angles the states $\langle J, +\tfrac{1}{2} |$ contribute to the cross-section

$$\sigma_{+\tfrac{1}{2}} = \frac{1}{2} \frac{1}{v} \frac{2\pi}{\hbar} \left\{ \frac{2}{3} \left| A_{\tfrac{3}{2}} \right|^2 + \frac{1}{3} \left| A_{\tfrac{1}{2}} \right|^2 \right\} \varrho$$

and the $\langle J, -\tfrac{1}{2} |$ states contribute the same, leaving

$$\sigma = \frac{1}{v} \frac{2\pi}{\hbar} \left\{ \frac{2}{3} \left| A_{\tfrac{3}{2}} \right|^2 + \frac{1}{3} \left| A_{\tfrac{1}{2}} \right|^2 \right\} \varrho$$

on summing over the initial states for $L = 1$.

If the forces are spin independent, $A_{\tfrac{3}{2}} = A_{\tfrac{1}{2}} = A_1$ where A_1 is the amplitude for scattering in the $L = 1$ state and then

$$\sigma = \frac{1}{v} \frac{2\pi}{\hbar} |A_1|^2 \varrho$$

which is exactly what we would have obtained for spinless particles.

If we want to look at the angular distribution of particles with $m_c = +\tfrac{1}{2}$ then the summation is taken only over those states with $m_c = +\tfrac{1}{2}$ but if the beam is unpolarized, we still average over m_a.

The effect of spin, in either elastic or inelastic reactions, may be summarized as follows:

1. If the forces are spin dependent, the problem no longer has circular symmetry about the beam axis, and polarization in general results.
2. There exists a matrix element connecting each completely specified initial state with each completely specified final state: these matrix elements are zero if conservation of J, J_z (energy and momentum) and, in strong or electromagnetic interactions, P are violated.
3. The cross-section is obtained by making the appropriate sum over final states of the matrix element and the appropriate average over the initial states.

Very detailed discussions of the modifications to a spinless formalism necessitated by the introduction of spin may be found in, for example, Blatt and Weisskopf, chap. VIII.[†]

[†] J. M. Blatt and V. Weisskopf, *Theoretical Nuclear Physics* (Wiley, 1952).

4.7. Time-independent potentials: the optical model

Consider a nucleon approaching a nucleus. At a very great distance the only forces acting are the long-range electromagnetic forces. As the nucleon comes within range of the nuclear forces, it is exposed to a potential which is the sum of the individual potentials generated by all the nucleons in the nucleus, and which therefore fluctuates with a time scale of $\sim 10^{-20}$ sec. A nuclear reaction cannot be represented by a potential: and strictly speaking even an elastic scattering process cannot be represented by the effect of a time-independent potential, just because of the fluctuations of the instantaneous potential at a point due to the motion of the individual nucleons in the nucleus. Thus a potential formalism is only exact for the (non-relativistic) scattering of two elementary particles. Indeed, if nucleons are themselves composite structures, the use of a potential in the two-nucleon problem is in itself an approximation.

The optical model of elastic scattering assumes that the scattering is insensitive to the rapid fluctuations in potential and can be adequately represented by the action of an averaged time-independent potential acting between the pair of objects which are scattering. Thus in electron scattering, at large distances local concentrations of charge produce fluctuating high-order multipoles of the electrostatic field which at large distances can be neglected. They show up in inelastic electron scattering, in which the nucleus is raised to an excited state. This is a process which depends on the detailed nuclear structure; and the relation with the fluctuating potential comes in through the uncertainty relation $\Delta E \, \Delta t \sim \hbar$ which says that the product of a transferred energy and the interaction time Δt is of order $\hbar$.

A model of scattering which represents the interaction by a time independent potential is incapable of describing any effects due to detailed nuclear structure (such as structure resonances)[†] but rather is used to correlate the trend from nucleus to nucleus of those features of the scattering that depend on gross effects: nuclear size, the range and depth of the average potential and the variation with radius of the nuclear density. Such a model is in no sense a theory, but is used to extract parameters that a real theory of nuclear matter must be able to reproduce. Because of its insensitivity to detailed structure, this optical model is best suited to scattering in the intermediate and high-energy regions, where structure effects are unimportant anyway. In the intermediate and high-energy regions the outgoing waves are depleted by inelastic processes and a substantial proportion of the elastic scattering comes from diffraction around the absorbing nucleus. This is taken into account by taking a complex time-independent potential, the real part giving rise to true potential scattering, and the imaginary part to diffraction scattering. Models of this type have been given the generic name of *optical models* because of the analogy with the scattering of light by a partially transparent dielectric sphere (they are also occasionally referred to as "cloudy crystal ball models") and they are conceptually identical with the independent particle model of nuclear structure.

In addition to central real and imaginary potentials, spin–orbit potentials may also be included in such a model to describe the polarization of scattered particles. For nucleon–

---

† See Section 4.12.

nucleus scattering, the real parts of the fitted potentials should be capable of reproducing the observed level sequence in the independent particle model, as indeed they are found to be. Thus the strong spin–orbit coupling which is built into the independent particle model to give the observed level sequence finds an independent confirmation from the fitting of optical model potentials to data on polarization in nucleon–nucleus scattering.

When the model is applied to nucleus–nucleus scattering, the parameters of the fitted potentials, while of less intrinsic interest than the nucleon–nucleus parameters, may be used to help in the understanding of nuclear reactions, to which the optical model cannot be applied directly. Suppose we consider a reaction $a+A \to b+B$. The reaction mechanism will be complicated, but much of the complexity may be removed by the use of optical potentials. Suppose we try to represent the interaction by a matrix element between two plane waves. Because the interactions are very strong, as soon as the incident particle enters the domain of nuclear forces, it is distorted from the plane-wave form by the nuclear potential. Similarly, as the secondary nucleon, or nucleus, is leaving the region of nuclear force, its wave function is distorted by the effective nuclear potential that it experiences. These distortions, in addition to a term describing the interaction, are part of any matrix element between two plane waves. However, if the appropriate potentials are known from optical model studies of elastic scattering, then these distortions can be calculated and the reactions data analysed to find the matrix element between the two distorted waves. This is known as the distorted wave Born approximation.

An optical model calculation proceeds as follows. A form for the potential is chosen. It must be short range, and is usually characterized by a range, a depth and a parameter giving the surface thickness of the nucleus, that is, a measure of how quickly the potential goes to zero in the region of the nuclear radius. Scattering is not very sensitive to the detailed form of such a potential, but the form chosen of course fixes the interpretation of the parameters. An absorption part must be included, not necessarily with the same parameters as the real part. A form very frequently used is the Woods–Saxon potential, mentioned in Chapter 1. Finally, if polarization is also to be fitted, a spin–orbit coupling term must go in. If the incident particles are charged, the real part of the potential must include a coulomb potential, with a distributed source. A set of values for the parameters in the potential is chosen, and the scattering which would be produced by this potential is then compared with the observed scattering and the parameters varied until the discrepancy between the observed scattering and the calculated scattering is minimized. The scattering that would be produced by an assumed potential is obtained by numerical integration of the Schrödinger equation with the assumed potential included, and the whole procedure is carried out using digital computers, although a qualitative picture of the properties of a potential may be obtained by calculating the matrix element between two plane waves (see Section 1.4).

For further details of the optical model we refer you to P. Hodgson, *The Optical Model of Elastic Scattering* (Oxford, 1963).

Nuclear Physics

4.8. Time-independent potentials: the two-nucleon system

The two nucleon system is the one place in nuclear physics where we can expect a time-independent potential to tell us everything, so long as we limit our discussion to energies below the threshold for meson production. Even so, there will be some absorption in neutron–proton scattering because of the existence of the deuteron bound state which may be reached from the continuum with emission of a γ-ray. At the end of Chapter 1, we remarked that potentials have been found which reproduce all features of the two nucleon system below ~ 400 MeV, but are very complex, it "employs a hard core and is different for singlet-even, singlet-odd, triplet-even and triplet-odd states. It consists of central, tensor, spin–orbit and quadratic spin–orbit parts. . . ."

Here we shall be primarily concerned with the s-wave two-nucleon system, at low energies.

The two-nucleon system may consist of two protons, two neutrons, a proton and neutron. It is possible to talk about a single potential for the two-nucleon system, regardless of the composition of the system, because of the observed charge independence of nuclear forces. If this were not so, we would need a potential for proton–proton scattering, another for neutron–neutron scattering and yet a third for proton–neutron scattering. As it is, the strong interactions between two nucleons in the same space and spin states are independent of the nucleon charge. This feature of the strong interactions is discussed in more detail in Chapter 6.

Consider two nucleons in an s-state. Because of the exclusion principle for particles with half integral spin, two protons or two neutrons must have their spins opposite—that is, they can exist only in a singlet s-state. A proton and a neutron are not so restricted, but can exist in either a singlet s-state or a triplet s-state. Because of charge independence, all three singlet s-states have the same interaction, while the triplet state may have a different interaction: the deuteron is bound, while the singlets have no bound states. Now consider a p-state. Two protons or two neutrons must have parallel spins, since the total wave function for two identical fermions must be totally antisymmetric. The triplet neutron–proton state will have the same interaction, the singlet state only exists for neutron and proton and so any value is all right for charge independence. The singlet s-states are examples of singlet-even states, the triplet s-state an example of a triplet-even state, the triplet p-states examples of triplet-odd states and the singlet p-state an example of a singlet-odd state. The nucleon–nucleon potentials thus depend on the spin orientations and the parity, but not otherwise on the relative orbital angular momentum.

The primary characteristic of the two-nucleon system at low energies is the existence of a single bound state in the triplet s configuration, the deuteron, with a binding energy of -2.225 MeV.

If we assume a square well form for the interaction in this state, with a depth of U and a radius R, then a relation between the depth and radius results. Let the total energy of the system be E, the reduced mass m and define

$$k_1^2 = \frac{2m}{\hbar^2} E \qquad (r > R), \qquad (4.8.1)$$

$$k_2^2 = \frac{2m}{\hbar^2}(E+U) \quad (r < R). \qquad (4.8.2)$$

Nuclear Reactions

For a bound state of energy E, k_1^2 is negative and k_2^2 positive. The appropriate solutions of the Schrödinger equation are:

$$\left.\begin{array}{l} u_1 = A\,\dfrac{e^{-|k_1|r}}{r} \quad \text{outside the well,} \\[2mm] u_2 = B\,\dfrac{\sin k_2 r}{k_2 r} \quad \text{inside the well} \end{array}\right\} \tag{4.8.3}$$

($e^{+|k_1|r}$ is excluded because it blows up as $r \to \infty$ and $\cos kr/kr$ because it blows up as $r \to 0$).

Applying the boundary conditions

$$\left.\begin{array}{l} u_1|_R = u_2|_R, \\[2mm] \dfrac{\partial u_1}{\partial r}\bigg|_R = \dfrac{\partial u_2}{\partial r}\bigg|_R, \end{array}\right\} \tag{4.8.4}$$

together with the normalization condition

$$\int_R^\infty 4\pi r^2 \,|u_1|^2\, dr + \int_0^R 4\pi r^2\, |u_2|^2\, dr = 1 \tag{4.8.5}$$

fixes the complete wave function, and yields

$$\tan k_2 R = -\frac{k_2}{|k_1|}$$

or

$$\tan R\sqrt{\frac{2m}{\hbar^2}(E+U)} = -\sqrt{\frac{E+U}{|E|}} \tag{4.8.6}$$

and the logarithmic derivative of the internal wave function ru_2 is negative for a bound state. If the bound state is at zero energy, then

$$R\sqrt{\frac{2mU}{\hbar^2}} = \frac{n\pi}{2}$$

and for $n = 1$

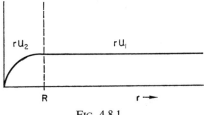

Fig. 4.8.1.

Nuclear Physics

With E negative and one bound state only we have

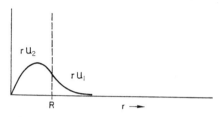

Fig. 4.8.2.

while if two bound states existed we would have

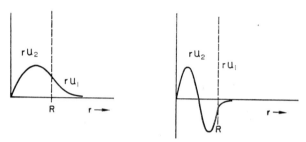

Fig. 4.8.3.

With only one bound state, we get only a relation between the parameters R and U, and need extra information in order to pin them down. This may be obtained from very low-energy s-wave scattering, the cross-section for which is, as the energy goes to zero, 20.36 barns. The situation is complicated by the singlet s and triplet s interactions: the singlet well is not deep enough to produce a bound state and so scattering in the triplet state will be different from scattering in the singlet.

For s-state scattering alone, in the spinless case

$$\sigma_s = \frac{4\pi}{k^2} \sin^2 \delta_s.$$

From eq. (4.4.21) we may note that in the expression

$$\sigma = \frac{2\pi}{\hbar} \frac{1}{v} |\mathcal{M}|^2 \varrho$$

the velocity terms in the flux factor and the phase space cancel, leaving all the energy dependence in the matrix element $\mathcal{M}$. If this is slowly varying near zero, then $\delta \sim k$ and we can write

$$\sigma_s = 4\pi a_s^2$$

where

$$a_s = \lim_{k \to 0} \frac{\delta(k)}{k}$$

and has the dimensions of a length, and is called the scattering length.†

† The scattering length is frequently defined as $-\lim_{k \to 0} \delta/k$.

Nuclear Reactions

There will be a different scattering length for the triplet and the singlet states—and because these correspond to different angular momentum, scattering in the two states is not mixed. We must sum over all final spin states for each initial spin state, and then complete the average over the initial states. There are three magnetic substates for the triplet, only one for the singlet, and so

$$\sigma_S = 4\pi\{\tfrac{1}{4}a_1^2 + \tfrac{3}{4}a_3^2\}. \tag{4.8.7}$$

In a little more detail, consider the spins of the nucleons before interaction:
The state with magnetic substates $(+\tfrac{1}{2}, +\tfrac{1}{2})$ has $J, J_z = 1, +1$.
The state with magnetic substates $(-\tfrac{1}{2}, -\tfrac{1}{2})$ has $J, J_z = 1, -1$.
The states with $(+\tfrac{1}{2}, -\tfrac{1}{2})$ and $(-\tfrac{1}{2}, +\tfrac{1}{2})$ are half each $J, J_z = 1, 0$ and $J, J_z = 0, 0$. Each of the $J = 1$ states has a scattering length a_3 and each of the $J = 0$ states a scattering length a_1. Summing over all these states and dividing by the weighting factor $(2S_a+1)(2S_b+1)$ with $S_a = S_b = \tfrac{1}{2}$ gives the result (4.8.7). Note that if a_1, was equal to a_3 we would just get $\sigma_s = 4\pi a_s^2$ in agreement with the spinless case.

The phase shift δ for a square well was obtained in Section 4.5:

$$\tan k_2 R = \frac{k_2}{k_1} \tan(k_1 R + \delta) \tag{4.5.15}$$

where for scattering both k_1 and k_2 are positive.

As $E \to 0$, $k_1 \to 0$ and $\tan(k_1 R + \delta) \to \tan \delta + k_1 R$, so

$$\frac{\delta}{k_1} \to \frac{\tan k_2 R}{k_2} - R$$

or

$$a^2 \to \left\{ \frac{\tan R \sqrt{\frac{2mU}{\hbar^2}}}{\sqrt{\frac{2mU}{\hbar^2}}} - R \right\}^2. \tag{4.8.8}$$

So

$$\sigma_s \to 4\pi \left\{ \frac{1}{4}\left[\frac{\tan R_1 \sqrt{\frac{2mU_1}{\hbar^2}}}{\sqrt{\frac{2mU_1}{\hbar^2}}} - R_1 \right]^2 + \frac{3}{4}\left[\frac{\tan R_3 \sqrt{\frac{2mU_3}{\hbar^2}}}{\sqrt{\frac{2mU_3}{\hbar^2}}} - R_3 \right]^2 \right\} \tag{4.8.9}$$

where the singlet potential has depth U_1 and radius R_1, the triple potential depth U_3 and radius R_3.

This gives two relations (the relation between U_1 and R_1 from the bound state of the deuteron, and the above expression for the zero energy cross-section) and four unknownse $U_1, R_1, U_3,$ and R_3. In order to solve the problem we need other relations, and these ar: provided by the requirement that

$$\sigma_s = 4\pi\{\tfrac{1}{4}a_1^2(k) + \tfrac{3}{4}a_3^2(k)\} \tag{4.8.10}$$

should reproduce the observed neutron–proton scattering cross-section at all energies low enough for s-wave scattering to predominate—that is, all energies such that $\lambda \gg R$. The

223

Nuclear Physics

radius R is $\sim 2\times 10^{-13}$ cm so this should work for energies of up to ~ 10 MeV. It would appear at first sight as though this requirement would probably be badly satisfied, for one thing we can be sure of is that no physical potential will have the infinite derivatives associated with a square well form. However, as we have earlier remarked, scattering is not very sensitive to the detailed form of a short-range potential.

If we write

$$a(k) = \frac{\sin \delta(k)}{k}$$

and expand $a(k)$ as a function of k

$$a(k) = a(0) + bk + ck^2 \ldots$$

and if $a(k)$ has zero gradient as $k \to 0$ then b is equal to zero, when

$$a(k) = a(0) + ck^2 \ldots$$

where c has the dimensions of length^3. If this form describes the scattering adequately, all that can be done is to determine the parameters $a(0)$ and c and the detailed shape is not accessible. We may check the applicability of such an approximation by returning to the square well:

$$a = \frac{\sin \delta}{k_1}$$

and for the square well

$$\tan k_2 R = \frac{k_2}{k_1} \tan (k_1 R + \delta)$$

as $\quad k_1 \to 0, \quad k_2 \to k_0 \quad$ and $\quad \dfrac{\tan \delta}{k_1} \to \dfrac{\tan k_0 R}{k_0} - R = a(0).$

Taking into account the next term in powers of k_1 we have on expanding $\tan (k_1 R + \delta)$ the result

$$\frac{\tan \delta}{k_1} \to \frac{\dfrac{\tan k_2 R}{k_2} - (R + \tfrac{1}{3} k_1^2 R^3)}{1 + \dfrac{k_1^2}{k_2} R \tan k_2 R} \qquad k_2 = \sqrt{k_1^2 + k_0^2}$$

and expanding k_2 yields

$$k_1 \cot \delta \to \frac{1}{a(0)} + \tfrac{1}{2} k_1^2 \left\{ R - \tfrac{1}{3} \frac{R^3}{a^2(0)} + \frac{1}{k_0^2 a(0)} \right\}$$

$$= \frac{1}{a(0)} + \tfrac{1}{2} k_1^2 \varrho. \qquad (4.8.11)$$

Now

$$\sin \delta = \frac{1}{\sqrt{1 + \cot^2 \delta}}.$$

So that

$$\frac{\sin^2 \delta}{k_1^2} = a^2(k) = \frac{a^2(0)}{a^2(0) k_1^2 + (1 + \frac{1}{2}k_1^2 a(0) \varrho)^2} \qquad (4.8.12)$$

and indeed $a(k)$ has no term linear in k, as we would expect—when the wavelength is very large in comparison with the size of the scattering potential the variation with energy of the scattering cross-section must be negligible.

This expansion will be valid provided $k_1 R \ll 1$ and $k_1 \ll k_2$. The form is independent of the detailed form of the potential, as shown by the following argument.[†]

We restrict ourselves to a short-range potential, which we define as being a potential for which the phase shift δ has an asymptotic value as $r \to \infty$ and write down the radial equations for zero energy, and for some energy E. For s-waves, the wave functions are $u(r)$ and we write $ru(r) = \psi(r)$ when $\psi(r)$ satisfies the equation

$$\frac{d^2\psi}{dr^2} + k_1^2 \psi - \frac{2m}{\hbar^2} V(r) \psi = 0$$

where $V(r)$ is a short-range potential. Then for $k_1 = 0$, $\psi(r) = \psi_0(r)$ and for arbitrary k_1, $\psi(r) = \psi_E(r)$ and these functions obey equations

$$\left.\begin{aligned}\frac{d^2\psi_0}{dr^2} - \frac{2m}{\hbar^2} V(r)\psi_0 &= 0, \\ \frac{d^2\psi_E}{dr^2} + k_1^2 \psi_E - \frac{2m}{\hbar^2} V(r)\psi_E &= 0,\end{aligned}\right\} \qquad (4.8.13)$$

whence

$$\psi_E \frac{d^2\psi_0}{dr^2} - \psi_0 \frac{d^2\psi_E}{dr^2} = k_1^2 \psi_E \psi_0 \qquad (4.8.14)$$

and integrating from zero to some finite radius r gives

$$\left[\left\{\psi_E \frac{d\psi_0}{dr} - \psi_0 \frac{d\psi_E}{dr}\right\}\right]_0^r = k_1^2 \int_0^r \psi_E \psi_0 \, dr.$$

The asymptotic solution is

$$\psi(r) \xrightarrow[r \to \infty]{} \Psi(r) = e^{i\delta} \frac{\sin(kr+\delta)}{k}$$

and

$$\frac{d^2\Psi}{dr^2} + k_1^2 \Psi = 0$$

so that

$$\left[\Psi_E \frac{d\Psi_0}{dr} - \Psi_0 \frac{d\Psi_E}{dr}\right]_0^r = k_1^2 \int_0^r \Psi_E \Psi_0 \, dr.$$

Now as $r \to \infty$, $\psi \to \Psi$ and as $r \to 0$ $\psi \to 0$ because of the requirement of regular behaviour of any wave function describing a physical system. Then subtracting these two

[†] For further details see J. M. Blatt and V. Weisskopf, *Theoretical Nuclear Physics*, chap. II (Wiley, 1952).

Nuclear Physics

equations, in the limit of large r, yields

$$\Psi_E(0)\left.\frac{d\Psi_0}{dr}\right|_{r=0} - \Psi_0(0)\left.\frac{d\Psi_E}{dr}\right|_{r=0} = k_1^2 \int_0^r \{\psi_E\psi_0 - \Psi_E\Psi_0\}\, dr$$

or

$$\frac{\sin\delta}{k_1} - a_0 \cos\delta = e^{-i\delta} k_1^2 \int_0^r \{\psi_E\psi_0 - \Psi_E\Psi_0\}\, dr, \qquad (4.8.15)$$

again showing that linear terms in the expansion of $a(k)$ are absent.

$$\Psi_E = e^{i\delta}\frac{\sin(k_1 r + \delta)}{k_1} \simeq \frac{\sin\delta}{k_1}$$

in the region where $k_1 r \ll 1$, that is in the region where ψ_E is different from Ψ_E. As we go outside this region, the integrand vanishes anyway. It would clearly be advantageous to have the integral as slowly varying with k_1 as possible, and to this end it is convenient to change the normalization of the wave functions so that

$$\psi' \xrightarrow[r\to\infty]{} \Psi = \frac{\sin(kr+\delta)}{\sin\delta}$$

(this is perfectly legitimate—for any single energy δ and k_1 are constants). This means that for $kr \ll 1$, $\Psi \sim 1$ regardless of the variation of δ with k_1. Then with this normalization we obtain a slightly different equation

$$k_1 \cot\delta - \left\{\lim_{k_1 \to 0} k_1 \cot\delta\right\} = -k_1^2 \int_0^r (\psi'_E\psi'_0 - \Psi'_E\Psi'_0)\, dr \qquad (4.8.16)$$

where the integrand in eq. (4.8.16) will be more slowly varying with k_1 than the integrand in (4.8.15) over the region where it is large, and hence (4.8.16) should have a larger region of applicability than (4.8.15), if we assume an energy independent integral. Our change of normalization has extracted some of the unknown variation of the integral and put it instead on the left-hand side of the equation. Equations (4.8.15) and (4.8.16) are exact: we now use our new normalization to assert that $\Psi'_E \simeq \Psi'_0$ and because the spatial variation of ψ_E and ψ_0 is controlled by the depth of the potential which is much greater than any value of E we are considering over most of the range where the integrand is large, we also approximate $\psi'_E = \psi'_0$ and so write

$$k_1 \cot\delta - \frac{1}{a_0} = \tfrac{1}{2} k_1^2 \varrho \qquad (4.8.17)$$

where ϱ is a constant and is called the *effective range*.† This expression for the variation of

---

† If a_0 is defined as

$$-\left\{\lim_{k_1 \to 0}\frac{\delta(k_1)}{k_1}\right\}$$

then (4.8.17) becomes

$$k_1 \cot\delta + \frac{1}{a_0} = \tfrac{1}{2} k_1^2 \varrho.$$

Nuclear Reactions

δ with k_1 is known as the shape independent, or effective range approximation, for we have nowhere injected the precise form of the potential. It is worth going over again the assumptions under which this is valid:

1. The potential is short range, in the sense that an asymptotic value of the phase shift δ exists.
2. The wavelengths considered are sufficiently large that almost all the scattering comes from a region within a radius r such that $k_1 r \ll 1$.
3. The depth of the potential over much of this region is very much greater than the energies considered.

It should also be clear that for the approximation to work, the scattering from the region in which the local depth of the potential is less than or of the order of the energy E must be an unimportant part of the total. Thus we should summarize this treatment by saying that *if empirically it is found that scattering over a certain range of energy is well fitted by the parametrization of eq. (4.8.17), then no scattering experiments in this region of energy can yield the detailed shape of the potential.*

It is in this sense that it is legitimate to employ the highly unphysical square well potential for low-energy nucleon–nucleon scattering and it is in this sense that we remarked in the last section that scattering is relatively insensitive to the detailed shape of a potential.

These results should hardly cause surprise. We know from classical electromagnetism that if we examine light scattered off a glob of dielectric with dimensions much smaller than the wavelength of the light, then we get no information on the shape of the glob—this is contained in the high-order multipole scattering and the high orders are negligible. The glob radiates as an electric dipole and all we can extract is the effective electric dipole moment, yielding an effective size and an effective refractive index.

The shape-independent approximation can be extended to bound states,[†] and we can finally conclude that if scattering is well described by eq. (4.8.17) and if the potential is not deep enough to produce a whole sequence of bound states, then the whole system can only be treated under the shape-independent approximation and any further information can only be obtained from scattering at energies so high that the shape-independent approximation breaks down.

The neutron–proton system with its one bound state is well described by the shape-independent approximation for energies at which only s-wave scattering is important, and so a square well treatment should reproduce all the low-energy scattering properties. It is important to note that the effective range ϱ is NOT equal to the radius parameter of a square well, as shown by eq. (4.8.11).

The effective range approximation can be improved by adding a term quartic in the momentum k_1, and this term is sensitive to the shape. For *predominantly* low-energy s-wave scattering, a term in k_1^4 is not necessary. If such a term is included, however, the scattering length and effective range parameters will change a little.

We take here parameters given by Wilson.[‡]

[†] See L. I. Schiff, *Quantum Mechanics*, 3rd ed., Sect. 50 (McGraw-Hill, 1968).
[‡] R. Wilson, *Nucleon–Nucleon Scattering* (Wiley, 1963).

Nuclear Physics

For the singlet neutron–proton state,

$$a^1(0) = 23.68 \pm 0.028 \times 10^{-13} \text{ cm}, \quad \varrho^1 = 2.46 \pm 0.12 \times 10^{-13} \text{ cm}.$$

For the triplet neutron–proton state,

$$a^3(0) = -5.399 \pm 0.011 \times 10^{-13} \text{ cm}, \quad \varrho^3 = 1.732 \pm 0.07 \times 10^{-13} \text{ cm}.$$

Substitution of these values for $a^1(0)$ and $a^3(0)$ gives the zero energy cross-section on a free proton

$$\sigma = 4\pi\{\tfrac{1}{4}a^1(0)^2 + \tfrac{3}{4}a^3(0)^2\}$$

equal to 20.36 barns and the cross-section for a pure triplet at zero energy is 3.68 barns while for the singlet it is 71 barns.

The parameters for square well potentials may be found by identifying the terms in (4.8.17) with the equivalent terms in (4.8.11).

If the triplet potential is represented by a radius R_3 and a depth U_3, the singlet potential by R_1, U_1 then those square well parameters are determined by the relations

$$a^3(0) = \frac{\tan\sqrt{\dfrac{2mU_3}{\hbar^2}} R_3}{\sqrt{\dfrac{2mU_3}{\hbar^2}}} - R_3, \tag{4.8.18a}$$

$$\varrho_3 = R_3 - \tfrac{1}{3}\frac{R_3^2}{a^3(0)^2} + \frac{1}{\dfrac{2mU_3}{\hbar^2} a^3(0)}, \tag{4.8.18b}$$

$$a_1^0(0) = \frac{\tan\sqrt{\dfrac{2mU_1}{\hbar^2}} R_1}{\sqrt{\dfrac{2mU_1}{\hbar^2}}} - R_1, \tag{4.8.18c}$$

$$\varrho^1 = R_1 - \tfrac{1}{3}\frac{R_1^2}{a^1(0)^2} + \frac{1}{\dfrac{2mU_1}{\hbar^2} a^1(0)}, \tag{4.8.18d}$$

(where m is the reduced mass of the two nucleon system, one-half of the nucleon mass). These equations must be solved numerically.

You should note at this point that $a(0)$ is positive if $k_0 R < \pi/2$ and negative if $k_0 R > \pi/2$. Thus with our definition of the scattering length, if the zero energy wave function turns over *before* $r > R$ then the scattering length is negative, while if it does not turn over, the scattering length is positive. The condition $k_0 R > \pi/2$ implies a bound state.

While the effective range must be determined from scattering over a range of energies, the scattering length is a zero energy property and so can be obtained from the scattering of very low-energy neutrons on protons. Indeed the values given above come predominantly from scattering at such low energies that the neutron wavelength is very much greater than the separation between two neighbouring protons in the hydrogen molecule. Under these circumstances scattering from adjacent protons is coherent and in addition to the contribution from each proton individually, the scattering also contains interference between the two. Thus if the scattering from a proton is in the singlet state, and from its neighbour in the triplet state, then the opposite sign of the scattering lengths implies destructive interference, while if the scattering is triplet–triplet or singlet–singlet, the interference is constructive. It was in this way that the relative sign of the scattering lengths was determined, showing conclusively that there is no bound singlet state of the deuteron.

This method is well illustrated by the difference in the molecular scattering cross-section on ortho- and para-hydrogen. In the diatomic hydrogen molecule, the spins of the two protons may be either parallel or antiparallel, with the total wave function antisymmetric under interchange of the two. Thus either the rotational or electronic states must be different, introducing an energy difference between the singlet and triplet states and there is also a difference in the energy from the magnetic moment interaction between the two protons, which is negligible. Ortho-hydrogen, the state in which the spins are parallel, has a higher energy than para-hydrogen in which the spins are antiparallel. The magnetic energy difference Δ_M is $\sim \mu^2/r^3$ where μ is the proton magnetic moment and r the mean separation. The proton magnetic moment is $\sim 10^{-20}$ erg-gauss^{-1} and $r \sim 10^{-8}$ cm so

$$\Delta_M \sim 10^{-16} \text{ erg} \sim 10^{-4} \text{ eV}.$$

The difference in rotational energy is

$$\Delta_R \sim \frac{\hbar^2}{2Mr^2} \quad \Delta_R \sim 10^{-14} \text{ erg} \sim 10^{-2} \text{ eV} \gg \Delta_M.$$

In the electronic ground states, two protons with parallel spin must have a rotational state antisymmetric under exchange, while two protons with antiparallel spins must have a rotational state symmetric under exchange. Thus the ground state of para-hydrogen has no rotational angular momentum and is lower in energy by $\sim 10^{-2}$ eV than the ground state of ortho-hydrogen, with one unit of rotational angular momentum. Room temperature corresponds to energies $\sim 2.5 \times 10^{-2}$ eV and so at room temperature hydrogen is a mixture of ortho- and para-hydrogen in the ratio of 3:1 (because of the three spin states accessible to orthohydrogen in comparison with the one state for para-hydrogen). If the temperature is reduced, this proportion is frozen in: the relaxation time to nearly pure para-hydrogen is very long because of the weak coupling between the two states. However, nearly pure para-hydrogen can be produced at low temperatures catalytically, and the molecular cross-sections for pure ortho- and para-hydrogen inferred from the results on parahydrogen and a mixture. Experimentally nearly pure para-hydrogen (purity $\sim 99.8\%$) gas can be prepared at 20°K, and because of the long relaxation time a mixture of 75% ortho-hydrogen and 25% para-hydrogen can be prepared at any such temperature. (If it is desired to compare ortho- and para-hydrogen scattering cross-sections, the hydrogen in the two cases should be at the same temperature to make easier corrections due to the details of molecular structure and thermal motion.)

The scattering in the two cases is easily related to scattering on a free proton, apart from detailed corrections depending on the structure of molecular hydrogen. The difference

Nuclear Physics

between the singlet and triplet states is determined by the values of the quantity $\sigma_n \cdot \sigma_p$. Now

$$(\sigma_n + \sigma_p)^2 = \sigma_n^2 + \sigma_p^2 + 2\sigma_n \cdot \sigma_p = 2 \text{ (triplet)}$$
$$= 0 \text{ (singlet)}.$$

So

$$\sigma_n \cdot \sigma_p = \tfrac{1}{4} \text{ (triplet)}$$
$$= -\tfrac{3}{4} \text{ (singlet)}.$$

So the operator $(\tfrac{3}{4} + \sigma_n \cdot \sigma_p)$ operating on a neutron–proton state has

eigenvalues 1 for a triplet state
0 for a singlet state,

while $(\tfrac{1}{4} - \sigma_n \cdot \sigma_p)$ has

eigenvalues 1 for a singlet state
0 for a triplet state.

Thus if we have two protons which are scattering coherently, the molecular scattering amplitude is proportional to

$$(\tfrac{3}{4} + \sigma_n \cdot \sigma_{p_1}) a_3 + (\tfrac{1}{4} - \sigma_n \cdot \sigma_{p_1}) a_1 + (\tfrac{3}{4} + \sigma_n \cdot \sigma_{p_2}) a_3 + (\tfrac{1}{4} - \sigma_n \cdot \sigma_{p_2}) a_1$$
$$= 2(\tfrac{3}{4} a_3 + \tfrac{1}{4} a_1) + (a_3 - a_1) \sigma_n \cdot \Sigma$$

where $\Sigma = \sigma_{p_1} + \sigma_{p_2}$. If $\Sigma = 0$ as for para-hydrogen, then the scattering length f for molecular para-hydrogen, sometimes called the coherent scattering length, is defined as

$$f = 2(\tfrac{3}{4} a_3 + \tfrac{1}{4} a_1)$$

and because a_1 is roughly three times a_3 and of opposite sign, this quantity is very small. Indeed, unless molecular hydrogen is nearly pure para-hydrogen, the second term with $\Sigma = 1$ predominates. Measurements of f using para-hydrogen are thus very sensitive to contamination from ortho-hydrogen.

For ortho-hydrogen $|\sigma_n + \Sigma| = \tfrac{3}{2}$ or $\tfrac{1}{2}$
corresponding to $\sigma_n \cdot \Sigma = \tfrac{1}{2}$ or -1.

The state with total angular momentum $\tfrac{3}{2}$ has weight $\tfrac{2}{3}$ and the state with total angular momentum $\tfrac{1}{2}$ has weight $\tfrac{1}{3}$. Thus the scattering cross-section for para-hydrogen is proportional to f^2 while the scattering cross-section for ortho-hydrogen is proportional to

$$f^2 + 2f(a_3 - a_1) \langle \sigma_n \cdot \Sigma \rangle + (a_3 - a_1)^2 \langle (\sigma_n \cdot \Sigma)^2 \rangle = f^2 + \tfrac{1}{2}(a_3 - a_1)^2.$$

This kind of experiment must be done at energies such that the neutron wavelength is large in comparison with the separation of the protons, $\sim 10^{-8}$ cm. The effective cross-section for a mixture of ortho- and para-hydrogen will be the appropriate average of the cross-sections for the two kinds: there is no interference between the scattering from different molecules, only between scattering from two protons in the same molecule. A wavelength of 10^{-8} cm, 1 Å unit, corresponds to a neutron energy of 10 eV. Measurements of f at energies $\lesssim \frac{1}{100}$ eV have been made on para-hydrogen, and measurements in this region on a mixture (3 : 1 ortho-hydrogen) can be combined with this to give a_1 and a_3 directly. In the region 1–15 eV neutron energy, coherence becomes steadily less important and the measurements can be extrapolated to yield the free proton cross-section

$$\sigma = 4\pi(\tfrac{3}{4}a_3^2 + \tfrac{1}{4}a_1^2) = 20.36 \text{ barns}$$

which remains essentially constant up to energies ~3 keV. (In relating the molecular cross-sections to the scattering lengths a_1 and a_3 it must be remembered that a_1 and a_3 are defined in terms of the centre of mass momentum, and the reduced mass is changed for molecular scattering because the two protons hang together provided the neutron energy is sufficiently low that excited states of the molecule are not produced (and we have seen this takes ~0.01 eV). For a given momentum in the centre of mass of a proton and a neutron, the matrix elements are the same, but referring back to (4.4.21) we see that in the product of phase space and flux factor the reduced mass (which enters squared) of

$$\left(\frac{M^2}{M+M}\right) = \frac{M}{2}$$

for free proton scattering is replaced by

$$\frac{M\,2M}{M+2M} = \tfrac{2}{3}M$$

where M is the nucleon mass. This means, for example, that the molecular scattering cross-section for para-hydrogen becomes

$$(\tfrac{4}{3})^2 4\pi f^2.$$

At very low energies it is necessary to average over thermal molecular velocities, and over the effects of rotation and vibration in the molecule. In addition, while neutron energies of $\lesssim 0.01$ eV cannot excite the molecule, molecular de-excitation with transfer of energy *to* the neutron can take place. The capture cross-section for $n+p \to d+\gamma$ obeys a $1/v$ law[†] and so also is important at very low energies. There are thus a large number of corrections, very dependent on the detailed molecular structure, to be included in an attempt to extract the scattering lengths from the coherent cross-sections. Finally, the sensitivity of the effective cross-section to an admixture of ortho-hydrogen in nearly pure para-hydrogen makes this method of finding the coherent scattering length f less reliable than a different method in which neutrons are reflected from a liquid hydrocarbon mirror.

Suppose that we can arrange to have coherent scattering from not just two protons, but from protons in a large number of molecules, so that the sum of the proton spins averages to zero. Then the appropriate scattering length is clearly the coherent scattering length f. Reflection of a beam of neutrons at a grazing angle from the surface of a liquid containing hydrogen fulfils these conditions: this is just Bragg reflection. Because the scattering from individual molecules is very close to the forward direction, the momentum transfer to the molecule is small and molecular structure effects are thus negligible. Absorption of neutrons and incoherent scattering reduce the coherent scattered intensity but do not affect the scat-

[†] See Section 4.9.

tering angle. If neutrons with a wavelength less than or comparable with the molecular separation are used, they will go straight through the liquid until a critical angle is reached at which total reflection takes place. This critical angle depends on the neutron refractive index, which in turn depends on f for hydrogen, and on the sum of f and the scattering length for carbon when hydrocarbons are used. This method provides the determination of f which is least susceptible to systematic error, when the contribution of f and the free proton scattering cross-section yields the values for the scattering lengths.

The scattering lengths from the very low-energy experiments, the variation of the scattering cross-section on free protons at energies of up to 15 MeV and the deuteron binding energy combine to give the scattering lengths and effective ranges characterizing the low-energy properties of the neutron–proton potential. For more details and reference to the original papers, we refer you to R. Wilson, *The Nucleon–nucleon Interaction: Experimental and Phenomenological Aspects* (Wiley, 1963).

Returning to eq. (4.8.18), numerical solution yields the following values for the square well parameters:

$$R_3 \sim 2.05 \times 10^{-13} \text{ cm} \quad R_1 \sim 2.4 \times 10^{-13} \text{ cm},$$
$$U_3 \sim 35 \text{ MeV} \quad U_1 \sim 16 \text{ MeV}.$$

The depth of the two potentials differs by a factor of 2, while the ranges are comparable. Note the difference between the square well radius and the effective range, which is small for R_1 because of the very large value of $a_1(0)$.

The two-proton state is easier to study because for each value of the orbital angular momentum there is only one value of the sum of the proton spins, so that for the s-state only the singlet parameters determine the scattering. However, it is complicated by the presence of the coulomb interaction, which at very low energies predominates, and prevents the protons getting within range of each other. The coulomb energy of two protons separated by 2×10^{-13} cm is ~ 0.8 MeV, so for the nuclear forces to influence the scattering appreciably the centre of mass kinetic energy must be ~ 0.8 MeV, implying a laboratory energy for the incident proton of ~ 3 MeV for strong nuclear scattering. However, below 1 MeV the angular distribution of scattered protons exhibits interference between the nuclear and the coulomb scattering amplitude, thus yielding not only the modulus of the scattering length but, from the sign of the interference term, the sign of the scattering length. The effective range parametrization of the scattering of two protons yields a scattering length and an effective range appropriate to the combination of the nuclear potential plus an e^2/r coulomb potential. Extraction of the parameters that would describe pp scattering at low energies in the limit $e \to 0$ is dependent on the form assumed for the strong interaction potential, and this limits the accuracy with which the strong scattering length can be obtained. A recent compilation[†] gives the following values for the pp singlet scattering length effective range $a(E+S)$, $\varrho(E+S)$ and the parameters obtained after correction for the coulomb part, $a(S)$, $\varrho(S)$:

pp shape-independent parameters $\times 10^{-13}$ cm

$$a(E+S) = 7.817 \pm 0.007 \quad \varrho(E+S) = 2.81 \pm 0.018$$
$$a(S) = 17 \pm 2 \quad \varrho(S) = 2.83 \pm 0.03$$

[†] E. M. Henley in *Isotopic Spin in Nuclear Physics*, Ed. D. H. Wilkinson (North Holland, 1969).

to be compared with

$$\text{Singlet } np \text{ shape-independent parameters} \times 10^{-13} \text{ cm}$$
$$a = 23.715 \pm 0.013 \qquad \varrho = 2.76 \pm 0.07$$

The scattering length and some estimate of *nn* scattering have also been obtained, although indirectly because free neutron targets are non-existent. A scattering length for the *nn* system of $\sim 20 \times 10^{-13}$ cm implies an attractive potential almost strong enough to produce a bound state, the di-neutron (which has never been observed). Suppose the di-neutron D existed as a stable bound state. Then it would be produced in reactions like

$$n + d \rightarrow D + p$$

and would be observed by finding a spike in the energy of the recoiling proton, corresponding to recoil against a mass slightly less than two neutron masses.

For an unbound *nn* system, the very strong *nn* interaction can still produce a spike in the proton recoil spectrum, corresponding to a mass slightly greater than two neutron masses: the strong interaction, although not strong enough to bind two neutrons, none the less makes them stick together on the nuclear time scale, and from the shape of the two neutron mass spectrum the scattering length can be extracted. The relation between the neutron–neutron *final state interaction* and neutron–neutron scattering is illustrated in Fig. 4.8.4 in which we visualize an almost free neutron in the outer region of the deuteron being scattered by the incident neutron. The deuteron itself is so weakly bound that this is a fair approximation.

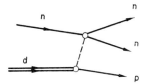

Fig. 4.8.4. Scattering of a bound (virtual) neutron into a continuum state. The lower vertex is well understood and so a study of this process yields the scattering amplitude at the upper vertex.

Some experimental results are given in Fig. 4.8.5 and from such data the following parameters for the neutron–neutron interaction have been extracted (Henley, *loc. cit.*):

$$a = 17.6 \pm 1.5 \qquad \varrho = 3.2 \pm 1.6.$$

The properties of the singlet *s*-wave nucleon–nucleon system may then be summarized as

	scattering length	effective range	$\times 10^{-13}$ cm
pp	$+17 \quad \pm 2$	2.83 ± 0.03	
np	$+23.715 \pm 0.013$	2.76 ± 0.07	
nn	$+17.6 \quad \pm 1.5$	3.2 ± 1.6	

These results are clearly consistent with charge symmetry of nuclear forces, and the effective ranges are clearly consistent with charge independence. There is, however, a substantial

Nuclear Physics

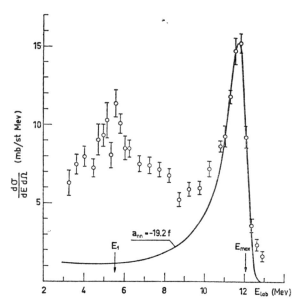

FIG. 4.8.5. Proton energy spectrum at 4° in the laboratory for the reaction $d(n, p) 2n$. The strong enhancement near the maximum possible energy is indicative of a strong attractive interaction between the two neutrons: the values of the parameters describing this interaction are, however, sensitive to the details of the theoretical analysis. [From K. Ilacovac et al., Phys. Rev. **124**, 1923 (1961).]

difference between the singlet np scattering length and the pp and nn scattering lengths. This difference was at one time attributed to the different magnetic moment scattering in the pp system and the np system (something which is clearly *not* charge independent) but this explanation depends on the $1/r^3$ singularity in a dipole–dipole interaction. A repulsive core in nucleon–nucleon scattering would prevent the magnetic moments from getting close enough to affect the scattering lengths, and anyway a study of magnetic moment scattering of electrons from nucleons shows that the magnetic moments are distributed throughout the nucleon, and so there is no singularity in the magnetic nucleon–nucleon interaction. The difference is really very small when expressed in terms of a potential—one-quarter wavelength so nearly fits into the singlet potential well (and if for a square well it fits in exactly, the scattering length becomes infinite) that a small variation in well depth produces large variations in the scattering length. For a square well potential, if the radius is taken as being equal for pp, np and nn interactions, then the difference of $\Delta a \sim 7 \times 10^{-13}$ cm in the scattering lengths implies that the quantity $k_0 R$ for the two cases is $\sim 1.5\%$ different, that is, the np well depth is ~ 0.4 MeV (3%) greater than the pp and nn well depth, or that the np radius is $\sim 0.4 f$ greater than the pp radius. The pp and nn (charge symmetric) square well parameters derived from these quantities are

$$R_{pp} = 2.66 \times 10^{-13} \text{ cm}, \quad U_{pp} = 12.8 \text{ MeV}$$

while the np parameters are

$$R_{np} = 2.64 \times 10^{-13} \text{ cm}, \quad U_{np} = 13.4 \text{ MeV}$$

but they are consistent with both depths equal and radii

$$R_{pp} = 2.60 \times 10^{-13} \text{ cm},$$
$$R_{np} = 2.64 \times 10^{-13} \text{ cm},$$

for example, if an effective range of 2.76 f is taken for both systems. Nowadays the difference in scattering lengths is regarded as being real and due to a lack of charge independence in the strong interactions at the 1% level. This can be qualitatively understood in terms of the forces expected from meson exchange. These are dependent on the meson masses, and the charged pion has a mass of 139.6 MeV to be compared with the neutral pion mass of 135.0 MeV, an $\approx 3\%$ difference, which is presumably electromagnetic in origin. However, there is an inverse relation between the range of meson exchange forces and the mass of the exchanged meson, and the light pion only controls the periphery of the nucleon–nucleon interaction, while the mass differences between the charged and neutral states of massive mesons are not known, and in any case no one really knows how to calculate meson exchange forces. Discussions of the theoretical "explanation" of nucleon–nucleon scattering are given by Wilson and Moravcik.[†]

For discussion of charge-dependent features of the nucleon–nucleon interaction we refer you to the article by Henley.

We have one last point in connection with low-energy nucleon–nucleon scattering. The s-wave phase shifts start falling in the MeV region, and the singlet s-wave phase shift goes through zero at ~ 200 MeV. This is the evidence for the presence of a repulsive core in the nucleon–nucleon potential. An attractive potential gives a positive phase shift, a repulsive potential a negative phase shift, and if the range of the repulsive potential is small in comparison with the range of the attractive potential, phase shifts will start out positive and decrease steadily as the wavelength decreases with increasing energy and the s-waves become progressively more sensitive to the short-range repulsion. It is of course the s-waves that are most sensitive to such a repulsive core: higher waves are kept clear by the centrifugal term $l(l+1)/r^2$ in the Schrödinger equation, which makes the wave functions behave like $(kr)^l$ near the origin.

As shown in Fig. 4.8.6, as the wavelength decreases a hard core pushes the zero of the wave function a larger and larger number of wavelengths away from the origin, and as the energy gets very large in comparison with the depth of the attractive part of the potential, this results in a negative phase shift, even though at low energies the influence of the repulsive core is negligible.

In our discussion of the time-independent nucleon–nucleon potential, we have neglected two residual effects: the presence of spin–orbit coupling and of tensor forces. Both these terms are non-central and so in their presence the orbital angular momentum **L** and the spin angular momentum **S** are not conserved separately: only the sum $\mathbf{J} = \mathbf{L} + \mathbf{S}$ is conserved. The evidence for the existence of such terms comes from attempts to reproduce all nucleon–nucleon scattering data with a phenomenological potential and is summarized by Moravcik (*loc. cit.*).

[†] R. Wilson, *The Nucleon–Nucleon Interaction* (Wiley, 1963); M. J. Moravcik, *The Two Nucleon Interaction* Oxford, 1963).

Nuclear Physics

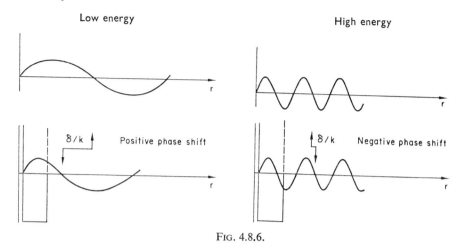

FIG. 4.8.6.

The observation of polarization in *pp* scattering by itself demands the existence of non-central forces (other evidence comes from the strong spin–orbit coupling needed to account for the level sequence in the independent particle model, and from polarization in nucleon–nucleus scattering when the nucleus has spin zero). In a system with two identical nucleons there is only one spin state for each value of the orbital angular momentum. However, polarization is also observed in *np* scattering, and is sufficiently different from the polarization in *pp* scattering to require the inclusion of non-central forces in all relative spin states of two nucleons. We shall not pursue this topic further, but we will discuss briefly the effects of non-central forces on the only two-nucleon bound state, the ground state of the deuteron.

A spin–orbit potential is the simplest non-central potential, of form $f(r)\mathbf{L}\cdot\mathbf{S}$. Such a potential does not mix into the $^3S_1$ dominant component of the deuteron wave function any $^3D_1$ component. The reason is that the operator $\mathbf{L}\cdot\mathbf{S}$ operating on a state of well-defined angular momentum can only mix in angular momentum states differing by zero or one unit from the ground state. Expanding the operator,

$$\mathbf{L}\cdot\mathbf{S} = L_xS_x+L_yS_y+L_zS_z = \tfrac{1}{2}(L_x+iL_y)(S_x-iS_y)+L_zS_z+\tfrac{1}{2}(L_x-iL_y)(S_x+iS_y)$$

and the factors (L_x+iL_y), L_z, (L_x-iL_y) only change the value of L by one unit, or by zero. Parity conservation prevents a change of one unit, and there is no spin–orbit splitting when $L=0$ as for the ground state of the deuteron. However, the deuteron possesses a small electric quadrupole moment and the magnetic moment of the deuteron is not precisely the sum of the neutron and proton magnetic moments. The values for these quantities are[†]

$$\left.\begin{array}{l}Q_d = 2.82\times 10^{-27} \text{ cm}^2,\\ \mu_d = 0.857\,41\pm 0.000\,08,\\ \mu_p = 2.792\,680\pm 0.000\,011,\\ \mu_n = -1.913\,04\pm 0.000\,10\end{array}\right\} \text{ nuclear magnetons}$$

so the value of the magnetic moment of the deuteron appropriate to a $^3S_1$ state, $\mu_p+\mu_n$, differs from the observed value by

$$\mu_d-(\mu_p+\mu_n) = -0.022\,23 \quad \text{nuclear magnetons}.$$

[†] See R. Wilson, *The Nucleon–Nucleon Interaction* (Wiley, 1963).

Nuclear Reactions

This value by itself suggests a few per cent of $^3D_1$ admixture, but does not absolutely necessitate it, for the magnetic moment of two bound nucleons need not be precisely the sum of the free magnetic moments: if exchange of charged mesons contributes to the potential, as is undoubtedly the case, internal mesonic currents can modify μ_d and the effect cannot be reliably calculated. The quadrupole moment, however, necessitates an admixture of an orbital angular momentum state other than zero, and the $^3D_1$ state is the only one allowed by conservation of total angular momentum and parity. A tensor force can produce this admixture: it is clear from the foregoing discussions that we need a potential that contains terms more complicated than the spin–spin interaction $\sigma_n \cdot \sigma_p$ and more complicated than the spin–orbit interaction $(\sigma_n + \sigma_p) \cdot \mathbf{L}$. Something which transforms under rotations as a second-rank tensor is an obvious candidate for mixing orbital angular momentum states differing by two units. A potential containing a term $(\sigma_1 \cdot \mathbf{r})(\sigma_2 \cdot \mathbf{r})$ is dependent on the orientation of each spin to the radius vector connecting the two nucleons, and has the desired properties: the form adopted for the tensor potential is

$$V_T(r) \left\{ \frac{3(\sigma_1 \cdot \mathbf{r})(\sigma_2 \cdot \mathbf{r})}{r^2} - \sigma_1 \cdot \sigma_2 \right\}$$

where $V_T(r)$ is some radial dependence. This operator, applied to a $^3S_1$ state, generates a $^3D_1$ state. This can be calculated directly using the properties of the angular momentum wave functions, and a discussion may be found in Blatt and Weisskopf. We shall not go into details here, for the relation of the observed quadrupole moment to the proportion of $^3D_1$ in the deuteron is dependent on the detailed form of the wave functions and estimates have ranged from $\sim 3\%$ to $\sim 7\%$. A value of $\sim 7\%$ is implied by the assumption of a pion exchange potential, which has a long tail (as opposed to something like a square well) and is also required by observations of deuteron photodisintegration: further details are given in Wilson, *loc. cit.*

4.9. Time-independent potentials: neutron capture by protons

We conclude our discussion of time-independent potentials with an example of a reaction: the process of deuteron formation through the reaction $n + p \rightarrow d + \gamma$. This is also our *sixth application of the Fermi Golden Rule.*

The initial state consists of a plane wave describing the unbound neutron–proton system. The interaction between neutron and proton is dominated by the appropriate potentials in the singlet and triplet states, and these potentials are time independent. At low energies only the incident *s*-waves of the neutron–proton system are appreciably scattered, or have an appreciable probability of interacting. Then there are two initial *s*-wave states to consider: the $^3S_1$ and $^1S_0$ states of the unbound proton–neutron system. In order for a deuteron to be formed, the excess energy must be carried off by some other particle coupled to the proton–neutron system, and at low energy only a photon is available. If we ignore the small $^3D_1$ admixture in the deuteron, there is no change of orbital angular momentum in the capture process, and so no change of parity. The transition is either

$$J^P = 1^+ \rightarrow J^P = 1^+ \qquad \text{(triplet initial state)}$$

Nuclear Physics

$$J^P = 0^{P+} \rightarrow J = 1^+ \qquad \text{(singlet initial state)}$$

and so the transition will be of magnetic dipole character,

$$\Delta J = \pm 1, 0 \; (0 \nrightarrow 0) \; N \, 0.$$

It is easy to see that the triplet s-state cannot go to a deuteron (also a triplet s-state) with emission of a photon. There are no currents corresponding to a change of orbital angular momentum, and so a photon can only be generated by a fluctuating magnetic moment corresponding to a reorientation of the nucleon spins—which does not occur in the process $^3S_1 \rightarrow {}^3S_1$. The spins have to keep the same relative orientation, and the sum of the spins cannot accelerate because of conservation of orbital angular momentum.†

Thus we have to write down just the matrix element for an M1 transition from the unbound $^1S_0$ state to the bound $^3S_1$ state, the deuteron.

Now the dominant process at all but the very lowest neutron energies is elastic scattering. The initial wave function is only well represented as a plane wave at very large separations, and it is better to take into account the modifications of the singlet initial state which are introduced by the presence of the singlet even potential. For a square well we have an initial wave function

$$\left. \begin{array}{ll} \psi_1(r) = \dfrac{e^{i\delta} \sin(k_1 r + \delta)}{k_1 r} & r \geqslant R, \\[1em] \psi_2(r) = \dfrac{\sin k_2 r}{k_2 r} & r < R \end{array} \right\} \qquad (4.9.1)$$

where δ is the singlet phase shift, whereas had we taken an initial plane wave we would have had

$$\psi(r) = \frac{\sin k_1 r}{k_1 r} \quad \text{for all } r.$$

The final state is just the deuteron ground state and for a square well is

$$\left. \begin{array}{ll} \psi_d(r) \propto \dfrac{e^{-|k_{1d}|r}}{r} & r > R, \\[1em] \psi_d(r) \propto \dfrac{\sin k_{2d} r}{k_{2d} r} & r < R. \end{array} \right\} \qquad (4.9.2)$$

We now want the transition matrix element between these two states—the transitory magnetic moment. Referring back to Chapter 3, eq. (3.5.10), we see that this is given by

$$\langle {}^3S_1 | \mu | {}^1S_0 \rangle = \frac{e\hbar}{2Mc} \langle {}^3S_1 | g_p \sigma_p + g_n \sigma_n | {}^1S_0 \rangle \qquad (4.9.3)$$

with $g_p = 5.58$ and $g_n = -3.82$.

---

† For a more technical discussion of this point, see J. M. Blatt and V. Weisskopf, *Theoretical Nuclear Physics*, p. 604 (Wiley, 1952).

The operator $g_p\boldsymbol{\sigma}_p + g_n\boldsymbol{\sigma}_n$ can be rewritten as

$$\tfrac{1}{2}(g_p+g_n)(\boldsymbol{\sigma}_p+\boldsymbol{\sigma}_n) + \tfrac{1}{2}(g_n-g_p)(\boldsymbol{\sigma}_n-\boldsymbol{\sigma}_p). \tag{4.9.4}$$

The spin part of the singlet wave function can be written as

$$\frac{1}{\sqrt{2}}\left\{\begin{pmatrix}1\\0\end{pmatrix}_p\begin{pmatrix}0\\1\end{pmatrix}_n - \begin{pmatrix}0\\1\end{pmatrix}_p\begin{pmatrix}1\\0\end{pmatrix}_n\right\}$$

and the components of $\boldsymbol{\sigma}_p$ operate only on the proton spinor and the components of $\boldsymbol{\sigma}_n$ only on the neutron spinor.

The operator $\boldsymbol{\sigma} = \mathbf{i}\sigma_x + \mathbf{j}\sigma_y + \mathbf{k}\sigma_z$ (where $\mathbf{i}, \mathbf{j}, \mathbf{k}$ are unit vectors along the x-, y-, z-axes) and the Pauli spin matrices are

$$\sigma_x = \frac{1}{2}\begin{pmatrix}0 & 1\\1 & 0\end{pmatrix} \quad \sigma_y = \frac{1}{2}\begin{pmatrix}0 & -i\\i & 0\end{pmatrix} \quad \sigma_z = \frac{1}{2}\begin{pmatrix}1 & 0\\0 & -1\end{pmatrix}$$

whence

$$\sigma^2 = \sigma_x^2 + \sigma_y^2 + \sigma_z^2 = \frac{3}{4}\begin{pmatrix}1 & 0\\0 & 1\end{pmatrix} = \sigma(\sigma+1)\begin{pmatrix}1 & 0\\0 & 1\end{pmatrix}.$$

The first term applied to the singlet state gives for the x-component:

$$\mathbf{i}\left\{\begin{pmatrix}0\\1\end{pmatrix}_p\begin{pmatrix}0\\1\end{pmatrix}_n - \begin{pmatrix}1\\0\end{pmatrix}_p\begin{pmatrix}1\\0\end{pmatrix}_n + \begin{pmatrix}1\\0\end{pmatrix}_p\begin{pmatrix}1\\0\end{pmatrix}_n - \begin{pmatrix}0\\1\end{pmatrix}_p\begin{pmatrix}0\\1\end{pmatrix}_n\right\} = 0 \tag{4.9.5}$$

and similarly for the y- and z-components. The magnetic moment transition matrix element for an initial singlet state is thus

$$\frac{e\hbar}{2Mc}\langle {}^3S_1 | \tfrac{1}{2}(g_n-g_p)(\boldsymbol{\sigma}_n-\boldsymbol{\sigma}_p) | {}^1S_0\rangle. \tag{4.9.6}$$

The operator $(\boldsymbol{\sigma}_n-\boldsymbol{\sigma}_p)$ applied to the singlet spin state gives

$$\left[\mathbf{i}\left\{\begin{pmatrix}1\\0\end{pmatrix}_p\begin{pmatrix}1\\0\end{pmatrix}_n - \begin{pmatrix}0\\1\end{pmatrix}_p\begin{pmatrix}0\\1\end{pmatrix}_n - \begin{pmatrix}0\\1\end{pmatrix}_p\begin{pmatrix}0\\1\end{pmatrix}_n + \begin{pmatrix}1\\0\end{pmatrix}_p\begin{pmatrix}1\\0\end{pmatrix}_n\right\}\right.$$

$$+ \mathbf{j}\left\{i\left[-\begin{pmatrix}1\\0\end{pmatrix}_p\begin{pmatrix}1\\0\end{pmatrix}_n - \begin{pmatrix}0\\1\end{pmatrix}_p\begin{pmatrix}0\\1\end{pmatrix}_n - \begin{pmatrix}0\\1\end{pmatrix}_p\begin{pmatrix}0\\1\end{pmatrix}_n - \begin{pmatrix}1\\0\end{pmatrix}_p\begin{pmatrix}1\\0\end{pmatrix}_n\right]\right\}$$

$$\left. + \mathbf{k}\left\{-\begin{pmatrix}1\\0\end{pmatrix}_p\begin{pmatrix}0\\1\end{pmatrix}_n - \begin{pmatrix}0\\1\end{pmatrix}_p\begin{pmatrix}1\\0\end{pmatrix}_n - \begin{pmatrix}1\\0\end{pmatrix}_p\begin{pmatrix}0\\1\end{pmatrix}_n - \begin{pmatrix}0\\1\end{pmatrix}_p\begin{pmatrix}1\\0\end{pmatrix}_n\right\}\right]\frac{1}{2\sqrt{2}}$$

$$= [\mathbf{i}\{2|1,+1\rangle - 2|1,-1\rangle\} + \mathbf{j}\{-i[2|1,+1\rangle + 2|1,-1\rangle]\} + \mathbf{k}\{-2\sqrt{2}|1,0\rangle\}]\frac{1}{2\sqrt{2}}$$

$$= \frac{1}{\sqrt{2}}[\{(\mathbf{i}-i\mathbf{j})|1,+1\rangle\} - \{(\mathbf{i}+i\mathbf{j})|1,-1\rangle\} - \mathbf{k}\{\sqrt{2}|1,0\rangle\}].$$

Thus we have matrix elements connecting the initial singlet state with a final state in which the deuteron has $m = \pm 1, 0$ with respect to the initial direction of the neutron. However, the physics is contained in the spin flip terms and since for the initial state there is no pre-

Nuclear Physics

ferred spin direction, we must get an isotropic distribution of radiation in space. Referring again to (3.5.7) we see that we must take the scalar product of the nuclear matrix element with the vector product of the photon polarization vector ϵ_0 and the propagation vector $\mathbf{k}_\gamma$. Defining polarization vectors ϵ_1 in the plane of $\mathbf{k}_\gamma$ and the z-axis and ϵ_2 normal to this plane we have

$$\epsilon_1 \times \mathbf{k}_\gamma = \epsilon_2 k_\gamma,$$

and so
$$\epsilon_2 \times \mathbf{k}_\gamma = -\epsilon_1 k_\gamma,$$

$$\langle {}^3S_1 | \tfrac{1}{2}(g_n - g_p)(\sigma_n - \sigma_p) \cdot (\epsilon_0 \times \mathbf{k}_\gamma) | {}^1S_0 \rangle \tag{4.9.7}$$

gives, on summing over the two polarization directions for the photon,

$$\frac{1}{2\sqrt{2}}(g_n - g_p) k_\gamma [\langle {}^3S_1 | (\epsilon_2 - \epsilon_1) \cdot \{(\mathbf{i} - i\mathbf{j}) | 1, +1\rangle - (\mathbf{i} + i\mathbf{j}) | 1, -1\rangle - \sqrt{2}\mathbf{k} | 1, 0\rangle\}]. \tag{4.9.8}$$

The two different polarizations, being orthogonal states, do not interfere. The deuteron spin states $|1, +1\rangle$, $|1, -1\rangle$ and $|1, 0\rangle$ are also orthogonal and other interferences are removed on squaring by the complex coefficients. We get then for the square of the summed matrix elements

$$\frac{(g_n - g_p)^2}{8} k_\gamma^2 \{2(\epsilon_2 \cdot \mathbf{i})^2 + 2(\epsilon_2 \cdot \mathbf{j})^2 + 2(\epsilon_2 \cdot \mathbf{k})^2$$
$$+ 2(\epsilon_1 \cdot \mathbf{i})^2 + 2(\epsilon_1 \cdot \mathbf{j})^2 + 2(\epsilon_1 \cdot \mathbf{k})^2\} | \int \Psi_d^*(r) \Psi(r) d^3r |^2. \tag{4.9.9}$$

If the angles between the axes and the photon propagation vector $\mathbf{k}_\gamma$ are θ, ϕ then

$$\epsilon_2 \cdot \mathbf{i} = \sin \phi, \qquad \epsilon_1 \cdot \mathbf{i} = \cos \theta \cos \phi,$$
$$\epsilon_2 \cdot \mathbf{j} = -\cos \phi, \qquad \epsilon_1 \cdot \mathbf{j} = \cos \theta \sin \phi,$$
$$\epsilon_2 \cdot \mathbf{k} = 0, \qquad \epsilon_1 \cdot \mathbf{k} = \sin \theta,$$

giving isotropy and

$$\frac{(g_n - g_p)^2 k_\gamma^2}{2} \left| \int \Psi_d^*(r) \Psi(r) d^3r \right|^2$$

(if we were only considering one deuteron magnetic substate an anisotropic angular distribution would result). On integrating over all angles, we get

$$(g_n - g_p)^2 k_\gamma^2 2\pi \left(\frac{e\hbar}{2Mc}\right)^2 \left| \int \Psi_d^*(r) \Psi(r) d^3r \right|^2 \tag{4.9.10}$$

for the transition matrix element squared. Putting in the factor of $2\pi/\hbar$, the normalization factor for the electromagnetic field $2\pi\hbar c^2/\omega$ and the phase space $\omega^2/\hbar(2\pi c)^3$ we find for the transition rate,

$$\begin{aligned}&\frac{\omega k_\gamma^2}{\hbar c}(g_n - g_p)^2 \left(\frac{e\hbar}{2Mc}\right)^2 \left| \int \Psi_d^*(r) \Psi(r) d^3r \right|^2 \\ &= \frac{\omega^3}{\hbar c^3}(g_n - g_p)^2 \left(\frac{e\hbar}{2Mc}\right)^2 \left| \int \Psi_d^*(r) \Psi(r) d^3r \right|^2 \end{aligned} \tag{4.9.11}$$

which indeed has the dimensions of sec^{-1} provided the wave functions have been normalized so that the integral is dimensionless. Finally, in order to convert this into a cross-section, it is necessary to divide by the relative velocity of the proton and the neutron in the centre of mass. The relative velocity in the centre of mass is $\hbar k_1/\tfrac{1}{2}M$ ($\tfrac{1}{2}M$ is the reduced mass) and so we get for the cross-section for capture of a neutron by a proton from the singlet state

$$\frac{M}{2\hbar k_1} \frac{\omega^3}{\hbar c^3} (g_n-g_p)^2 \left(\frac{e\hbar}{2Mc}\right)^2 \left| \int \Psi_d^*(r)\, \Psi(r)\, d^3r \right|^2 \qquad (4.9.12)$$

which must be divided by a factor of 4 to allow for the statistical weight of the singlet state —we have already summed over all possible initial and final states—yielding

$$\sigma(n+p \to d+\gamma) = \frac{1}{8}\frac{M}{\hbar k_1}\frac{\omega^3}{\hbar c^3}(g_n-g_p)^2\left(\frac{e\hbar}{2Mc}\right)^2\left|\int \Psi_d^*(r)\,\Psi(r)\,d^3r\right|^2. \qquad (4.9.13)$$

The value of the integral $\int \Psi_d^*(r)\Psi(r)\,d^3r$ depends on the form of the potential, but because of the long tail on the deuteron wave function (which arises because of its small binding energy) we will get as reasonable a result by assuming that the exterior wave functions apply from $r=0$ to $r=\infty$—this is the zero range approximation—as by assuming a square well. Then we write

$$\Psi(r) \simeq \frac{e^{i\delta}\sin(k_1 r+\delta)}{k_1 r} \qquad 0 \leqslant r \leqslant \infty,$$

$$\Psi_d(r) \simeq \sqrt{\frac{2|k_{1d}|}{4\pi}}\,\frac{e^{-|k_{1d}|r}}{r} \qquad 0 \leqslant r \leqslant \infty$$

where $\Psi_d(r)$ has been normalized to unity—the normalization of $\Psi(r)$ has already been fixed by the plane wave normalization. Note that the squared integral, with this normalization, now has the dimensions of cm^3, as required for a cross-section calculation.

The integral then becomes (apart from the irrelevant phase factor $e^{i\delta}$)

$$\int \sqrt{\frac{2|k_{1d}|}{4\pi}}\,\frac{e^{-|k_{1d}|r}}{r}\,\frac{\sin(k_1 r+\delta)}{k_1 r}\,d^3r = 4\pi\sqrt{\frac{2|k_{1d}|}{4\pi}}\,\frac{1}{k_1}\int_0^\infty e^{-|k_{1d}|r}\sin(k_1 r+\delta)\,dr$$

$$= 4\pi\sqrt{\frac{2|k_{1d}|}{4\pi}}\,\frac{1}{k_1}\left[\frac{k_1\cos\delta+|k_{1d}|\sin\delta}{k_1^2+|k_{1d}|^2}\right]. \qquad (4.9.14)$$

So that squaring and substituting into the cross-section gives

$$\sigma(n+p \to d+\gamma) = \pi\frac{M}{\hbar k_1}\frac{\omega^3}{\hbar c^3}(g_n-g_p)^2\left(\frac{e\hbar}{2Mc}\right)^2|k_{1d}|\left[\frac{\cos\delta+|k_{1d}|\sin\delta/k_1}{k_1^2+k_{1d}^2}\right]^2.$$

As $k_1 \to 0$, k_1 may be neglected in comparison with $|k_{1d}|$ and $\sin\delta/k_1 \to a_0$. The transition energy $E_\gamma = \hbar\omega$ and we also have

$$E_\gamma = \frac{\hbar^2}{M}(k_1^2+|k_{1d}|^2), \qquad \omega = \frac{\hbar}{M}(k_1^2+|k_{1d}|^2)$$

so that

$$\sigma(n+p \to d+\gamma)_{k_1 \to 0} \cdot \frac{\pi|k_{1d}|}{k_1}\frac{\hbar|k_{1d}|^2}{M^2 c^3}\left(\frac{e\hbar}{2Mc}\right)^2(g_n-g_p)^2[1+|k_{1d}|a_0]^2,$$

$$k_{1d}^2 = \frac{MB}{\hbar^2}, \qquad k_1^2 = \frac{E_{cm}M}{\hbar^2}$$

where B is the deuteron binding energy, so

$$\sigma(n+p \to d+\gamma) \to \pi \sqrt{\frac{B}{E_{cm}} \frac{\hbar}{M^2 c^3}} \left(\frac{MB}{\hbar^2}\right) \left(\frac{e\hbar}{2Mc}\right)^2 (g_n - g_p)^2 \left[1 + \sqrt{\frac{MB}{\hbar^2}} a_0\right]^2$$

$$= \pi \sqrt{\frac{2B}{E_L}} \frac{e^2}{\hbar c} \left(\frac{B}{Mc^2}\right) \left(\frac{\hbar}{Mc}\right)^2 (\mu_n - \mu_p)^2 \left[1 + \sqrt{\frac{MB}{\hbar^2}} a_0\right]^2 \quad (4.9.15)$$

where E_L is the laboratory energy of the neutron (for a stationary proton target) and μ_n, μ_p are the nucleon magnetic moments measured in nuclear magnetons, $\mu_n = -1.91$, $\mu_p = 2.79$.

This is a nice example of an exothermic reaction obeying a $1/v$ law in the low energy region where the matrix element is only slowly varying and the phase space is almost independent of the incident energy. In this example, as soon as $k_1 \ll |k_{1d}|$ the whole structure of the matrix element is controlled only by the properties of the deuteron. The expression is particularly easy to evaluate in this form: the only factor with dimensions is $\hbar/Mc$, the compton wavelength of the nucleon equal to 2.1×10^{-14} cm.

The scale factor for the deuteron wave function $\sqrt{MB}/\hbar = 2.32 \times 10^{12}$ cm^{-1} so

$$1 + \frac{\sqrt{MB}}{\hbar} a_0 = 1 + 5.5 = 6.5.$$

Note that if the singlet np system had a bound state characterized by a scattering length of -23.68×10^{-13} cm then $1 + a_0 \sqrt{MB}/\hbar$ would be -4.5 and the resulting cross-section would be down by a factor of 2.

Substituting these numbers into the expression for the cross-section we obtain

$$\sigma(n+p \to d+\gamma) = 0.477 \times 10^{-28} E_N^{-1/2} \quad (E_N \text{ in MeV})$$

in the low-energy limit, $k_1 \ll |k_{1d}|$.

At thermal energy $E_N = 0.025$ eV, 2.5×10^{-8} MeV, $E_N^{1/2} = 1.58 \times 10^{-4}$ (MeV)$^{1/2}$ and the laboratory velocity of the neutron is 2.2×10^5 cm/sec,

$$\sigma(n+p \to d+\gamma)_{\text{thermal}} = 0.3 \times 10^{-24} \text{ cm}^2 = 0.3 \text{ barn}.$$

This is the cross-section for neutrons diffusing through a bucket of water at room temperature, and this calculated value compares very well with the measured value of 0.3314 ± 0.0019 barn given by Wilson (*loc. cit.*) where the actual experiments and further refinements to the calculation are also discussed.

The inverse endothermic process, photodisintegration of the deuteron, can be calculated similarly. For magnetic dipole photodisintegration the matrix element is determined by deuteron structure only at low energy and consequently the cross-section rises from threshold as k_1. The magnetic dipole term is, however, rapidly overtaken as the photon energy increases by electric dipole transitions to the triplet p-state, where the matrix element contains a centrifugal barrier effect and consequently the cross-section of electric dipole disintegration rises as k_1^3 near threshold. The calculations may be found in, for example, Blatt and Weisskopf, *Theoretical Nuclear Physics*.

Nuclear Reactions

4.10. Resonances. Introduction

Resonance is a phenomenon very familiar from classical physics: if a note is played on a violin at a frequency that corresponds to the natural frequency of a wine glass, the glass resonates and the amplitude of the induced ringing may become sufficiently large to shatter the glass. Resonance corresponds to driving a physical system capable of vibration at one of its natural frequencies, and always corresponds to large induced amplitudes. In quantum systems, resonance corresponds to driving a physical system, with a number of energy levels, at a frequency appropriate to induce transitions from the original state of the system, energy E_0, to another state, energy E'—thus the resonant frequency ω is given by $\hbar\omega = E' - E_0$. In the terminology of nuclear and particle physics, a *resonant state* (or just a *resonance* for short) refers to such an excited state. Before discussing resonant phenomena in nuclear physics, we will briefly give one example of resonance scattering in classical physics—resonant scattering of light.

Consider one-dimensional motion of an electron, characterized by an angular frequency ω_0 and in the absence of incident radiation by an equation of motion

$$\frac{d^2x}{dt^2} + \gamma \frac{dx}{dt} + \omega_0^2 x = 0. \tag{4.10.1}$$

This equation has a solution $x = x_0 e^{-i\Omega t}$

where
$$\Omega = -\frac{i\gamma}{2} \pm \sqrt{\omega_0^2 - \frac{\gamma^2}{4}}$$

$$\approx -\frac{i\gamma}{2} \pm \omega_0 \quad \text{if} \quad \gamma \ll \omega_0$$

and corresponds to a fluctuating electric dipole moment d with exponential damping

$$d = ex \simeq ex_0 e^{-(\gamma/2)t} e^{-i\omega_0 t}.$$

(In the absence of collisional damping, radiation from the oscillating dipole itself damps the motion, although since the instantaneous power radiated is proportional to the square of the acceleration, the equation of motion is more complicated.)

If an oscillating electric field $E = E_0 e^{-i\omega t}$ is applied to this system, the equation of motion becomes

$$\frac{d^2x}{dt^2} + \gamma \frac{dx}{dt} + \omega_0^2 x = \frac{eE_0}{m\omega_0} e^{-i\omega t}$$

where m is the mass of the electron. After any transients have disappeared, the time dependence of the forced oscillation is $x = x_0 e^{i\omega t}$ and this gives an expression for the dipole moment induced by the radiation

$$d = \frac{\frac{e^2 E_0}{m} e^{-i\omega t}}{\omega_0^2 - \omega^2 - i\omega\gamma} \simeq \frac{\frac{1}{2}\frac{e^2 E_0}{m\omega_0} e^{-i\omega t}}{(\omega_0 - \omega) - i\frac{\gamma}{2}} \tag{4.10.2}$$

Nuclear Physics

for $\omega \sim \omega_0$. The power radiated is proportional to the square of the acceleration and so the fraction of incident energy scattered is proportional to

$$\omega^4 |d_0|^2 = \frac{\left(\dfrac{e^2}{m}\right)^2 \omega^4}{(\omega_0^2-\omega^2)^2+\omega^2\gamma^2}$$

which reaches a maximum when $\omega \simeq \omega_0$ and at $\omega = \omega_0$ corresponds to a purely imaginary scattering amplitude—meaning that the electron is vibrating out of phase with the exciting radiation. In the region where $\omega \sim \omega_0$ this may be written in the following approximate form provided $\gamma \ll \omega_0$:

$$\frac{\frac{1}{4}\left(\dfrac{e^2}{m}\right)^2 \omega_0^2}{(\omega_0-\omega)^2+\gamma^2/4} \tag{4.10.3}$$

which has the famous Breit–Wigner shape. You should note at this point that in the event of a large damping factor γ the shape of the scattering cross-section departs significantly from this form, being skewed towards high frequencies. This provides a (classical) illustration of our remarks in Section 3.2. In the region $\omega \sim \omega_0$ we have resonant scattering, and the resonant frequency is the undamped natural frequency of the oscillator.

In a quantized system, we have a ground state with a well-defined energy and an excited state which is unstable and so does not have a well-defined energy, but rather a spread characterized by a width $\hbar\gamma$ and a central energy $E_1 = \hbar\omega_1$. The resonant frequency at which maximum scattering occurs is thus given by

$$\hbar\omega_0 = E_1 - E_g$$

and resonant scattering of radiation occurs for photon energies other than $\hbar\omega_0$ because of the spread of energy in the upper state. The picture we have is of a photon being absorbed in the transition from the state with energy E_g to the state with energy E_1 followed by decay of the state E_1 back to ground with emission of a photon. The lifetime of the state is $1/\gamma$ with the definitions we have used here. All these features have their analogues in nuclear physics, and indeed in particle physics.

To begin our discussion of resonances in nuclear physics, we return to eq. (4.5.14) for the cross-section for elastic scattering, with no absorption, of two spinless particles:

$$\frac{d\sigma}{d\Omega} = \frac{1}{k^2}\left|\sum_{l=0}^{\infty} e^{i\delta_l}(2l+1)\sin\delta_l P_l(\cos\theta)\right|^2 \tag{4.5.14); (4.10.4}$$

which gives for the total scattering cross-section

$$\sigma = \frac{4\pi}{k^2}\sum_{l=0}^{\infty}(2l+1)\sin^2\delta_l$$

where k is the propagation constant at large distances. The phase shift in the lth partial wave is δ_l and the scattering amplitude $e^{i\delta_l}\sin\delta_l$. If $\sin\delta_l = 1$ the phase shift $\delta_l = (n+\frac{1}{2})\pi$ where n is an integer, and the cross-section in the lth partial wave, is given by

$$\sigma_l^U = (2l+1)\frac{4\pi}{k^2}.$$

This is the *unitarity limit* to σ_l and expresses conservation of probability. The amount of inward travelling lth partial wave in the total wave function must be greater than or equal to the amount of outward travelling lth partial wave—you cannot scatter what is not there—and this results in the above limit. The outward going lth partial wave cannot be boosted by contributions from a different partial wave because l is conserved in the scattering of spinless particles. If the particles have spin, we would write

$$\sigma_J = \frac{4\pi}{k^2} \alpha_J \sin^2 \delta_J$$

where α_J is a number depending on the total angular momentum and its composition, and the unitary limit for scattering in the system with angular momentum J becomes

$$\sigma_J = \frac{4\pi}{k^2} \alpha_J.$$

Then if $\delta = (n+\tfrac{1}{2})\pi$ the cross-section is close to a maximum (and reaches the unitary limit). The scattering amplitude $e^{i\delta} \sin \delta$ also becomes purely imaginary at this value, and these are two of the characteristics of resonant scattering in our classical example. It should also be clear that the faster δ goes through $(n+\tfrac{1}{2})\pi$ with k the sharper the resonance will be.

It is worth while at this stage pointing out that if the phase shift δ decreases through $(n+\tfrac{1}{2})\pi$ instead of increasing through $(n+\tfrac{1}{2})\pi$ as k increases, then although the cross-section in the appropriate angular momentum state reaches its unitary maximum, resonant scattering does not occur. Formation of an excited state only corresponds to δ increasing through $(n+\tfrac{1}{2})\pi$ and there is a limit on the rate at which δ decreases through $(n+\tfrac{1}{2})\pi$ but no such limit on the rate at which δ increases. The reason is to be found in the assumption of *causality*: no output signal before an input signal. Since plane waves being scattered correspond to a steady state, it is necessary to go back to the scattering of a wave packet to discuss this point.

Consider an incoming spherical wave packet, defined by

$$\psi_{\text{in}} \simeq \frac{1}{r} \int A(k) \, e^{-ikr - i\omega t} \, dk \tag{4.10.5}$$

at a great distance from the interaction. The outgoing wave is

$$\psi_{\text{out}} \simeq \frac{1}{r} \int A(k) \, e^{+i(kr + 2\delta) - i\omega t} \, dk \tag{4.10.6}$$

where δ is a function of k.

If we make an expansion of k and ω, as is always done to derive the group velocity, then we write

$$\left. \begin{aligned} k &= k_0 + (k - k_0) \ldots \\ \omega &= \omega_0 + \frac{\partial \omega}{\partial k}\bigg|_{k_0} (k - k_0) \ldots \\ \psi_{\text{in}} &\simeq \frac{1}{r} e^{-i(k_0 r + \omega_0 t)} \int A(k) \, e^{-i(k-k_0)\left\{ r + \frac{\partial \omega}{\partial k}\big|_{k_0} t \right\}} dk \end{aligned} \right\} \tag{4.10.7}$$

Nuclear Physics

where the information about the shape of the wave packet travels with velocity $\partial\omega/\partial k$, the group velocity.

Similarly the outgoing wave becomes

$$\psi_{out} = \frac{1}{r} e^{i(k_0 r - \omega_0 t + 2\delta_0)} \int A(k) e^{i(k-k_0)\left\{r + 2\frac{\partial\delta}{\partial k}\big|_{k_0} - \frac{\partial\omega}{\partial t}\big|_{k_0} t\right\}} dk \qquad (4.10.8)$$

if we neglect dispersion, that is terms in $(k-k_0)^2$ and higher powers.

Again information about the shape and position is contained in the exponent under the integral. If we define the origin of time so that $t = 0$ when $r = 0$, then the ingoing pulse passes a particular radius R at a time

$$t^R_{in} = -R \Big/ \frac{\partial\omega}{\partial k}\bigg|_{k_0}$$

and the outgoing pulse reaches the same radius R at

$$t^R_{out} = R \Big/ \frac{\partial\omega}{\partial k}\bigg|_{k_0} + 2\frac{\partial\delta}{\partial k}\bigg|_{k_0} \Big/ \frac{\partial\omega}{\partial k}\bigg|_{k_0}.$$

If $t^R_{out} = t^R_{in}$ this means that the incoming wave is reflected at a radius $r = R$. In general, t^R_{out} must be greater than or equal to t^R_{in} and so the minimum requirement we can impose from causality is that

$$t^R_{out} \geqslant t^R_{in}$$

or

$$R + 2\frac{\partial\delta}{\partial k}\bigg|_{k_0} \geqslant -R$$

or

$$\frac{\partial\delta}{\partial k}\bigg|_{k_0} \geqslant -R.$$

This limit, of course, only becomes stringent if R is very small, and we must take it to be approximately the radius of interaction. The quantity t^R_{out} can be as much greater than t^R_{in} as it likes—if a very long-lived excited state is formed, this corresponds to a very large positive value of $\frac{\partial\delta}{\partial k}$ and a narrow resonance. However, if δ is decreasing with R, then the limit is

$$\frac{\partial\delta}{\partial k} \geqslant -R$$

and this corresponds to no trapping of the particles inside the potential, and no excited state, even though δ happens to be passing through $(n+\frac{1}{2})\pi$, and a sharp peak in the cross-section does not result.

If we have a large probability for the formation of an excited state, which has a number of different decay modes, then we may expect maxima in the cross-section for inelastic reactions at the resonant energy for the elastic reaction. The criteria for the existence of resonance are then

(1) Sharp peaks in both elastic and inelastic reactions,
(2) The elastic phase shift passes upwards through $(n+\frac{1}{2})\pi$.

Nuclear Reactions

We may distinguish fairly clearly between two different kinds of resonance in nuclear physics, shape resonances and structure resonances. Shape resonances are those resonances adequately described by a time-independent potential treatment and thus are essentially a phenomenon of elastic scattering. They correspond to a potential acting between the initial particle and the nucleus which is strong enough that at certain energies the elastic phase shift goes through $\pi/2$ (or any odd multiple of $\pi/2$ of course). The depth and range of the effective potential determine these energies, and so different nuclei which are neighbours in the nuclear periodic table will be expected to exhibit substantially the same shape resonances, since the effective potential will change only slowly from one nucleus to the next. These resonances correspond to single-particle excited states in an independent particle model, and so the separation in energy between successive shape resonances is expected to be roughly equal to the separation between the highest single-particle levels in the nuclear ground state; shape resonances are thus natural to the optical model in which they are handled with great success: we have chosen to discuss them here rather than explicitly as part of the optical model because of the important features that all resonance phenomena have in common.

Structure resonances present a great contrast. On entering the nucleus an individual nucleon may collide with nucleons in bound states, raising these nucleons to excited states, and thus forming a many-particle excited state of the *compound nucleus* that mediates the reaction. The precise energies at which these many-particle excitations occur are not given by a model in which all nucleons move in a time-independent potential, because of all the fluctuations in instantaneous potential which go on in the hot nucleus. The number, position and density of these many particle excitations is a sensitive function of nuclear structure and we do not expect any obvious relationship between the structure resonances of neighbouring nuclei.

4.11. Resonances in scattering from a potential hole

The origin of shape resonances is easily illustrated by the scattering from a square well potential

$$V(r) = -U \quad r < R,$$
$$V(r) = 0 \quad r \geq R.$$

If the kinetic energy in the centre of mass is E, then we define

$$k_1^2 = \frac{2ME}{\hbar^2}, \quad k_2^2 = \frac{2M(E+U)}{\hbar^2}$$

and

$$k_0^2 = \frac{2MU}{\hbar^2}, \quad k_2^2 = k_1^2 + k_0^2$$

where M is the reduced mass of the system.

We will consider here scattering in two partial waves, s- and p-waves, for spinless particles.

For s-waves, assuming we have no absorption, the radial wave function inside the potential boundary R is the regular function

$$\frac{\sin k_2 r}{k_2 r}$$

Nuclear Physics

while outside the boundary the solution is

$$\frac{\sin(k_1 r + \delta_0)}{k_1 r}$$

where δ_0 is the s-wave phase shift. This phase shift δ_0 is found by matching these functions and their first derivatives across the potential boundary yielding

$$\frac{\tan(k_1 R + \delta_0)}{k_1} = \frac{\tan k_2 R}{k_2}$$

or (4.11.1)

$$\frac{\tan(a_1 + \delta_0)}{a_1} = \frac{\tan a_2}{a_2}$$

where $a = kR$.

Expanding the left-hand side of eq. (4.11.1) gives

$$\tan \delta_0 = \frac{a_1 \tan a_2 - a_2 \tan a_1}{a_2 + a_1 \tan a_1 \tan a_2}. \qquad (4.11.2)$$

If we are searching for resonances at low energy where $\tan a_1 \simeq a_1$ then we have

$$\tan \delta_0 \simeq \frac{a_1 \{\tan a_2 - a_2\}}{a_2 + a_1^2 \tan a_2}.$$

If $\tan a_2 \to \infty$ as $a_1 \to 0$, then $\tan \delta_0 \to 1/a_1 \to \infty$ and we have a resonance at zero energy for $\tan k_0 R = (n + \frac{1}{2})\pi$. This is also the condition for a bound state at zero energy—the two are indistinguishable. Such a resonance does not of course exhibit a Breit-Wigner shape. In Fig. 4.11.1 we have plotted the variation of the phase shift δ_0 and the s-wave cross-section σ_0 as a function of a_1 for various values of a_0, where a_0 characterizes the potential. Figure 4.11.1(a) shows the variation of δ_0 and the quantity σ_0/R^2 as a function of a_1 for $a_0 = \pi/4$. The potential well characterized by this value contains no bound state, and no resonance. The phase shift δ_0 rises to a maximum value of $\sim 12°$ and then decreases according to eq. (4.11.2) as a_1 approaches a_2. The cross-section σ_0 is a monotonically decreasing function of a_1. Figure 4.11.1(b) shows what happens when $a_0 = \pi/2$. This is the condition for a bound state at zero energy, and also the condition for a resonance at zero energy. The phase shift decreases steadily from $\pi/2$ and the cross-section, infinite at $a_1 = 0$, decreases monotonically. Since the resonance and the bound state are the same phenomena, and the bound state is stable, the resonance follows a curve appropriate to zero width, and the form of σ_0 for $a_1 > 0$ in this case is analogous to the theoretical response of an undamped L–C circuit in the region $\omega > \omega_0$ where ω_0 is the resonant frequency. Figure 4.11.1(c) for $a_0 = 3\pi/4$ corresponds to scattering from a square well containing one bound state in the s-wave. The phase shift decreases from π (because the exterior wave has to join on to an interior wave that has turned over) and the cross-section drops monotonically. Note that δ_0 passes downwards through $\pi/2$ slowly and this produces no bump in the cross-section. Finally in Fig. 4.11.1(d) we show the cross-section for $a_0 = 3\pi/2$ corresponding to a second bound state in the square well, just going in at zero energy.

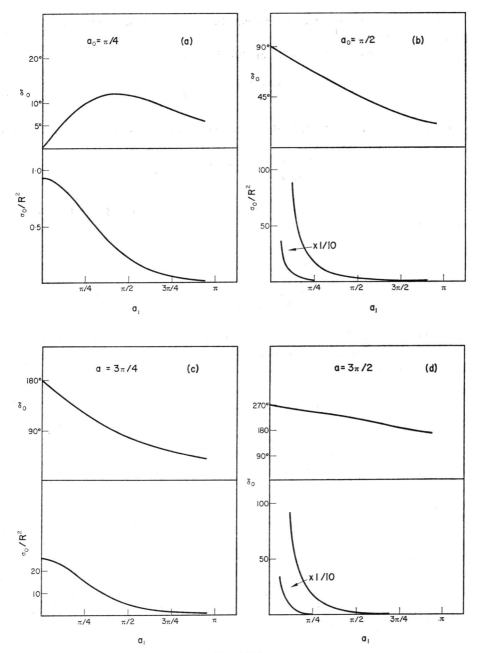

Fig. 4.11.1.

s-wave scattering on a square well only shows resonance at zero energy, when $a_0 = (n+\tfrac{1}{2})\pi$ although very large zero energy cross-sections may be obtained when a_0 is close, but not equal, to $\pi/2$. This is the origin of the singlet np scattering cross-section of 68 barns.

For p-wave scattering the centrifugal repulsion ensures that at low energy the effect of the potential is not felt by the p-wave, and the centrifugal barrier, resulting from the super-

Nuclear Physics

position of a square well potential and the centrifugal repulsion, introduces striking differences from the s-wave case.

Inside the potential the regular p-wave solution has a radial part

$$3i\left\{\frac{\sin k_2 r}{(k_2 r)^2} - \frac{\cos k_2 r}{k_2 r}\right\} \tag{4.11.3}$$

and, for elastic scattering only, the exterior wave function is shifted by a real phase δ_1 and is

$$3i\left\{\frac{\sin (k_1 r + \delta_1)}{(k_1 r)^2} - \frac{\cos (k_1 r + \delta_1)}{k_1 r}\right\}. \tag{4.11.4}$$

Matching these radial functions and their first derivatives with respect to r across the boundary at $r = R$ yields

$$\tan(\delta_1 + a_1) = \frac{a_1 a_1^2 \tan a_2}{a_0^2 \tan a_2 + a_1^2 a_2}$$

or

$$\tan \delta_1 = \frac{a_1 a_2 [a_2 \tan a_2 - a_1 \tan a_1] - a_0^2 \tan a_2 \tan a_1}{a_0^2 \tan a_2 + a_1 a_2 [a_1 + a_2 \tan a_1 \tan a_2]}, \tag{4.11.5}$$

a complicated expression, the content of which is by no means obvious: but it is easy to track the variation of δ_1 by numerical calculation. For small a_0 the phase shift and hence the cross-section rise from zero, reach a maximum and then decrease without reaching resonance. In Fig. 4.11.2 we show the behaviour that results when the potential becomes strong enough to drive the phase shift δ_1 upwards through $\pi/2$. Figure 4.11.2(a) shows the variation of δ_1 and the quantity $\sigma_1/3R^2$ as a function of a_1 for $a_0 = 7\pi/8$, a value just large enough for the phase shift δ_1 to pass through $\pi/2$. The phase shift δ_1 reaches a maximum of $\sim 93°$ and then drops slowly back towards zero. The cross-section exhibits the p-wave resonant shape, peaking *well below* the point at which $\delta_1 = \pi/2$ and skewed from the simple Breit–Wigner shape towards larger values of k_1. This is because the centrifugal barrier decreases in importance as k_1 increases, and the resonance is broad enough for this effect to skew the shape. Figure 4.11.2(b) shows what happens when a_0 is increased, in this case to $a_0 = \frac{15}{16}\pi$. The resonance has moved lower in energy and has become narrower, because the centrifugal barrier has become more important and inhibits decay to a greater extent. This is reflected in the greater slope $\partial \delta_1/\partial k_1$ near $\delta_1 = \pi/2$. The peak in the cross-section and the energy at which $\delta = \pi/2$ are now closer. Notice that δ_1 rises to $\sim 115°$ and descends slowly through $\pi/2$ producing no maximum in the cross-section at this point, where the quantity $\partial \delta_1/\partial a_1$ may be read from Fig. 4.11.2(b) and is

$$\frac{\partial \delta_1}{\partial a_1} = -0.49$$

when δ is measured in radians.

Our causality argument required $\partial \delta_1/\partial a_1 > -1$ and is well satisfied.

Figure 4.11.2(c) is for an even more extreme case. The well has been deepened to $a_0 = \frac{39}{40}\pi$, resulting in a very narrow resonance where $\delta_1 = \pi/2$, with δ_1 reaching a maximum of $\sim 130°$. The characteristic p-wave distortion of the simple Breit–Wigner formula is still

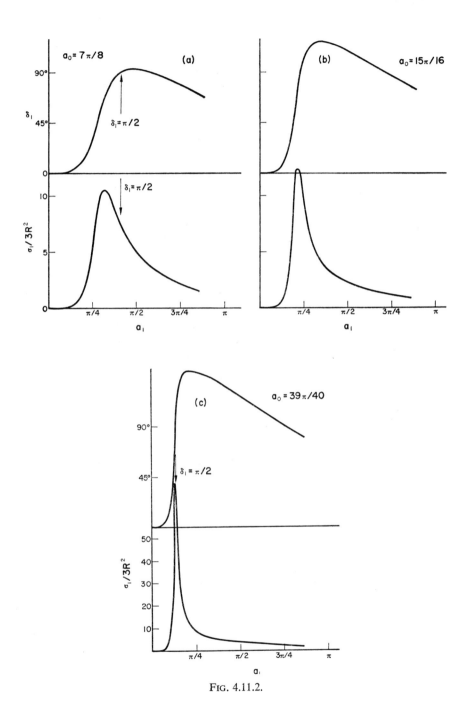

Fig. 4.11.2.

noticeable. It is clear that a further strengthening of the potential will result in the resonance disappearing below zero energy and becoming the first *p*-wave bound state: expansion of eq. (4.11.5) shows that this occurs at $a_0 = \pi$.

The difference between *s*- and *p*-wave resonant scattering from a square well is as follows: the *s*-wave is only ever in resonance at zero energy, and then only when the well parameter $a_0 = (n+\frac{1}{2})\pi$ and the scattering cross-section decreases monotonically from zero energy. In *p*-wave scattering there is a range of values of the well parameter a_0 for which resonance can occur, the value of a_1 at resonance is determined by the value of a_0 within this range, and sharp peaks in the cross-section result when $\delta_1 = (n+\frac{1}{2})\pi$ at non-zero values of a_1. The difference is due to the centrifugal barrier for *p*-waves, which is capable of containing the interior wave function over an extended period for certain values of the total energy. A real potential barrier—the only important example of which in nuclear physics is the coulomb barrier—can do the same thing and in the time-independent potential treatment resonances at other than zero energy imply a barrier—a real or centrifugal repulsion outside the attractive region of the effective nuclear potential. The effect of the centrifugal barrier is (for a square well) easily calculable in terms of trigonometric functions: the effect of a coulomb barrier is more complicated.

The characteristic of shape resonances is thus that inside the range of the potential k_2 is approximately constant and that the interior solution is regular at the origin. The latter condition, when coupled with the slow variation of k_2 with r determines the interior wave function at the boundary very simply and it is this pinning of the interior wave function at the origin that requires that shape resonances occur roughly every time an extra interior half wavelength can be inserted into the nuclear radius, and not with closer spacing.

4.12. Structure resonances: the compound nucleus and low-energy reactions

In the optical model the scattering of a particle incident on a nucleus is assumed to be given by the effect of a time-independent potential. Such an approximation is quite good for the ground state of nuclei, in which variation of individual nucleon wave functions with time is suppressed by the Pauli exclusion principle. However, if even a zero energy nucleon is dropped into a nucleus, the resulting complex has ∼ 8 MeV excitation and this energy may be distributed among many nucleons. With many nucleons raised to single-particle excited states, the Pauli exclusion principle no longer inhibits collective effects and an independent particle approach promptly becomes invalid. We need a phenomenological approach to nuclear reactions under these conditions.

In Section 4.11 we approximated a short-range potential by the effect of a square well with a definite boundary. We retain this feature for our discussion of structure effects, and suppose outside a critical radius R the incident particle experiences no nuclear potential at all. On passing the radius R the particle is acted upon by a time-independent potential so that its wavelength changes but it still retains its original identity. Then, at a radius perhaps ∼ one nucleon diameter less than R, the incident particle plunges into the nuclear surface and any treatment with a time-independent potential loses its validity. The consequences of these assumptions are that, provided the separation between the nucleus and the incident or scattered particle is greater than R, we may apply the same treatment as for a time-independent square well. At a separation just less than R the same treatment is valid but the interior wave function is now no longer constrained to be regular at the origin, since the whole treatment breaks down long before the origin is reached. Consider s-wave neutrons. For an optical model approximation we have

Interior wave function	Exterior wave function
$\dfrac{\sin k_2 r}{k_2 r}$	$\dfrac{\sin(k_1 r + \delta)}{k_1 r}$
valid for $0 \leqslant r \leqslant R$	valid for $R \leqslant r \leqslant \infty$

and for the phase shift δ to reach $90°$ we require

$$\frac{\tan k_2 R}{k_2 R} = -\frac{\cot k_1 R}{k_1 R}$$

which is only satisfied by $k_1 = 0$, $k_0 R = \pi/2$ giving the peculiar zero energy resonance discussed in Section 4.11. However, if we abandon our optical model approximation for the assumptions discussed above, we may write

Interior wave function	Exterior wave function
$\dfrac{\sin(k_2 r + \eta)}{k_2 r}$	$\dfrac{\sin(k_1 r + \delta)}{k_1 r}$
valid only for	valid for
$r \pm R$	$R \leqslant r < \infty$

and the condition for δ to reach $90°$ is now

$$\frac{\tan(k_2 R + \eta)}{k_2 R} = -\frac{\cot k_1 R}{k_1 R}$$

and so resonance can occur more or less anywhere, depending on the value of η. The interior phase shift η is determined by the detailed nuclear structure and so is a sensitive function of k_1. A rapid and large variation of η with k_1 can generate a lot of closely spaced s-wave resonances, and the closer the spacing the narrower they will clearly be.

Since we have no way of determining η, or the corresponding interior phase shifts for other partial waves, it is preferable to make a phenomenological description purely in terms of the exterior wave functions, comparing their properties with the logarithmic derivative of the interior wave function which is the parameter transmitting to the exterior all information about the nuclear structure relevant to the behaviour of the exterior wave function. If in general the exterior radial wave function is $\psi_{\text{ext}} = U_{\text{ext}}/r$ and the interior wave function is near the boundary

$$\psi_{\text{int}} = \frac{U_{\text{int}}}{r}$$

Nuclear Physics

then matching the functions and their first derivatives at $r = R$ requires

$$\frac{1}{U_{int}} \frac{\partial U_{int}}{\partial r}\bigg|_R = \frac{1}{U_{ext}} \frac{\partial U_{ext}}{\partial r}\bigg|_R. \quad (4.12.1)$$

The left-hand side is the logarithmic derivative of U_{int} with respect to r and we define the parameter

$$f = \frac{R}{U_{int}(R)} \frac{\partial U_{int}}{\partial r}\bigg|_R \quad (4.12.2)$$

which is determined by the detailed nuclear structure. The exterior wave function is then subject to the constraint

$$\frac{R}{U_{ext}(R)} \frac{\partial U_{ext}}{\partial r}\bigg|_R = f. \quad (4.12.3)$$

It should be remembered that although there is no nuclear potential for $r > R$, both centrifugal and coulomb repulsion may still operate.

The exterior wave function is made up of a combination of ingoing and outgoing waves, U_{in} and U_{out} and may alternatively be viewed as a combination of the incident wave (comprising both ingoing and outgoing portions) and a scattered wave (which is outgoing only). The incident wave must be regular at the origin, the exterior wave may be shifted in phase and may have a reduced outgoing part if inelastic scattering can also take place.

If we consider just one angular momentum state at a time, so that we do not have to bother with keeping track of the relative amplitudes and phases of the different states, then we write

$$U_{incident} = U_{in} - U_{out} = U_{out}^* - U_{out}. \quad (4.12.4)$$

EXAMPLE: for s-waves, no coulomb potential, we have

$$U_{incident} = \sin kr = \frac{e^{ikr} - e^{-ikr}}{2i}$$

and defining the time dependence as $e^{-i\omega t}$, e^{ikr} is an outgoing term and e^{-ikr} an ingoing s-wave term. The particular combination $U_{in} - U_{out}$ for a wave which is not acted upon by a nuclear potential is necessary to form a solution regular at the origin. The p-wave term, for no coulomb barrier, is

$$U_{incident} = \frac{\sin kr}{kr} - \cos kr$$

which obviously breaks up in just the same sort of way. The total wave function, U_{ext}, must be given by

$$U_{ext} = U_{in} - \xi U_{out} \quad (4.12.5)$$

where ξ is a complex number describing any reduction in intensity of the outgoing wave and the shift in phase of U_{out} relative to U_{in} as a result of the specifically nuclear effects. The limit on ξ is $|\xi| \leq 1$. The scattered wave function is given by

$$U_{scattered} = U_{ext} - U_{incident} = (1 - \xi) U_{out}. \quad (4.12.6)$$

Nuclear Reactions

Since we wish only to match the logarithmic derivative of U_{ext} to the parameter f, our neglect of the proper numerical coefficients does not matter, and we get

$$f = \frac{R \left.\frac{\partial U_{\text{ext}}}{\partial r}\right|_R}{U_{\text{ext}}(R)} = \frac{R\left\{\frac{\partial U_{\text{in}}}{\partial r} - \xi \frac{\partial U_{\text{out}}}{\partial r}\right\}\bigg|_R}{U_{\text{in}}(R) - \xi U_{\text{out}}(R)} \tag{4.12.7}$$

and since $U_{\text{in}} = U_{\text{out}}^*$

$$f = \frac{R\left\{\frac{\partial U_{\text{out}}^*}{\partial r} - \frac{\partial U_{\text{out}}}{\partial r}\right\}\bigg|_R}{U_{\text{out}}^* - \xi U_{\text{out}}}.$$

So

$$\xi = \frac{fU_{\text{out}}^* - R\left.\frac{\partial U_{\text{out}}^*}{\partial r}\right|_R}{fU_{\text{out}} - R\left.\frac{\partial U_{\text{out}}}{\partial r}\right|_R} = \frac{f - f_{\text{out}}^*}{f - f_{\text{out}}} e^{i\chi} \tag{4.12.8}$$

where

$$f_{\text{out}} = \frac{R\left.\frac{\partial U_{\text{out}}}{\partial r}\right|_R}{U_{\text{out}}(R)}$$

and

$$e^{i\chi} = \frac{U_{\text{out}}^*(R)}{U_{\text{out}}(R)}.$$

The two quantities f_{out} and χ are known quite independently of the details of nuclear structure. For an s-wave, with no coulomb barrier,

$$\left.\begin{array}{l} f_{\text{out}} = ikR \\ e^{i\chi} = e^{-2ikR}; \quad \chi = -2kR \end{array}\right\} \text{ s-wave}, \tag{4.12.9}$$

for a p-wave with no coulomb barrier

$$\left.\begin{array}{l} f_{\text{out}} = \frac{-1 + ik^3R^3}{1 + k^2R^2} \\ e^{i\chi} = \frac{ikR+1}{-ikR+1} e^{-2ikR} \end{array}\right\} \text{ p-wave}. \tag{4.12.10}$$

The quantity f_{out} thus has in general both real and imaginary parts. However, if there is no absorption, the interior wave function contains both incoming and outgoing waves in the same proportion and so for pure elastic scattering f is real. If there is absorption, however, the amount of outgoing wave can become small and so when there is absorption f becomes complex.

We may now identify the complex parameter ξ with the quantity $\eta e^{i\delta}$ of eq. (4.5.17) which allows us to write the scattering and reaction cross-sections for a particular angular

Nuclear Physics

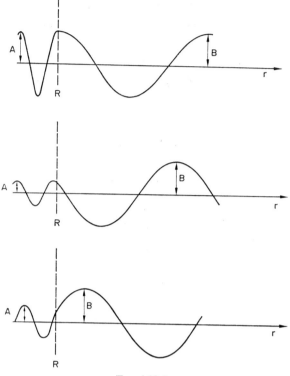

FIG. 4.12.1.

momentum

$$\sigma^J_{\text{elastic}} = \frac{\pi}{k^2}\alpha_J |\xi - 1|^2,$$

$$\sigma^J_{\text{reaction}} = \frac{\pi}{k^2}\alpha_J |1 - |\xi||^2$$

where α_J is a spin factor for the channel with angular momentum J and is $2l+1$ for spinless scattering with orbital angular momentum l.

$$\xi = \frac{f^* - f^*_{\text{out}}}{f - f_{\text{out}}} e^{i\chi} = \frac{\{\text{Re}\,f - \text{Re}\,f_{\text{out}}\} + i\{\text{Im}\,f + \text{Im}\,f_{\text{out}}\}}{\{\text{Re}\,f - \text{Re}\,f_{\text{out}}\} + i\{\text{Im}\,f - \text{Im}\,f_{\text{out}}\}}, \qquad (4.12.11)$$

and so $|\xi|^2 = 1$ unless f contains an imaginary part. We thus see once more that reactions other than elastic scattering imply an imaginary part to f.

We now make one further reduction, and ask what the scattering would be if the incident wave could not pass the boundary at R. This must mean for matching that $U_{\text{ext}}(R) = 0$ and $f \to \infty$. Such a situation could arise in principle, dictated by the vagaries of detailed nuclear structure, and the scattering would then be that due to an impenetrable sphere of radius R with $\xi = e^{i\chi}$ (clearly if the projectile cannot get into the nucleus it cannot initiate a reaction!). Then

$$\sigma^J_{\text{potential}} = \frac{\pi}{k^2}\alpha_J |e^{i\chi} - 1|^2$$

for hard sphere scattering, and we divide the scattering amplitude $\xi-1$ into two portions:

$$\frac{f-f^*_{\text{out}}}{f-f^*_{\text{out}}} e^{i\chi} - 1 = \{e^{i\chi}-1\} - e^{-i\chi}\frac{2i\,\text{Im}\,f_{\text{out}}}{f-f_{\text{out}}} \tag{4.12.12}$$

which we may write as $A_{\text{potential}} + A_{\text{nuclear}}$ where $A_{\text{potential}}$ is the scattering amplitude for a hard sphere (plus the effect of a coulomb potential if one is present) and A_{nuclear} is the scattering amplitude due to penetration of the nucleus (in the literature, the quantity we have chosen to call A_{nuclear} is frequently called $A_{\text{resonance}}$ even though the scattering may be at an energy very far from a resonance). Then

$$\sigma^J_{\text{elastic}} = \frac{\pi}{k^2}\alpha_J \left| \{e^{i\chi}-1\} - e^{i\chi}\frac{2i\,\text{Im}\,f_{\text{out}}}{f-f_{\text{out}}} \right|^2, \tag{4.12.13(a)}$$

$$\sigma^J_{\text{reaction}} = \frac{\pi}{k^2}\alpha_J \left\{ \frac{-4\,\text{Im}\,f\,\text{Im}\,f_{\text{out}}}{|f-f_{\text{out}}|^2} \right\}. \tag{4.12.13(b)}$$

This is a formal parametrization and an enormous ignorance of what goes on inside the nucleus is summed up in the parameter f. A detailed theory of nuclear structure could predict f, but if we had such a theory we would hardly need to employ such a parametrization. The value of this treatment is that by making plausible assumptions about the behaviour of f we can gain some insight into the physics of various kinds of nuclear process, and by measuring f we can find out about the detailed structure of the nucleus.

So far our treatment has been general and the only assumption built in has been the assumption of a sharp nuclear boundary. It should, however, be evident that a parametrization in terms of f is most appropriate to low-energy reactions in which the nucleus acts as a whole. The parameter f is sensitive to detailed nuclear structure in the appropriate angular momentum state, and at high energy where many partial waves contribute such a parametrization will be far too complicated. For low energies where a single partial wave in the nucleus-projectile system is predominant, the only effect of dropping the assumption of a sharp nuclear boundary is to confuse inextricably the contribution of nuclear and potential scattering, although the phenomenological form for the scattering cross-section will remain valid.

With this framework established, we will now investigate the nature of resonances and their relation to unstable excited states of the nucleus in more detail, independently of the details of nuclear structure. Since we expect a resonance to correspond to a maximum in the nuclear scattering we first discover the condition for a maximum amount of penetration into the nucleus. Let the interior wave function be $AU_{\text{int}}(\varkappa r+\eta)$ (just inside the boundary) and the exterior function be $BU_{\text{ext}}(kr+\delta)$. $\varkappa^2 = k_0^2+k^2$; η is determined by the detailed nuclear structure and δ by matching the logarithmic derivatives. The penetration of the nucleus is greatest when

$$\left|\frac{A}{B}\right|^2 \text{ is a maximum,}$$

$$\left|\frac{A}{B}\right|^2 = \frac{U^*_{\text{ext}}(kR+\delta)}{U^*_{\text{int}}(\varkappa R+\eta)}\frac{U_{\text{ext}}(kR+\delta)}{U_{\text{int}}(\varkappa R+\eta)}. \tag{4.12.14}$$

Nuclear Physics

The condition for a maximum as a function of k is obtained by setting the differential with respect to k equal to zero

$$\frac{\left(R+\frac{\partial \delta}{\partial k}\right)\{U_{ext}^{*'}U_{ext}+U_{ext}^{*}U_{ext}'\}}{U_{int}^{*}U_{int}} - \frac{U_{ext}^{*}U_{ext}\left(R\frac{\partial \varkappa}{\partial k}+\frac{\partial \eta}{\partial k}\right)\{U_{int}^{*'}U_{int}+U_{int}^{*}U_{int}'\}}{U_{int}^{*}U_{int}} = 0$$

at $R = r$, (4.12.15)

$$\left(R+\frac{\partial \delta}{\partial k}\right)\left\{\frac{U_{ext}^{*'}}{U_{ext}^{*}}+\frac{U_{ext}'}{U_{ext}}\right\}\bigg|_R = \left(R\frac{k}{\varkappa}+\frac{\partial \eta}{\partial k}\right)\left\{\frac{U_{int}^{*'}}{U_{int}^{*}}+\frac{U_{int}'}{U_{int}}\right\}\bigg|_R \quad (4.12.16)$$

which requires $f+f^* = 0$ or $\mathrm{Re} f = 0$.

This is very reasonable: if we match two oscillatory functions where the gradient is not zero, the interior function, oscillating faster than the exterior function, has no chance to achieve its maximum (see Fig. 4.12.1).

Let us now expand f about the zero

$$f = (E-E_0)\frac{\partial \mathrm{Re} f}{\partial E}\bigg|_{E_0} + i\,\mathrm{Im} f(E_0) \quad (4.12.17)$$

taking the leading terms only in the expansion. (It is of course perfectly possible for the real part f to be zero more or less everywhere—this would correspond to a wave going into the nucleus having a negligible probability of ever coming out. Under these circumstances you do not get resonant phenomena.)

Then the nuclear scattering term

$$\frac{-2i\,\mathrm{Im} f_{out}}{f-f_{out}}$$

becomes

$$\frac{-2i\,\mathrm{Im} f_{out}}{(E-E_0)\frac{\partial \mathrm{Re} f}{\partial E}\bigg|_{E_0} + i\,\mathrm{Im} f(E_0) - \mathrm{Re} f_{out} - i\,\mathrm{Im} f_{out}}. \quad (4.12.18)$$

Since an outgoing wave behaves like $\sim e^{ikr}$ the imaginary part of the logarithmic derivative, $\mathrm{Im} f_{out}$, is a positive quantity. Since reaction cross-sections are positive quantities, $\mathrm{Im} f$ must be negative. This means that the ingoing flux is always greater than or equal to the outgoing flux—you cannot get out more than you put in, another expression of unitarity.

Define

$$\left.\begin{array}{l}\Gamma_{el} = \dfrac{-2\,\mathrm{Im} f_{out}}{\dfrac{\partial \mathrm{Re} f}{\partial E}\bigg|_{E_0}} \\[2em] \Gamma_R = \dfrac{+2\,\mathrm{Im} f}{\dfrac{\partial \mathrm{Re} f}{\partial E}\bigg|_{E_0}}\end{array}\right\} \quad \Gamma = \Gamma_{el}+\Gamma_R, \quad (4.12.19)$$

$$E_R = E_0 + \frac{\mathrm{Re} f_{out}}{\dfrac{\partial \mathrm{Re} f_{out}}{\partial E}\bigg|_{E_0}} \quad (4.12.20)$$

and the nuclear scattering term becomes

$$\frac{i\Gamma_{el}}{(E-E_R)+i(\Gamma/2)}$$

where the parameters Γ_{el}, Γ_R are positive quantities because $\partial \operatorname{Re} f/\partial E \leqslant 0$. This is easy to see for the case of shape resonances where the origin of the interior wave is pinned, because the interior wave function contracts inwards as the energy is increased (see Fig. 4.12.2).

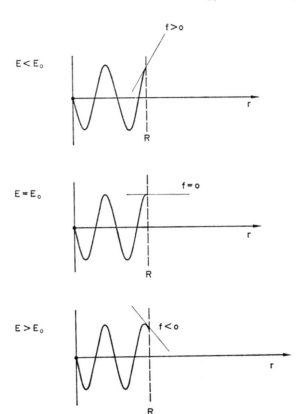

FIG. 4.12.2.

If instead of

$$\frac{1}{E-E_R+i\Gamma/2}$$

we had a nuclear scattering term

$$\frac{1}{E-E_R-i\Gamma/2}$$

then although the identical nuclear scattering cross-section results (that is, the cross-section for negligible potential scattering) there are nonsensical implications.

Nuclear Physics

We can now write

$$\sigma_{\text{elastic}}^J = \frac{\pi}{k^2} \alpha_J \left| \{e^{i\chi} - 1\} + \frac{i e^{i\chi} \Gamma_{\text{el}}}{E - E_R + i\Gamma/2} \right|^2, \quad (4.12.21)$$

$$\sigma_{\text{reaction}}^J = \frac{\pi}{k^2} \alpha_J \left\{ \frac{\Gamma_{\text{el}} \Gamma_R}{(E - E_R)^2 + \Gamma^2/4} \right\}. \quad (4.12.22)$$

If potential scattering can be neglected

$$\sigma_{\text{elastic}}^J = \frac{\pi}{k^2} \alpha_J \left\{ \frac{\Gamma_{\text{el}}^2}{(E - E_R)^2 + \Gamma^2/4} \right\} \quad (4.12.23)$$

and these are the familiar single level Breit–Wigner formulae that we have introduced in many places in this book.

These formulae have been reached by an expansion of the parameter f and so are clearly only approximately valid. They may be expected to be most accurate for large values of $\partial f/\partial E$, that is, narrow widths.

Even if the expansion

$$f \simeq (E - E_0) \frac{\partial f}{\partial E}$$

is valid, variation of $\text{Im} f$ and $\text{Im} f_{\text{out}}$ with energy will distort the above Breit–Wigner shapes if Γ is not very much smaller than E_R. Indeed, E_R itself may be a function of the energy E.

For an s-wave, $\text{Im} f_{\text{out}} = kR$ and so

$$\Gamma_s \simeq \Gamma_s(E_0) \frac{k}{k_0} \quad (4.12.24)$$

and the energy variation is given only by the two-particle phase space.

As another example, consider p-wave elastic scattering.

$$\Gamma = \Gamma_{\text{el}} = \frac{-2 \, \text{Im} f_{\text{out}}}{\left. \dfrac{\partial f}{\partial E} \right|_{E_0}}$$

and for a p-wave in the absence of any coulomb barrier we found

$$\text{Im} f_{\text{out}} = \frac{(kR)^3}{1 + k^2 R^2}$$

so that

$$\Gamma_p(E) \simeq \Gamma_p(E_0) \left(\frac{k}{k_0}\right)^3 \quad (4.12.25)$$

in the next approximation.

The extra factor of k^2 is the centrifugal barrier effect and this gives the long high-energy tail to a p-wave resonance. For low-energy structure resonances in nuclear physics these effects are usually not important. At higher energies for broad resonances they cannot be

Nuclear Reactions

neglected. As an example we show in Fig. 4.12.3 the cross-section for $\pi^+ p$ in the neighbourhood of a *p*-wave resonance which occurs at a centre of mass energy of 1.236 GeV and has a width of 0.116 GeV. The distortion from the simple Breit–Wigner shape is very marked.

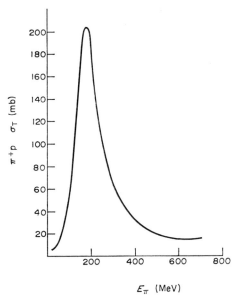

FIG. 4.12.3. The total cross-section for $\pi^+ p$ as a function of pion laboratory kinetic energy. In the energy range shown the total cross-section is almost entirely elastic and reaches the unitarity limit at about 190 MeV. The width of the *p*-wave resonance is 116 MeV/c² and the departure from a Breit–Wigner formula with an energy independent width is very marked.

If a resonance occurs at the same energy and angular momentum in a system with the *same* number of protons and neutrons, but with a different projectile and target nucleus a pair of formulae similar to (4.12.21) and (4.12.22) would result, although the relation between k and the total energy of the system would be different. Thus if we are forming a genuine excited state of the nucleus, we must interpret Γ and E_R as properties of this state which are independent of the way in which it is formed. Then if Γ is independent of the particular way in which the state is formed, $\Gamma_{el} + \Gamma_R$ is independent of the way in which the state is formed and we write

$$\Gamma = \sum_i \Gamma_i \qquad (4.12.26)$$

where the Γ_i are the *partial widths* for each possible decay mode of the excited state.

This assumption that the scattering and reaction cross-sections can be written in terms of such parameters which do not depend on the formation process contains all the physics of the *compound nucleus model* of nuclear reactions. While it is clear that this assumption will hold for structure resonances, it may also hold well away from resonance.

The treatment of resonance we have so far adopted tells us nothing about the way in which the excited state of the nucleus decays in time, although everyone knows the mean lifetime is $\hbar/\Gamma$. In order to establish a direct connection between the width of an observed

Nuclear Physics

resonance and the lifetime of the excited state to which this resonance corresponds, time-independent scattering must be abandoned and we must construct an argument similar to the one we used for establishing the limit on the rate at which a phase shift can fall. We must consider a pulse of radiation fairly well defined in time going into the target nucleus and study the time dependence of what comes out.

For a fixed frequency, if the ingoing wave in a particular angular momentum state is $\sim e^{-i(kr+\omega t)}$ the outcoming wave is $\sim e^{2i\delta} e^{i(kr-\omega t)}$. The effect of the nuclear potential is to shift the phase of the outgoing wave relative to the incoming wave by 2δ (where δ may be complex if absorption is present). Let us suppose for simplicity that potential scattering may be neglected, that is $\chi \sim 0$. Then

$$e^{2i\delta} = \xi = \frac{f - f^*_{\text{out}}}{f - f_{\text{out}}}$$

and in the extreme resonance approximation we can write this as

$$\begin{aligned}\xi &= \frac{E - E_R + i(\Gamma_R/2) - i(\Gamma_{\text{el}}/2)}{E - E_R + i(\Gamma/2)} \\ &= 1 - \frac{i\Gamma_{\text{el}}}{E - E_R + i(\Gamma/2)}.\end{aligned} \qquad (4.12.27)$$

We now define a pulse of radiation

$$\sim \int A(\omega) e^{-i(kr+\omega t)} d\omega$$

coming in.

This represents a pulse localized in space and time and built up by adding together incoming spherical waves of all frequencies, with appropriate relative phases and amplitudes. If $k = \omega/v$ where v is a frequency-independent phase velocity, then, for example,

$$\int e^{-i(kr+\omega t)} d\omega = \delta(r + vt)$$

and represents an infinite spike of radiation travelling inwards with velocity v. In non-relativistic wave mechanics, k^2 is proportional to ω and so any pulse suffers dispersion, but this effect is not important for the argument we are constructing here, and we leave $A(\omega)$ as an arbitrary function which merely has the property of generating a pulse in time.

The outgoing wave is constructed by shifting the phases of each frequency component as a result of passing through the nucleus, and then recombining all these frequencies to give the outgoing pulse, which is thus

$$\sim \int A(\omega) \xi e^{i(kr-\omega t)} d\omega = \int A(\omega) e^{i(kr-\omega t)} d\omega - i \int A(\omega) \frac{\Gamma_{\text{el}} e^{i(kr-\omega t)} d\omega}{(E - E_R + i(\Gamma/2))}. \qquad (4.12.28)$$

The first term just reconstructs the original pulse, going out instead of coming in. The second term is proportional to

$$\Gamma_{\text{el}} e^{i(k_R r - \omega_R t)}$$

where

$$\hbar \omega_R = E_R - i \frac{\Gamma}{2}$$

$$k_R^2 = \frac{2m \left\{ E_R - i \frac{\Gamma}{2} \right\}}{\hbar^2}$$

Nuclear Reactions

and so becomes

$$\sim A(\omega_R)\, \Gamma_{\rm el} e^{i({\rm Re}\, k_R r - (E_R/\hbar)t)}\, e^{-({\rm Im}\, k_R r + (\Gamma/2\hbar)t)}. \tag{4.12.29}$$

At a particular value of r, long after the unscattered outgoing pulse has passed, a signal oscillating with frequency $E_R/\hbar$ and damping out with increasing time is observed. This signal is the resonant scattered pulse which will only interfere with the unscattered pulse in the local region of overlap. The fact that this signal is the scattered pulse is easily seen. We have just calculated the total outgoing wave. The scattered wave is

$$(1-\xi)U_{\rm out}$$

where $U_{\rm out}$ is the outgoing wave in the absence of nuclear scattering and

$$1-\xi = \frac{i\Gamma_{\rm el}}{E-E_R+i(\Gamma/2)}.$$

At the beginning of this chapter we already saw that the effect of a phase shift which is a function of frequency is to introduce a delay into a pulse. In the region of a resonance the effect is more complicated and results in an outgoing signal totally different from the incoming pulse, and characterized not by a delay time but by an exponential damping, with a fall off in the amplitude given by

$$e^{-(\Gamma/2\hbar)t}$$

and so a mean lifetime in the intensity of

$$\tau = \frac{\hbar}{\Gamma}.$$

The imaginary part of k_R merely ensures that at a fixed time the outgoing signal grows with r, for the larger r at a fixed time, the earlier the signal originated.

We can now see why $\partial f/\partial E|_{E_0}$ must be a negative quantity. If it were positive, then instead of a denominator in $1-\xi$ of $E-E_R+i(\Gamma/2)$ (Γ positive) we would get instead a denominator $E-E_R-i(\Gamma/2)$ (Γ positive) and this would give an output signal growing exponentially with time, which we discard as unphysical.

Our time-independent study, based entirely on boundary conditions at the nuclear surface has thus told us that a resonant peak in either elastic or inelastic cross-section corresponds to excitation of an unstable nuclear state with an exponential decay. For on sending a pulse of radiation into the nucleus, we get out an unscattered pulse (and in general a distorted pulse from potential scattering) followed by a signal due to the nuclear scattering that decays exponentially with time. If $\tau \sim$ seconds the target can be picked up, moved across the laboratory and put into a separate detector, and the scattering process still continues. However, the mode of formation of the nuclear excited state is irrelevant and it is in this sense that treating nuclear decay by itself in the previous chapter is justified.

The existence of resonances in nuclei, then, occurs when conditions in the nuclear interior are such as to allow an incident particle to have a high probability of being found inside the nucleus over a limited range of energy. This is not primarily a problem of potential barrier penetration, for transmission probability is a monotonic function of energy. It is a problem of nuclear structure which can only be easily tackled for the special case of shape resonances,

Nuclear Physics

and in general would require solution of the nuclear many-body problem. It must be emphasized that the existence of resonances is not caused by it being easy for a particle to get in through a barrier, but difficult to get out, for transmission probabilities are symmetric and easy one way means easy the other. Rather, when penetration into the nucleus is possible only over a narrow resonant band, the single-particle wave functions in the region of the nuclear surface conspire together to minimize the amplitude of the net outgoing signal. It is perhaps misleading to contemplate resonance phenomena in the framework of a fixed frequency, even though our very crude wave packet treatment has shown that this may be done formally by introducing a complex energy

$$E_0 - i\frac{\Gamma}{2}$$

but this is just an economical way of summarizing the conspiracy among the infinite number of functions of real energy that is induced by the possibility of exciting the compound nucleus over a narrow band of energies.

There is no classical particle model of shape resonances. Classically a particle directed at a potential barrier is either deflected off it or alternatively goes over the top and comes out of the nucleus in roughly the time it takes to cross a nuclear diameter. However, we can make a semiclassical model of structure resonances, in which we have discreet or nearly discreet single-particle energy levels below zero energy and yet still talk of a particle coming in. If we take a schematic picture of the nucleus as an assembly of nucleons confined in a square well potential, then we have a sequence of single-particle energy levels in this well, filled up to ~ 8 MeV below the zero of energy for a single nucleon.

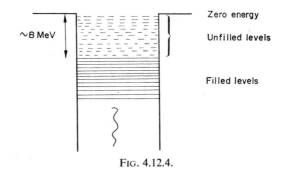

Fig. 4.12.4.

Any nucleon in a level below zero energy cannot escape from the nucleus, and in the absence of electromagnetic interactions all levels below zero energy would have zero width. However, a nucleon in a level above the filled levels can drop down to a lower level by emitting a photon and so any level with unfilled levels below it has a non-zero width. Consider a nucleon entering the nuclear volume from outside and encountering the high density of nucleons. It will collide with some of these nucleons and if it has the right energy to excite one to a single-nucleon state, then the incident nucleon will lose energy and (if the original nucleon had $\ll 8$ MeV kinetic energy) we will have a situation in which a compound nucleus has been formed with two nucleons in unfilled levels. Similarly, if the incident nucleon has just the right energy to excite two nucleons into the unfilled single particle

Nuclear Reactions

levels, and end up in one itself, it will again be trapped below zero energy. If the incident nucleon does not have the right energy to excite some nucleons into unfilled levels and end up in one itself, then because of the quantization of these levels collisions of the incident nucleon with internal nucleons cannot change the state of the system and the incident nucleon comes out again fast, scattered only by the optical potential. This makes us suspect that the incident nucleon only gets trapped for fixed values of energy, but we must remember that (a) the unfilled single-particle levels below zero energy have a width because of the electromagnetic interaction, and (b) that once the energy of the incoming nucleon has been communicated to the compound nucleus, then the whole system is unstable against particle emission. Now because of the strong nucleon–nucleon interaction, and the lack of suppression of these interactions by the Pauli principle when nucleons are moving around in unfilled levels, it is clear that in general these excited nuclear states may not be represented as a superposition of nucleons in single particle levels, but will involve collective motion of all the excited nucleons. Thus our picture is of an incident nucleon being scattered only by an optical potential unless it has the right energy to excite a collective mode of the compound nucleus (which will only rarely be approximated by the independent particle model with several excited nucleons). This energy is not discrete, because any such states are unstable. Once such a state has been formed, no single nucleon is likely to have sufficient energy to get out of the nucleus and the state can decay only electromagnetically (or by the weak interactions in principle) or by waiting until the collective motion results in a return to a configuration consisting of an outgoing nucleon and the original target ground state (or of course some other configuration which allows one or more nuclear fragments to get out). This is essentially a statistical process in such a semiclassical model, giving rise to an exponential decay for an assembly of such states.

The density of single-particle levels near the top of a short-range well is easily estimated by treating it as a container, with impenetrable walls, of volume $(4\pi/3) R_0^3 A$ where A is the number of nucleons. The number of neutron levels, say, corresponding to momentum $\leqslant p$ in such a box is

$$2 \times \frac{4\pi}{3} \frac{p^3}{(2\pi\hbar)^3} \frac{4\pi}{3} R_0^3 A$$

where $R_0 A^{1/3}$ is the nuclear radius and the factor 2 comes from the two spin states of the nucleon. Then the nucleon momentum $p_{\max}$ in the highest single-neutron state filled in the nuclear ground state is given by

$$N = 2 \times \frac{4\pi}{3} \frac{p_{\max}^3}{(2\pi\hbar)^3} \frac{4\pi}{3} R_0^3 A$$

where N is $\sim A/2$. The maximum momentum $p_{\max}$ is thus roughly independent of the nuclear number A and on differentiating, the level density for one kind of nucleon is approximately

$$\left.\frac{dn}{dE}\right|_{E_{\max}} \simeq \frac{3}{2} A \frac{1}{2E_{\max}} \qquad (4.12.30)$$

Nuclear Physics

a result we already used in Section 1.5. E_{max} is ~ 12 MeV and the effective potential well depth ~ 20 MeV. The level density near the top of the well is thus

$$\frac{dn}{dE} \sim \frac{3A}{40} \text{ per MeV.}$$

Thus single-particle levels are $\sim 1/\text{MeV}$ for light nuclei, and $\sim 1/0.1$ MeV for $A \sim 100$. This separation will continue above zero energy, centrifugal or coulomb barriers confining all except the occasional s-wave excited states, and so these numbers give the approximate spacing of shape resonances.

Now let us calculate approximately the separation of the structure resonances near zero energy (for neutrons, since zero energy protons have an enormous coulomb barrier in heavy nuclei). If we suppose the single-particle levels to be uniformly spaced, each with four nucleons in it, how many ways are there of dividing up the energy of the incoming neutron among excitations to single-particle levels? For light nuclei a neutron coming in near zero energy gives the compound nucleus an excitation energy of approximately 8 times the single-particle spacing—the neutron binding energy divided by the level spacing. It can stay where it is, exciting no other single-particle states. It can drop one level, raising one other nucleon one level. It can drop two levels, raising one nucleon two levels, or two nucleons one level each. It can drop three levels, raising one nucleon three levels, three nucleons one level or one nucleon two levels and one nucleon one level, and so on. The total number of ways of distributing this energy among different configurations of single-particle levels is a combinational problem. If the incident particle conveys to the compound nucleus an excitation energy equal to an integral number n of single-particle excitations, then the number of ways of dividing up this energy is given by the number of different ways of partitioning the integer n which for large n is equal to approximately[†]

$$\frac{1}{\sqrt{48n}} \exp\left\{\pi \sqrt{\frac{2n}{3}}\right\}.$$

Thus in this very simple model, when an incident nucleon enters the nucleus with enough energy to reach a many particle excited state, while the allowed incident energies are separated by ~ 1 MeV for light nuclei and ~ 0.1 MeV for heavy nuclei, each possible energy level is very highly degenerate. Thus a neutron entering a heavy nucleus at $\sim$ zero energy and conveying to it an excitation $n \times 0.1$ MeV will form a many-excitation excited state with a degeneracy

$$\sim \frac{1}{\sqrt{48n}} \exp\left\{\pi \sqrt{\frac{2n}{3}}\right\}$$

where n will be ~ 80 and so the degeneracy of these levels near zero energy for a heavy nucleus is $\sim 10^7$.

Finally, the fact that nucleon–nucleon scattering in these highly excited states is not inhibited by the Pauli principle means that individual nucleon–nucleon interactions will be

---

[†] See S. A. Moszkowski, Models of nuclear structure, *Handbuch der Physik*, XXXIX, 437 (Springer-Verlag, 1957).

Nuclear Reactions

more important that the overall time-independent component of the potential in determining the energy levels, and so the degeneracy implied in a simple independent particle model with equal spacing will be removed and we are left with the expectation of $\sim 10^7$ levels in the 0.1-MeV interval between single particle states, a density of $10^8/\text{MeV}$. This model is of course far too naïve, and even within the framework of this model all these states will not be observed. For low-energy neutrons, for example, we would expect to pick out only the s-wave resonances since the capture cross-section is $\sim \Gamma_{el}$ and the scattering cross-section to Γ_{el}^2 and Γ_{el} will be suppressed by the centrifugal barrier for p and higher waves. However, it is clear how it is possible to find resonances at low energy with spacings and widths $\sim$ electron volts, when the optical model spacing is ~ 0.1 MeV.

Our semiclassical picture also shows us why it is a reasonable approximation to divide the scattering amplitude into a potential part (hard sphere scattering for neutrons) and a nuclear, or resonant part. If a low-energy nucleon enters the nucleus, it has a high probability of colliding with at least one nucleon. If the incident energy is right for excitation of the compound nucleus, then this and subsequent collisions can change the internal state of the system and retain the incident particle. If the energy is not right, the quantized nature of the system means that the struck nucleon cannot change its state of motion relative to other nucleons in the target nucleus, and all that can happen is that the target nucleus as a whole recoils, the incident nucleon bouncing off what is effectively a hard sphere.

Our treatment of resonances through the matching of particle wave functions across the nuclear boundary relied on the expansion

$$f = (E-E_0) \left.\frac{\partial \operatorname{Re} f}{\partial E}\right|_{E_0} + i \operatorname{Im} f(E_0).$$

If the real part of f is everywhere zero, or nearly zero, this corresponds to the internal wave consisting of an ingoing component only, which implies strong absorption in the nuclear scattering. This in turn implies a large number of outgoing channels. In the neighbourhood of a resonance we may pass to this situation by letting

$$\left.\frac{\partial \operatorname{Re} f}{\partial E}\right|_{E_0} \to 0 \qquad (4.12.31)$$

and because the widths are proportional to the reciprocal of this quantity, the widths get larger and larger as

$$\left.\frac{\partial \operatorname{Re} f}{\partial E}\right|_{E_0} \to 0$$

and neighbouring resonances blur together into a *continuum*, the nuclear scattering term becoming

$$\frac{-2i \operatorname{Im} f_{\text{out}}}{-\operatorname{Re} f_{\text{out}} + i \operatorname{Im} f(E_0) - i \operatorname{Im} f_{\text{out}}}. \qquad (4.12.32)$$

However, if in spite of the many exit channels and the overlapping of the resonances, the quantities in this expression still remain characteristic only of the energy of the system and do not depend on the particular channel through which the reaction was initiated, the

Nuclear Physics

factorizability of the scattering and reaction amplitudes, which we noted for the formation of a single excited state, are still maintained. This factorizability is the characteristic property of the compound nucleus hypothesis. The observation of a density of excited states very much greater than the optical model density demands a compound nucleus model at low energies, at higher energy the persistence of compound nucleus behaviour is manifested by the factorizability of reaction cross-sections, in which resonant structure is not apparent.

The treatment we have followed so far is not applicable to incident photons. However, the behaviour of γ-ray scattering by nuclei is analogous to, and the treatment is no different in principle from, our brief study of the scattering of light at the beginning of this chapter. If an incident γ has the right energy to induce a transition to an excited state of the compound nucleus, then it is resonant scattered, not necessarily through the electric dipole mechanism responsible for the scattering of light. If the wavelength of the photon is comparable with or smaller than the nuclear size, then the variation of the photon phase with position in the nucleus must be taken into account and absorption may proceed via the excitation of higher multipoles, the inverse of the decay process. In the resonance region, the electromagnetic decay of an excited state is determined by the properties of the state, and whether this state has been formed by a γ, a slow neutron or by coulomb excitation or any other method has no effect on the decay probabilities. The factorizability of reaction cross-sections in the low-energy region will thus hold for photon-initiated or terminated reactions just as for reactions involving particles. The analogue of potential scattering for photons is compton scattering of the photon by the nucleus, with no disturbance of the nuclear structure. If the photon energy is small in comparison with the mass of the nucleus, the cross-section is given approximately by the familiar Thompson cross-section, and because of the great mass of the nucleus is negligible. The elastic scattering cross-section for photon scattering may thus be written as proportional to

$$\left| \frac{\Gamma_{el}}{E - E_R + i(\Gamma/2)} \right|^2 \qquad (4.12.33)$$

and for those states where particle emission is energetically forbidden, this becomes

$$\frac{\Gamma_{el}^2}{(E - E_R)^2 + \Gamma_{el}^2/4}.$$

On integrating this quantity over energy, the scattering cross-section becomes

$$\int \sigma_{sc} \, dE \propto \Gamma_{el} \qquad (4.12.34)$$

so that a measurement of the integrated cross-section in the region of an excited state determines Γ_{el} even if the energy resolution of the detector is too poor to allow an accurate measurement of the line shape. The lifetime is then given by

$$\tau_\gamma = \frac{\hbar}{\Gamma_{el}}$$

and it is in this way that the lifetimes of states decaying by γ emission may be determined in the region $10^{-16} \lesssim \tau \lesssim 10^{-14}$ sec as we mentioned in Chapter 3.

Nuclear Reactions

Resonance phenomena and compound nucleus effects in general may be treated in the framework of the Fermi Golden Rule by an extension of the methods of Sections 3.2 and 4.4. We consider an initial state ψ_1 feeding a state ψ_3 through an intermediate state ψ_2 as well as directly. In the framework of a reflecting box we write the amplitudes as

$$\left. \begin{aligned} i\hbar \frac{\partial A_1}{\partial t} &= \langle 1|H|1\rangle A_1 + \langle 1|H|2\rangle A_2 + \langle 1|H|3\rangle A_3, \\ i\hbar \frac{\partial A_2}{\partial t} &= \langle 2|H|1\rangle A_1 + \langle 2|H|2\rangle A_2 + \langle 2|H|3\rangle A_3, \\ i\hbar \frac{\partial A_3}{\partial t} &= \langle 3|H|1\rangle A_1 + \langle 3|H|2\rangle A_2 + \langle 3|H|3\rangle A_3. \end{aligned} \right\} \quad (4.12.35)$$

ψ_1 is an initial state in the continuum, ψ_2 is an intermediate compound nuclear state, and ψ_3 is a final continuum state. We ignore any possible effects of ψ_1 being fed from ψ_2 or ψ_3, and we ignore any possible effects of ψ_2 being fed from ψ_3 and obtain the approximate set for the case where $\psi_1 \not\rightarrow \psi_3$ directly

$$\left. \begin{aligned} i\hbar \frac{\partial A_1}{\partial t} &= \langle 1|H|1\rangle A_1, \\ i\hbar \frac{\partial A_2}{\partial t} &= \langle 2|H|1\rangle A_1 + \langle 2|H|2\rangle A_2, \\ i\hbar \frac{\partial A_3}{\partial t} &= \langle 3|H|2\rangle A_2 + \langle 3|H|3\rangle A_3. \end{aligned} \right\} \quad (4.12.36)$$

If we now drive the whole system under steady state conditions, allowing ψ_3 to escape from the normalization volume, we obtain (see Section 4.4)

$$i\hbar\lambda A_2 = \langle 2|H|1\rangle e^{\lambda t} + \langle 2|H|2\rangle A_2,$$
$$i\hbar\lambda A_3 = \langle 3|H|2\rangle A_2 + \langle 3|H|3\rangle A_3 - i\alpha A_3$$

and so

$$A_3 = \frac{\langle 3|H|2\rangle A_2}{i\hbar - \langle 3|H|3\rangle} = \frac{\langle 3|H|2\rangle\langle 2|H|1\rangle e^{\lambda t}}{\{i\hbar\lambda - \langle 3|H|3\rangle\}\{i\hbar\lambda - \langle 2|H|2\rangle\}} + i\alpha.$$

The calculation of the rate at which ψ_3 is fed from ψ_1 via the intermediate state ψ_2 follows directly from our treatment of a direct process in Section 4.4, yielding

$$T_{1\to 2\to f} = \frac{2\pi}{\hbar} \left| \frac{\langle f|H|2\rangle\langle 2|H|1\rangle}{i\hbar\lambda_1 - \langle 2|H|2\rangle} \right|^2 \frac{dn_f}{dE}. \quad (4.12.37)$$

Since ψ_2 is an unstable state,

$$\langle 2|H|2\rangle = E_2 - i\Gamma_2/2$$

and so we obtain

$$T_{1\to 2\to f} = \frac{2\pi}{\hbar} \frac{|\langle f|H|2\rangle\langle 2|H|1\rangle|^2}{(E_1-E_2)^2 + \Gamma_2^2/4} \frac{dn_f}{dE} \quad (4.12.38)$$

Nuclear Physics

where E_1 is the exciting energy. Note that the whole resonance ψ_2 is not excited under these steady state conditions.

If the direct matrix element $\langle 3 | H | 1 \rangle$ is not equal to zero then it is easy to see that we obtain

$$T_{1 \to f} = \frac{2\pi}{h} \left| \langle f | H | 1 \rangle + \frac{\langle f | H | 2 \rangle \langle 2 | H | 1 \rangle}{E_1 - E_2 + i(\Gamma_2/2)} \right|^2 \frac{dn_f}{dE} \qquad (4.12.39)$$

and if Γ_2 is small the resonance term dominates for $E_1 \sim E_2$. But if $E_1 - E_2 \sim \Gamma_2$ then the two terms may interfere. This is illustrated in Fig. 4.12.5. Another example may be found in Fig. 5.5.3.

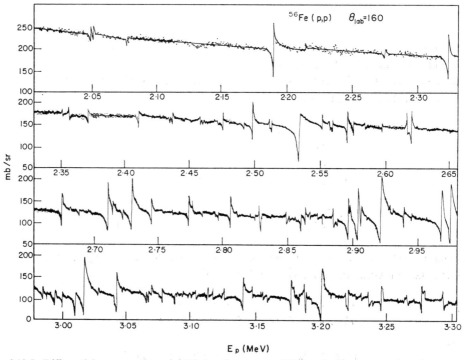

FIG. 4.12.5. Differential cross-section at 160° for the process ^{56}Fe(p, p). Very strong interference exists between the coulomb scattering and the scattering through resonant states of ^{57}Co. [From D. P. Lindstrom et al., Nucl. Phys. A **168**, 37 (1971).]

The advantage of this treatment is that it shows clearly the origin of the factorization property of cross-sections for reactions going through an excited state (in this connection it must be remembered, however, that the first-order terms are not necessarily zero and can interfere with a resonant amplitude, and also that when a great many excited states overlap the factorization property may be destroyed).

We have not specified the nature of the specific interaction I which is responsible for inducing transitions—it may be the sum of various strong interaction terms, an electromagnetic coupling and even weak interactions. Then if the state ψ_1 consists of a neutron and a nucleus, and the state ψ_3 of a photon and a nucleus, the matrix element $\langle 2 | I | 1 \rangle$ picks out the strong interaction coupling the neutron with the compound nucleus while the matrix

Nuclear Reactions

element $\langle 3|I|2\rangle$ picks out the electromagnetic interaction responsible for γ-decay of the compound nucleus and has the appropriate multipole characteristics.

This treatment also sheds light on the nature of the damping term. If we split the formation-decay chain as we did in Chapter 3 and just consider the decay of the compound nucleus ψ_2 regarded as an initial state prepared by some arbitrary means, then the partial decay rate into the state ψ_3 is

$$T_3 = \frac{2\pi}{\hbar} |\langle 3|I|2\rangle|^2 \varrho(E_3). \qquad (4.12.40)$$

The partial decay rate into the state ψ_1 is

$$T_1 = \frac{2\pi}{\hbar} |\langle 1|I|2\rangle|^2 \varrho(E_1) \qquad (4.12.41)$$

and so on. If we write

$$\frac{1}{\tau} = \frac{\Gamma}{\hbar} \quad \text{and} \quad \Gamma = \sum \Gamma_i$$

then

$$\Gamma = 2\pi \sum_i |\langle i|I|2\rangle|^2 \varrho(E_i) \qquad (4.12.42)$$

and the resonant cross-section σ_i may be expressed in terms of the partial widths.

$$\sigma_{1i} = \frac{1}{v_1} \frac{2\pi}{\hbar} \left| \frac{\langle i|I|2\rangle\langle 2|I|1\rangle}{E_1-E_2+i(\Gamma/2)} \right|^2 \varrho(E_i)$$

Now

$$\varrho(E_1) = \frac{4\pi p_1^2 \, dp_1}{(2\pi\hbar)^3 \, dE_1} = \frac{4\pi p_1^2}{(2\pi\hbar)^3 v_1}$$

so that

$$|\langle 2|I|1\rangle|^2 = \frac{\Gamma_1(E_1)}{2\pi} \frac{(2\pi\hbar)^3}{4\pi p_1^2} v_1$$

$$\sigma_{1i} = \frac{\pi}{k_1^2} \frac{\Gamma_1 \Gamma_i}{(E_1-E_2)^2 + \Gamma^2/4} \qquad (4.12.43)$$

where both Γ_i, Γ_1 are functions of the energy E, which enters not only through the energy dependence of the phase space factor but also through the energy dependence of the appropriate matrix element. Comparison with expression (4.12.22) for a reaction cross-section shows that the only difference is in the absence of the statistical factor. This is always introduced by summing over final spin states and averaging over initial spin states: a procedure which may be carried out on eq. (4.12.43) to yield an expression identical to (4.12.22).

The variation of the partial widths Γ_i cannot in general be neglected. For particle emission (two-body final states) the phase space is proportional to k_i. At low energies the final state wave function varies as $(k_i r)^l$ because of the centrifugal repulsion and so $\langle i|H|2\rangle \sim (k_i R)^l$ where R is a radius parameter. We may thus expect $\Gamma_i \sim k_i^{2l+1}$, a result which may be compared with the explicit calculation for a p-wave obtained from boundary conditions at

271

Nuclear Physics

the nucleus, eq. (4.12.25). At momenta sufficiently high that the overlap of the outgoing particle wave function with the nucleus is substantially independent of l, the phase space variation only will be important—the explicit p-wave calculation also exhibits this feature. It is worth noting that for an exothermic reaction like (n, γ) slow neutron capture, where the slow neutron is in an s-wave and so $\Gamma_n \sim k_n$, the cross-section goes as $1/k_n$ well below the resonance energy—the $1/v$ law again. Finally, we should note that the partial width Γ_{ch} for any channel in which both particles are charged will contain a coulomb barrier factor of precisely the kind we calculated for α-decay rates.

$$\Gamma_{ch} = \Gamma_0 \left(\frac{k}{k_0}\right)^{2l+1} \frac{e^{-G(v)}}{e^{-G(v_0)}}. \tag{4.12.44}$$

This is discussed further in the next section.

4.13. The coulomb barrier

We have already encountered the coulomb barrier in α-decay, and in this section give a fairly detailed discussion of the physics of the coulomb barrier and its influence upon nuclear reactions involving charged particles. The results of this section are of enormous importance since it is the properties of the coulomb barrier that control thermonuclear reaction rates which we discuss in Chapter 5.

We will suppose here either that we are working only with s-waves, or that the coulomb barrier is sufficiently large in comparison with the centrifugal barrier that we may neglect the latter. We thus write the radial part of the Schrödinger equation as

$$\frac{1}{r^2}\frac{d}{dr}\left(r^2 \frac{d\mathcal{R}}{dr}\right) + \frac{2m}{\hbar^2}(E-V)\mathcal{R} = 0. \tag{4.13.1}$$

We now suppose that for $r > R$ the potential is given by

$$V = \frac{Zze^2}{r}$$

and for $r < R$

$$V = -V_0$$

so that as a charged particle approaches the nucleus it experiences first a repulsive coulomb potential. Then at a radius $r = R$ it comes within range of a strong attractive potential and on moving further in it dives into the nuclear surface and loses its identity.

On substituting $\mathcal{R} = \phi/r$ the equation (4.13.1) takes on the simple one-dimensional form

$$\frac{d^2\phi}{dr^2} + \frac{2m}{\hbar^2}(E-V)\phi = 0. \tag{4.13.2}$$

Thus in the region $r < R$ the solution is oscillatory with a constant propagation constant k_1 given by

$$k_1^2 = \frac{2m}{\hbar^2}(E+V_0). \tag{4.13.3}$$

Nuclear Reactions

In the region $R < r < R_c$, $R_c = Zze^2/E$, the propagation constant k_2 is a function of r and is imaginary,

$$k_2^2 = \frac{2m}{\hbar^2}\left(E - \frac{Zze^2}{r}\right) \qquad E < \frac{Zze^2}{r} \qquad (4.13.4)$$

and in the region $r > R_c$ the propagation constant k_3 is real once more and given by

$$k_3^2 = \frac{2m}{\hbar^2}\left(E - \frac{Zze^2}{r}\right) \qquad E > \frac{Zze^2}{r}. \qquad (4.13.5)$$

Our problem is to obtain solutions to eq. (4.13.2) in each of these three regions and match them to each other across the boundaries at $r = R, R_c$. For $r < R$ the solutions are standard

$$\phi \propto e^{\pm ik_1 r}.$$

In the other two regions the propagation constant is a function of r and in general the equation should be solved numerically. Here we will get approximate solutions by using the W.B.K. method (Wentzel, Brillouin, Kramers).

If k is a very slowly varying function of r, then locally

$$\phi \simeq e^{\pm ik(r)r}$$

will be a good approximation to the true solution of eq. (4.13.2). In the W.B.K. approximation we suppose the solution to be given locally by such a form, and allow both the amplitude and phase to be functions of r, so as to accommodate the effect of the changing potential. We try a solution

$$\phi = A e^{i\varkappa}$$

where A and $\varkappa$ are functions of r. Then

$$\frac{d^2}{dr^2}[A e^{i\varkappa}] + k^2 A e^{i\varkappa} = 0. \qquad (4.13.6)$$

This equation has both real and imaginary parts which must be separately equated to zero yielding

$$2\frac{dA}{dr}\frac{d\varkappa}{dr} + A\frac{d^2\varkappa}{dr^2} = 0, \qquad (4.13.7)$$

$$\frac{d^2 A}{dr^2} - \left(\frac{d\varkappa}{dr}\right)^2 A + k^2 A = 0. \qquad (4.13.8)$$

Equation (4.13.7) is easily solved by direct integration.

$$\frac{1}{A}\frac{dA}{dr} = -\frac{1}{2}\frac{1}{d\varkappa/dr}\frac{d}{dr}\left(\frac{d\varkappa}{dr}\right),$$

$$\ln A = -\frac{1}{2}\ln\left(\frac{d\varkappa}{dr}\right) + \text{const},$$

$$A = C\left(\frac{d\varkappa}{dr}\right)^{-1/2}.$$

Nuclear Physics

Equation (4.13.8) has a simple solution in any region where the potential varies slowly enough for the effects to be important only in the phase and not the amplitude. Thus if

$$\left|\frac{d^2 A}{dr^2}\right| \ll \left|\left(\frac{d\varkappa}{dr}\right)^2 A\right|$$

then

$$\frac{d\varkappa}{dr} \simeq k$$

and

$$\varkappa \simeq \int^r k \, dr.$$

The solution in this approximation of eq. (4.13.2) is thus

$$\phi \simeq \frac{c}{k^{1/2}} e^{\pm i \int^r k \, dr} \tag{4.13.9}$$

(and you can easily show that as $\partial V/\partial r \to 0$ this approaches the standard oscillatory solution with constant amplitude). We now write (4.13.2) for the three regions as

$$\begin{aligned}
\frac{d^2\phi_1}{dr^2} + k_1^2 \phi_1 = 0 \quad & k_1^2 = \frac{2m}{\hbar^2}(E+V_0) & r < R, \\
\frac{d^2\phi_2}{dr^2} - k_2^2 \phi_2 = 0 \quad & k_2^2 = -\frac{2m}{\hbar^2}\left(E - \frac{Zze^2}{r}\right) & R < r < R_c, \\
\frac{d^2\phi_3}{dr^2} + k_3^2 \phi_3 = 0 \quad & k_3^2 = \frac{2m}{\hbar^2}\left(E - \frac{Zze^2}{r}\right) & r > R_c
\end{aligned} \tag{4.13.10}$$

(where we have changed a sign in the definition of k_2 so that all k are real) and have the solutions

$$\phi_1 = \frac{c_1^+}{k_1^{1/2}} e^{ik_1 r} + \frac{c_1^-}{k_1^{1/2}} e^{-ik_1 r} \qquad r < R, \tag{4.13.11}$$

$$\phi_2 = \frac{c_2^+}{k_2^{1/2}} e^{\int_R^r k_2 \, dr} + \frac{c_2^-}{k_2^{1/2}} e^{-\int_R^r k_2 \, dr} \qquad R < r < R_c, \tag{4.13.12}$$

$$\phi_3 = \frac{c_3^+}{k_3^{1/2}} e^{i \int_{R_c}^r k_3 \, dr} + \frac{c_3^-}{k_3^{1/2}} e^{-i \int_{R_c}^r k_3 \, dr} \qquad r > R_c. \tag{4.13.13}$$

Provided that $E \ll Zze^2/R$ (that is, the height of the barrier is very much greater than the energy) we may expect (4.13.12) and (4.13.13) to be good approximations *except* in the region $r \approx R_c$. Matching these solutions at $r = R$ is thus straightforward, but matching at $r = R_c$ is complicated by the fact that this is precisely the region in which the solutions break down.

Nuclear Reactions

Rather than make a general solution of the problem, we will specialize to three cases of physical interest, characterized by different boundary conditions as $r \to \infty$ or $r \to 0$. The first corresponds to an outgoing wave only at $r \gg R_c$. This is appropriate to α-decay, or indeed to the emission of any charged particle through a coulomb barrier in nuclear decay. Thus for this case our boundary condition at infinity gives $c_3^- = 0$.

Let us now proceed to match ϕ_2 and ϕ_3 at $r = R_c$.

$$\phi_2(r) \simeq \frac{c_2^+}{k_2^{1/2}} e^{\int_R^r k_2 dr} + \frac{c_2^-}{k_2^{1/2}} e^{-\int_R^r k_2 dr},$$

$$\frac{\partial \phi_2(r)}{\partial r} \simeq c_2^+ k_2^{1/2} e^{\int_R^r k_2 dr} - c_2^- k_2^{1/2} e^{-\int_R^r k_2 dr},$$

$$\phi_3(r) \simeq \frac{c_3^+}{k_3^{1/2}} e^{i\int_R^r k_3 dr},$$

$$\frac{\partial \phi_3(r)}{\partial r} \simeq c_3^+ i k_3^{1/2} e^{i\int_R^r k_3 dr},$$
(4.13.14)

where we have neglected $\partial A/\partial r$ in comparison with $\partial x/\partial r$. This neglect is justified in the region where ϕ_2 and ϕ_3 are adequate approximations. Now ϕ_2 and ϕ_3 are not valid solutions at $r = R_c$. Therefore let us suppose that ϕ_2 is valid for $r \leq R_c - x_0$ and that ϕ_3 is valid for $r \geq R_c + x_0$. If the quantity x_0 is small, then we should expect

$$\phi_2(R_c - x_0) \approx \phi_3(R_c + x_0),$$

$$\frac{\partial \phi_2}{\partial r}(R_c - x_0) \approx \frac{\partial \phi_3}{\partial r}(R_c + x_0).$$

Now

$$k_2^2(R_c + x) = -\frac{2m}{\hbar^2}\left(E - \frac{Zze^2}{R_c + x}\right) \approx -\frac{2m}{\hbar^2}\frac{Zze^2}{R_c^2}x,$$

$$k_3^2(R_c + x) = \frac{2m}{\hbar^2}\left(E - \frac{Zze^2}{R_c + x}\right) \approx \frac{2m}{\hbar^2}\frac{Zze^2}{R_c^2}x$$

and so for $x/R_c \ll 1$ we may write

$$\left.\begin{array}{l}\dfrac{\partial^2 \phi_2}{\partial x^2} + \alpha^2 x \phi_2 = 0 \\[6pt] \dfrac{\partial^2 \phi_3}{\partial x^2} + \alpha^2 x \phi_3 = 0\end{array}\right\} \quad \alpha^2 = \frac{2m}{\hbar^2}\frac{Zze^2}{R_c^2}.$$
(4.13.15)

Thus in the region $r \sim R_c$ both ϕ_2 and ϕ_3 are solutions of the same equation, and their second derivatives are zero at $x = 0$. The function ϕ_2 is the solution for x negative and the function ϕ_3 for x positive. Thus we may expect both solutions to become valid at $|x| = x_0$ where

$$k_2(-x_0) = k_3(x_0).$$
(4.13.16)

Nuclear Physics

Let us now match the functions and their first derivatives at $x = \pm x_0$.

$$c_2^+ e^{\int_R^{R_c-x_0} k_2\, dr} + c_2^- e^{-\int_R^{R_c-x_0} k_2\, dr} = c_3^+ e^{i\int_{R_c}^{R_c+x_0} k_3\, dr},$$

$$c_2^+ e^{\int_{R_c}^{R_c-x_0} k_2\, dr} - c_2^- e^{-\int_R^{R_c-x_0} k_2\, dr} = ic_3^+ e^{i\int_{R_c}^{R_c+x_0} k_3\, dr}.$$

k_2 and k_3 are very nearly zero in the region $r \sim R_c$ and so the integrals are changed very little by replacing the upper limits by R_c to obtain

$$c_2^+ e^{\int_R^{R_c} k_2\, dr} + c_2^- e^{-\int_R^{R_c} k_2\, dr} = c_3^+,$$

$$c_2^+ e^{\int_R^{R_c} k_2\, dr} - c_2^- e^{-\int_R^{R_c} k_2\, dr} = ic_3^+,$$

whence

$$c_3^+ = \frac{2}{1-i} e^{-\int_R^{R_c} k_2\, dr} c_2^- \tag{4.13.17}$$

and

$$c_2^+ = c_2^{-*} e^{-2\int_R^{R_c} k_2\, dr}. \tag{4.13.18}$$

However, since both ϕ_2 and ϕ_3 have a finite slope, these results are not exact—what we have done is represented schematically by

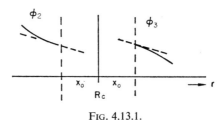

Fig. 4.13.1.

We can improve things by interpolating linearly across the intermediate region

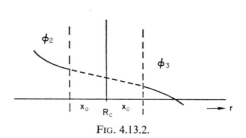

Fig. 4.13.2.

by still matching the gradient but setting

$$\phi_2(R_c - x_0) = \phi_3(R_c + x_0) - 2x_0 \frac{\partial \phi_3}{\partial x}\bigg|_{x_0}$$

yielding

$$c_3^- = \frac{2c_2^- e^{-\int_R^{R_c} k_2 \, dr}}{1-i-2x_0 ik(x_0)}.$$

The fact that this result depends on the value of x_0 in first order shows that our interpolation is not that good. However, we can estimate the size of the correction from the condition that our approximate solutions are valid at $x = \pm x_0$, namely that

$$\left|\frac{\partial^2 A}{\partial x^2}\right| \ll \left|A\left(\frac{\partial \varkappa}{\partial x}\right)^2\right|$$

if the linear expansion of $1/r$ is a reasonable approximation in the region where our approximate solutions obtain.

$$A = \frac{c}{k^{1/2}} \qquad \frac{\partial \varkappa}{\partial x} = k,$$

$$k = \alpha x^{1/2}$$

so the condition is

$$\alpha^2 x_0^3 = k^2 x_0^2 \gg \tfrac{5}{16}$$

(which must be valid simultaneously with $x_0/R_c \ll 1$).

Thus if we set $kx_0 \sim 1$

$$c_3^- \sim \frac{2}{1-3i} c_2^- e^{-\int_R^{R_c} k_2(r) \, dr} \qquad (4.13.19)$$

(we may also note that the condition that we can drop $\partial A/\partial r$ in the derivative of ϕ is

$$\frac{\partial}{\partial x}\left(\frac{c}{k^{1/2}}\right) \ll ck^{1/2}$$

or

$$\alpha x^{3/2} \gg \tfrac{1}{4}.$$

Thus we have evaluated the derivatives correctly).

There will always be an energy low enough that

$$k^2 x_0^2 \gg \tfrac{5}{16} \quad \text{or} \quad \frac{2m}{\hbar^2} Zze^2 R_c \left(\frac{x_0}{R_c}\right)^3 \gg \tfrac{5}{16}$$

and $x_0/R_c \ll 1$ are both satisfied and so at low energy

$$c_3^- \sim c_2^- e^{-\int_R^{R_c} k_2 \, dr}. \qquad (4.13.20)$$

This is the fundamental coulomb barrier result. The form must be correct: the residual uncertainty lies in the precise numerical factor. This can be obtained from more elaborate

Nuclear Physics

continuation procedures than we have considered, which depend upon solving eq. (4.13.15) for the intermediate region and then matching this solution to our solutions. The details may be found in many standard texts on quantum mechanics.[†]

We have already found the constant $c_2^+ \sim c_2^- e^{-2\int_R^{R_c} k_2\, dr}$ when ϕ_3 contains only an outgoing part, and so at $r \sim R$ the only part of the interior solution ϕ_2 that survives is

$$\frac{c_2^-}{k_2^{1/2}} e^{-\int_R^r k_2\, dr}. \tag{4.13.21}$$

This must be matched to the oscillatory solutions inside the barrier. Now because the barrier very effectively reflects a charged particle trying to get out, we may approximate by supposing

$$|c_1^+| \sim |c_1^-|$$

and write

$$\phi_1 = \frac{c}{k_1^{1/2}} \sin(k_1 r + \eta).$$

In many treatments it is supposed that the charged particle moves in a time-independent potential everywhere inside the barrier, $r < R$, and consequently ϕ_1 must be regular at the origin. You should note that this is not necessary. We have assumed only that there is a region between the nuclear surface and the barrier in which the strong interactions may be approximated by a time-independent potential, and consequently the phase η is controlled by the details of nuclear structure. Then

$$\frac{c}{k_1^{1/2}} \sin(k_1 R + \eta) = \frac{c_2^-}{k_2^{1/2}},$$

$$k_1^{1/2} \cos(k_1 R + \eta) = -c_2^- k_2^{1/2}$$

yielding

$$\tan(k_1 R + \eta) = -\frac{k_1}{k_2(R)}. \tag{4.13.22}$$

This fixes η in consequence of our assumption that the reflection is very efficient. If η does not approximately satisfy (4.13.22) we have a larger amplitude getting out through the barrier and $|c_1^-| < |c_1^+|$. However, for α-decay clearly the excited states will correspond to efficient reflection, eq. (4.13.22). As $k_2(R) \to \infty$ the internal solution approaches a standing wave.

Finally we may write for an outgoing wave at $r \gg R_c$

$$c_3^- = 0 \qquad c_3^+ \sim c\, \frac{k_2^{1/2}(R)}{k_1^{1/2}} \sin(k_1 R + \eta)\, e^{\int_R^{R_c} k_2\, dr}$$

[†] For example, L. I. Schiff, *Quantum Mechanics*, 3rd ed., sect. 34 (McGraw-Hill, 1968).

and so since $k_1 \ll k_2(R)$

$$\phi_3(r) \approx \frac{c}{k_1^{1/2}} \frac{k_1}{k_2(R)} \frac{k_2(R)^{1/2}}{k_3^{1/2}(r)} e^{-\int_R^{R_c} k_2 \, dr} e^{ik_3(r)r}. \tag{4.13.23}$$

$c/k_1^{1/2}$ is the amplitude of the interior function ϕ_1, and at low energy both k_1 and $k_2(R)$ are approximately independent of the energy E. The outward going flux is proportional to

$$|\phi_3(\infty)|^2 v(\infty) \quad \text{and} \quad v(\infty) = \frac{\hbar}{m} k_3(\infty).$$

The rate of decay T of an already prepared state through a coulomb barrier is thus

$$T \propto e^{-2\xi} = e^{-G} \tag{4.13.24}$$

where

$$\xi = \int_R^{R_c} k_2 \, dr.$$

The factor e^{-G} is known as the Gamow factor. This result is applied to α-decay in Section 3.16. In addition to this application, it is at once clear that the partial width of a resonance for a charged particle channel is controlled by a factor

$$e^{-G},$$

a result which we shall obtain more directly in the next section.

Pictorially, the situation we have considered corresponds to

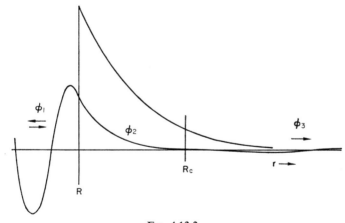

Fig. 4.13.3.

We now go on to apply this method to nuclear reactions, and in this case a charged particle is approaching from outside the coulomb barrier. Most of the intensity will be reflected from the coulomb barrier before the incident particle even penetrates into the shielded nucleus and therefore our boundary condition at $r \gg R_c$ is that

$$|c_3^+| \approx |c_3^-|$$

Nuclear Physics

and so we write ϕ_3 in the form

$$\phi_3 \approx \frac{c_3}{k_3^{1/2}} \sin\left\{\int_{R_c}^r k_3 \, dr + \delta\right\}.$$

We now match at $r \sim R_c$ to the interior function ϕ_2:

$$c_2^+ e^{\int_R^{R_c} k_2 \, dr} + c_2^- e^{-\int_R^{R_c} k_2 \, dr} \sim c_3 \sin \delta,$$

$$c_2^+ e^{\int_R^{R_c} k_2 \, dr} - c_2^- e^{-\int_R^{R_c} k_2 \, dr} \sim c_3 \cos \delta$$

(since $k_2(-x) = k_3(x)$ at $r \sim R_c$). Inside the nucleus we may consider two separate cases. In the first case the charged particle, if it ever gets through the barrier, has a negligible probability of ever emerging after diving into the nuclear surface—it ejects a neutron or something. Then inside the barrier there will only be an incoming wave and

$$c_1^+ = 0 \qquad \phi_1 \sim \frac{c_1^-}{k_1^{1/2}} e^{-ik_1 r} \tag{4.13.25}$$

so that on matching at $r = R$

$$\frac{c_1^-}{k_1^{1/2}} e^{-ik_1 R} = \frac{c_2^+}{k_2(R)^{1/2}} + \frac{c_2^-}{k_2(R)^{1/2}},$$

$$-ik_1^{1/2} c_1^- e^{-ik_1 R} = k_2(R)^{1/2} c_2^+ - k_2(R)^{1/2} c_2^-$$

or

$$c_2^+ = \tfrac{1}{2} c_1^- e^{-ik_1 R} \left\{ \frac{k_2(R)^{1/2}}{k_1^{1/2}} - \frac{ik_1^{1/2}}{k_2(R)^{1/2}} \right\}$$

and if $k_2(R) \gg k_1$ then

$$c_2^+ \sim \tfrac{1}{2} c_1^- e^{-ik_1 R} \left\{ \frac{k_2(R)}{k_1} \right\}^{1/2}. \tag{4.13.26}$$

Now

$$|c_1^-| \simeq |c_2^+|$$

and so at $r = R_c$ we may neglect the term

$$c_2^- e^{-\int_R^{R_c} k_2(r) \, dr}$$

and so obtain $\tan \delta = 1$ and

$$c_2^+ \simeq \frac{c_3}{\sqrt{2}} e^{-\int_R^{R_c} k_2(r) \, dr}$$

whence

$$c_1^- \simeq \sqrt{2} \, c_3 e^{-\int_R^{R_c} k_2(r) \, dr} e^{-ik_1 R} \left[\frac{k_1}{k_2(R)}\right]^{1/2} \tag{4.13.27}$$

Nuclear Reactions

and in this case the coulomb barrier hinders a charged particle getting into the nucleus by roughly the same factor as it prevents a charged particle established inside getting out.

Pictorially we have for this case

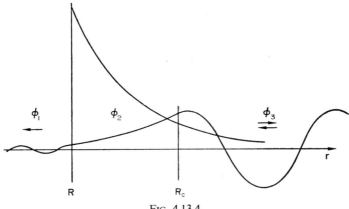

FIG. 4.13.4.

Finally, we consider the case in which the charged particle, having got into the nucleus, eventually comes out again. In this case, we write

$$\phi_1 = \frac{c_1}{k_2^{1/2}} \sin(k_1 r + \eta),$$

$$\phi_2 = \frac{c_2^+}{k_2^{1/2}} e^{\int_R^r k_2(r)\,dr} + \frac{c_2^-}{k_2^{1/2}} e^{-\int_R^r k_2(r)\,dr}, \qquad (4.13.28)$$

$$\phi_3 = \frac{c_3}{k_3^{1/2}} \sin(k_3 r + \delta).$$

This case is more complicated and because of the factorization property is the one which contains resonant scattering. Matching the solutions gives us

$$\frac{c_1}{k_1^{1/2}} \sin(k_1 R + \eta) = \frac{c_2^+}{k_2(R)^{1/2}} + \frac{c_2^-}{k_2(R)^{1/2}},$$

$$k_1^{1/2} c_1 \cos(k_1 R + \eta) = k_2(R^{1/2}) c_2^+ - k_2(R)^{1/2} c_2^-,$$

$$c_2^+ e^{\int_R^{R_c} k_2(r)\,dr} + c_2^- e^{-\int_R^{R_c} k_2(r)\,dr} = c_3 \sin \delta, \qquad (4.13.29)$$

$$c_2^+ e^{\int_R^{R_c} k_2(r)\,dr} - c_2^- e^{-\int_R^{R_c} k_2(r)\,dr} = c_3 \cos \delta$$

and to estimate nuclear reaction rates we are again interested in c_1 in terms of c_3, for we would normalize ϕ_3.

$$2c_2^+ = c_3(\sin \delta + \cos \delta)\, e^{-\int_R^{R_c} k_2(r)\,dr}, \qquad (4.13.30)$$

$$2c_2^- = c_3(\sin \delta - \cos \delta)\, e^{\int_R^{R_c} k_2(r)\,dr}.$$

Nuclear Physics

For the case where $\sin \delta = -\cos \delta$, $c_2^+ = 0$ and a small amplitude outside corresponds to a large one inside the nucleus. This condition gives rise to resonance, but it is very rare. Eliminating $c_2^\pm$ gives us

$$\frac{c_1}{k_1^{1/2}} \sin(k_1 R + \eta) = \frac{c_3}{k_2(R)^{1/2}} \left\{ (\sin \delta + \cos \delta) e^{-\int_R^{R_c} k_2(r)dr} + (\sin \delta - \cos \delta) e^{\int_R^{R_c} k_2(r)dr} \right\},$$

$$k_1^{1/2} c_1 \cos(k_1 R + \eta) = c_3 k_2(R)^{1/2} \left\{ (\sin \delta + \cos \delta) e^{-\int_R^{R_c} k_2(r)dr} - (\sin \delta - \cos \delta) e^{\int_R^{R_c} k_2(r)dr} \right\}.$$

To simplify the notation let $s = \sin \delta$,
$$c = \cos \delta,$$
$$\xi = \int_R^{R_c} k_2(r)dr$$

when

$$\frac{k_2(R)}{k_1} \tan(k_1 R + \eta) = \frac{(s+c)e^{-\xi} + (s-c)e^{\xi}}{(s+c)e^{-\xi} - (s-c)e^{\xi}}; \qquad (4.13.31)$$

writing

$$K = \frac{k_2(R)}{k_1} \tan(k_1 R + \eta)$$

we have

$$K = \frac{(s+c)e^{-\xi} + (s-c)e^{\xi}}{(s+c)e^{-\xi} - (s-c)e^{\xi}}$$

where K is controlled primarily by the details of nuclear structure through the parameter η. Then

$$\frac{s-c}{s+c} = -e^{-2\xi} \frac{1-K}{1+K}$$

and so $s-c \approx 0$ unless K is very close to -1.

Then for
$$s - c \simeq 0$$
$$c_1 \sim c_3 e^{-\xi}$$

and we have

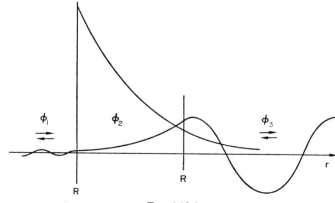

Fig. 4.13.5.

Nuclear Reactions

If

$$\frac{s-c}{s+c} = B \quad \text{then} \quad s+c = \frac{\sqrt{2}}{\sqrt{1+B^2}}, \quad s-c = \frac{\sqrt{2}B}{\sqrt{1+B^2}}$$

so that

$$\begin{aligned}
\frac{c_1}{k_1^{1/2}} \sin(k_1 R + \eta) &= c_3 \frac{\sqrt{2}}{k_2(R)^{1/2}} \left\{ \frac{2K}{1+K} \right\} \frac{e^{-\xi}}{\sqrt{1+B^2}} \\
&= c_3 \frac{\sqrt{2}}{k_2(R)^{1/2}} \frac{2Ke^{-\xi}}{[(1+K)^2 + e^{-4\xi}(1-K)^2]^{1/2}}
\end{aligned} \quad (4.13.32)$$

which is indeed $\sim e^{-\xi}$ unless $1+K \approx 0$ when $K = -1$ and

$$\frac{c_1}{k_1^{1/2}} \sin(k_1 R + \eta) = \frac{\sqrt{2} c_3}{k_2(R)^{1/2}} e^{\xi} \quad (4.13.33)$$

and a very large amplitude inside results—this is resonance. The condition for resonance is $K = -1$ or $\tan(k_1 R + \eta) = -k_1/k_2(R)$. Thus in this case we have

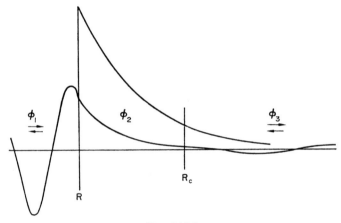

FIG. 4.13.6.

and on cutting off the incident wave we are left with our prepared state decaying through the coulomb barrier.

If we now set $K+1 = \Delta K$ and choose ΔK such that c_1 has dropped to $1/\sqrt{2}$ of the resonant value we have

$$\frac{1}{2\sqrt{2}} e^{\xi} = \frac{e^{-\xi}}{[(\Delta K)^2 + 4e^{-4\xi}]^{1/2}} \quad (4.13.34)$$

or $\Delta K = 2e^{-2\xi}$.

Now

$$\Delta K = \frac{\partial K}{\partial E} \bigg|_{K=-1} \Delta E$$

and so if the energy is shifted from the peak value by an amount

$$\frac{\Gamma}{2} = 2e^{-2\xi} \left(\frac{\partial K}{\partial E} \right)^{-1} \quad (4.13.35)$$

Nuclear Physics

the interior amplitude drops by $\sqrt{2}$. Thus the width of a resonance observed in charged particle scattering contains a factor

$$e^{-2\xi} = e^{-G}$$

which ties in very nicely with this same factor that we found in the decay rate of an already prepared state. The quantity $\partial K/\partial E$ is of course controlled by the detailed nuclear structure.

The resonances corresponding to $K = -1$ are thus very narrow because of the coulomb barrier (and would not exist in its absence). Except in the resonant peak

$$\frac{c_1}{k_1^{1/2}} \sin(k_1 R + \eta) \approx \frac{c_3 \sqrt{2}}{k_2(R)^{1/2}} \left(\frac{2K}{1+K}\right) e^{-\xi}$$

to be compared with

$$\frac{|c_1^-|}{k_1^{1/2}} e^{-ik_1 R} \approx \frac{c_3 \sqrt{2}}{k_2(R)^{1/2}} e^{-\xi}$$

for the case of total absorption of an incident charged particle that has penetrated the coulomb barrier.

The rate of formation of a compound nucleus is determined by the amplitude $c_1/k_1^{1/2}$ of the interior wave function, and it is the external amplitude $c_3/k_3^{1/2}$ that is normalized. Thus an inhibiting factor proportional to $e^{-2\xi}$ enters the cross-section for production of a compound nucleus through a charged channel and enters the decay of a compound nucleus through a charged channel, a result we quoted in Section 4.4.

If there were no coulomb barrier then we would eliminate the intermediate solution ϕ_2 and match at $r = R$ the functions

$$\phi_1 = \frac{N_1}{k_1^{1/2}} \sin(k_1 r + \eta), \qquad (4.13.36)$$

$$\phi_3 = \frac{N_3}{k_3(\infty)^{1/2}} \sin(k_3(\infty) r + \delta_N) \qquad (4.13.37)$$

for those cases when the incident charged particle has a high probability of re-emerging, whence

$$\frac{N_1}{k_1^{1/2}} \sin(k_1 R + \eta) = \frac{N_3}{k_3(\infty)^{1/2}} \sin(k_3(\infty) R + \delta_N)$$

with

$$\frac{1}{k_1} \tan(k_1 R + \eta) = \frac{1}{k_3(\infty)} \tan(k_3(\infty) R + \delta_N). \qquad (4.13.38)$$

Writing

$$K_N = \frac{k_3(\infty)}{k_1} \tan(k_1 R + \eta)$$

gives

$$\frac{N_1}{k_1^{1/2}} \sin(k_1 R + \eta) = \frac{N_3}{k_3(\infty)^{1/2}} \frac{K_N}{\sqrt{1+K_N^2}}.$$

Thus the amplitude on the inside of the coulomb barrier $A_{1c} = c_1/k_1^{1/2}$ is related to the amplitude in the absence of a coulomb barrier $A_{1N} = N_1/k_1^{1/2}$ by

$$\frac{A_{1c}}{A_{1N}} \sim \frac{k_3^{1/2}(\infty)}{k_2(R)^{1/2}} e^{-\xi} \frac{K}{K_N} \left\{ \frac{1+K_N^2}{(1+K)^2 + e^{-4\xi}(1-K)^2} \right\}^{1/2} \quad (4.13.39)$$

(where the normalization conditions at $r \to \infty$ set $c_3 = N_3$).

Now $\tan(k_1 R + \eta)$ is controlled by the nuclear structure and in our approximation will not depend on the presence or absence of a barrier. Then

$$K = \frac{k_2(R)}{k_3(\infty)} K_N$$

and

$$\frac{A_{1c}}{A_{1N}} \approx \left\{ \frac{k_2(R)}{k_3(\infty)} \right\}^{1/2} e^{-\xi} \left\{ \frac{1+K_N^2}{(1+K)^2 + e^{-4\xi}(1-K)^2} \right\}^{1/2}. \quad (4.13.40)$$

At low energy $K_N \ll K$ and provided that $\tan(k_1 R + \eta)$ is slowly varying then

$$\frac{A_{1c}}{A_{1N}} \approx \left\{ \frac{k_2(R)}{k_3(\infty)} \right\}^{1/2} e^{-\xi} \frac{1}{[(1+K)^2 + e^{-4\xi}(1-K)^2]^{1/2}}.$$

Thus provided that we are not in the middle of a resonant peak the cross-section for the formation of a compound nucleus through a coulomb barrier may be written

$$\sigma_c \approx \frac{k_2(R)}{k_3(\infty)} e^{-2\xi} \sigma_N$$

$$= \left(\frac{E_B - E}{E} \right)^{1/2} e^{-2\xi} \sigma_N, \quad (4.13.41)$$

$$E \ll E_B = Zze^2/R.$$

This expression is of great importance in dealing with thermonuclear reactions. These reactions are exothermic and in consequence would follow approximately a $1/v$ law in the absence of the coulomb barrier. The result (4.13.41) allows us to write for such reactions the standard parametrization

$$\sigma = \frac{S(E)}{E} e^{-2\xi} \quad (4.13.42)$$

where $S(E)$ will be a slowly varying function of E except in a resonant peak.

As far as our treatment of the coulomb barrier goes, it now remains only to evaluate the integral

$$\left. \begin{array}{l} G = 2\xi = \displaystyle\int_R^{R_c} k_2(r)\, dr, \\[2mm] \xi = \displaystyle\int_R^{R_c} k_2(r)\, dr = \displaystyle\int_R^{R_c} \sqrt{\frac{2m}{\hbar^2}\left(\frac{Zze^2}{r} - E\right)}\, dr. \end{array} \right\} \quad (4.13.43)$$

Nuclear Physics

Set

$$\frac{rE}{Zze^2} = \cos^2\theta,$$

$$dr = \frac{Zze^2}{E} 2\cos\theta\, d\cos\theta,$$

$$\sqrt{\frac{Zze^2}{r} - E} = \sqrt{E}\sqrt{\frac{1}{\cos\theta} - 1} = \frac{\sqrt{E}}{\cos\theta}\sqrt{1-\cos^2\theta}.$$

So

$$\int k_2(r)dr = \sqrt{\frac{2m}{\hbar^2}}\int\sqrt{\frac{Zze^2}{r} - E}\, dr = -\sqrt{\frac{2m}{\hbar^2}}\int\frac{Zze^2}{\sqrt{E}} 2\sin^2\theta\, d\theta,$$

$$2\sin^2\theta = 1 - \cos 2\theta,$$

so

$$\int_R^{R_c} k_2(r)\, dr = Zze^2\sqrt{\frac{2m}{\hbar^2 E}}\int_{\theta_c}^{\theta_R}[1-\cos 2\theta]\, d\theta$$

where

$$\theta_c = \cos^{-1}\sqrt{\frac{R_c E}{Zze^2}} \qquad \theta_R = \cos^{-1}\sqrt{\frac{RE}{Zze^2}},$$

$$\int_\theta^\theta [1-\cos 2\theta]\, d\theta = [\theta - \theta_c] - \tfrac{1}{2}[\sin 2\theta - \sin 2\theta_c].$$

Now

$$R_c = \frac{Zze^2}{E} \quad\text{so}\quad \theta_c = 0 \quad\text{and}$$

$$\xi = \int_R^{R_c} k_2(r)\, dr = Zze^2\sqrt{\frac{2m}{\hbar E}}\left\{\cos^{-1}\sqrt{\frac{RE}{Zze^2}} - \sqrt{\frac{RE}{Zze^2}}\sqrt{1-\frac{RE}{Zze^2}}\right\}. \quad (4.13.44)$$

The exponential term is thus

$$\begin{aligned}e^{-G} = e^{-2\xi} &= e^{-2Zze^2\sqrt{\frac{2m}{\hbar^2 E}}\left\{\cos^{-1}\sqrt{\frac{RE}{Zze^2}} - \sqrt{\frac{RE}{Zze^2}}\sqrt{1-\frac{RE}{Zze^2}}\right\}} \\ &= e^{-\frac{4Zze^2}{\hbar v}\left\{\cos^{-1}\sqrt{\frac{RE}{Zze^2}} - \sqrt{\frac{RE}{Zze^2}} - \sqrt{1-\frac{RE}{Zze^2}}\right\}}.\end{aligned} \quad (4.13.45)$$

This is the Gamow Factor and is encountered in all problems involving either absorption or emission of a charged particle from a nucleus. The argument of the exponential may be expanded to give the dependence on the parameter R to any order desired, and it is interesting to note that for a very high potential barrier

$$\frac{Zze^2}{R} \gg E \quad \cos\theta_R \sim 0 \quad \theta_R \sim \frac{\pi}{2}$$

286

Nuclear Reactions

so that the Gamow penetration factor becomes

$$e^{-2\pi Zze^2/\hbar v} \tag{4.13.46}$$

where v is the velocity of the particle at infinity and is independent of R.

4.14. Examples of direct reactions: peripheral scattering processes

In Section 4.12 we have considered an extreme set of assumptions most appropriate to low-energy nuclear reactions. We considered the incident particle entering the target nucleus and forming an intermediate compound nucleus, the principal characteristic of which is that the breakup of a specified excited compound nucleus is a function only of the excitation energy and is independent of the mode of formation: the compound nucleus has no memory of its genesis. A compound nucleus reaction $a+A \rightarrow B+b+c\ldots$ may be represented diagrammatically as

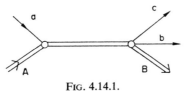

Fig. 4.14.1.

Even if an excited state of well-defined angular momentum is not produced, the presence of a large number of outgoing channels (which makes the width parameter Γ very large, the quantity Re $\partial f/\partial E$ very small) may still generate this feature.

As the energy goes up and the wavelength of the incident particle decreases, reactions take place more and more with individual nucleons or small clusters of nucleons. If the collision is central, the incident particle ploughs through the sea of nucleons, and if the energy is not too large, a compound nucleus treatment will still work because of the effect of secondary collisions. In this section we consider a fairly well-defined class of direct reactions in which the reaction does not take place through a well-defined angular momentum state, and in which we cannot suppose the number of outgoing channels to be large. These are reactions which in the tens of MeV region involve preferentially a relatively loosely bound nucleon which because of the loose binding has a wave function relatively large near the nuclear surface: we call these peripheral reactions. An example is the knock-on $(p, 2p)$ reaction. In a central collision, the incident proton forms a compound nucleus which decays through the emission of two protons (among its other modes). In a peripheral interaction a loosely bound proton in the target is struck by the incident proton and knocked out of the target nucleus, which is nearly a spectator of what is almost free proton–proton scattering.

We will discuss several examples of peripheral reactions and the model we assume is that the nucleus consists of an inert core plus one active peripheral nucleon with which the interaction takes place. We will also assume that the core is very heavy and so can take up momentum without recoiling away. The calculations of this section are our *seventh application of the Fermi Golden Rule*.

Nuclear Physics

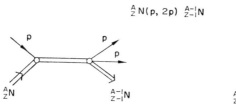
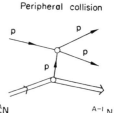

FIG. 4.14.2.

If we further neglect the interaction of the core with a free nucleon, we may write the matrix element for a reaction like $(p, 2p)$ in the following way.

Let the wave function of the incident nucleon be $\psi_i(\mathbf{r}_1)$ and the outgoing wave function $\psi_f(\mathbf{r}_1)$. The other nucleon is initially bound in the nucleus with a wave function $\psi_i(\mathbf{r}_2)$ and has a final wave function $\psi_f(\mathbf{r}_2)$. The core wave function we represent by χ and the product of volume elements associated with it by d^3r_χ. Then if the core wave function remains unchanged we have for the matrix element

$$M = \int \psi_f^*(\mathbf{r}_1)\, \psi_f^*(\mathbf{r}_2)\, \chi^* V(\mathbf{r}_1 - \mathbf{r}_2)\, \psi_i(\mathbf{r}_1)\, \psi_i(\mathbf{r}_2)\, \chi\, d^3r_1\, d^3r_2\, d^3r_\chi$$

and with

$$\left. \begin{array}{c} \int \chi^*\chi\, d^3r_\chi = 1, \\ M = \int \psi_f^*(\mathbf{r}_1)\, \psi_f^*(\mathbf{r}_2)\, V(\mathbf{r}_1 - \mathbf{r}_2)\, \psi_i(\mathbf{r}_1)\, \psi_i(\mathbf{r}_2)\, d^3r_1\, d^3r_2 \end{array} \right\} \quad (4.14.1)$$

where V is the potential acting between the incident nucleon and the loosely bound peripheral nucleon. Injection of this matrix element into the expression for the cross-section (4.4.9) gives the cross-section for that part of the process which goes by this peripheral mechanism (in general there will be more complicated mechanisms simultaneously in operation and the different amplitudes can interfere). This matrix element is only to first order in V, but illustrates the principles and and can easily be generalized.

It is now convenient to perform a change of variables and write

$$\mathbf{r} = \mathbf{r}_1 - \mathbf{r}_2.$$

Then using plane waves for nucleon 1, the matrix element

$$\int e^{-i\mathbf{k}_1'\cdot\mathbf{r}_1}\, \psi_f^*(\mathbf{r}_2)\, V(\mathbf{r}_1-\mathbf{r}_2)\, e^{i\mathbf{k}_1\cdot\mathbf{r}_1}\, \psi_i(\mathbf{r}_2)\, d^3r_1\, d^3r_2$$

becomes

$$\left. \begin{array}{c} \int e^{-i(\mathbf{k}_1'-\mathbf{k}_1)\cdot\mathbf{r}}\, V(\mathbf{r})\, d^3r \int e^{-i(\mathbf{k}_1'-\mathbf{k}_1)\cdot\mathbf{r}_2}\, \psi_f^*(\mathbf{r}_2)\, \psi_i(\mathbf{r}_2)\, d^3r_2 \\ = \left\{ \int e^{-i\mathbf{q}\cdot\mathbf{r}}\, V(\mathbf{r})\, d^3r \right\} \left\{ \int e^{-i\mathbf{q}\cdot\mathbf{r}_2}\, \psi_f^*(\mathbf{r}_2)\, \psi_i(\mathbf{r}_2)\, d^3r_2 \right\} \end{array} \right\} \quad (4.14.2)$$

where $\mathbf{k}_1$ and $\mathbf{k}_1'$ are the propagation vectors of nucleon 1 before and after scattering, and $\mathbf{k}_1' - \mathbf{k}_1 = \mathbf{q}$, the momentum transfer to the incident nucleon 1.

The first integral

$$\int e^{-i\mathbf{q}\cdot\mathbf{r}}\, V(\mathbf{r})\, d^3r$$

Nuclear Reactions

with $\mathbf{r} = \mathbf{r}_1 - \mathbf{r}_2$ is nothing other than the matrix element for free elastic scattering of nucleon 1 on nucleon 2. The second integral contains the modifications due to nucleon 2 being bound in a nucleus, and contains three cases all of which are of interest.

(i) If $\psi_f(\mathbf{r}_2) = \psi_i(\mathbf{r}_2)$ then the matrix element describes the contribution to elastic scattering of the incident nucleon scattering only on the peripheral nucleon. The second integral is then

$$\int e^{-i\mathbf{q}\cdot\mathbf{r}_2} \varrho(\mathbf{r}_2)\, d^3r_2$$

where $\varrho(\mathbf{r}_2)$ is the probability density of nucleon 2 in the struck nucleus. This term is thus precisely analogous to the electromagnetic form factor we studied first in Chapter 1. This particular peripheral process is, however, of little interest, for it will interfere with and be swamped by all the other contributions to elastic scattering. The diagram for this process is

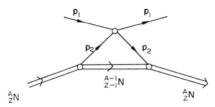

Fig. 4.14.3.

This process should show up if at the upper vertex a proton was swapped with a neutron and the nucleus made a transition between two members of an isospin doublet. It is clearly only important for small q^2; the form factor

$$\int e^{-i\mathbf{q}\cdot\mathbf{r}_2} \varrho(\mathbf{r}_2)\, d^3r_2$$

gives the probability that the nucleus will remain intact when a peripheral nucleon is struck, and this falls off rapidly as the momentum transfer q increases.

(ii) If $\psi_f(\mathbf{r}_2)$ represents a bound nucleon, but in a different state from $\psi_i(\mathbf{r}_2)$ then the matrix element (4.14.2) represents the contribution to inelastic scattering due to exciting nucleon 2 from one single particle level to a second. This process may be expected to be fairly clean if we select scattered nucleons of the right energy, i.e. if we sweep through the energy band of the inelastically scattered nucleons to find the peaks in the cross-section as a function of incident energy which correspond to exciting a state of definite quantum numbers and then sit on the peak. The diagram is

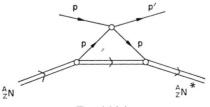

Fig. 4.14.4.

Nuclear Physics

Clearly the principle is the same whether nucleon 1 is a proton or a neutron, whether nucleon 2 is a proton or a neutron or if we have charge exchange scattering at the upper vertex, so that in the final nucleus we have swapped a proton for a neutron:

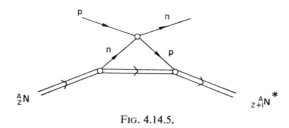

FIG. 4.14.5.

Let us suppose that the nucleus makes a transition between two spherically symmetric states which in the peripheral region tail off exponentially

$$\psi_f(\mathbf{r}_2) \sim \frac{e^{-\alpha_f r_2}}{r_2},$$

$$\psi_i(\mathbf{r}_2) \sim \frac{e^{-\alpha_i r_2}}{r_2}.$$

Then

$$\int e^{-i\mathbf{q}\cdot\mathbf{r}_2} \psi_f^*(\mathbf{r}_2) \psi_i(\mathbf{r}_2) d^3 r_2$$
$$= \int e^{-i\mathbf{q}\cdot\mathbf{r}_2} e^{-\alpha_f r_2} e^{-\alpha_i r_2} d\cos\theta \, d\phi \, dr_2$$

and we may further suppose that the main contribution will come from a radius $r_2 > R$: compound nucleus effects predominating where $r_2 < R$. The nuclear structure integral is then

$$\int_R^\infty e^{-i\mathbf{q}\cdot\mathbf{r}_2} e^{-\alpha_f r_2} e^{-\alpha_i r_2} d\cos\theta \, d\phi \, dr_2. \tag{4.14.3}$$

Measuring the angle θ from the vector $\mathbf{q}$ this becomes

$$2\pi \int_R^\infty e^{-iqr_2\cos\theta} e^{-\alpha_f r_2} e^{-\alpha_i r_2} d\cos\theta \, dr_2$$

$$= -4\pi \int_R^\infty \frac{\sin qr_2}{qr_2} e^{-\alpha_f r_2} e^{-\alpha_i r_2} dr_2$$

and this is, of course, the s-wave radial part of the term $e^{i\mathbf{q}\cdot\mathbf{r}_2}$ folded into the overlap of the nuclear initial and final state.

This integral

$$\int_R^\infty \frac{\sin qr_2}{qr_2} e^{-\alpha r_2} dr_2 \quad (\alpha = \alpha_i + \alpha_f) \tag{4.14.4}$$

Nuclear Reactions

is very nasty—and if we put in wave functions involving transitions to different angular momentum states the equivalent integral becomes even nastier. We can get some feeling for the form, however, by making some approximations. The most drastic approximation we can make is to suppose that α is sufficiently large to result in almost all the contribution to the integral coming from $r_2 \sim R$ in which case

$$\int_R^\infty \frac{\sin qr_2}{qr_2} e^{-\alpha r_2} \, dr_2 \approx \frac{\sin qR}{qR} e^{-\alpha R} \qquad (4.14.5)$$

and the inelastic scattering cross-section is proportional to

$$\frac{\sin^2 qR}{(qR)^2}$$

a function very familiar from physical optics. This function has a principal maximum at $q = 0$ corresponding to forward inelastic scattering. As another approximation, if we suppose that the variation of $\sin qr_2$ and $e^{-\alpha r_2}$ may not be treated so cavalierly, but that the denominator $1/qr_2$ may be approximated by $1/qR$ then we get

$$\int_R^\infty \frac{\sin qr_2}{qr_2} e^{-\alpha r_2} \, dr_2 \approx \frac{1}{qR} \int_R^\infty \sin qr_2 \, e^{-\alpha r_2} \, dr_2$$

$$\approx \frac{1}{qR} \left\{ \frac{q \cos qR + \alpha \sin qR}{[q^2 + \alpha^2]} \right\} \qquad (4.14.6)$$

and this again has a principal maximum at $q = 0$.

Now consider what happens if instead of a transition between two s-states in the nucleus we have a transition between an s-state and a p-state. This introduces a term $\cos \theta$ into the angular part of the integral and the matrix element is now

$$\int_R^\infty e^{i\mathbf{q} \cdot \mathbf{r}_2} e^{-\alpha r_2} |\cos \theta \, d\cos \theta \, d\phi \, dr_2 \approx \frac{\sin qR}{(qR)^2} - \frac{\cos qR}{qR} \qquad (4.14.7)$$

and this function, familiar to us from partial wave analysis, has a zero at $q = 0$ and rises away from $q = 0$ as $\sim qR$. It is again an oscillatory function, but the first maximum does not come at $q = 0$.

It is clear from this very simplified treatment that the angular distribution of nucleons inelastically scattered via the peripheral process (and regardless of whether the scattering involves charge exchange or not) is an oscillatory function of q for which the positions of the maxima are dependent on the change of angular momentum of the nucleus, and the differential cross-section is also to a lesser extent a measure of the radial distribution of the target nucleon.

(iii) The third case is found when the function $\psi_f(\mathbf{r}_2)$ represents a free nucleon that has been knocked out of the nucleus: the original peripheral process that we considered. If we

Nuclear Physics

approximate $\psi_f(\mathbf{r}_2)$ by a plane wave

$$\psi_f(\mathbf{r}_2) = e^{i\mathbf{k}_2 \cdot \mathbf{r}_2}$$

and the matrix element becomes

$$\left\{ \int e^{-i\mathbf{q} \cdot \mathbf{r}} V(r) \, d^3r \right\} \left\{ e^{-i(\mathbf{q}+\mathbf{k}_2) \cdot \mathbf{r}_2} \psi_i(\mathbf{r}_2) \, d^3r_2 \right\}. \tag{4.14.8}$$

The subsequent treatment of this matrix element is almost identical to the previous case for inelastic scattering. The study of such reactions is however a more direct measure of the properties of the single-particle level ψ_i. If for a particular level

$$\psi_i(\mathbf{r}_2) = Y_l^m(\theta, \phi) f(r_2)$$

then measuring θ from the vector $(\mathbf{q}+\mathbf{k}_2)$ the integral over θ projects out that part of $e^{-i(\mathbf{q}+\mathbf{k}_2) \cdot \mathbf{r}_2}$ with angular momentum l and $m = 0$, and the differential cross-section expressed in terms of the variable $(\mathbf{q}+\mathbf{k}_2)$ will be oscillatory, the position of the maxima being determined by the single quantity l rather than by two angular momenta l_i and l_f; as is the cross-section for inelastic scattering in terms of the variable q (in our examples we only considered $l_i = 0$, a particularly simple case). While it is this factor that makes peripheral knock-on processes a particularly useful probe of single particle states, experimentally the study of such reactions is more difficult, for the correlation between both nucleons in the final state must be measured.

4.15. Examples of direct reactions: pickup and stripping reactions

The processes discussed above were essentially the scattering of one nucleon on another loosely bound in the nucleus, and differences come in according to what happened to the target nucleon after the scattering. The deuteron pickup reaction (p, d) is a similar peripheral process in which instead of the incident proton bouncing off the target nucleus, it instead picks it up and the two go off together as a deuteron

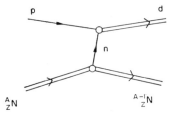

FIG. 4.15.1.

The angular distribution of the outgoing deuteron depends on the angular momentum of the single-particle level from which the neutron was snatched, and so the pickup reaction, and its inverse, the stripping reaction, are again sensitive probes of the higher single-particle states in nuclei. In the crude approximation scheme we have been using, the matrix element for the pickup process is easily written down. The incident proton has a wave function $e^{i\mathbf{k}_1 \cdot \mathbf{r}_1}$.

Nuclear Reactions

The neutron bound in the nucleus has a wave function $\psi_i(\mathbf{r}_2)$. The core wave functions we again suppress, as we neglect the p-core and d-core interactions. The outgoing deuteron contains both nucleons and so its wave function contains both $\mathbf{r}_1$ and $\mathbf{r}_2$ as variables; we write the wave function of the outgoing deuteron as

$$\psi_d(\mathbf{r}_1-\mathbf{r}_2)\, e^{i\mathbf{k}_d\cdot\{(\mathbf{r}_1+\mathbf{r}_2)/2\}}$$

where ψ_d is the internal wave function of the deuteron and the second factor $e^{i\mathbf{k}_d\cdot\{(\mathbf{r}_1+\mathbf{r}_2)/2\}}$ describes the motion of the centre of mass of the two nucleons, that is, of the deuteron. We take the interaction generating the deuteron as $V\delta(\mathbf{r}_1-\mathbf{r}_2)$ as an approximation, and only let it act if $(\mathbf{r}_1+\mathbf{r}_2)/2$ is greater than some cutoff R, for if the deuteron is formed within the nucleus it will collide with other nucleons, break up, and the process will be more like compound nucleus formation than a peripheral reaction. We then write for the matrix element

$$\begin{aligned} M_{pd} &= \iint e^{-i\mathbf{k}_d\cdot\{(\mathbf{r}_1+\mathbf{r}_2)/2\}}\, \psi_d^*(\mathbf{r}_1-\mathbf{r}_2)\, V\delta(\mathbf{r}_1-\mathbf{r}_2) \\ & \quad \psi_i(\mathbf{r}_2)\, e^{i\mathbf{k}_1\cdot\mathbf{r}_1}\, d^3r_1\, d^3r_2 \\ &= \int_R^\infty e^{-i\mathbf{k}_d\cdot\mathbf{r}_1}\, \psi_d(0)\, V\psi_i(\mathbf{r}_2)\, e^{i\mathbf{k}_1\cdot\mathbf{r}_1}\, d^3r_1 \end{aligned} \qquad (4.15.1)$$

where we take the integral from a lower limit of R to eliminate non-peripheral processes. Writing $\mathbf{k}_d-\mathbf{k}_1 = \mathbf{Q}$ this becomes proportional to

$$\int_\infty^R e^{-i\mathbf{Q}\cdot\mathbf{r}_1}\, \psi_i(\mathbf{r}_1)\, d^3r_1 \qquad (4.15.2)$$

and so the oscillatory properties of the differential scattering cross-section are directly controlled by the angular momentum properties of the initial bound neutron wave function, just as for knock-on reactions. Indeed, the pickup reaction may be regarded as a special case of a knock-on reaction, in which the need to measure the correlation between the two nucleons is eliminated: the formation of a deuteron makes them always come out together.

You should be able to convince yourself by mere inspection of expressions (4.15.1) and (4.15.2) that the matrix element for stripping, in which the deuteron is the incident particle and a neutron is torn from it and absorbed into a peripheral nuclear state, is just the complex conjugate of expression (4.15.2). In stripping reactions it is, of course, the angular distribution of the outgoing solitary nucleon that carries the information about the state to which its partner has been transferred.

Pickup and stripping processes make it possible to get at neutron wave functions without producing beams of neutrons or having to detect scattered neutrons. You cannot get at proton wave functions via deuterium stripping or pickup reactions without handling neutrons somewhere—the $(p, 2p)$ process must be used. However, it is possible to get at proton wave functions without handling neutrons or the correlation problems associated with the $(p, 2p)$ process by stripping a triton (t, d) or allowing a deuteron to pick up a proton (d, t).

In the preceding two sections we have been concerned only to give you a feeling for how direct reactions proceed, and how the relevant properties can be calculated, and what they can be used for. We did not work out anything properly. We made no attempt to put in

Nuclear Physics

realistic radial wave functions for the bound states, and in particular we completely ignored the fact that nucleons have spin and that states containing two like nucleons are antisymmetric under interchange of all variables. In an attempt to apply these peripheral methods to a real experiment, all these omissions must be rectified, and there is yet a further refinement. We have represented incoming and outgoing nuclei with definite momenta by plane waves. In reality, not only will coulomb forces distort the wave functions of charged particles, but the nuclear optical potential will also distort the wave functions of both charged and uncharged strongly interacting particles. If the optical parameters are known from the scattering experiments, the matrix elements can be corrected for this distortion—the distorted wave Born approximation. All these points are for the professionals; we shall pursue this subject no further.

4.16. Examples of direct reactions: neutrino interactions

The interaction of neutrinos with nuclei is another example of a direct reaction. The first neutrino reaction to be observed was the inverse of neutron decay

$$\bar{\nu}+p \rightarrow e^+ +n.$$

The cross-section for this process is easily estimated from the elementary theory of weak interactions, using the Fermi Golden Rule. The transition rate is

$$T = \frac{2\pi}{\hbar} |M|^2 \varrho$$

where ϱ is the density of two particle e^+n final states. At neutrino energies $\lesssim 100$ MeV the flux factor is just the reciprocal of the neutrino velocity and so

$$\sigma = \frac{2\pi}{\hbar c} |M|^2 \varrho.$$

If the four-fermion interaction is pointlike (that is, we ignore nucleon structure effects) then the square of the matrix element is just g^2, where g is the weak coupling constant, $g = 1.41 \times 10^{-49}$ erg cm^3. In the low-energy region the laboratory frame and the centre of mass frame are essentially identical and the two-particle phase space is then

$$\frac{4\pi p_e^2 \, dp_e}{(2\pi\hbar)^3 dE_e} \tag{4.16.1}$$

where p_e is the laboratory electron momentum and $dp_e/dE_e = E_e/p_e c^2$.

$$\sigma = \frac{g^2 p_e E_e}{\pi \hbar^4 c^3}. \tag{4.16.2}$$

Equation (4.16.2) is our *eighth application of the Fermi Golden Rule*.

This is an endothermic reaction, the threshold for which is given approximately by

$$E_\nu = m_e c^2 + (m_n - m_p)c^2 = 1.8 \text{ MeV}$$

so that neglecting the neutron recoil

$$E_e = E_\nu - 1.3 \text{ MeV}, \quad E_\nu > 1.8 \text{ MeV}.$$

For $E_e = 1$ MeV$_e$, $p_e \sim 1$ MeV/c^2 and the cross-section given by eq. (4.16.2) is

$$\sigma(E_e = 1 \text{ MeV}) \sim 2\times 10^{-44} \text{ cm}^2 = 2\times 10^{-20} \text{ barn}.$$

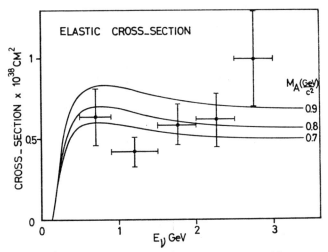

FIG. 4.16.1. Cross-section for the "elastic" reaction $\nu_\mu + n \to p + \mu^-$ measured in a propane bubble chamber. The solid curves correspond to the theoretical prediction for various form factor parameters. [From an article by D. H. Perkins in *Proc. Topical Conf. on Weak Interactions*, CERN 69-7 (1969).]

This is to be compared with cross-sections for strong interaction processes which are typically $\sim$ barns in this energy region. The content of this result is strikingly indicated by calculating the interaction length λ_ν—that is, the distance in which a beam of neutrinos is attenuated by a factor e

$$\lambda_\nu = \frac{1}{N\sigma_\nu}$$

where N is the number of protons per cm^3 of absorber: if we take water as an absorber with two target protons per molecule, then $N \approx 10^{23}$ and

$$\lambda_\nu \sim 10^{21} \text{ cm} \simeq 1000 \text{ light years}.$$

Alternatively, the diameter of the earth is $\sim 10^9$ cm and so a beam of $\sim$ MeV neutrinos passing through the earth along a diameter will only be attenuated by $\approx$ one part in 10^{12}.

More realistically, if a neutrino flux f is incident upon a tank of water, a 1-metre cube, the number of reactions per second is $\sim 10^{-15} f$ and to achieve a reaction rate of one per second in such a tank requires a flux $\approx 10^{15}$ neutrinos per second per cm^2. A flux of monoenergetic neutrinos of this sort of intensity has never been achieved, but a nuclear reactor is an intense source of antineutrinos. The reason is that the fragments produced in nuclear fission are neutron rich and achieve stability by conversion of the excess neutrons into

Nuclear Physics

protons through the process

$$n \to p + e^- + \bar{\nu}.$$

The resulting antineutrino energy spectrum is a superposition of the spectra from the decay of the different fragments formed in fission, peaks in the region of 1 MeV and then drops exponentially

$$N(E)_{\bar{\nu}} dE_{\bar{\nu}} \approx e^{-0\cdot 6 E_{\bar{\nu}}} dE_{\bar{\nu}}$$

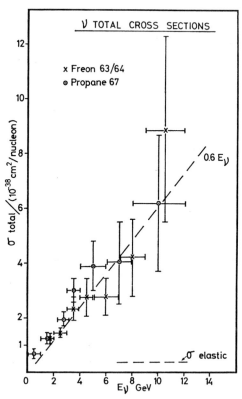

FIG. 4.16.2. Total cross-section as a function of neutrino energy, taken from CERN heavy liquid bubble chamber data. [From an article by D. H. Perkins in *Proc. Topical Conf. on Weak Interactions*, CERN 69-7 (1969).]

where $E_{\bar{\nu}}$ is in MeV. The average value of the cross-section is thus

$$\langle \sigma \rangle \approx \frac{g^2}{\pi \hbar^4 c^4} \langle E_e^2 \rangle$$

where

$$\langle E_e^2 \rangle = \frac{\int_{1\cdot 8}^{\infty} (E_{\bar{\nu}} - 1.3)^2 \, e^{-0\cdot 6 E_{\bar{\nu}}} \, dE_{\bar{\nu}}}{\int_{1\cdot 8}^{\infty} e^{-0\cdot 6 E_{\bar{\nu}}} \, dE_{\bar{\nu}}} = 7.5 \text{ MeV}^2.$$

This will be a slight overestimate because below $E_e \sim 1$ MeV the approximation $p_e c \sim E_e$ becomes inadequate. The experiment of Reines and Cowan, the first in which neutrino

Nuclear Reactions

(strictly speaking, antineutrino) interactions were detected, was carried out with just such a set-up, and the calculated cross-section $\langle\sigma\rangle$ was $\sim 6\times10^{-20}$ barns, 6×10^{-44} cm². In this first experiment[†] the detector was a 10-ft³ tank of liquid scintillator, loaded with cadmium. After the interaction the neutrons diffused through the liquid until captured by the $^{113}Cd(n, \gamma)^{114}Cd$ reaction and the γ produced a pulse of light from the scintillator. The tank was surrounded by counters to catch the two γ from the annihilation of the positron and the combination of signals provided a sufficiently characteristic signature for the reaction $\bar{\nu}+p \rightarrow n+e^+$ for detection in a flux of $\sim 10^{13}$ $\bar{\nu}$ cm^{-2} sec^{-1}. In a later and more elaborate experiment[‡] a counting rate of 36 ± 4/hr was achieved.

The neutrino (as opposed to antineutrino) reaction

$$^{37}Cl+\nu \rightarrow {}^{37}A+e^-$$

is of considerable interest. The ground state of ^{37}A is unstable and decays by electron capture

$$^{37}A+e^- \rightarrow {}^{37}Cl+\nu.$$

Because ^{37}A is the nucleus of a noble gas, ^{37}A atoms may be readily separated from a liquid containing chlorine and the decays individually counted. This possibility has been utilized by Davis and his collaborations in an attempt to detect the neutrino flux from the sun: neutrino astronomy. The particular interest of this study is that neutrinos are created in the energy-producing core of the sun. Because of the enormous mean free path they are in no way moderated in their passage through the outer regions and so constitute a direct measure of nuclear activity in the centre of the sun, with which theoretical models based on much less direct evidence may be compared. The nucleus ^8B is believed to be produced in the course of hydrogen burning in the sun, and is the source of neutrinos with energy ≈ 15 MeV

$$^8B \rightarrow {}^8Be+e^++\nu \quad 14.06 \text{ MeV},$$
$$^8B+e^- \rightarrow {}^8Be+\nu \quad 15.08 \text{ MeV}.$$

The importance of these energetic neutrinos is that they are capable of inducing transitions not only to the ground state of ^{37}A but also the excited states, in particular via a superallowed transition to an excited state at about 5 MeV. Thus a measurement of the rate of production of ^{37}A from ^{37}Cl would yield the ^8B concentration in the centre of the sun, and it is the ^8B concentration that is to be compared with theoretical models.

Davis and his collaborators[§] used a tank of (liquid) C_2Cl_4 containing 520 metric tons of chlorine, 4850 ft underground in a mine in South Dakota (the reason for going deep underground is to cut production of ^{37}A from nuclear reactions involving the cosmic radiation). Any argon formed was swept out of the liquid by purging with helium, concentrated, and allowed to decay in a counter sensitive to the 2.8 keV Auger electrons produced after the electron capture. The measured rate of production of ^{37}A was < 0.5 nuclei per day, to be compared with a rate expected of $\sim$ 2–7 per day. This very sensitive experiment set a limit

---

[†] F. Reines and C. L. Cowan, *Phys. Rev.* **92**, 830 (1953).
[‡] F. Reines and C. L. Cowan, *Phys. Rev.* **113**, 273 (1959); see also the following paper.
[§] R. Davis, D. S. Harmer and K. C. Hoffman, *Phys. Rev. Lett.* **20**, 1205 (1968).

on the flux of neutrinos from ^{37}A decay at the earth of $< 2 \times 10^{16}$ cm^{-2} sec^{-1}. It is well worth your while to read the original paper.

The four-fermion interaction may be a point interaction, and if not is certainly of very much shorter range than the strong interaction processes, and is very weak. We might therefore expect the transition rate from the Golden Rule, calculated with plane wave initial and final states, to work better for high-energy neutrino interactions than for anything else. At high energy it is convenient to calculate the cross-section in the overall centre of mass and in this system at high energies the incident neutrino and the nucleon are travelling with velocity c. The flux factor becomes $1/2c$. The phase space is $\propto E_e^2 \approx (E_{cm}/2)^2$ where E_{cm} is the centre of mass energy, and if $E_{\bar{\nu}}$ is the laboratory energy of the incident neutrino,

$$E_{cm}^2 \simeq 2Mc^2 E_{\bar{\nu}}$$

where M is the mass of the proton. Then

$$\sigma(E_{\bar{\nu}}) = \frac{g^2 Mc^2 E_{\bar{\nu}}}{4\pi \hbar^4 c^4} \qquad (4.16.3)$$

which increases linearly with incident energy. Now if the four-fermion coupling is really a point interaction, the cross-section must be wholly s-wave.

Without making a proper relativistic treatment, we can see that this s-wave nature of the interaction implies

$$\sigma(E_{\bar{\nu}}) \lesssim \frac{\pi \hbar^2 c^2}{E_{\bar{\nu}cm}^2} = \frac{2\pi \hbar^2 c^2}{Mc^2 E_{\bar{\nu}}}$$

(where $E_{\bar{\nu}cm}$ is the neutrino energy in the centre of mass) and so at high energies the cross-section should fall at least linearly with laboratory energy: our calculation shows that it rises linearly. Thus the calculated cross-section for the single process

$$\bar{\nu} + p \rightarrow e^+ + n$$

will violate the unitarity limit at an energy given by

$$\frac{g^2 Mc^2 E_{\bar{\nu}}}{4\pi \hbar^4 c^4} \simeq \frac{2\pi \hbar^2 c^2}{Mc^2 E_{\bar{\nu}}} \qquad (4.16.4)$$

or

$$E_{\bar{\nu}} \simeq 10^3 \text{ erg} \simeq 10^6 \text{ GeV}.$$

Such energies are well beyond present-day experiment and so the resolution of this inconsistency cannot be found from experiment.

The present (1971) generation of proton synchrotrons have made available relatively intense beams of neutrinos, $\sim 10^5$ cm^{-2} sec^{-1}, in the energy range of 1–10 GeV. This has been done by focusing intense beams of pions and then letting them decay:

$$\pi^+ \rightarrow \mu^+ + \nu_\mu.$$

After passing through enough material to stop the muons, the neutrinos alone remain, and although the flux is down by $\sim 10^8$ on reactor fluxes, the cross-section at ~ 1 GeV is $\sim 10^{-38}$ cm^2 which makes experiments feasible.

Nuclear Reactions

Neutrino reactions have been observed in both spark chambers and bubble chambers. It is experimentally observed that there are two kinds of neutrino: a neutral muon as well as a neutral electron, and it is necessary to associate with leptons a further internal quantum number which specifies either muonic or electronic nature. The reaction

$$\nu_\mu + n \to p + \mu^- \tag{4.16.5}$$

is observed: the reaction

$$\nu_\mu + n \to p + e^-$$

is not observed.

The cross-section for the reaction (4.16.5) does not rise linearly as implied by our calculation. The reason is that nucleons do not behave like point fermions but have a spatial structure: manifestations of this are their anomalous magnetic moments and the form factors observed in electron–nucleon scattering. If the density in space of the fermionic element of a nucleon is $\varrho(r)$ then our matrix element for the reaction (4.16.5) is

$$g \int e^{-i\mathbf{k}_\mu \cdot \mathbf{r}} \varrho(r) e^{i\mathbf{k}_\nu \cdot \mathbf{r}} \, d^3r = gF(p^2) \quad \text{where} \quad F(q^2) = \int e^{-i\mathbf{q} \cdot \mathbf{r}} \varrho(r) \, d^3r,$$

$$\mathbf{q} = \mathbf{k}_\nu - \mathbf{k}_\mu.$$

The electromagnetic form factors are approximately represented by

$$F(q^2) = \frac{1}{(1+\alpha^2 q^2)^2} \tag{4.16.6}$$

and if this form also represents the weak interaction form factors then the reaction is damped out for large q^2, that is, large production angles in the centre of mass. The neutrino form factors, although as yet not accurately determined, indeed seem to be compatible with the electromagnetic form factors. The reaction is peaked forward more and more as the neutrino energy increases, and the total cross-section for the "elastic" process

$$\nu_\mu + n \to \mu^- + p$$

flattens off. This result is not specific to the form (4.16.6) but applies to any form factor falling fast enough for $\int F(q^2) dq^2$ to be asymptotically flat. However, the nucleon form factors associated with strong interactions do not flatten off the cross-sections for purely leptonic processes, and a flat cross-section does not necessarily remove the problem of unitarity violation. If $F(q^2) \propto (1+\alpha^2 q^2)^{-1}$ the cross-section is flat, but the contribution of the s-wave has a logarithmic divergence and so eventually violates unitarity. This is the kind of form factor an intermediate boson would contribute.

The simple point interaction model of weak interactions cannot be right and the construction of a weak interaction theory which does not exhibit violations of unitarity is a major problem of weak interaction physics.

CHAPTER 5

Self-sustaining Nuclear Reactions and Nuclear Energy Sources

5.1. Introduction

Our discussion of the physics of the nucleus and in particular nuclear reactions has so far been academic and confined to the laboratory. In this chapter we step outside the laboratory and discuss some situations in which nuclear reactions take place on a grand scale. We shall first consider the physics of the nuclear reactor, of increasing relevance to the ordinary world as the number of fission reactors in operation multiplies. Our second topic will be thermonuclear reactions and thermonuclear reactors—there are no terrestrial thermonuclear reactors yet (1971), but each star in the universe is a thermonuclear reactor, and there are 10^{11} in our galaxy alone, a mere 10^5 light years in diameter. These topics are not only interesting and important in their own right, but also illustrate the principles we discussed in Chapter 4.

If we return to the curve of mean binding energy versus mass number for stable nuclei, Fig. 1.3.1, we note that this curve has a maximum in the region of the iron group of elements, with a rapid rise on the low mass side and a more gradual fall off on the high mass side. In our development of the semiempirical mass formula in Chapter 1, we saw that the rise is associated with the decreasing ratio of nuclear surface area to volume as the number of nucleons increases, while the gradual fall off is due to the increasing coulomb repulsion as the nuclear charge Z grows with the mass number.

If light nuclei can be combined to form a stable nucleus in which the mean binding energy is greater than the mean binding energy of the ingredients, energy is evolved. These are exothermic reactions and it is the evolved energy from such reactions that powers the stars. Apart from effects due to the detailed structure of the binding energy curve, building elements up to iron from lighter components releases energy—the ultimate source of this energy must be sought in the origin of the universe.

If a massive nucleus can be persuaded to fragment into pieces that have a mean binding energy greater than the mean binding energy of the original nucleus, again energy is evolved as the fragments rush apart driven by their mutual coulomb repulsion. The ultimate source of this energy is the process which prepared massive nuclei in the first instance—possibly a dense hot glob of nucleons in the early stages of the evolution of the universe, although as we shall see there is every reason to believe that these massive nuclei can be formed during the last stages of the evolution of many stars.

Self-sustaining Nuclear Reactions and Nuclear Energy Sources

5.2. Nuclear fission[†]

This is the process in which a nucleus splits into two fragments of approximately equal mass, with the evolution of energy. We may distinguish two kinds of fission: spontaneous fission, in which a nucleus decays into fission fragments without any prodding from outside (α-decay is an extreme example, but not usually classified as fission), and induced fission in which the splitting takes place only after the nucleus has been excited by, for example, dropping in a slow neutron, or an inelastic scattering process. For a discussion of nuclear reactors, we are concerned only with induced fission, but the nature of this process becomes clearer if we first discuss spontaneous fission.

Suppose we have a nucleus of charge Z and mass number A, more or less stable against β-decay. Then $Z \sim A/2$ and the binding energy is given approximately by the volume energy, surface energy and coulomb energy terms of the semiempirical mass formula:

$$\alpha A - \beta A^{2/3} - \varepsilon \frac{Z^2}{A^{1/3}}.$$

If this binding energy is compared with the sum of the binding energies of two fragments having charges Z_1 and Z_2, mass numbers A_1 and A_2 such that

$$Z_1 + Z_2 = Z, \qquad A_1 + A_2 = A,$$

then if the sum of the binding energies of the fragments is greater than the binding energy of the nucleus Z, A then work must be done to persuade the fragments to coalesce, while if the sum of the binding energies is less than the binding energy of the nucleus Z, A then energy is evolved if the fragments coalesce. In the former case the nucleus Z, A is unstable against decay into the two fragments, in the latter case it is stable. The condition for a nucleus to be just stable is then

$$\alpha A - \beta A^{2/3} - \varepsilon \frac{Z^2}{A^{1/3}} \simeq \alpha A_1 - \beta A_1^{2/3} - \varepsilon \frac{Z_1^2}{A_1^{1/3}} + \alpha A_2 - \beta A_2^{2/3} - \varepsilon \frac{Z_2^2}{A_2^{1/3}} \qquad (5.2.1)$$

where in eq. (5.2.1) we have neglected the asymmetry energy (assuming $Z \sim A/2$, $Z_1 \sim A_1/2$ and $Z_2 \sim A_2/2$) and the pairing energy term which fluctuates from nucleus to nucleus. The volume energy terms are the same on both sides of the equation, and we may start our investigation by finding the way in which the energy balance varies with the disparity in fragment size. The energy released, E_R, when the nucleus (Z, A) splits into the two fragments is given by

$$E_R = -\beta(A_1^{2/3} + A_2^{2/3} - A^{2/3}) - \varepsilon \left(\frac{Z_1^2}{A_1^{1/3}} + \frac{Z_2^2}{A_2^{1/3}} - \frac{Z^2}{A^{1/3}} \right) \qquad (5.2.2)$$

$$\approx \beta(A^{2/3} - A_1^{2/3} - (A-A_1)^{2/3}) + \frac{\varepsilon}{4}(A^{5/3} - A_1^{5/3} - (A-A_1)^{5/3}). \qquad (5.2.3)$$

[†] A comprehensive review of all aspects of fission may be found in an article by I. Halpern, *Ann. Rev. Nucl. Sci.* **9**, 245 (1959).

Nuclear Physics

We maximize the energy released with respect to A_1 by setting $\partial E_R/\partial A_1 = 0$ which gives

$$\tfrac{2}{3}\beta(A-A_1)^{-1/3} + \tfrac{5}{12}\varepsilon(A-A_1)^{2/3} = \tfrac{2}{3}\beta A_1^{-1/3} + \tfrac{5}{12}\varepsilon A_1^{2/3} \tag{5.2.4}$$

with the solution

$$A - A_1 = A_1. \tag{5.2.5}$$

The maximum energy is released in fission when the two fragments have equal masses and charges. There will, of course, be local fluctuations due to the terms we have neglected and also the effects of those variations in binding energy that never appear in the semiempirical mass formula, but roughly speaking the maximum energy is released in fission when the two fragments have equal mass. If we wish to calculate the point at which nuclei become unstable against spontaneous fission as the mass increases, we inject this result into eq. (5.2.2) and obtain on setting E_R equal to zero

$$\beta\left\{A^{2/3} - 2\left(\frac{A}{2}\right)^{2/3}\right\} + \varepsilon\left\{\frac{Z^2}{A^{2/3}} - 2\frac{(Z/2)^2}{(A/2)^{1/3}}\right\} = 0$$

or

$$\left.\begin{array}{l} \dfrac{\beta}{\varepsilon}\cdot\dfrac{\{2^{1\cdot 3}-1\}}{\{1-2^{-2/3}\}} = \dfrac{Z^2}{A}, \\[6pt] \dfrac{Z^2}{A} = 0.7\dfrac{\beta}{\varepsilon}. \end{array}\right\} \tag{5.2.6}$$

Inserting the values of β and ε quoted in eq. (1.5.5)

$$\beta = 18.33 \text{ MeV}, \quad \frac{\varepsilon}{4} = 0.1785 \text{ MeV}$$

gives

$$\frac{Z^2}{A} \simeq 18$$

as the condition governing the limit of stability of a nucleus Z, A against decay into two fission fragments. If we put $Z \sim A/2$ then $A \sim 70$ while if we feed this condition into the actual stability curve we find otherwise stable nuclei with $A \gtrsim 90$ are unstable against spontaneous decay into two fragments. In Table 5.1 we summarize a few spot checks on even–even nuclei.

These calculations have shown us that elements of quite moderate mass number, like silver and tin, have nuclei at the edge of the stability limit for spontaneous fission into two symmetric fragments. Yet it is well known that spontaneous fission does not become an important decay mode for nuclei until the transuranic region, $A \sim 250$ is reached. The explanation of this must be sought in the dynamics of nuclear fission. Suppose we consider the two symmetric fragments of $^{80}_{36}$Kr, $^{40}_{18}$A and plot the energy of such a system as a function of the separation of the two fragments. At large distances the only interaction will be the long-range coulomb potential (Fig. 5.2.1).

The description breaks down at a separation $\sim 2R_1$ where R_1 is the radius of the fragment. At some such distance the strong interactions start to distort the fragments, nucleons may be exchanged between them and the description of the system becomes unmanageably

Self-sustaining Nuclear Reactions and Nuclear Energy Sources

TABLE 5.1

Nucleus	Binding energy (MeV)	Symmetric fragment and binding energy (MeV)		Energy E_R released in fission
$^{80}_{36}$Kr	695	$^{40}_{18}$A	344	-7
$^{90}_{40}$Zr	784	$^{45}_{20}$Ca	388	-8
$^{102}_{44}$Ru	878	$^{52}_{22}$Ti	444	$+10$
$^{112}_{50}$Sn	954	$^{56}_{25}$Mn	489	$+22$

As the mass number goes up, the ratio of neutrons to protons for the β-stable isotopes increases so as to combat the growing coulomb repulsion. The result of this is that the symmetric fragments rapidly become very neutron rich as the mass number of the parent nucleus increases, and so difficult to make except through fission: consequently binding energies of massive fragments will not be found in the tables.

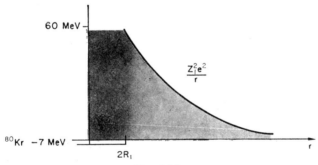

FIG. 5.2.1.

complicated. The intimate mixture formed has a ground state at negative energy (referred to a zero in which the two $^{40}_{18}$A fragments are at an infinite distance from each other) which may be reached from any positive level by the emission of radiation. It is clear, however, that to make the stable $^{80}_{36}$Kr out of two fragments, an enormous coulomb barrier must be surmounted before the nuclear forces get a grip: the height of this barrier is $\sim Z_1^2 e^2 / 2R_1$ ~ 60 MeV for $^{80}_{36}$Kr. If we now go to $^{112}_{50}$Sn, which has a positive energy of 22 MeV referred to a zero when the fragments are at infinite separation, then starting the fragments off with 11 MeV each allows us in principle to form $^{112}_{50}$Sn, but the cross-section is negligible because of the coulomb barrier.

Conversely, if this nucleus attempts to dissociate into two fragments, the impossibility of penetrating the enormous coulomb barrier, before the fragments emerge into a region in which the total energy minus the potential energy is positive, reflects the fragments straight back to reform the original nucleus (Fig. 5.2.2).

The effect of the coulomb barrier will only become negligible when the ground state of

Nuclear Physics

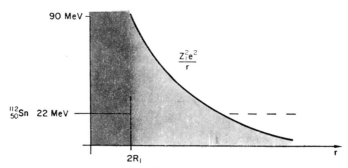

FIG. 5.2.2.

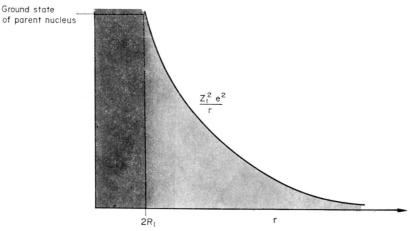

FIG. 5.2.3.

the parent nucleus is near the top of the coulomb barrier, so that the energy released is approximately equal to the height of the barrier (Fig. 5.2.3).

Thus we may expect the condition that spontaneous fission take place without inhibition to be

$$E_R \simeq \frac{Z_1^2 e^2}{2R_0 A_1^{1/3}} \tag{5.2.7}$$

or

$$\left\{ \beta(A^{2/3} - A_1^{2/3} - (A-A_1)^{2/3}) + \varepsilon \left(\frac{Z^2}{A^{1/3}} - \frac{Z_1^2}{A_1^{1/3}} \right. \right.$$
$$\left. \left. - \frac{(Z-Z_1)^2}{(A-A_1)^{1/3}} \right) \right\}_{\substack{Z_1 = Z/2 \\ A_1 = A/2}} \simeq \frac{Z_1^2 e^2}{2R_0 A_1^{1/3}},$$

$$\beta A^{2/3} \{1 - 2^{1/3}\} + \varepsilon \frac{Z^2}{A^{1/3}} \{1 - 2^{-2/3}\} = 1/4 \frac{e^2}{R_0} \frac{Z^2}{A^{1/3}} 2^{-2/3},$$

or the new condition

$$\frac{\beta \{2^{1/3} - 1\}}{\varepsilon \{1 - 2^{-2/3}\} - 2^{-2/3} (e^2/R_0)/4} = Z^2/A, \tag{5.2.8}$$

$$\frac{Z^2}{A} = \frac{0.26\beta}{0.37\varepsilon - 0.16 e^2/R_0}.$$

Self-sustaining Nuclear Reactions and Nuclear Energy Sources

If we take $R_0 \sim 1.3 \times 10^{-13}$ cm

$$\frac{e^2}{R_0} \sim 1.1 \text{ MeV} = 1.54\,\varepsilon.$$

So the condition that spontaneous fission be not suppressed by the coulomb barrier is

$$\frac{Z^2}{A} \simeq 2.1 \beta/\varepsilon \qquad (5.2.9)$$

to be compared with $Z^2/A \sim 0.7\, \beta/\varepsilon$ for the stability limit. Thus

$$\frac{Z^2}{A} \approx 55 \qquad (5.2.10)$$

is the condition for spontaneous fission to be important. For $Z = A/2$ this gives $A = 220$, $Z = 110$ while taking the actual relationship between Z and A for maximum stability gives $A = 340$, $Z = 136$. This places uninhibited spontaneous fission deep in the transuranic region, and the calculation is not very reliable because the denominator contains the difference of two quantities neither of which is very well known. We can attempt to remove some of the errors by noting that

$$\varepsilon \frac{Z^2}{A^{1/3}} = \frac{3}{5} \frac{e^2 Z^2}{R_0 A^{1/3}}$$

in the liquid drop model. Then we have consistency between the two coulomb energies if we set

$$\frac{e^2}{R_0} = \frac{5}{3} \varepsilon$$

when the condition on Z^2/A becomes $Z^2/A \sim 2.5\, \beta/\varepsilon$.

The condition

$$\frac{Z^2}{A} = 2\beta/\varepsilon \qquad (5.2.11)$$

may be derived from the liquid drop model in a more rigorous way, although the greater rigour hardly adds to the reliability of the result. So far we have applied the model to the initial nucleus (assumed spherical) and the final fission fragments. The liquid drop model gives us a means of dealing with the initial stages of breakup in which the strong interactions are important. Since the nuclear density is observed to be about constant throughout the nuclear periodic table, it is reasonable to suppose that the density remains constant under a small deformation of the parent nucleus. Since a sphere is the geometrical shape with the smallest ratio of surface area to volume, under a small ellipsoidal deformation at constant density the surface area increases, and so does the surface energy. Because such a deformation separates charges, the coulomb energy decreases. Classically such a nucleus will be stable against a small deformation if the increase in surface energy is greater than the decrease of coulomb energy, and unstable if the decrease in coulomb energy wins out.

Nuclear Physics

A small symmetric deformation may be represented by writing the length of the radius vector from the centre to the surface as

$$R(\theta) = R(0)\{1 + \alpha\, P_2(\cos\theta)\ldots\}.$$

With the constant density constraint applied, the change of coulomb energy is

$$\Delta E_c = -\tfrac{1}{5}\alpha^2 E_c \ldots$$

and the change of surface energy is

$$\Delta E_S = +\tfrac{2}{5}\alpha^2 E_S \ldots$$

so that the classical limit is found by setting $2E_S - E_c = 0$ or

$$\frac{Z^2}{A} = \frac{2\beta}{\varepsilon}.$$

This is the original treatment of Bohr and Wheeler;[†] the details of these tedious but entirely classical calculations may be found in a monograph by Wilets.[‡] If we plot the deformation energy of a nucleus as a function of this distortion parameter α then we find as seen in Fig. 5.2.4:

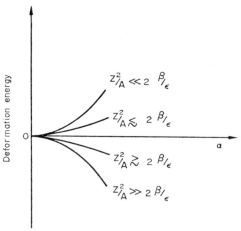

Fig. 5.2.4.

Classically a nucleus with a positive value of the deformation energy for small values of α is absolutely stable. In quantum mechanics a positive deformation energy taken in conjunction with the coulomb repulsion at large distances creates a potential barrier through which the fragments must tunnel in order for fission to take place. Thus if we plot the potential energy as a function of a single parameter which for small deformations is α and which beyond the *scission point* becomes the separation of the two fragments, we have Fig. 5.2.5.

[†] N. Bohr and J. A. Wheeler, *Phys. Rev.* **56**, 426 (1939).
[‡] L. Wilets, *Theories of Nuclear Fission* (Oxford, 1964).

Self-sustaining Nuclear Reactions and Nuclear Energy Sources

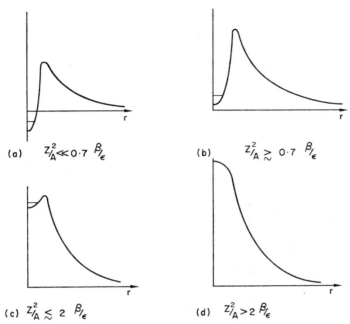

FIG. 5.2.5. This figure shows schematically the potential energy of a nucleus as a function of some deformation parameter which beyond the scission point becomes the separation of two fragments. In case (a) the ground state of the nucleus lies below the zero of energy for two fragments infinitely separated, and the nucleus is absolutely stable against spontaneous fission. In case (b) the ground state lies above zero energy and spontaneous fission is energetically possible but suppressed by an enormous potential barrier. In case (c) the ground state lies near the top of the potential and fission takes place spontaneously, inhibited by the barrier which determines the lifetime of the nucleus against spontaneous fission. In case (d) the nucleus is totally unstable.

These considerations based on the semiempirical mass formula have allowed us to see why spontaneous fission is not observed for elements with mass numbers ~ 100 and lead us to expect that spontaneous fission of ground state nuclei will only become important among the transuranic elements. Thus ^{235}U has a half-life against spontaneous fission of 2×10^{17} years, ^{234}U, ^{236}U, ^{238}U $\sim 10^{16}$ years, ^{240}Pu $\sim 10^{11}$ years, ^{250}Cf $\sim 10^4$ years while the isotope ^{254}Cf has a half-life of only 55 days.

We must emphasize that the variation of energy with distortion is not a simple phenomenon. Many nuclei are significantly distorted in their ground state, the number of ways in which a nucleus may distort is unlimited and details not contained in the liquid-drop model will also be important. The form of the barrier may even be double humped. A number of isotopes of americium are observed to have isomeric states at excitations $\sim$ MeV which decay via spontaneous fission with lifetimes $\sim$ milliseconds. With such a long lifetime against spontaneous fission one would expect these isomeric states to decay electromagnetically unless very large spin charges are involved, which is not the case. A potential barrier is then presumably preventing electromagnetic decay to the ground state: the nucleus is trapped in a local minimum of the potential energy.

Nuclear Physics

5.3. Induced fission

We have so far only discussed spontaneous fission of nuclei in their ground state. If a nucleus with a negligible probability of spontaneous fission is excited, the probability of the excited state undergoing fission depends on the excitation energy: if the energy of the excited state is about equal to the coulomb energy of the two fragments at the scission point, then we may expect fission to be an important mode of decay of the excited nucleus. We may define the *activation energy* as that excitation energy that is sufficient for fission to compete with other decay modes of the excited nucleus: the activation energy will be roughly the height of the barrier presented to the fragments (see Fig. 5.3.1).

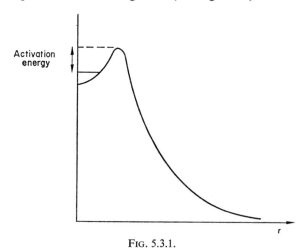

FIG. 5.3.1.

While a calculation of the barrier shape in the region between small deformation and the scission point is bound to be unreliable, we can find the activation energy directly. For example, ^{235}U is fissionable on absorption of a slow neutron: this makes ^{236}U in an excited state with an excitation energy ~ 7 MeV, which decay by spontaneous fission with a probability of 0.84, by γ-emission to the ground state of ^{236}U with a probability of 0.16. On the other hand, the nucleus ^{238}U is not fissionable by slow neutrons: the excited state of ^{239}U formed by slow neutron capture decays electromagnetically to the ground state. Now ^{236}U is an even–even nucleus, while ^{239}U is even–odd. Thus on adding a slow neutron to ^{235}U the excitation energy of ^{236}U is increased over the liquid-drop form by the pairing energy term, while on adding a neutron to ^{238}U there is no such term. This is reflected in the neutron-separation energies: the separation energy for ^{236}U is 7 MeV; for ^{239}U, 5 MeV. The activation energy in the region of uranium must therefore be about 6 MeV.

The importance of the pairing energy is illustrated by the sequence of events that follows neutron capture by ^{238}U in a slow neutron reactor.

$$^{238}_{92}\text{U} + n \rightarrow {}^{239}_{92}\text{U}^* \rightarrow {}^{239}_{92}\text{U} + \gamma,$$
$$^{239}_{92}\text{U} \rightarrow {}^{239}_{93}\text{Np} + e^- + \bar{\nu},$$
$$^{239}_{93}\text{Np} \rightarrow {}^{239}_{94}\text{Pu} + e^- + \bar{\nu}.$$

Self-sustaining Nuclear Reactions and Nuclear Energy Sources

$^{239}_{94}$Pu has an odd number of neutrons and on capture of a slow neutron the excited state of $^{240}_{94}$Pu thus formed decays by fission. This is how a breeder reactor works: it is not true to say that a breeder reactor makes its own fuel: rather the processes following capture of a neutron by ^{238}U convert the ^{238}U which a thermal neutron reactor is unable to consume into ^{239}Pu which it can. As we might expect from the work of the previous pages, the probability of induced fission of ^{238}U only becomes significant for neutrons of energy > 1 MeV, and thermal reactors operate at a mean neutron energy ~ 0.1 eV.

At those bombarding energies of practical interest ($\lesssim$ a few MeV) induced fission may be regarded as proceeding in two stages. First, an excited state of the nucleus is formed, by neutron capture, photon absorption, coulomb excitation, or some other mechanism, and this excited compound nuclear state then decays by fission. For most fission reactors, the excited compound nucleus is formed by thermal neutron capture. Since the excitation energy is ~ 6 MeV, the probability of fission is independent of the energy of the absorbed thermal neutron, and so the cross-section $\sigma(n, f)$ will follow the $1/v$ law characteristic of exothermic reactions. The cross-section for ^{235}U is shown in Fig. 5.5.1.

5.4. Secondary features of fission

For nuclei in the region of uranium, symmetric fission (i.e. fission into two equal fragments) is relatively unlikely, unless the excitation energy is $\gtrsim 100$ MeV. The distribution of fission fragments as a function of mass number is shown in Fig. 5.4.1 for fission of ^{235}U by thermal neutrons. The symmetry of the curve is assured by the requirement that $A_1 + A_2 \simeq A$ (not exactly equal because of prompt neutron emission) but symmetric fission

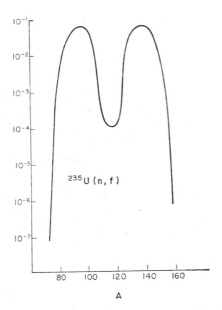

FIG. 5.4.1. The approximate distribution of fission fragment mass number in the fission of ^{235}U by slow neutrons.

occurs with a frequency only $\sim 10^{-3}$ of the most probable configuration. This feature of fission is not understood, and an explanation is presumably only to be found in the detailed dynamics of the fission process. It will be noted that the most probable configuration is in the region of the closed neutron shells at 50 and 82, but it is not clear why such closed shells should retain their influence in so disruptive a process as nuclear fission.

Because the coulomb energy $\sim Z^2$ the fragments formed at scission are neutron rich and are also highly excited. The initial stage by which the fragments shed their neutron excess is the evaporation of so-called prompt neutrons.

The average number of neutrons given off in fission is a function of the compound nucleus formed and the excitation energy: on average an extra 8.5 MeV of excitation results in one more neutron. If ^{236}U is formed by thermal neutron capture, ~ 2.5 neutrons are produced per fission. Their average energy is ~ 2 MeV in the laboratory, and the laboratory energy spectrum together with the laboratory angular distribution measured with respect to the line of flight of the fission fragments shows that these prompt neutrons are emitted isotropically in the centre of mass of the fragments, and with an energy spectrum in the fragment centres of mass consistent with that expected for neutrons evaporating from a Fermi gas, with an average energy ~ 1 MeV.

The angular distribution of these prompt neutrons in the laboratory is not affected by whether the fragments are moving through a gas or solid material. In Chapter 3 we estimated the slowing-down time for completely ionized carbon moving through carbon, using the expression

$$t \sim \frac{v^3 M m_e}{12\pi Z^2 e^4 N}$$

for an ion of charge Ze, mass M, moving with velocity v through a medium containing N electrons/cm^3. For carbon moving through carbon, $t \sim 10^{-12}$ sec. The fission fragment energy is ~ 1 MeV/nucleon, so we may take v as $\sim 10^9$ cm sec^{-1}. M is increased from ~ 12 nucleon masses to ~ 120 nucleon masses, and Z from 6 to ~ 50. The slowing-down time is thus reduced from 10^{-12} sec to 10^{-12} $120/50^2 \sim 5 \times 10^{-14}$ sec for a fission fragment moving through carbon. The fact that the angular distribution of prompt neutrons is unaffected by a fragment moving through solid matter thus implies that the prompt neutrons are evaporated in a time less than $< 5 \times 10^{-14}$ sec as we would expect: even in the thermal region the widths of neutron resonances imply lifetimes $\sim 10^{-16}$ sec, and these fission fragments are excited by at least 1 MeV above the excitation achieved in slow neutron capture.

The energy released in fission of ^{235}U by thermal neutron capture is 193 MeV. In the most probable configuration the fragments, after evaporation of prompt neutrons, have 166 MeV, leaving ~ 27 MeV for excitation of the fragments. The binding energy per nucleon in the fission fragment region of the nuclear periodic table is ~ 8 MeV for nuclei stable against β-decay, and will be reduced below this for the neutron-rich fragments. Thus of the 27 MeV of excitation energy available in the fission of ^{235}U by slow neutrons, we may expect 20 MeV to disappear in evaporating the prompt neutrons and the rest to be radiated as photons as the cooler fragments, having discarded all the neutrons they can, de-excite electromagnetically.

Self-sustaining Nuclear Reactions and Nuclear Energy Sources

The shake-off of ~ 1 neutron each leaves the fragments stable against further nucleon emission, but they are still neutron rich. For example, if we take fragments of mass number 94 and 142, in the region of the most probable configuration, and suppose the charge to be divided in the same ratio we expect atomic numbers of 37 and 55 respectively. The stable isotope of $_{37}$Rb is $^{85}_{37}$Rb (and $^{87}_{37}$Rb has a half-life of 4.3×10^{10} years). The stable isotope of $_{55}$Cs is $^{133}_{55}$Cs. Thus the fragment $^{94}_{37}$Rb must undergo β-decay to reach stability: three successive transitions take it to $^{94}_{40}$Zr which is stable. The fragment $^{142}_{55}$Cs would end up at $^{142}_{58}$Ce, again after three transitions. It is the neutron-rich character of the fission fragments that is responsible for the highly radio-active ash that remains after fission in a nuclear reactor.

Another very interesting phenomenon, of great importance to the practical problems of making a useful nuclear reactor, is the emission of delayed neutrons. Delayed neutrons are emitted from certain fission fragments in their approach to stability with a time scale $\sim$ seconds or minutes. This very long lifetime cannot be associated directly with a state unstable against neutron emission: the neutron energy is ~ 0.1 MeV. The lifetime against delayed neutron emission is characteristic of the weak interactions, and the delay in introduced through β-decay. A fission fragment which has emitted prompt neutrons has to make several consecutive β-transitions in order to reach stability. If in the course of these transitions a nucleus neutron rich, but stable against neutron emission, decays into a second nucleus which is less neutron rich but happens to be unstable against neutron emission, then the product of the β process will emit a prompt neutron, delayed from the initial fission by time taken for the intermediate β processes.

A well-known example is the fission fragment $^{87}_{35}$Br$_{52}$ with a half-life of 55 sec. ^{87}Br decays $\sim 30\%$ of the time to the ground state of ^{87}Kr which while β unstable is stable against neutron emission. In the other 70% of cases it decays to one or more levels of ^{87}Kr at around 5.4 MeV and these excited levels decay via neutron emission 3% of the time. (It is not clear whether one or more excited levels are formed in the β-decay of ^{87}Br: the 3% of neutron emission could come from one level with an $n:\gamma$ branching ratio of 3%, or in some 2% of β-decays another level with a very large $n:\gamma$ ratio might be formed.) The delayed neutron has an energy ~ 0.3 MeV.

^{87}Br has two neutrons outside a closed shell at $N = 50$; ^{87}K has only one neutron beyond the closed shell.

5.5. Physics of the fission reactor

If a nucleus of ^{235}U absorbs a slow neutron and undergoes fission, on average 193 MeV of energy is released of which 166 MeV appears in the form of the kinetic energy of the fission fragments, moving with velocities $\sim 10^9$ cm sec^{-1}. These fragments collide with both electrons and nuclei in the lattice containing the uranium, and are stopped within 10^{-5} cm. Most of their energy is converted into heat in the process, but some is stored in the lattice as a result of displacement of atoms. This radiation damage can ruin the mechanical properties of a fuel element.

Several neutrons accompany the process, on average 2.5 prompt neutrons per fission and

Nuclear Physics

0.018 delayed neutrons per fission. If things can be so arranged that on average one of these neutrons is captured by a further nucleus and induces fission, then we have a self-sustaining chain reaction with constant power output. If more than one fission is induced by the neutrons of the primary fission, then the power output will rise exponentially—the extreme example is a bomb.

Consider what may happen to neutrons produced in the fission process:

1. They may escape from the reactor.
2. They may be absorbed by non-fissile material.
3. They may be absorbed by fissile material but not cause fission.
4. They may be absorbed by fissile material and induce fission.

Loss of neutrons from the active core of a reactor can be minimized by making the assembly big, reducing the ratio of surface area to volume, and surrounding the core with reflecting material, which should have a low cross-section for neutron capture. The inactive parts of the reactor core must also have a low neutron capture cross-section: that is, the cans containing the fuel elements, the control system, and whatever fluid is pumped through the hot core to transport out heat. However, ^{238}U is not fissionable by neutrons of energy less than 1 MeV and makes up 99.28% of naturally occurring uranium. In the region of 1 MeV the cross-section for ^{238}U(n, γ) ^{239}U is ~ 10 times the ^{235}U(n, f) cross-section and so in naturally occurring uranium most of the neutrons produced in the fission of ^{235}U are soaked up by ^{238}U without producing further fission. Thus in order to produce a core capable of sustaining a chain reaction, the alternatives are either to enrich the uranium with ^{235}U or to interfere with the neutron energy spectrum so as to maximize the probability of neutron capture by ^{235}U. This is done by thermalizing the neutrons. Since neutrons are uncharged and the (n, f) reaction is highly exothermic, the cross-section $\sigma(n, f)$ follows a $1/v$ law (with superimposed resonant peaks) and so to maximize $\sigma(n, f)$ the neutrons must be reduced in velocity from $\sim 10^9$ cm sec^{-1} to as small a value as possible. The best that can be done is to bring the neutrons into thermal equilibrium with their environment.

At thermal energies the following cross-sections apply (see Fig. 5.5.1):

Process	σ (barns) at $v = 2.2 \times 10^5$ cm sec^{-1}, $E = 0.025$ eV
^{235}U(n, f)	577
^{235}U$(n, \gamma)^{236}$U	101
^{238}U$(n, \gamma)^{239}$U	273

Thus in naturally occurring uranium a thermal neutron has a probability of 0.54 of capture on ^{235}U leading to fission, 0.36 of radiative capture on ^{238}U and 0.1 of radiative capture on ^{235}U: it takes 1.9 thermal neutrons to make a ^{235}U fission in natural uranium. Since a mean of 2.5 neutrons per fission are produced a self-sustaining chain reaction in naturally occurring uranium is possible, provided that the neutrons can be thermalized without incurring a loss greater than 24% in the process. The 1-MeV neutrons produced in fission can only be brought down to thermal energies by collision with atoms in the core of the reactor. This process of moderation must be done as quickly as possible, for the greater the path length in the reactor at non-thermal energies, the greater the probability of the neutron

Self-sustaining Nuclear Reactions and Nuclear Energy Sources

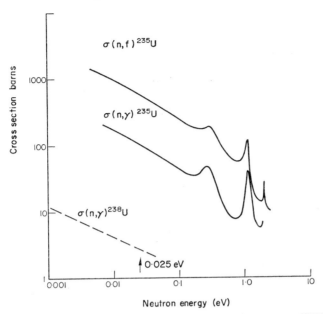

FIG. 5.5.1. The figure shows the low-energy neutron cross-sections for the processes $^{235}U(n,f)$, $^{235}U(n,\gamma)^{236}U$ and $^{238}U(n,\gamma)^{239}U$. Note the $1/v$ dependence of all these cross-sections below $E_n = 0.1$ eV. [Data taken from *The Barn Book*.]

being captured by the ^{238}U. It is easy to see that the moderator should consist of high atoms if the energy is to be lost by elastic collision. At energies ~ 1 MeV we expect s-wav scattering to dominate:† in the centre of mass of the neutron and the moderating nucleut the scattered neutron will be scattered isotropically. The mean energy lost may then be calculated very simply, using non-relativistic mechanics.

We work initially in the centre of mass system. The scattered neutron has the same energy as the incident neutron and velocity v. The target has velocity V.

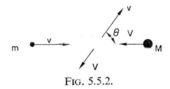

FIG. 5.5.2.

In this centre of mass the number of neutrons scattered through an angle θ is

$$dN(\theta) = \frac{N}{2} d\cos\theta. \tag{5.5.1}$$

The velocity of the centre of mass in the laboratory is V and that of the incident neutron $v+V = v_L$. The velocity of the scattered neutron in the laboratory is

$$\{(v\cos\theta + V)^2 + v^2\sin^2\theta\}^{1/2} = \{v^2 + V^2 + 2vV\cos\theta\}^{1/2}. \tag{5.5.2}$$

† See Section 4.2.

Nuclear Physics

Thus the energy of a neutron in the laboratory after being scattered through an angle θ in the centre of mass is

$$E_L = \tfrac{1}{2}m \{v^2 + V^2 + 2vV \cos \theta\} \tag{5.5.3}$$

and the average laboratory energy after one collision is

$$\bar{E}_L = \frac{\tfrac{1}{2}m \int_{-1}^{+1} \{v^2 + V^2 + 2vV \cos \theta\} N d \cos \theta}{N \int_{-1}^{+1} d \cos \theta}. \tag{5.5.4}$$

So that the fraction of the initial laboratory energy E_{in} still carried by the emergent neutron after the collision is, on average,

$$\frac{\bar{E}_L}{E_{in}} = \frac{m^2 + M^2}{(m+M)^2}. \tag{5.5.5}$$

As $M \to \infty$ this ratio tends to unity: if the struck nucleus is of infinite mass, the centre of mass and laboratory system coincide. This function also has a minimum at $m = M$ when it is equal to $\tfrac{1}{2}$. Thus colliding with protons, the neutrons would lose on average $\tfrac{1}{2}$ of their energy per collision and to reduce their energy from 1 MeV to 0.1 eV, a factor of 10^7 will take n collisions where

$$2^n = 10^7 \quad \text{or} \quad n = \frac{7}{\ln_{10} 2} = 23.$$

If $M = 12\,m$ (for a carbon moderator) then the fractional energy is 0.86 and

$$n = \frac{7}{\ln_{10} 1.16} = 108$$

while if we left it to the uranium to do the moderating the fractional emergent energy is 0.99 and it would take ~ 2000 collisions to reduce the neutron energy to ~ 0.1 eV (and the neutrons would never get there).

There is a reason other than the need to minimize path length for thermalizing the neutrons in as big steps as possible. If the cross-section for radiative capture by ^{238}U varied smoothly with energy, it would make little difference whether the neutrons were moderated with small energy loss at each stage by a moderator with a large scattering cross-section, or by fewer collisions with a light moderator, with smaller cross-sections: the only requirement would be to minimize the path length. However, the cross-section for radiative capture on ^{238}U has a cluster of resonances, superimposed on relatively smooth behaviour, between 5 and 100 eV (Fig. 5.5.3). These resonances are a few electron volts wide, and the cross-section rises to 5000 barns. If a neutron on entering this region lost energy only in steps of $\lesssim 1$ eV it could not avoid being swallowed up by the resonant (n, γ) process. If, on the other hand, the energy lost at each step was large in comparison with the widths of the resonances, then most neutrons would dodge the resonant peaks, and only by misfortune would a neutron come out of a scattering process with its energy matched to a resonance. This is shown schematically in Fig. 5.5.4.

Self-sustaining Nuclear Reactions and Nuclear Energy Sources

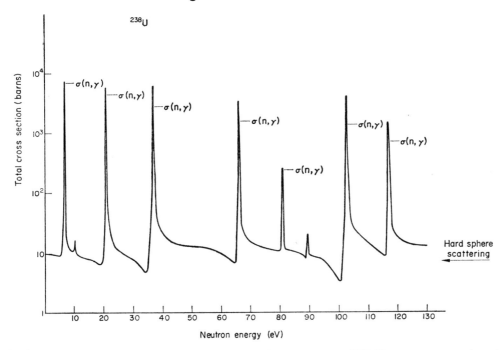

FIG. 5.5.3. The figure shows the total cross-section for slow neutrons on ^{238}U. The resonances reach up to total cross-sections ~ 5000 barns and $\Gamma_n \lesssim \Gamma_\gamma$. The background is predominantly s-wave neutron elastic scattering and interference between the background and the elastic part of the resonances can be seen. The peak cross-sections for the process ^{238}U$(n, \gamma)^{239}$U are included, and we have also indicated the expected cross-section for elastic s-wave scattering of neutrons from a hard sphere of radius $R = 1.2 A^{1/3} \times 10^{-13}$ cm. [Data taken from *The Barn Book—Neutron Cross Sections*, BNL-325 2nd ed., Supplement 2, Vol. III (1965).]

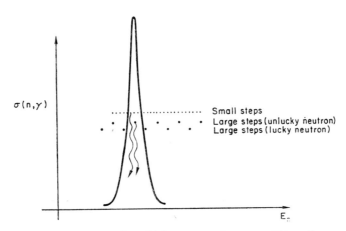

FIG. 5.5.4. This figure shows how moderation with large energy loss per collision allows a neutron to dodge narrow resonances. A neutron losing energy in small steps has a very high probability of being captured by a resonance of width larger than the mean energy loss. If moderation is effected in steps large in comparison with the resonance width, the probability of capture is greatly reduced.

Nuclear Physics

A 100-eV neutron in collision with a free proton will lose on average 50 eV. In collision with a free carbon nucleus the energy loss is only 14 eV, reduced to 7 eV on average for a 50 eV neutron. Thus neutrons moderated by hydrogen have a good chance of dodging the resonances, while the energy loss for carbon-moderated neutrons is matched rather too closely to the widths of the ^{238}U resonances. Hydrogen is apparently the best choice for a moderator. However, if the light nucleus is to do most of the moderating, the number of such nuclei should be large in comparison with the number of uranium nuclei. We saw in Section 4.9 that the cross-section for neutron capture by a proton $p(n, \gamma)d$ has a value of 0.3 barn at thermal energies. This is too large for hydrogen to be used as moderator in a natural uranium reactor. With 10 times as many protons as uranium nuclei in the reactor, sufficient thermal neutrons are absorbed by the hydrogen to prevent a self-sustaining reaction—even if the neutrons were thermalized without loss. Hydrogen, in the form of water or an organic compound, can be used in a reactor fuelled with enriched uranium.

Heavy water, D_2O, can be used as moderator in a natural uranium assembly, but it is expensive and ill suited to high-temperature operation. Graphite is the material used as moderator in natural uranium piles producing power on a commercial scale. ^{12}C is a light nucleus, tightly bound and with all nucleons paired off reluctant to pick up another. The capture cross-section at thermal energies is only $\sim 5 \times 10^{-3}$ barn and falls off according to the $1/v$ law. Finally, graphite is both cheap and a refractory material suitable for use at high temperatures as both moderator and a reflector outside the uranium-containing core. The disadvantages of a relatively small average energy loss, which in the region of 100eV energy is dangerously close to the resonance widths, may be overcome by distributing the uranium through the reactor in lumps rather than constructing an approximation to a homogeneous mixture of graphite and uranium. (This has obvious advantages for the construction, control and refuelling of the reactor as well.) If we disregard any moderation that may take place within a lump of uranium, any neutrons with an energy matching the resonances in ^{238}U will have to approach the uranium from the moderator. Such a neutron will be trapped by the ^{238}U in the surface region of the lump and will never penetrate deep into the interior. Thus a ^{238}U nucleus deep within a lump of uranium will never see a neutron of the right energy for resonance capture and so cannot act as a neutron trap.

Suppose that we make a simple one-dimensional model in which a flux of neutrons N_R at resonant energies and a flux N_T at thermal energies are incident on the face of a slab of uranium. At a depth of x inside the fluxes are

$$\left. \begin{array}{l} N_R(x) = N_R e^{-x/L_R}, \\ N_T(x) = N_T e^{-x/L_T}. \end{array} \right\} \quad (5.5.6)$$

The number of resonant neutrons absorbed per gram of material is then proportional to

$$N_R \frac{[1-e^{-x/L_R}]}{x} \quad (5.5.7)$$

and the number of thermal neutrons leading to fission on ^{235}U per gram of material is proportional to

$$N_T \frac{[1-e^{-x/L_T}]}{x}. \quad (5.5.8)$$

Self-sustaining Nuclear Reactions and Nuclear Energy Sources

As $x \to \infty$ both these quantities $\to 0$, and both have a maximum at $x = 0$ corresponding to a homogeneous mixture of moderator and uranium. Thus it is not possible to minimize the first and simultaneously maximize the second, but because $L_R \ll L_T$ any dimension $x \sim L_T$ is in the right ball park. The absorption length for thermal neutrons is given by

$$L_T = \frac{1}{n_{235}\{\sigma_{235}(n,\gamma) + \sigma_{235}(n,f)\} + n_{238}\,\sigma_{238}(n,\gamma)} \tag{5.5.9}$$

where n_{235}, n_{238} are the number of nuclei per cm³ in naturally occurring metallic uranium, of density 18.7. This yields $L_T \sim 0.3$ cm. The quantity L_R depends violently on the precise energy of the neutron: a cross-section of 1000 barns yields a value of $L_R \sim 2 \times 10^{-3}$ cm. The function $(1 - e^{-x/L_T})/x$ is slowly varying up to a value of $x \sim L_T$ and has fallen to half the value for x by 0 at $x = 1.5\,L_T$. Thus in this one-dimensional example, there is nothing to be gained by making the thickness of the uranium slab more than ~ 0.5 cm thick. In a three-dimensional problem the flux is coming at a lump from all sides and a lump or bar ~ 1–2 cm across will be about optimum. The best value will depend on the detailed history of a given neutron in the reactor: how long it spends at each energy in both moderator and fuel. The detailed history in turn depends on the configuration.

A reactor fuelled with natural uranium, then, will consist of an assembly of bars of metallic uranium, or a uranium compound, and graphite moderator. A reflector of graphite will extend beyond the uranium bearing core to reflect neutrons back in and the core must be large enough that the unavoidable loss of neutrons from the surface does not prevent a self-sustaining chain reaction—there is a critical size for such an assembly which must be exceeded. Through this core must be pumped a coolant with a low neutron capture cross-section. (Quite apart from the need to keep neutron capture in non-fissile material down, it is desirable that the coolant shall not become violently radioactive by neutron capture). Carbon dioxide gas under high pressure, D_2O and liquid sodium are all coolants which have been used. Around the whole assembly must be a gas-tight shell to keep radioactive fission products in, and outside this a biological shield against the radiation produced.

Most of the energy from the fission of a ^{235}U nucleus appears as kinetic energy of the fragments. These fragments are stopped, almost always within the fuel elements, and their kinetic energy is in the process degraded to heat, while some is stored in the lattice of the fuel element as a result of radiation damage. The β-active fission fragments make ~ 4 transitions to achieve stability. The decay energy is turned into heat again, except for the leak of energy carried by the neutrinos, which is unstoppable. The neutrons carry only a few MeV. In the moderation this energy is also degraded to heat, but again some is stored in the graphite lattice when atoms are displaced in the collision with a neutron. The β activity of the fission fragments means that the reactor will continue to produce heat after the chain reaction has been shut down. The energy accumulated in the graphite must be annealed out periodically, for a spontaneous release of this energy could melt parts of the reactor.

The heat energy produced is transported out of the reactor core by a coolant and then used via a heat exchanger to raise steam. The nuclear energy sources of the twentieth century are thus coupled to the electricity network by techniques two centuries old—and known to Hero of Alexandria!

Nuclear Physics

5.6. Time constant of a fission reactor

The neutron flux in a reactor is a function of position, time and the energy of the neutron. The density of neutrons at any point in the assembly is given by the solution of a diffusion equation with energy-dependent coefficients and boundary conditions determined by the size and shape of the assembly. The physical separation of moderator and fuel element means that the equation must be solved separately for the moderator and for the inserted fuel elements, and the solutions matched at the boundary. In this section we shall ignore these complications and assume only an homogeneous assembly and a neutron flux independent of position.

Suppose we have a population of thermal neutrons in the core of such an assembly. The capture processes for such neutrons are dominated by the $1/v$ law which simply says that the faster the neutrons travel the smaller the chance of capture. The capture rate is independent of neutron velocity. Then in the absence of any replacement

$$dn = -n\frac{dt}{\tau}; \quad n(t) = n_0 e^{-t/\tau}. \tag{5.6.1}$$

At $1/40$ eV, a velocity of 2.2×10^5 cm sec^{-1}, the cross-sections are

$^{238}U(n, \gamma)^{239}U$	2.73 barns
$^{235}U(n, f)$	577 barns
$^{235}U(n, \gamma)^{236}U$	101 barns

We can calculate τ directly for metallic uranium

$$\frac{1}{\tau} = \sum_i \sigma_i v N_i$$

where σ_i is the cross-section for the ith process taking place on a nucleus of density N per cm^3 and v is the neutron velocity. With $\sigma_i v$ constant for the $1/v$ law we find

$$\frac{1}{\tau} \simeq 6 \times 10^5 \text{ sec}^{-1},$$

$$\tau \simeq 1.6 \times 10^{-6} \text{ sec}.$$

If the natural uranium is diluted by ~ 100 carbon nuclei to each uranium nucleus, the time scale for capture will become $\sim 10^{-4}$ sec.

If each disappearing thermal neutron produced on average k fresh thermal neutrons, then

$$dn = -n\frac{dt}{\tau} + kn\frac{dt}{\tau}.$$

and

$$n(t) = n_0 e^{-(1-k)t/\tau} = n_0 e^{(k-1)t/\tau}. \tag{5.6.2}$$

If $k = 1$ the neutron flux and the fission rate remain constant: the chain reaction is just self sustaining. If $k < 1$ the chain reaction dies out exponentially with a time constant $\tau/|k-1|$ and if $k > 1$ then the chain reaction grows exponentially with a time scale $\tau/|k-1|$.

Self-sustaining Nuclear Reactions and Nuclear Energy Sources

In the case of a volume of pure ^{235}U there is no moderator, and in the MeV region the cross-section for radiative capture is only $\sim \frac{1}{10}$ the cross-section for fast fission, and so neutrons either escape or induce fission. The fission cross-section in the MeV region is ~ 1 barn, the velocity $\sim 10^9$ cm sec^{-1} and so the time scale for capture is $\sim 2\times 10^{-8}$ sec. If escape from the assembly is unimportant then $k \sim 2.5$ and the reaction grows exponentially with a time scale of $\sim 10^{-8}$ sec. The fission fragments travel no faster than the initiating neutrons and so the chain reaction consumes all the fuel before the energy released can disrupt the reaction mass: this is a nuclear fission explosion.[†] The interaction length for a cross-section of 1 barn in ^{235}U is ~ 20 cm, and a sphere of uranium of diameter 20 cm weighs ~ 60 kg. (This simple calculation gives only the order of magnitude of the critical mass of ^{235}U.) An explosion is made by smacking together a number of pieces below the critical mass so as reduce the surface area and boost k from below 1 to substantially greater than 1.

In a thermal neutron reactor, the assembly is built greater than the critical size, with dampers in to keep k down. Such dampers are rods of material with a large cross-section for the absorption of thermal neutrons, such as cadmium. As these dampers are pulled out k approaches unity and on exceeding unity the reaction starts to grow. In order to control a reactor it is necessary to be able to pull dampers in and out in a time short compared with the time constant of the reactor—it would be irritating to have the pile go out and fatal to have parts start melting. The time scale $\sim 10^{-4}$ sec which we have just calculated is clearly dangerously short: if $k = 1 + 10^{-4}$ the overall time scale is 1 sec. The situation is saved by the presence of the delayed neutrons. These delayed neutrons can affect the time scale of a reactor drastically. If the reactor is critical on prompt neutrons alone, then the delayed neutrons are irrelevant. However, if the reactor is just subcritical on the prompt neutrons, but just critical when the delayed neutrons are added in, then the time constant of the reactor will reflect the delay.

Let n_p be the density of prompt neutrons,
n_d the density of delayed neutrons,
k_p the average number of prompt neutrons produced in fission which reach thermal energies.

Then

$$\frac{dn_p}{dt} = \frac{k_p}{\tau}\{n_p + n_d\} - \frac{1}{\tau} n_p. \tag{5.6.3}$$

Let the density of pregnant nuclei (that is, of those nuclei that may emit a delayed neutron) be p, f the probability that a pregnant fission fragment is produced when a thermal neutron is absorbed, and g the probability that a pregnant nucleus will be delivered of a neutron which is subsequently thermalized (rather than de-exciting electromagnetically). To keep things simple we suppose only one species of pregnant nucleus to be important, with decay constant λ_β. Then we can write

$$\frac{dp}{dt} = \frac{f}{\tau}(n_p + n_d) - \lambda_\beta p \tag{5.6.4}$$

[†] Such a device is usually called, inaccurately, an atomic bomb.

Nuclear Physics

and
$$\frac{dn_d}{dt} = \lambda_\beta gp - \frac{n_d}{\tau}. \tag{5.6.5}$$

$n_p + n_d = n$, the total thermal neutron density, and on adding eqs. (5.6.3) and (5.6.5) we find
$$\frac{dn}{dt} = \frac{(k_p - 1)}{\tau} n + \lambda_\beta gp. \tag{5.6.6}$$

The pair of equations (5.6.4) and (5.6.6) must now be solved simultaneously. This is most easily done by substituting $p = p_0 e^{\lambda t}$, $n = n_0 e^{\lambda t}$ when two solutions are obtained:

$$2\lambda = -\left\{\lambda_\beta - \frac{(k_p-1)}{\tau}\right\} \pm \sqrt{\left\{\lambda_\beta + \frac{(k_p-1)}{\tau}\right\}^2 + \frac{4fg\lambda_\beta}{\tau}};$$

writing the product $fg = k_d$ and $k = k_p + k_d$ yields

$$2\lambda = -\left\{\lambda_\beta - \frac{(k_p-1)}{\tau}\right\} \pm \sqrt{\left\{\lambda_\beta - \frac{(k_p-1)}{\tau}\right\}^2 + \frac{4(k-1)\lambda_\beta}{\tau}}. \tag{5.6.7}$$

If $k_d = 0$, i.e. either f or $g = 0$, then the two solutions correspond to $\lambda = -\lambda_\beta$ and $\lambda = (k_p - 1)/\tau$: the pregnant nuclei and the neutron flux are decoupled. If $k_p > 1$ the reaction will run away, but if $k > 1$ and $k_p < 1$ then the time constant is sensitive to λ_β. Qualitatively, if we start with a closed-down reactor and pull out the dampers, the reactor first goes critical on both the prompt and delayed neutrons. A small change in damping then has a delay depending on λ_β and in this region the time constant may be hours. The dampers must not be pulled out far enough for the reaction to become "prompt critical"!

There are 0.018 delayed neutrons per ^{235}U fission, and 2.5 prompt neutrons. In the region where the reactor is just critical the negative term $(k_p - 1)/\tau$ is most important, and we may expand eq. (5.6.7) to give

$$\lambda = \frac{(k-1)\lambda_\beta}{\tau\left\{\lambda_\beta - \frac{(k_p-1)}{\tau}\right\}}$$

(the other root is large and negative). If we set

$$k_d = \frac{0.018}{2.5} k_p = 0.007 k_p,$$

then for $k = 1$, $k_p - 1 = -7 \times 10^{-3}$ and so $(k_p - 1)/\tau = -70$ which is indeed large in comparison with λ_β ($\sim 10^{-1} - 10^{-2}$).

If $k - 1 = 10^{-4}$ then $k_p - 1 = -6.9 \times 10^{-3}$ and if $\lambda_\beta = 10^{-1}$

$$\lambda = 1.45 \times 10^{-3} \text{ sec}^{-1}.$$

The time scale of the reactor will then be given by $1/\lambda = 690$ sec ~ 11 min, while without delayed neutrons a value $k - 1 = 10^{-4}$ gives a time scale of 1 sec. A reactor subcritical on prompt neutrons but critical on prompt plus delayed neutrons can be controlled.

Self-sustaining Nuclear Reactions and Nuclear Energy Sources

5.7. A note on breeding and fast neutron reactors

In Section 5.3 we pointed out that absorption of a neutron by ^{238}U leads via two successive β-decay processes to the even–odd nucleus $^{239}_{94}$Pu which is fissile by slow neutrons. Similarly capture of a neutron by $^{232}_{90}$Th leads to the fissile nucleus ^{233}U. Material which becomes fissile as a result of seeding by neutron capture is called fertile material, and the conversion of non-fissile but fertile nuclei into fuel which can be consumed in a fission reactor is called breeding. This process is of great importance in making a reactor which can produce power competitively with chemically fuelled plants: it may be likened to catalytic cracking of heavy petroleum fractions to produce a fuel of sufficient volatility to be used in petrol engines. Suppose that for each fission one neutron went to sustain the reaction, while a second fertilized ^{238}U. Such a system would have constant power output and a constant number of fissile nuclei until most of the ^{238}U originally present was converted into ^{239}Pu. Now ^{235}U produces on average 2.5 neutrons per fission. If they were thermalized instantaneously for every one neutron absorbed by ^{235}U leading to fission, 0.18 will be absorbed by ^{235}U through radiative capture. The maximum number of neutrons η produced per absorption by a ^{235}U nucleus is thus only 2.1 at thermal energies. In order to sustain the chain reaction and replace the fuel burned this number must be equal to or greater than 2, after allowing for capture of neutrons by the moderators and other materials neither fissile nor fertile, and leakage from the core. A figure of 2.1 neutrons produced per neutron absorbed on ^{235}U is too marginal to allow a ^{235}U–^{238}U thermal breeder reactor. (A ^{233}U–^{232}Th mixture $\eta = 2.28$, would breed at thermal energies.)

In a fast reactor the neutrons have an energy spectrum close to the original spectrum resulting from fission, softened somewhat by the moderating effect of coolant, and indeed of the fuel itself. The neutrons are thus ~ 1 MeV in energy and can cause fission directly in the fertile material, and the number of neutrons produced per absorption on fissile material increases in the MeV region and so makes breeding practicable. Such a reactor must be fed initially with enriched fuel, for an assembly of naturally occurring uranium will not go critical. This would not be economical were it not for the prospect of breeding far more fissile material than the initial loading.

5.8. Thermonuclear reactions and stellar evolution. Introduction[†]

The fusion of light nuclei is a process which also releases energy until nuclei in the neighbourhood of iron have been built up: the mean binding energy has a maximum in this region of the nuclear periodic table. If four protons can be assembled to make a ^{4}He nucleus ~ 28 MeV will be given off in the process, an energy release of ~ 7 MeV per nucleon, to be compared with ~ 1 MeV per nucleon in fission. The light elements, and in particular hydrogen, are stable, however, under terrestrial conditions, and the potential energy stored in

---

[†] Very detailed discussion of nuclear astrophysics may be found in D. D. Clayton, *Principles of Stellar Evolution and Nucleosynthesis* (McGraw-Hill, 1968) and references given there.

Nuclear Physics

a tank of hydrogen can only be released when the mean energy of the protons is already ~ 1 MeV. The reason is that light nuclei are kept apart by the coulomb repulsion between them so that the short-range nuclear forces never have a chance to force a fusion.

Suppose we wish to fuse two deuterium nuclei to form ^{4}He. The energy released in such a transition is ~ 24 MeV, but the two nuclei must be brought to within $\sim 2\times 10^{-13}$ cm before the reaction can take place. The energy required to overcome the electrostatic repulsion of the two deuterons is thus e^2/r where

$$e = 4.8\times 10^{-10} \text{ esu}$$

and

$$r = 2\times 10^{-13} \text{ cm}; \quad \frac{e^2}{r} \sim 10^{-6} \text{ erg} \simeq 0.7 \text{ MeV}.$$

In a head-on collision each deuteron must have ~ 0.35 MeV in the centre of mass. While this coulomb barrier can only inhibit a transition rate and cannot prevent the process, the inhibition is sufficiently great that for all practical purposes these processes do not occur except under conditions of extreme temperature and pressure. Suppose that at an energy of 1 MeV in the centre of mass the cross section is one barn and suppose further that the nuclear matrix element is more or less independent of incident energy. With no coulomb barrier, the cross-section would follow a $1/v$ law but this will be modified by a barrier factor for two incoming charged particles

$$e^{-G} \approx e^{-2\pi e^2/\hbar v}$$

where v is the relative velocity. (This expression only applies when the energy in the centre of mass is very much less than the barrier height.) For a velocity $\sim 10^9$ cm/sec (MeV region)

$$\frac{2\pi e^2}{\hbar v} = 1.4$$

and the barrier factor is unimportant. At room temperature, however, the mean kinetic energy is ~ 0.025 eV. If two singly charged deuterium nuclei approach each other with 0.025 eV each, the relative velocity is 3×10^5 cm/sec. The exponent

$$\frac{2\pi e^2}{\hbar v} = 4.6\times 10^3$$

and

$$e^{-G} = e^{-4.6\times 10^3} = 10^{-2000}.$$

A cross-section of ~ 1 barn in the MeV region is thus increased by a factor $\sim 10^4$ from the flux factor and reduced by a factor $\sim 10^{2000}$ by the coulomb barrier. For comparison, neutrino cross-sections in the MeV region are $\sim 10^{-20}$ barn. Another way of looking at this number is to suppose that if in the absence of a coulomb barrier two deuterons kept close to each other coalesced in $\sim 10^{-20}$ sec, then the coulomb barrier increases this time to $\sim 10^{1980}$ sec, and the age of the universe is only $\sim 10^{16}$ sec.

While it is relatively easy to cause exothermic nuclear reactions by bombarding nuclei with ions from an accelerator, if thermonuclear reactions are to take place in a large mass

Self-sustaining Nuclear Reactions and Nuclear Energy Sources

of material under circumstances approaching thermal equilibrium, then our simple calculation shows that the mean kinetic energy of the nuclei must be ~ 0.1 MeV, corresponding to a temperature of $\sim 10^9$ °K. Such temperatures are found in the interior of the stars.

The sun produces energy almost entirely by hydrogen burning, with helium as an end product. The mass of gas constituting the sun must have been raised to a temperature sufficient for the energy released in hydrogen burning to equal the radiated energy, and the initial input of energy to get the hydrogen burning started must have come from the conversion of gravitational energy. We can easily estimate the gravitational energy released in the contraction of a diffuse cloud of gas to a body the size of the sun. Suppose the initial cloud of gas to have had the mass of the sun and to have been spread over a volume of space very large in comparison with the present volume. For simplicity suppose the density of the sun to be approximately constant and equal to ϱ. Then

$$E_{\text{released}} = \int_0^{R_\odot} \frac{G}{r} \left(\frac{4\pi r^3 \varrho}{3} \right) 4\pi r^2 \varrho \, dr = \frac{3}{5} G \frac{M_\odot^2}{R_\odot} \tag{5.8.1}$$

where $M_\odot$ is the solar mass and $R_\odot$ the radius.

$$M_\odot = 2 \times 10^{33} \text{ g,}$$
$$R_\odot = 7 \times 10^{10} \text{ cm at the photosphere,}$$
$$G \sim 6.6 \times 10^{-8} \text{ cgs units.}$$

So
$$E_{\text{released}} = 2 \times 10^{48} \text{ ergs.}$$

The energy of a mass of gas at temperature T is $\sim RT$ per mole, and one mole of ionized hydrogen is approximately $\frac{1}{2}$ g. If the sun is pure atomic hydrogen then it contains $\sim 2 \times 10^{33}$ moles and at a temperature T the thermal energy will be $2 \times 10^{33} RT$: the gas constant R has a value 8.31×10^4 erg mole^{-1} degree^{-1}. If all the gravitational energy released in the contraction of the sun to the present size was converted into heat, then the temperature is given by $2 \times 10^{48} = 2 \times 10^{33} RT$; $T \simeq 10^7$ °K. Despite the large number of dubious assumptions we have made in this calculation, the central temperature of the sun is indeed $\sim 10^7$ °K: a central temperature of this order is necessary for the internal pressure to support the weight of the overlying layers. This may be seen from an approximate solution to the equation of hydrostatic equilibrium for a star:

$$\frac{dP}{dr} = -\frac{GM(r)\varrho(r)}{r^2} \tag{5.8.2}$$

where P is the pressure at a radius r, $M(r)$ is the mass contained within radius r, and $\varrho(r)$ is the local density. Then approximating

$$\frac{P_c}{R_\odot} \simeq \frac{GM_\odot \bar{\varrho}}{R_\odot^2} \tag{5.8.3}$$

where $\bar{\varrho}$ is the mean density and the central pressure $P_c = NRT_c$ where T_c is the central temperature and $N = \varrho_c/\mu$, ϱ_c being the central density and μ the mean molecular weight.

Nuclear Physics

Then
$$T_c \simeq G \frac{M_\odot}{R_\odot} \frac{\mu}{R}, \qquad (5.8.4)$$

$$R = 8.3 \times 10^7 \text{ erg/mole/°K},$$
$$M_\odot = 2 \times 10^{33} \text{ g},$$
$$R_\odot = 7 \times 10^{10} \text{ cm},$$
$$G = 6.6 \times 10^{-8} \text{ cgs},$$

whence $T_c \approx 10^7$ °K once more. The exact number is dependent on the detailed structure: both the variation of density and temperature with radial distance from the centre of a star depend on the type and distribution of the nuclear energy source and on the energy transport properties of the material making up the star. In the central regions the temperature and pressure of a star will be at a maximum, and because of the continual loss of radiation from the surface a temperature gradient will exist in the star, and the energy which is generated in the core will transport outward and become degraded until it is radiated from the surface. The surface temperature of the sun is 6000°K and so the mean temperature gradient is 1°K per 100 metres.

If the temperature of $\sim 10^7$ °K in the interior of the sun was insufficient to drive the thermonuclear processes at a rate capable of sustaining the sun in equilibrium, then the sun would contract, releasing more gravitational energy and hence raising the interior temperature until the hydrogen burning went fast enough to prevent further gravitational collapse. We must conclude that a temperature of $\sim 10^7$ °K is sufficient to generate thermonuclear power at a rate that will balance the energy loss.

5.9. Hydrogen-burning processes in the stars

The most important reactions occurring in the sun, and all main sequence stars of solar mass or lighter are those of the proton–proton chains, which in initially pure hydrogen take the PP I form:
$$p+p \rightarrow d+e^++\nu,$$
$$p+d \rightarrow {}^3\text{He}+\gamma,$$
$${}^3\text{He}+{}^3\text{He} \rightarrow {}^4\text{He}+2p+\gamma,$$

which provides the most direct mechanism for conversion of hydrogen into helium. Because two out of four protons must be converted into neutrons to build a ^{4}He nucleus from hydrogen, the weak interactions play a crucial role. If there were no weak interactions the conversion of pure hydrogen into helium could not take place.

The rate of burning in the p–p chain is controlled by the first reaction, the rate of which is suppressed by both a coulomb barrier and the weak coupling: the other two are only suppressed by coulomb barriers. If a p–p bound state existed, then the cross-section for formation of this state would be controlled by the coulomb barrier, and it could decay at leisure. In the absence of a bound state, the decay must take place within $\sim 10^{-20}$ sec which is at least 10^{10} shorter than a characteristic β-decay time scale.

Self-sustaining Nuclear Reactions and Nuclear Energy Sources

We may use the results of the previous two chapters to form a crude estimate for the cross-section for the process

$$p+p \to d+e^+ +\nu.$$

Restricting ourselves to s-waves, the $^3S_1$ state of two protons does not exist, and the reaction is an allowed Gamow–Teller process. In the absence of any coulomb repulsion, we could approximate the s-wave function $\sin kr/kr$ by unity, and the deuteron wave function by the zero range approximation result (Section 4.9)

$$\Psi_d(r) \simeq \sqrt{\frac{2|k_{1d}|}{4\pi}} \frac{e^{-|k_{1d}|r}}{r}.$$

As usual, we approximate the plane wave electron and neutrino wave functions by unity in the interaction region. The matrix element for the transition would then be

$$\mathcal{M} \sim g \int \sqrt{\frac{2|k_{1d}|}{4\pi}} \frac{e^{-|k_{1d}|r}}{r} d^3r, \qquad (5.9.1)$$

$$|\mathcal{M}|^2 \sim g^2 \frac{8\pi}{k_{1d}^3} \quad \text{where} \quad k_{1d} = 2.32 \times 10^{12} \text{ cm}^{-1}.$$

The cross-section becomes

$$\frac{d\sigma}{dp}(p+p \to d+e^+ +\nu) \simeq \frac{2\pi}{\hbar v_{pp}} \left\{ g^2 \frac{8\pi}{|k_{1d}|^3} \right\} \frac{(4\pi)^2 p^2(E-E_0)^2}{(2\pi\hbar)^6 c^3} \qquad (5.9.2)$$

for electron momentum p. Integrating the phase space over the electron momentum yields approximately

$$\frac{(4\pi)^2}{(2\pi\hbar)^6} \frac{1}{c} \frac{p_0^5}{30}$$

if we set $pc \sim E$. In the absence of a coulomb barrier, then, we have

$$\sigma(p+p \to d+e^+ +\nu) \simeq \frac{1}{v_{pp}} \frac{2}{\pi^2 \hbar^7 c} \frac{g^2 p_0^5}{15|k_{1d}|^3}. \qquad (5.9.3)$$

For a centre of mass energy of 1 MeV, $v_{pp} \sim 2 \times 10^9$ cm sec^{-1} and $E_0 = 1.9$ MeV. Then at an energy ~ 1 MeV we expect

$$\sigma \approx 10^{-48}\text{--}10^{-47} \text{ cm}^2 = 10^{-24}\text{--}10^{-23} \text{ barn}.$$

At low-incident energies E_0 is approximately constant and equal to 0.9 MeV. Then

$$v_{pp}\sigma \approx \frac{2}{\pi^2\hbar^7 c} \frac{g^2 p_0^5}{15|k_{1d}|^3} \approx 0.66 \times 10^{-40} \text{ cm}^3 \text{ sec}^{-1} \qquad (5.9.4)$$

at energies $\ll 1$ MeV in the absence of a coulomb barrier. Replacing our approximate integration over phase space by an exact integration reduces this value by $\sim 40\%$.

This value will be modified by a coulomb barrier factor

$$\approx \left(\frac{E_B}{E}\right)^{1/2} e^{-G(v)} \qquad (4.13.41)$$

Nuclear Physics

where

$$G(v) = \frac{2\pi Z_1 Z_2 e^2}{\hbar v} \qquad (4.13.46)$$

and E_B is the barrier height at the nuclear radius.

$$E_B = \frac{e^2}{R} \simeq 0.7 \text{ MeV}.$$

At the mean thermal energy corresponding to a temperature $\sim 10^7 \,°\text{K}$, $E \sim 1\text{ KeV}$ and the exponential factor $\sim e^{-24}$

$$v_{pp}\sigma \approx 3 \times 10^{-49} \text{ cm}^3 \text{ sec}^{-1},$$
$$\sigma \sim 5 \times 10^{-33} \text{ barn}.$$

Even at energies ~ 1 MeV where the coulomb barrier has become insignificant, the cross-section $\sim 10^{-23}$ barn is too low to be observed experimentally with present-day techniques. However, the pp phase shifts are known sufficiently well and our theoretical understanding of the weak interactions is sufficiently good that a refined version of the calculation we have just sketched may be made with complete confidence.

While we expect a cross-section $\sim 10^{-33}$ barn at the mean thermal energy corresponding to $10^7 \,°\text{K}$, the average cross-section at $10^7 \,°\text{K}$ will substantially larger. The coulomb barrier is such a rapidly varying function of velocity that the high velocity tail of the Maxwell–Boltzmann distribution will be most effective in causing interactions. This distribution is given by

$$N(v) \propto v^2 e^{-mv^2/2kT} \qquad (5.9.5)$$

where v is the relative velocity of the nuclei and m the reduced mass of the system. $N(v)\, dv$ is the number of particles between v and $v+dv$.

Since the phase space in the final state of these exothermic reactions is almost constant in the keV region the appropriate average is

$$\frac{\int_0^\infty v^2 e^{-mv^2/2kT} \left\{ \frac{1}{v^2} e^{-G(v)} \right\} dv}{\int_0^\infty v^2 e^{-mv^2/2kT}\, dv} \qquad (5.9.6)$$

where we have written

$$\sigma \propto \frac{1}{v^2} e^{-G(v)}.$$

One factor of $1/v$ is the flux factor and the second factor comes from the coulomb barrier. Hopefully the fall off in $N(v)$ with increasing v will kill the integral in the numerator before the approximate expression we have used for the barrier factor becomes invalid.

For problems involving nuclear reaction rates, it is more useful to evaluate the average rate rather than the average cross-section. If we write

$$\sigma \simeq \frac{1}{v} \left(\frac{E_B}{E}\right)^{1/2} \alpha e^{-G(v)} \qquad (5.9.7)$$

Self-sustaining Nuclear Reactions and Nuclear Energy Sources

then

$$\langle \sigma v \rangle = \alpha C(T) \tag{5.9.8}$$

where

$$C(T) = \frac{\int_0^\infty \left(\frac{E_B}{E}\right)^{1/2} v^2 e^{-mv^2/2kT} e^{-G(v)} dv}{\int_0^\infty v^2 e^{-mv^2/2kT} dv}. \tag{5.9.9}$$

Since $\frac{1}{2} mv^2 = E$, the centre of mass energy of the two interacting nuclei, we may write

$$C(T) = \frac{\int_0^\infty E_B^{1/2} e^{-E/kT} e^{-\beta/E^{1/2}} dE}{\int_0^\infty E^{1/2} e^{-E/kT} dE} \tag{5.9.10}$$

where

$$\beta = \frac{\sqrt{2m\pi e^2}}{\hbar}. \tag{5.9.11}$$

The approximately constant factor α is equal to the value the product σv would have in the absence of a coulomb barrier and is $\approx 10^{-40}$ for the pp reaction. For reactions where the cross-section has been measured at a particular energy E_s the factor α is given by

$$\alpha = v_s \sigma_s e^{\beta/E_s^{1/2}} \left(\frac{E_s}{E_B}\right)^{1/2}. \tag{5.9.12}$$

The integral in the numerator of eq. (5.9.10) must be evaluated numerically. The biggest contribution will come from close to the maximum of the exponential factor, given by

$$\frac{d}{dE} \{-E/kT - \beta/E^{1/2}\}|_{E_{\max}} = 0$$

or

$$E_{\max} = \left(\beta \frac{kT}{2}\right)^{2/3}. \tag{5.9.13}$$

The mean centre of mass energy of the nuclei is $\bar{E} = \frac{3}{2} kT$ (and $kT = 0.025$ eV at 300°K). At 10^7 °K, $\bar{E} = 1.25$ keV. For two protons the constant $\beta = 7.3 \times 10^2$ (eV)$^{1/2}$ [$\beta/E^{1/2} = 1$ when $E = 0.55$ MeV] and so at 10^7 °K $E_{\max} = 4.4$ keV.

The numerator of the expression for $C(T)$ may now be evaluated by approximating to some appropriate shape centred on $E_{\max}$. We suppose this to be

$$E_B^{1/2} e^{-E_{\max}/kT} e^{-\beta/E_{\max}^{1/2}} e^{-\frac{(E-E_{\max})^2}{\varepsilon^2}}. \tag{5.9.14}$$

This function has the correct height at $E = E_{\max}$ and we fix ε by requiring that not only the first derivative of $e^{-E/kT} e^{-\beta/E_{\max}^{1/2}}$ and $e^{-(E-E_{\max})^2/\varepsilon^2}$ match at $E = E_{\max}$ (which is included

Nuclear Physics

in the definition of E_{max}) but also that the second derivatives match, which yields

$$\varepsilon^2 = \frac{8}{3} \frac{E_{max}^{5/2}}{\beta}. \tag{5.9.15}$$

Now

$$\int_0^\infty e^{-(E-E_{max})^2/\varepsilon^2} \, dE = \frac{\sqrt{\pi}\varepsilon}{2}$$

and

$$\int_0^\infty E^{1/2} e^{-E/kT} \, dE = \frac{\sqrt{\pi}}{2} (kT)^{3/2}.$$

Then the dimensionless quantity $C(T)$ becomes

$$C(T) = \frac{\varepsilon E_B^{1/2}}{(kT)^{3/2}} e^{-E_{max}/kT} e^{-\beta/E_{max}^{1/2}}$$

$$= \sqrt{8/3} \left(\frac{E_B}{kT}\right)^{1/2} \left(\frac{E_{max}}{kT}\right) \left\{\frac{E_{max}^{1/2}}{\beta}\right\}^{1/2} e^{-E_{max}/kT} e^{-\beta/E_{max}^{1/2}}. \tag{5.9.16}$$

This is to be compared with

$$\left(\frac{E_B}{kT}\right)^{1/2} e^{-\beta/(kT)^{1/2}}$$

which obtains if the coulomb barrier factor is evaluated at $E = kT$. At 10^7 °K for two protons the values are respectively $\sim 10^{-5}$ and $\sim 10^{-10}$ taking $E_B = 0.7$ MeV.

The effect of folding together the coulomb barrier and the Maxwell–Boltzmann factor thus can hardly be ignored: for two protons at 10^7 °K the reaction rate is $\sim 10^5$ times the value obtained by merely substituting the mean energy. (We have laboured this point because it is not only important in the context of stellar energy sources but also in the problems of terrestrial thermonuclear reactors.)

The solar constant is the radiant energy flux at the earth's orbit and is 1.93 cal cm^{-2} min^{-1} = 1.34×10^6 erg cm^{-2} sec^{-1}. The energy generated by the sun is thus 3.8×10^{33} erg sec^{-1}. The energy generated in the proton–proton chain is easily computed. The binding energy of deuterium is 2.225 MeV. In assembling a deuteron from two protons, one proton has to be turned into a neutron and a positron, so the energy released is $2.225 - (M_n + m_e - M_p)c^2 = 0.42$ MeV. A bit over half this energy will go to the neutrino and so escapes from the star without contributing to the internal energy, but another 1.02 MeV is released on annihilation of the positron. The useful energy is thus

$$p + p \, (+e^-) \rightarrow d + \nu + (e^+ + e^-) \sim 1.2 \text{ MeV}.$$

The difference in binding energy alone gives the energy released in the other two steps of the chain

$$d + p \rightarrow {}^3He + \gamma \quad 5.49 \text{ MeV}$$

and

$$^3He + {}^3He \rightarrow {}^4He + 2p \quad 12.86 \text{ MeV}.$$

Self-sustaining Nuclear Reactions and Nuclear Energy Sources

Since the energy released in these two steps appears as either photons or kinetic energy of nucleons and nuclei, it all contributes to the useful energy. The useful energy released per deuteron formed is thus 13.12 MeV on completion of the last two stages of the proton–proton chain. (It is interesting to note that the last stage of this sequence is

$$^3He + {}^3He \rightarrow {}^4He + 2p + \gamma$$

rather than

$$d + d \rightarrow He^4 + \gamma.$$

The deuterium abundance remains very small because it is consumed in the reaction

$$p + d \rightarrow He^3 + \gamma$$

but ^4Li is unbound so that the overwhelming preponderance of protons does not destroy ^3He as it destroys d.)

The rate at which deuterium is formed in the sun is the ratio of energy radiated per second to the energy released per deuteron formed and is 1.8×10^{38} reactions/sec.

The reaction rate for deuterium formation is $\approx 10^{-5} \times 10^{-40}$ sec^{-1} for unit density of protons. The total reaction rate is then

$$10^{-45} n_p^2 V = 1.8 \times 10^{38} \tag{5.9.17}$$

where there are n_p protons/cm^3 and the V is the volume in which protons are interacting. Thus

$$n_p^2 V \simeq 10^{83}.$$

Even at 10^7 °K the weight of the overlying layers will crush the central regions to a density ~ 100 g cm^{-3}. The proton density n_p is thus $\sim 6 \times 10^{25}$ whence $V \simeq 10^{31}$ cm^3 corresponding to a volume circumscribed by $\sim \frac{1}{10}$ of the solar radius. Thus despite the very low reaction rate for the controlling weak interaction process

$$p + p \rightarrow d + e^+ + \nu$$

the solar energy output can be supplied by the PP I reactions taking place in a central core of the sun containing only $\sim 10^{-3}$ of the solar volume at the photosphere. To do better than our rough estimates is a job for the professional astrophysicist. In Fig. 5.9.1 we plot the temperature and the integrated rate of energy production in the sun as a function of radius. Our crude calculations of the last few pages have demonstrated not only the ability of the proton–proton chain to maintain the sun's power output but have also come surprisingly close to giving the radius of the energy-producing core.

It is worth mentioning that with ^4He present in the core of the star, either generated by hydrogen burning or present in the original formation, there are other possible paths that the hydrogen-burning chain may take after the formation of ^3He:

$$^3He + {}^4He \rightarrow {}^7Be + \gamma$$

followed by either

$$^7Be + e^- \rightarrow {}^7Li + \nu$$

Nuclear Physics

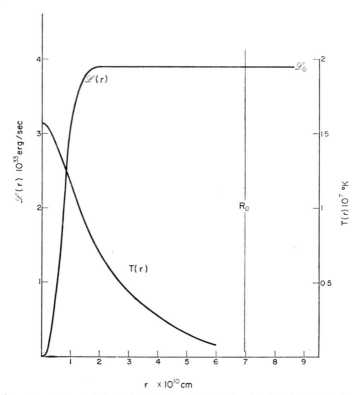

FIG. 5.9.1. The figure shows the variation of the temperature and luminosity of the sun with radius. The later quantity is the total energy passing outward through a sphere of radius r per second and so for a star in equilibrium corresponds to the energy generated per second in a volume of radius r. Our approximate calculation yielding the result that the energy-producing core has a radius $\sim 0.1 R_\odot$ is in excellent agreement with these curves which are calculated from the model of the present sun. [See, for example, D. D. Clayton, *Principles of Stellar Evolution and Nucleosynthesis*, pp. 481 ff. (McGraw-Hill, 1968).]

(^{7}Be can *only* decay by electron capture),

$$^7\text{Li} + p \rightarrow \text{He}^4 + \text{He}^4 \quad \text{PP II}$$

or

$$^7\text{Be} + p \rightarrow {}^8\text{B} + \gamma$$
$$^8\text{B} \rightarrow {}^8\text{Be} + e^+ + \nu$$
$$^8\text{Be} \rightarrow {}^4\text{He} + {}^4\text{He} \quad \text{PP III}$$

(the ground state of ^{8}Be is unstable).

The relative importance of the PP I, PP II and PP III chains depends on the temperature, density and composition of the energy-producing core, and so on the age of the star and the initial composition. It is the reactions of the PP III chain which produce the energetic neutrinos recently searched for by Davis in the reaction[†]

$$^{37}\text{Cl} + \nu \rightarrow {}^{37}\text{A} + e^-.$$

[†] See Section 4.16.

Self-sustaining Nuclear Reactions and Nuclear Energy Sources

The rival process for the consumption of hydrogen in the stars is the catalytic carbon–nitrogen cycle, which may be entered at any of the four points indicated:

$$\to {}^{12}C + p \to {}^{13}N + \gamma$$
$$ {}^{13}N \to {}^{13}C + e^+ + \nu$$
$$\to {}^{13}C + p \to {}^{14}N + \gamma$$
$$\to {}^{14}N + p \to {}^{15}O + \gamma$$
$$ {}^{15}O \to {}^{15}N + e^+ + \nu$$
$$\to {}^{15}N + p \to {}^{16}O^* \to {}^{12}C + {}^{4}He$$

Again two positron–neutrino pairs are produced for the conversion of four protons into helium, and the catalytic nucleus emerges at the end of the cycle. The weak interaction stages consist of decay from nuclei stable against all except the weak interactions, in contrast to the PP chains in which the crucial positron decay must take place within $\sim 10^{-20}$ sec before the interacting protons bounce out of range of each other. In the CN cycle the rate is controlled by the coulomb barriers.

If we take the maximum charge encountered for a (p, γ) reaction in the CN cycle, $Z = 7$, then the constant β is given by

$$\beta = \sqrt{2m}\, \frac{7\pi e^2}{\hbar}.$$

The reduced mass m is approximately the nucleon mass for these reactions involving heavy targets, and so β is increased by $7\sqrt{2}$ over the value for PP reactions to 73×10^2 (eV)$^{1/2}$. The barrier height is

$$E_B \sim 3 \text{ MeV}$$

and

$$E_{\max} = \left(\beta \frac{kT}{2}\right)^{2/3} = 21 \text{ keV at } 10^7 \,{}^\circ K$$

to be compared with 4.4 keV for the PP chains.

To get an idea of the reaction rates in the CN cycle, suppose that at an energy high enough for the coulomb barrier not to matter much the (p, γ) processes have cross-sections $\sim 10^{-2}$ barn. This guess is made up from the typical nuclear cross-section of 1 barn for formation of the compound nucleus, coupled with an assumed branching ratio for γ-decay of $\sim e^2/\hbar c \sim 10^{-2}$.

This gives us our estimated value for the constant α in (p, γ) reactions of the CN cycle of

$$\alpha \approx 10^{-16} \text{ cm}^3 \text{ sec}^{-1}.$$

We may check this estimate by using as input a measured value of the cross-section. At a laboratory energy of 0.1 MeV (well below the coulomb barrier) the cross-section for the

Nuclear Physics

process $C^{12}(p, \gamma) N^{13}$ is $\sim 10^{-10}$ barn. Then

$$\alpha = \sigma_s v_s \left(\frac{E_s}{E_B}\right)^{1/2} e^{\beta/E_s^{1/2}},$$

$$\sigma_s \sim 10^{-10} \text{ barn} \quad E_s = 0.1 \text{ MeV},$$
$$v_s \sim 5 \times 10^8 \text{ cm sec}^{-1},$$

whence $\alpha = 10^{-16}$ cm^3 sec^{-1} in excellent agreement with our guess.
This value $\alpha = 10^{-16}$ cm^3 sec^{-1} for the CN cycle is to be compared with the value

$$\alpha \simeq 10^{-40} \text{ cm}^3 \text{ sec}^{-1}$$

for the PP chains. The reaction rate in the CN cycle is then given by

$$N_N N_p \, 10^{-16} \, C_{CN}(T) \text{ cm}^{-3} \text{ sec}^{-1}$$

to be compared with

$$N_p^2 \, 10^{-40} \, C_{PP}(T) \text{ cm}^{-3} \text{ sec}^{-1}$$

where N_p is the density of protons and N_N the density of the least abundant catalytic nucleus under conditions of equilibrium in the CN cycle.

At 10^7 °K

$$C_{CN}(T) \approx 3 \times 10^{-30},$$
$$C_{PP}(T) \approx 5 \times 10^{-6},$$

so that the rate of power output from the CN cycle R_{CN} is given by

$$R_{CN}(10^7 \text{ °K}) = 6 \times 10^{-1} \, R_{PP}(10^7 \text{ °K}) \, N_N/N_p.$$

If we go up to a temperature of 1.5×10^7 °K

$$C_{CN}(T) = 1.5 \times 10^{-26},$$
$$C_{PP}(T) = 4.5 \times 10^{-5}.$$

So $\quad R_{CN} = (1.5 \times 10^7 \text{ °K}) = 3 \times 10^2 \, R_{PP}(1.5 \times 10^7 \text{ °K}) \, N_N/N_p.$

Over this temperature range $R_{PP} \sim T^5$ and $R_{CN} \sim T^{22}$. The CN cycle is depressed to an inferior status at $\sim 10^7$ °K by the large coulomb barrier presented to incident protons, but because the coulomb barrier is larger than for the PP chains, the reaction rate increases much faster with temperature. As the temperature increases the CN cycle overtakes the PP chain at some point depending on the density of CNO nuclei in the star. Our crude calculation has shown that we expect the CN cycle to dominate hydrogen burning for modest admixtures of catalytic nuclei at temperatures $\gtrsim 1.5$–2×10^7 °K. Because of the very strong temperature dependence of the CN cycle this conclusion depends only weakly on the accuracy of our cross-section estimate.

The neutrino astronomy work of Davis and his colleagues is of great interest in comparing the PP chains and the CN cycle. While the neutrino energy spectrum from the PP chains is sensitive to both temperature and the admixture of ^{4}He in the burning hydrogen, the energy spectrum and absolute flux of neutrinos from the CN cycle is exactly calculable in terms of

Self-sustaining Nuclear Reactions and Nuclear Energy Sources

the power generated by the CN cycle. The upper limit given by Davis and his colleagues is sufficient to set an upper limit on the fraction of solar energy generated in the CN cycle: $\lesssim 9\%$ only.[†]

The CN cycle does have an extra branch to it. The excited ^{16}O formed by

$$^{15}N + p \rightarrow {}^{16}O^*$$

decays electromagnetically with a partial rate of $\sim 4 \times 10^{-4}$ of the dominant α-emitting mode. A sequence of reactions

$$^{16}O + p \rightarrow {}^{17}F + \gamma,$$
$$^{17}F \rightarrow {}^{17}O + e^+ + \nu,$$
$$^{17}O + p \rightarrow {}^{14}N + He^4$$

consumes the ^{16}O and returns ^{14}N to the cycle. Thus doping the stellar core with any of the nuclei ^{12}C, ^{13}C, ^{14}N, ^{15}N, ^{16}O, ^{17}O will get the carbon–nitrogen cycle started.

The initial stages of stellar evolution emerge as follows. A mass of gas, mostly hydrogen, contracts under the influence of its own gravitational field. Of the potential energy released, some is emitted as radiation and some is converted into internal heat, and in the process the protostar becomes self-luminous. In the absence of large internal pressures the initial stages of the collapse are rapid in comparison with the quasi-equilibrium of hydrogen burning, and because the gravitational energy is being released quickly the turbulent contracting mass of gas is highly luminous. In this phase of evolution the radiated energy is produced solely by the conversion of gravitational potential energy. The gravitational contraction continues to increase the density and temperature, is slowed down as the internal pressure builds up, but is only terminated by the onset of thermonuclear reactions in the core of the star. An equilibrium is set up when hydrogen burning in the core supplies enough energy to make good that lost from the surface as radiation and at the same time maintain the internal pressures needed to counter the gravitational collapse. The more massive a star, the greater the internal pressures needed to support the weight of the outer regions and hence the higher the central temperature needed for equilibrium. The higher the temperature, the faster the hydrogen burning goes: this is purely an effect of the coulomb barrier. Above $\sim 2 \times 10^7$ °K the CN cycle will be responsible for most of the energy generation (unless the core is undoped with C, N or O) and the CN cycle is much more temperature sensitive than the reactions of the PP chains. Thus when a star has reached an equilibrium maintained by hydrogen burning in the core, the greater the mass the faster the fuel is consumed. The luminosity increases as the third or fourth power of the mass, and the greater the mass, the faster the star will evolve. A star in an equilibrium maintained by hydrogen burning in the central core is said to lie on the main sequence. We should also note that in a protostar of very low mass the central temperature never rises high enough to ignite the hydrogen: the gravitational collapse is stopped by electron degeneracy pressure. (The planet Jupiter, mass $\sim 10^{-3}$ solar masses, is not self-luminous and has a surface temperature $\sim 100°K$.) The sun is a fairly typical main sequence star and is believed to be $\sim 4 \times 10^9$ years old. The sun is currently radiating 3.8×10^{33} ergs/sec which corresponds to the conversion of $\sim 5 \times 10^{14}$ g of hydrogen to helium

---

[†] J. N. Bahcall, N. A. Bahcall and G. Shaviv, *Phys. Rev. Lett.* **20**, 1209 (1969).

Nuclear Physics

each second. Over 4×10^9 years, $\sim 6\times 10^{31}$ g will have been converted to helium if the present rate of energy generation is typical of this period, and if this rate continued into the future the projected hydrogen-burning life of the sun is $\sim 10^{11}$ years.

5.10. Stellar evolution during hydrogen burning

The understanding of the hydrogen-burning processes as the principal source of stellar energy was established by the end of the 1930s, although work on details is still going on. The basic clue to stellar evolution off the main sequence was provided during the 1940s by

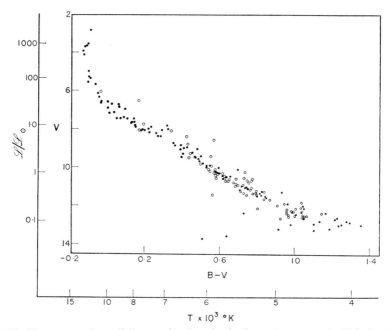

FIG. 5.10.1. The Hertzsprung–Russell diagram for the galactic cluster known as the Pleiades. The majority of stars lie on the main sequence, but the brightest stars have evolved significantly upwards and to the right. The ordinate is the visual magnitude of the stars and the abscissa the colour index. The figure is from R. I. Mitchell and H. L. Johnson, *Ap. J.* **125,** 418 (1957). [By permission of the University of Chicago Press. Copyright 1957 by the University of Chicago.] We have added approximate surface temperature and luminosity scales.

the study of clusters of stars. If we suppose that stars all evolve in much the same way, but with the rate at which this evolution proceeds being a sensitive function of the mass, then a cluster of stars of common origin will contain representatives of all the stages of stellar evolution. Given the distance of a star cluster from the solar system, the absolute luminosity of a star may be determined while the surface temperature may be inferred from the distribution of the radiated energy in the spectrum. A two-dimensional plot of these quantities is known as a Hertzsprung–Russell diagram, examples of which are shown in Figs. 5.10.1 and 5.10.2. On such a diagram the main sequence is a line running from the lower right-hand corner to the upper left-hand corner, the main sequence luminosity increasing with temperature. Clusters of stars lying within the galactic plane are mostly young and the majority of stars within them lie in the region of the main sequence: the Pleiades form such a cluster. The more populous and denser globular clusters are found well outside the galactic plane and are presumably older: if clusters are formed from hydrogen lying originally in the galactic plane then enough time must have elapsed since the formation of the globular clusters for them to spread through the galactic halo. The stars of these globular clusters, when plotted on the Hertzsprung–Russell diagram, are distributed very differently from those of the

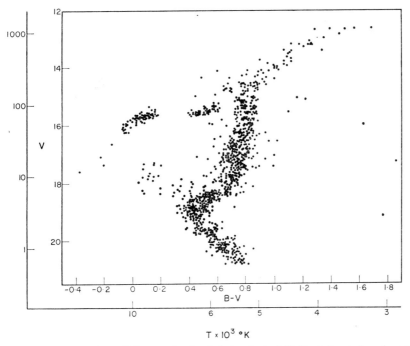

FIG. 5.10.2. The Hertzsprung–Russell diagram for the globular cluster M3. Evolution is far advanced and the scatter of stars represents an approximate evolutionary track. The gap on the horizontal branch contains the oscillating RR Lyrae stars. The ordinate is visual magnitude and the abscissa the colour index. The figure is from H. C. Johnson and A. Sandage, *Ap. J.* **124**, 379 (1956). [By permission of the University of Chicago Press. Copyright 1956 by the University of Chicago.] We have added approximate surface temperature and luminosity scales.

galactic clusters. A portion of the main sequence corresponding to relatively cool stars is populated, but the population breaks away from the main sequence, upwards and to the right into the region of high luminosity and low surface temperature, the red giant region. In addition there is a horizontal band containing stars of roughly the same luminosity but a wide range of surface temperatures. This continuous track is to be supposed representative of some average evolutionary path on the Hertzsprung–Russell diagram.

Evolution away from the main sequence, then, proceeds first by an increase in luminosity with little change of surface temperature, while further evolution results in a continuing growth in luminosity accompanied by a substantial drop in surface temperature. A large

Nuclear Physics

luminosity and low surface temperature can only mean that the energy is being radiated from a very large surface area: a red giant not only has giant luminosity but is also giant sized. Familiar red giants in the night sky of the northern hemisphere are Arcturus, Aldebaran and Betelgeuse in Orion. (The latter is the first star for which the radius was directly measured. In 1920 Michelson found the angular diameter interferometrically. The distance of Betelgeuse was already known from its annual apparent excursions relative to the very distant stars. The parallax of Betelgeuse is 0.018″ and with the angular diameter of 0.047″ this at once gives the diameter of Betelgeuse as 2.6 times the radius of the eath's orbit, or 3.9×10^8 km and so the radius of Betelgeuse is ~ 278 solar radii.)

The explanation of this evolutionary track off the main sequence and towards the region of the red giants has little to do with nuclear physics, being caused by the accumulation of a core of helium at the centre of hydrogen-burning stars. If a mechanism existed capable of so mixing the material in a star that the composition remained homogeneous, this evolution upward and to the right on the Hertzsprung–Russell diagram would not take place. In such a star evolution off the main sequence would be governed by the slowly changing molecular weight of both the burning *and* the unburning material. The nuclear energy source keeps the star in equilibrium, and so the effect of efficient mixing would be to take the star from its initial stable configuration slowly through other stable configurations appropriate to increasing molecular weight (since this material in a star will be highly ionized throughout most of the volume, the mean molecular weight must be interpreted as the mean particle mass, and so for a star composed of pure hydrogen will be half the mass of atomic hydrogen, for pure helium approximately three-quarters the mass of atomic hydrogen, while heavy elements with $A \sim 2Z$ each atom will contribute $Z+1$ particles but a weight $2Z$ and so for completely ionized heavy elements the mean molecular weight will be 2). The particle pressure P at a given temperature is given by

$$P = NkT$$

where the particle density is N, and N is reduced by conversion of hydrogen into helium. The opacity of the star is proportional to the free electron density, and is also reduced by conversion of hydrogen to helium at a constant temperature and mass density. For a star in which the helium is mixed with hydrogen throughout the volume as it is formed equilibrium can only be maintained by the star contracting throughout, getting both hotter and more luminous. Such a star would evolve upward on the Hertzsprung–Russell diagram and off to the left of the main sequence.

This does not happen. There is no known mechanism[†] capable of stirring a star at a rate sufficient to prevent the formation of a helium-rich core in the course of the hydrogen burning and it is this feature which drives the star upward into the subgiant and on into the red giant configuration. Consider what happens in the centre of a star like the sun as the hydrogen in the energy-burning region is depleted and replaced by helium. As the helium concentration increases the particle density is reduced, and so the pressure in the core tends to drop. This disturbs the initial main sequence equilibrium and so the core will undergo a gravitational contraction. The release of potential energy results in the slowly collapsing core getting hotter. No new equilibrium can be established, however, for the hydrogen is continually

---

† In *very* light stars convection currents throughout the body of the star will achieve this.

Self-sustaining Nuclear Reactions and Nuclear Energy Sources

being consumed. Now the increase in temperature of the core implies a greater flux of energy through the unburning hydrogen, which must also get hotter and as a result the outer regions of the star tend to expand and the luminosity increases at the same time. The star thus tends to evolve upwards off the main sequence. The big difference from the hypothesis of homogeneous mixing lies in the fact that conversion of hydrogen to helium results in a pressure deficiency throughout the homogeneous star, but only in the core of an unmixed star.

Now the hydrogen-burning rates depend on the square of the proton density which is being continually reduced as the hydrogen is converted into helium. Energy generation in the core gradually ceases as the helium accumulates and hydrogen burning continues in a shell source surrounding the inert helium core. The temperature continues to rise as the helium core contracts under its own weight and the rising flow of energy through the outer regions of the star blows them outwards to form the enormous distended envelopes of a red giant.

None of these features is particularly obvious and the details of evolution off the main sequence have only become established as a result of numerical calculations made by many workers over a period between the 1940s and 1960s. The detailed behaviour of a star is a sensitive function of the initial mass and of the composition. (The abundance of metals in the outer regions of a star affects the opacity of the envelope: the easily ionized metals act as sources of extra electrons and so increase the concentration of H^- ions which play a dominant role in the opacity.) Numerical computations of the evolution of stars with an accumulating core of helium yield the result that massive stars leave the main sequence almost horizontally as the envelope distends and in distending absorbs the extra energy, the luminosity remaining roughly constant. In less massive stars, the hydrogen burning takes place over a larger fraction of the stellar volume and the luminosity goes up more rapidly as the envelope expands.

In stars of around a solar mass the contraction of the helium core is delayed by the Pauli exclusion principle. The number of states available to a free electron of momentum p is

$$\frac{4\pi p^2 \, dp}{(2\pi \hbar)^3} \times 2 \text{ per unit volume}$$

(where the factor of 2 comes from the two spin states). As these states are filled electrons must sit in states of progressively higher energy. If the top of this Fermi sea is comparable with or well above the thermal energies appropriate to a Boltzmann distribution, the electron gas must be described by Fermi–Dirac statistics and the pressure of an electron gas rises over that obtaining classically. In stars of around solar mass the central temperatures in the core are sufficiently low for this degeneracy pressure to provide extra support which defers the collapse of the core.† The temperature of the core still increases as more helium is added in outer, less dense, regions and the hydrogen-burning shell works its way outwards driving the envelope before it.

Evolution off the main sequence thus takes a star into a configuration in which a central core of helium is getting hotter as gravitational energy is released. The core is surrounded by a shell of burning hydrogen and beyond this shell there is a relatively cool distended envelope. The stage is now set for the re-entry of nuclear physics into the problem of stellar

---

† This is discussed further in the next section.

Nuclear Physics

evolution. If the central temperature and density of the helium core become sufficiently elevated, helium burning will start up. We must point out, however, that this does not necessarily happen. Just as an aggregation of a relatively small mass of hydrogen can reach a stable configuration with the weight of the material balanced by electron degeneracy pressure at a temperature so low that hydrogen burning does not start, so with a lightweight helium core. The coulomb barrier for helium burning is greater than for the proton–proton chains which dominate hydrogen consumption in a light star, and helium burning will only take place at temperatures $\sim 10^8\,°K$. If in the consumption of the hydrogen and the growth of the helium core the temperature never becomes high enough for the helium to ignite and the helium is prevented from contraction by the degenerate electrons, then as the hydrogen is exhausted the star will be left as a homogeneous sphere of helium cooling to oblivion through the white dwarf phase. The energy such a white dwarf radiates into space cannot be taken from the electrons, for they have already filled all low-energy states available: it will come from the kinetic energy of the non-degenerate nuclei. In stars of mass less than 0.5 solar mass the temperature will not rise high enough to ignite the helium.

5.11. Helium burning

In contrast to the hydrogen-burning reactions of the PP chains and the CN cycle, the nuclear reactions which consume helium in a star are dominated by resonant processes. The first is the resonant α–α scattering which is responsible for building up an equilibrium concentration of the unstable ground state of ^{8}Be

$$^4He + {}^4He \rightleftharpoons {}^8Be.$$

The 0^+ ground state of ^{8}Be is only just unstable against breakup into two α's and occurs at 94 keV above the zero for two α. The width of this state is about 2.5 keV (the first two excited states are respectively 2^+ and 4^+ at 2.9 and 11.4 MeV: they may be neglected as far as helium burning is concerned). The first phase of helium burning is completed by a second resonant reaction

$$^8Be + {}^4He \rightarrow {}^{12}C + \gamma$$

which takes place through a 0^+ excited state of ^{12}C. This state lies 7.644 MeV above the ground state of ^{12}C but only 278 keV above the zero of $^8Be + {}^4He$. The width of this state is ~ 8 eV and on formation it usually decays straight back to $^4He + {}^8Be$. The branching ratio for photon emission to the ground state of ^{12}C is only $\sim 4 \times 10^{-4}$, when the 0^+ excited state decays via two E2 transitions going through a 2^+ excited state of ^{12}C at 4.433 MeV.

The cross-section for a resonant s-wave (α, γ) process will be given by

$$\sigma = \frac{4\pi}{k^2} \frac{\Gamma_\alpha \Gamma_\gamma}{(E-E_0)^2 + \Gamma^2/4}$$

where E_0 is the resonant energy. The factor Γ_γ contains the phase space for the final state and, since the reaction is exothermic, is independent of incident energy. The factor Γ_α describes the formation of the excited ^{12}C (and its decay back to $^4He + {}^8Be$) and so contains a

Self-sustaining Nuclear Reactions and Nuclear Energy Sources

phase space factor and a coulomb barrier factor

$$\Gamma_\alpha = \Gamma_{0\alpha} \frac{e^{-G(v)}}{e^{-G(v_0)}}$$

using the results of Section 4.13.

On the tail of such a resonance, then $\sigma \propto (1/v^2)e^{-G(v)}$ just as for a non-resonant reaction. Thus if the result of folding the Maxwell–Boltzmann distribution with the resonant shape gives a contribution mostly from the tail of the resonance, then

$$\langle \sigma v \rangle_{\text{tail}} \propto C(T). \tag{5.11.1}$$

As the temperature increases and the resonant peak moves deeper into the Maxwell–Boltzmann distribution, a point is reached at which the average value of $\langle \sigma v \rangle$ is well represented by replacing the varying Maxwell–Boltzmann distribution by its value at the resonant peak, when

$$\langle \sigma v \rangle_{\text{peak}} \propto \frac{1}{(kT)^{3/2}} e^{-E_0/kT} \int_0^\infty \frac{dE}{(E-E_0)^2 + \Gamma^2/4}. \tag{5.11.2}$$

Thus when the most effective energy

$$E_{\max} \ll E_0, \quad \langle \sigma v \rangle \propto C(T)$$

and we can write $C(T)$ in the form

$$C(T) = \frac{2}{\sqrt{3}} \left(\frac{E_B}{kT}\right)^{1/2} \left(\frac{E_{\max}}{kT}\right)^{1/2} e^{-3E_{\max}/kT} \tag{5.11.3}$$

where the dominant term is the factor

$$e^{-3E_{\max}/kT}.$$

As $E_{\max}$ works its way towards the peak of the resonance, the variation becomes

$$\propto \frac{1}{(kT)^{3/2}} e^{-E_0/kT} \tag{5.11.4}$$

where the dominant term is $e^{-E_0/kT}$. Let us take the energy at which these two exponential factors are equal

$$E_{\max} = 1/3\, E_0.$$

Above this value the peak expression takes over, for E_0/kT varies as $1/kT$ while $E_{\max}/kT$ varies as only $1/(kT)^{1/3}$. Thus the 3α process may be expected to take off at an energy such that

$$E_{\max} \approx \tfrac{1}{3} E_0$$

where $E_0 = 278$ keV for the $^8\text{Be}(\alpha, \gamma)\, ^{12}\text{C}$ reaction. (A less-educated guess could have given us $E_{\max} \approx E_0$.)

$$E_{\max} = \left(\frac{\beta kT}{2}\right)^{2/3},$$

$$\beta = \frac{Z_1 Z_2 \pi e^2 \sqrt{2m}}{\hbar}.$$

Nuclear Physics

For ^{8}Be–^{4}He reactions $Z_1 = 2$, $Z_2 = 4$ and the reduced mass m is approximately 2.7 nucleon masses. The constant β is thus $8\sqrt{2.7}$ times the value for proton–proton interactions, giving

$$\beta = 9600 \quad (\text{eV})^{1/2}.$$

Then if

$$E_{\max} \approx 100 \text{ keV}, \quad kT = 6.4 \text{ keV}$$

or

$$T = 8 \times 10^7 \, ^\circ\text{K}.$$

As kT goes above this value, the very rapidly varying factor $e^{-E_0/kT}$, $E_0 = 278$ keV, comes in to dominate the reaction rate.

We may expect helium in the core of a star to ignite at temperatures $\sim 10^8 \, ^\circ$K. The values of temperature and density at which helium burning maintains the equilibrium of the helium core depends, of course, on the mass of the star.

The effect of the ignition of helium in the central regions of the helium core is to arrest the contraction of the core and expand it outwards. The temperature of the hydrogen-burning shell surrounding the helium core falls, energy production via hydrogen burning decreases and the luminosity of the star drops. This is accompanied by the contraction of the diffuse hydrogen envelope since the energy flux is no longer sufficient to maintain it against the gravitational attraction of the central regions. The star thus decreases in luminosity, its surface temperature rises and on the Hertzsprung–Russell diagram the star descends the red giant branch and moves out to the left into the horizontal region.

In a low mass star the central temperatures are low enough and the central pressures high enough for the helium core to be supported by electron degeneracy pressure at the time of ignition. This means that during the onset of helium burning the increase in temperature only increases the pressure due to the helium nuclei and the electron pressure is not affected. There is thus little tendency for the core to expand until the temperature rises high enough to lift electrons into unpopulated energy levels. Since the $3\alpha \to {}^{12}$C process is very temperature sensitive, and the core is an excellent thermal conductor (the electrons in a metal are highly degenerate) the helium core detonates rather than ignites. This phenomenon is called the helium flash. The end product is still a star moving off to the left of the red giant branch, for the outer layers damp the explosion. The helium flash raises the temperature of the core, removes the electron degeneracy and again the star takes on a configuration with helium burning in the centre of a non-degenerate core, hydrogen burning in a shell around the helium core and the whole is surrounded by an unburning hydrogen envelope moderating the energy and transporting it to the stellar surface.

The $3\alpha \to {}^{12}$C reaction is backed up by a series of further α-capture processes:

$$^{12}\text{C} + {}^4\text{He} \to {}^{16}\text{O} + \gamma.$$

This is believed to go through a p-wave resonant interaction in the tail of a 1^- excited state of ^{16}O which lies just below the zero energy level for ^{12}C + ^{4}He.

Subsequent processes are

$$^{16}\text{O} + {}^4\text{He} \to {}^{20}\text{Ne} + \gamma,$$
$$^{20}\text{Ne} + {}^4\text{He} \to {}^{24}\text{Mg} + \gamma,$$
$$^{24}\text{Mg} + {}^4\text{He} \to {}^{28}\text{Si} + \gamma.$$

Self-sustaining Nuclear Reactions and Nuclear Energy Sources

At each stage the coulomb barrier increases and nuclei beyond ^{16}O will only be synthesized in helium burning in massive stars, burning at relatively high temperatures. It is, however, clear that ^{16}O at least will be formed in the course of helium burning. The rate of ^{12}C formation is proportional to the density of ^4He and the density of ^8Be. The density of the ^8Be in equilibrium is proportional to the square of the ^4He density. Thus if there are N ^4He nuclei per unit volume, the production rate falls as N^3 as the ^4He in the burning core is exhausted. A point will come at which the remaining helium is consumed by the carbon to form ^{16}O.

The evolution of model stars past the helium ignition point has been computed in some detail.† A helium-burning star works its way out to the left of the red giant branch in the Hertzsprung–Russell diagram, although it still remains above the main sequence. Its surface temperature and luminosity work towards those values characteristic of the pulsating stars known as Cepheid variables and RR Lyrae stars. These are stars vibrating in a breathing mode. The RR Lyrae stars have periods $\sim \frac{1}{2}$ day, and the Cepheid variables periods of days (the existence of a direct relationship between the period and the luminosity of these pulsating variables allows them to be used to find the distance of any stellar system containing examples). RR Lyrae stars have surface temperatures $\sim 7000°K$ and luminosities ~ 100 times that of the sun: Cepheids have surface temperatures $\sim 5000°K$ and luminosities ~ 1000 times that of the sun. The driving mechanism of the pulsations is believed to be as follows. Suppose that in a certain region of a stellar envelope the opacity is an increasing function of temperature. Then induce a small breathing mode oscillation. As the zone

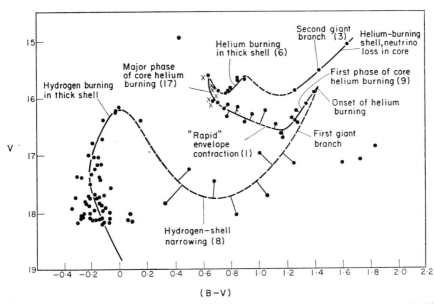

FIG. 5.11.1. Position in the colour-magnitude (H.R.) diagram of stars in the cluster NGC 1866. The curves represent a calculated average evolutionary track and the crosses indicate pulsating variables. [From I. Iben, *Ann. Rev. Ast. Ap.* **5**, 571 (1967).]

† For a review see the article by Icko Iben, Stellar evolution along and off the main sequence, *Ann. Rev. Ast. Ap.* **5**, 571 (1967).

Nuclear Physics

collapses inwards it gets hotter, the opacity grows and the zone is driven outwards again by the energy flux from the central regions of the star. On being driven outwards, the opacity of the zone drops, the dammed-up energy escapes and the now unsupported zone collapses inwards again. In the regions of a star where the temperature is just about right to ionize an abundant species of atom the opacity will indeed be an increasing function of temperature, for as the temperature goes up the atoms ionize and the increasing density of free electrons raises the opacity, thus increasing the temperature more. Numerical calculations give the result that in the hot but diffuse outer envelope of stars burning helium in the core, this effect can generate radial pulsation in the envelope with the right characteristics. The agent responsible for raising the opacity is the ionization of the He^+ ion. A striking piece of observational evidence is shown in Fig. 5.11.1.

A helium-burning star works its way through the pulsating variable zone (the pulsations only affect the diffuse envelope and do not penetrate to the energy-producing region) and then as a core of carbon and oxygen begins to accumulate within the helium and contract, the outer regions expand, the star moves back through the pulsating variable zone and eventually starts to climb the red giant branch for a second time.

5.12. Stellar evolution after helium burning

Most of the energy which may be liberated in the fusion of hydrogen into elements at the peak of the mean binding-energy curve has already been generated in the production of helium: the mean binding energy of helium is ~ 7.1 MeV and of iron ~ 8.8 MeV. All possible thermonuclear reactions beyond helium only release a further 1.7 MeV per nucleon. A star in which helium burning is going on is thus depleting the resources of energy seriously in the central regions and is in danger of going into the red. Evolution beyond the stage of helium burning in the central regions is governed by two phenomena which are worth a brief digression. The first is the Chandrasekhar limit: a star whose mass exceeds 1.45 solar masses can never be stabilized against gravitational collapse by the Pauli principle acting on degenerate electrons. The second phenomenon is radiation of neutrinos.

The Chandrasekhar limit is crucial to the evolution of massive stars beyond the helium-burning phase, and provides yet another example of the ubiquitous density of states factor. In quantum mechanics the number of states of a particle having momentum between p and $p+dp$ is given by

$$\frac{4\pi p^2 dp}{(2\pi\hbar)^3} \times 2$$

where the additional factor of two applies for particles of spin $\frac{1}{2}$. In a completely degenerate Fermi gas all levels up to some maximum momentum p_{max} are filled and so if the number of electrons per unit volume is N then

$$N = \frac{8\pi}{3} \left(\frac{p_{max}}{2\pi\hbar^3}\right)^2, \qquad (5.12.1)$$

Self-sustaining Nuclear Reactions and Nuclear Energy Sources

an expression we already encountered for nucleons in Section 1.5. The greater the electron density N, the greater the momentum of the electrons in the Fermi gas and so the greater the pressure they exert. This degeneracy pressure is just the uncertainty principle plus the Pauli exclusion principle at work.

Now the pressure exerted by particles of momentum p, velocity v, randomly directed is just

$$\tfrac{1}{3} pv.$$

Non-relativistically

$$v = \frac{p}{m_e},$$

and the pressure exerted by non-relativistic electrons is

$$P_D = \int_0^{p_{\max}} \frac{8\pi}{3} \frac{p^4 \, dp}{m_e (2\pi\hbar)^3} \propto p_{\max}^5 \propto N^{5/3}. \tag{5.12.2}$$

If, however, the material is so compressed that the momentum $p_{\max}$ is highly relativistic, then it will be a better approximation to write for the pressure P_D,

$$P_D = \int_0^{p_{\max}} \frac{8\pi}{3} \frac{cp^3 \, dp}{(2\pi\hbar)^3} = \frac{8\pi c}{12} \frac{p_{\max}^4}{(2\pi\hbar)^3} \propto N^{4/3} \tag{5.12.3}$$

and the degeneracy pressure due to relativistic electrons increases less rapidly with the electron density. For the problem of the Chandrasekhar limit it is convenient to express N in terms of the mass density ϱ of the material. For all compositions except hydrogen, the number of electrons is approximately half the number of nucleons, so that

$$N \approx \frac{1}{2} \frac{\varrho}{M_n}$$

where M_n is the nucleon mass. Then

$$P_D = K \varrho^{4/3}$$

where

$$K = \frac{\pi \hbar c}{8} 2^{-1/3} \left(\frac{3}{\pi}\right)^{1/3} \frac{1}{M_n^{4/3}} = 5.8 \times 10^{14} \quad \text{(cgs units)}. \tag{5.12.4}$$

A star in which the internal pressures at any radius r balance the gravitational attraction is described by the equation

$$\frac{dP}{dr} = -\varrho \frac{GM_r}{r^2} \tag{5.12.5, 5.8.2}$$

where P is the pressure at radius r, ϱ is the density at radius r, and M_r the mass enclosed within a radius r. Since

$$dM_r/dr = 4\pi r^2 \varrho$$

Nuclear Physics

we may rewrite (5.12.5) as

$$\frac{1}{r^2}\frac{d}{dr}\left(\frac{r^2}{\varrho}\frac{dP}{dr}\right) = -4\pi G\varrho. \tag{5.12.6}$$

Substitution of $K\varrho^{4/3}$ for P yields the equation describing a star in which the pressure is provided by relativistically degenerate electrons. The resulting equation must be solved numerically, but rather than do this we will tackle the problem approximately.

The existence of a limit follows from dimensional arguments without specifying the form of the density, $\varrho(r)$. Writing

$$\varrho(r) = \varrho_c f(r) \tag{5.12.7}$$

where $f(r)$ is dimensionless, the mass of the star is

$$M = \int_0^a 4\pi r^2 \varrho_c f(r)\, dr = x\varrho_c a^3. \tag{5.12.8}$$

The pressure at the centre due to the weight of the overlying layers is

$$P_W = \int_0^a \varrho_c f(r) \frac{G}{r^2}\, dr \int_0^r 4\pi r^2 \varrho_c f(r)\, dr = y\varrho_c^2 G a^2 \tag{5.12.9}$$

where x and y are dimensionless constants depending on the form of $f(r)$.

To prevent collapse, P_W must be balanced by the degeneracy pressure P_D at the centre

$$P_D = K\varrho_c^{4/3} = P_W = y\varrho_c^2\, Ga^2. \tag{5.12.10}$$

Substituting for the mass from eq. (5.12.8) *both ϱ_c and the stellar radius a cancel,*[†] yielding

$$K = yG\left(\frac{M}{x}\right)^{2/3} \tag{5.12.11}$$

and equilibrium is only possible with complete relativistic degeneracy for a unique value of the mass M_c

$$M_c = x\left(\frac{K}{yG}\right)^{3/2}. \tag{5.12.12}$$

If $M < M_c$ then the equilibrium configuration of the star will be one in which the degenerate electrons are not highly relativistic and the degeneracy pressure will support the star against collapse. If, on the other hand,

$$M > M_c$$

and the temperature has dropped low enough for the electrons to be both relativistic and degenerate, then the central pressure cannot sustain the star against gravitational collapse. Such a star must either shed its surplus mass, as might happen in a supernovae explosion with M only a little greater than M_c, or it will collapse in on itself until stopped by something else—like the repulsive core in nucleon–nucleon interactions.

[†] Note that this does NOT obtain for non-relativistic electrons, $P_D \propto \varrho^{5/3}$.

Self-sustaining Nuclear Reactions and Nuclear Energy Sources

The dimensionless quantities x and y must be of order unity, giving

$$M_c \approx \left(\frac{K}{G}\right)^{3/2} = 0.8 \times 10^{33} \text{ g } (0.4 M_\odot).$$

This simple calculation shows the Chandrasekhar limit to be of the order of 1 solar mass: to do better the proper form of $f(r)$ must be employed. For a sphere of helium in which the energy is transported by radiation the Chandrasekhar limit is[†]

$$M_c = 1.45 M_\odot.$$

Relativistic degeneracy will set in when the top of the Fermi sea is given by

$$p_{max} c \approx 1 \text{ MeV}$$

provided that the temperature corresponds to energies substantially less than this, that is temperatures $\ll 10^9$ °K,

$$p_{max} c = 2\pi\hbar c \left(\frac{3}{8\pi}\right)^{1/3} \left(\frac{\varrho}{2M_n}\right)^{1/3}$$

so that $p_{max} c = 1$ MeV corresponds to a density $\varrho \approx 10^7$ g cm^{-3} (cf. nuclear density $\approx 10^{15}$ g cm^{-3}).

The accumulating carbon–oxygen core in a star burning helium will be maintained at a temperature $\sim 10^8$ °K by the helium-burning shell with energy being slowly released as helium is converted into carbon and oxygen. As the mass of such a core rises above the Chandrasekhar limit, only the ignition of further thermonuclear processes can save the star from catastrophic collapse. However, at temperatures 10^8–10^9 °K there appears to exist an efficient mechanism for transporting energy out of the core without any heating of the stellar material: radiation of neutrinos.

In a carbon–oxygen core the origin of the neutrinos is not β processes but rather the direct Fermi interaction between four leptons. The only example studied in the laboratory is muon decay

$$\mu^+ \to e^+ \nu \bar{\nu}_\mu$$
$$\mu^- \to e^- \nu \bar{\nu}_\mu$$

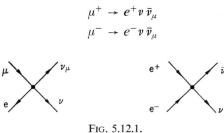

FIG. 5.12.1.

which goes through the standard Fermi interaction with coupling constant $g = 1.4 \times 10^{-49}$ erg cm^3. Similarly we expect

$$e^+ + e^- \to \nu + \bar{\nu}.$$

[†] S. Chandrasekhar, *An Introduction to the Study of Stellar Structure*, p. 423 (Dover, 1957), or see D. D. Clayton, *Principles of Stellar Evolution and Nucleosynthesis*, p. 161 (McGraw-Hill, 1968).

Nuclear Physics

Using the Fermi Golden Rule we expect the cross-section to be (see Section 4.16)

$$\sigma(e^+e^- \to \nu\bar{\nu}) \approx \frac{\pi}{\hbar c} g^2 \frac{4\pi p_\nu^2}{(2\pi\hbar)^3} \frac{dp_\nu}{dE_\nu} \tag{5.12.13}$$

where the neutrino momentum is ~ 1 MeV/c. Putting in numbers yields

$$\sigma(e^+e^- \to \nu\bar{\nu}) \approx 10^{-44} \text{ cm}^2$$

and the rate is

$$\approx 2c N_e^+ N_e^- \sigma$$
$$\approx 10^{-33} N_e^+ N_e^- \text{ cm}^{-3} \text{ sec}^{-1}.$$

The energy loss in neutrinos is thus

$$L_\nu \sim 10^{-33} N_e^+ N_e^- \text{ MeV cm}^{-3} \text{ sec}^{-1}$$

where N_e^+, N_e^- are the densities of electrons and positrons. At $\sim 10^9$ °K photons ~ 1 MeV will be present and so there will be an equilibrium concentration of positrons. At a density $\sim 10^7$

$$N_e^- \simeq \frac{10^7}{2 \times 1.6 \times 10^{-24}} \sim 10^{31}$$

when the energy loss in neutrinos becomes

$$L_\nu \approx 10^{-2} N_e^+ \text{ MeV cm}^{-3} \text{ sec}^{-1}.$$

The sun radiates $\sim 10^{39}$ MeV/sec. If a neutrino radiating core at $\sim 10^9$ °K and a density of 10^7 g cm^{-3} has the mass of the sun, then

$$L_\nu \approx 10^{24} N_e^+ \text{ MeV/sec}.$$

Thus N_e^+ need only be $\approx 10^{15}$ ($\sim 10^{-16} N_e^-$) for the neutrino luminosity to reach that of the sun.

The energy content of 1 cm^3 is $\sim 10^{31}$ MeV so that if $N_e^+ = N_e^-$ this energy could be radiated away in ~ 100 sec. If $N_e^+ \sim 10^{-2} N_e^-$ this becomes ~ 3 hr.

The equilibrium concentration of positrons may be calculated readily using the techniques of statistical mechanics. We shall not make this calculation here but merely note that N_e^+ will contain a factor

$$\approx e^{-2m_e c^2/kT}$$

and so the neutrino luminosity is a rapidly increasing function of both temperature and pressure.

Another important process is direct conversion of photons into $\nu\bar{\nu}$ pairs, through a virtual e^+e^- pair:

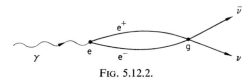

Fig. 5.12.2.

Self-sustaining Nuclear Reactions and Nuclear Energy Sources

This process does not take place for real photons. Neutrinos are completely left-handed and antineutrinos completely right-handed. To couple to spin 1 (and a real photon has spin 1) they must be produced back to back, but this configuration is forbidden by conservation of energy and momentum. However, the photons present in a very dense plasma are continually interacting with the electrons and are sufficiently virtual for this process to take place: the plasma itself can provide the extra momentum necessary.

Finally, the virtual electron generated when a photon is absorbed by a real electron may radiate its excess energy as a neutrino pair rather than by radiating a photon: this is the photoneutrino process which is like compton scattering with the scattered photon replaced by a neutrino pair.

Under the assumption that the universal Fermi coupling constant applies to these processes, the effects of neutrino loss are calculable. Which process is dominant depends on both the temperature and density of the material. Reines has recently attempted to measure the cross-section for the related neutrino-scattering process[†]

$$\bar{v} + e^- \rightarrow \bar{v} + e^-$$

yielding

$$g^2(ev \rightarrow ev) < 4g^2$$

and astrophysical observations themselves suggest that neutrino radiation indeed takes place with about the right coupling constant. If the coupling constant were much larger, white dwarf stars would cool very rapidly by neutrino loss, and there would be significantly less white dwarfs than are observed. If, on the other hand, the coupling constant were much less than the Fermi constant red supergiants would evolve more slowly than they appear to. A recent limit[‡] is

$$g^2(ev \rightarrow ev) = 10^{0 \pm 2} g^2.$$

It seems, then, that neutrino radiation provides an efficient refrigerating process in the cores of massive stars that have progressed beyond the helium-burning stage. The refrigeration will delay the ignition of carbon-burning processes.

The primary carbon burning reaction is

$$^{12}C + ^{12}C \rightarrow ^{24}Mg + \gamma.$$

Helium ignites at $\sim 10^8 \,^\circ K$. The barrier factor is $\propto Z_1 Z_2 / \sqrt{E}$ and so for the same barrier factor in carbon burning as in helium burning we need a mean energy ~ 9 times as large and so might expect carbon burning to commence at a temperature $\leqslant 10^9 \,^\circ K$. If carbon does ignite it will do so in a degenerate environment and the star will experience a carbon flash analogous to the earlier helium flash.

Although the energy released in $^{12}C + ^{12}C \rightarrow ^{24}Mg + \gamma$ is only 0.58 MeV per nucleon, the $^{12}C - ^{12}C$ system lies at an excitation of 14 MeV above the ^{24}Mg ground state. At such high excitations, particle emission competes with electromagnetic decay and introduces protons and α-particles into the carbon–oxygen mix. A complicated chain of subsequent reactions takes place, building ^{16}O, ^{20}Ne, ^{23}Na and ^{24}Mg. In this phase of core burning, the coulomb

[†] F. Reines and H. S. Gurr, *Phys. Rev. Lett.* **24**, 1448 (1970).
[‡] R. B. Stothers, *Phys. Rev. Lett.* **24**, 538 (1970).

Nuclear Physics

barrier is larger for ^{12}C–^{16}O reactions than for ^{12}C–^{12}C reactions, with the result that carbon is mostly consumed before the oxygen begins to burn through

$$^{16}O + {}^{16}O \rightarrow {}^{32}S + \gamma$$

and a number of associated reactions.

In any region of a star in which carbon and oxygen burning are going on, photons of ~ 1 MeV energy will be present. These photons can eject nucleons from the nuclei with mass numbers between 12 and 32 which are present in the mixture. The result is a hot soup of nuclei, nucleons, some alphas and photons, in which as the temperature increases, the tendency is for the light fragments to be soaked up by the heavier nuclei, the whole process tending to build up elements in the iron region. Beyond this point, no further energy can be released by thermonuclear processes. This terminal phase of thermonuclear processes is known as the *e* (equilibrium) process.

The synthesis of elements beyond the Fe group can only be accomplished through the input of energy, and the only available source in the central regions of a highly evolved star is the conversion of gravitational potential energy as the core collapses.

We can now sketch the evolution of a star beyond helium burning, although the details are hazy and still the subject of intensive study.

First, just as the temperature of a star of mass less than $0.5 M_\odot$ will never rise high enough to ignite the helium, so in a star of mass $<0.7 M_\odot$ the star will be stabilized without the central temperature ever rising high enough to ignite carbon-burning processes. Such a star will become degenerate in its central region and eventually evolve into a white dwarf, although as the outer layers collapse and heat up, explosive reactions may take place blowing off some of the outer layers.

A star of mass between 0.7 and 1.44 solar masses can eventually become stable against gravitational collapse, although the temperature may rise high enough to start carbon burning—the role of neutrino emission will be crucial in determining this. While a star with a mass below $0.7 M_\odot$ may be expected to evolve into a white dwarf composed mostly of carbon, a star with a mass between 0.7 and $1.44 M_\odot$ may be expected to enter on the terminal stage of its evolution as a white dwarf composed largely of iron.

A star of mass well in excess of $1.44 M_\odot$ is bound to end in catastrophe. Electron degeneracy pressure can never stop the gravitational collapse and as the density and pressure in the central regions rise, neutrino losses siphon off the energy liberated to continually increase the instability. These losses result in the central regions being at a lower temperature than a zone somewhat removed from the centre, in which the residual thermonuclear processes will continue. In those regions of a star in which the temperature continues to rise, photo-disintegration of the iron group of elements into α's and neutrons will take place providing an additional sink of energy with a consequent reduction of pressure. At this stage the core implosion is so advanced that (under highly non-equilibrium conditions) elements above iron may be built up by successive neutron capture followed by β-decay. If neutron capture has a time scale long compared with the β-decay lifetime, massive nuclei will be built up along the valley of β-stability. This is known as the *s* (slow) process. If the neutron capture time scale is short in comparison with the β-decay lifetimes, neutron-rich nuclei will be built up through this *r* (rapid) process into the transuranic region. In both cases, the β-decays follow-

Self-sustaining Nuclear Reactions and Nuclear Energy Sources

ing neutron capture increase the energy lost in neutrinos. At such temperatures and densities as are expected at the end of the evolution of a massive star, the neutrino loss will also be boosted by the *Urca process* in which nuclei successively capture electrons and then undergo β-decay:

$$(Z, A) + e^- \rightarrow (Z-1, A) + \nu,$$
$$(Z-1, A) \rightarrow (Z, A) + e^- + \bar{\nu}.$$

The final stages of the collapse of a massive star are thus characterized by the extraction of energy from the core. This results in a pressure deficiency, and the outer layers implode, heat up and explode. Nothing except mass loss can stop the central regions collapsing to the point where the electrons are squeezed into the protons to form neutrons, the subsequent collapse being halted by nucleon degeneracy and the short range repulsion between nucleons. Since this neutron formation is an endothermic process and is also accompanied by neutrino loss, there is no energy left for the enormous nucleus thus formed to bounce back, and the supernova explosion which marks the end of the evolution of a moderately massive star is expected to leave behind it a neutron star.

The neutrinos produced during the final collapse will have energies $\sim$MeV and so the cross-section for interaction with matter will be $\sim 10^{-44}$ cm^2 as we found in Section 4.16. While such a cross-section corresponds to an interaction length of ~ 1000 light years at terrestrial densities, in a medium of density $\sim 10^7$ g cm^{-3} the interaction length becomes $\sim 10^{13}$ cm decreasing to $\sim 10^{10}$ cm at a density $\sim 10^{10}$ g cm^{-3}. If in the final stages of the implosion of the outer layers the density rises to $\sim 10^{10}$ g cm^{-3} then much of the energy carried in the intense neutrino flux from the core of the star may be deposited in this outer material and it is this neutrino energy that may be the agent responsible for blowing off the outer layers in a supernova explosion.[†]

Whether the core of a star involved in a supernova explosion ends as a white dwarf or as a neutron star will depend on at what stage in the collapse of the core the neutrino pressure and detonation of unburned material blows the star apart. A star not very far above the Chandrasekhar limit may end as a white dwarf, but a massive (but not too massive) star seems likely to end as a neutron star at the centre of the expanding envelope generated in the supernova explosion. The discovery of a pulsar in the core of the crab nebula has provided strong circumstantial evidence for the formation of neutron stars in supernovae.

The crab nebula is the remnant of the expanding envelope generated in the supernova of A.D. 1054, and at present the gases are still expanding outwards with velocities $\sim 10^7$ cm sec^{-1} and the diameter of the nebula is $\sim 10^{18}$ cm.

A pulsar is a radio source emitting sharp pulses of radiation very regularly, with a period $\lesssim 1$ sec. The pulsar NP 0532 at the heart of the crab has a period of 33 msec and has been observed to emit optical pulses of the same frequency—it is through the optical pulses that the source has been precisely located.[‡]

It is generally believed at present (1971) that pulsars are neutron stars, and that the radiation is produced by the interaction of a magnetic field, rotating with the neutron star, with

---

[†] You should note that in a very massive star the core may reach the implosion point before it has grown to a sufficient size to drive the stars off the main sequence.

[‡] W. J. Cocke, M. J. Disney and D. J. Taylor, *Nature* **221**, 525 (1969).

Nuclear Physics

the surrounding gases.[†] The regularity of the pulses argues an enormous inertia, thus pointing to a stellar origin rather than to plasma oscillations, and the high frequency requires a small size. The point is that the finite signal velocity sets a limit on the size of any vibrating or rotating system $\sim c\tau$, where τ is the period of the oscillations. A frequency ~ 1 sec^{-1} requires linear dimensions $\ll 10^{10}$ cm. Now suppose a mass $M_\odot$ is compressed to nuclear densities $\sim 10^{14}$–10^{15} g cm^{-3}. $M_\odot = 2\times 10^{33}$ g so that the volume of such an object is $\sim 10^{18}$–10^{19} cm^3 and the radius $\sim 10^6$ cm, that is $\sim 10^{-2}$ earth radii. Now if the angular momentum characteristic of the sun is frozen into such an object in its collapse, the resulting period may be computed directly. The rotational period of the sun is

$$\tau_\odot = 26 \text{ days} = 2.2\times 10^6 \text{ sec},$$
$$R_\odot = 7\times 10^{10} \text{ cm} \sim 10^{11} \text{ cm},$$

so that a shrinkage in radius to 10^6 cm demands a rotational period $\sim 10^{-4}$ sec. (This corresponds to a surface velocity at the equator of $\sim 10^{10}$ cm sec^{-1} which is verging on relativistic.)

The reason why a neutron star can exist while a stable nucleus consisting of ~ 100 neutrons cannot is, of course, gravitation.

So light stars end up as white dwarfs, cooling towards a configuration in which the electrons are fully degenerate and the kinetic energy initially stored in the non-degenerate nuclei has leaked away. Stars a bit above the Chandrasekhar limit may end up as white dwarfs after a supernova outburst, while in the supernova phase of more massive stars, a neutron star is born.

There is one further possibility for which there is at present no observational evidence. If the collapsing core is sufficiently massive, the collapse may proceed to the point at which a star swallows itself and disappears from space.

To treat this phenomenon properly, we need general relativity. However, we can get a rough picture of what is involved from the following argument, using only special relativity. Consider a photon of angular frequency ω emitted at the surface of a body of mass M, radius R. The mass of the photon is $\hbar\omega/c^2$ and so this photon has kinetic energy $\hbar\omega$ and gravitational potential energy of $GM/R \cdot \hbar\omega/c^2$.

Thus when infinitely removed from the parent body, the energy of the photon will be

$$\hbar\omega' = \hbar\omega\left\{1 - \frac{GM}{Rc^2}\right\}$$

and the gravitational red shift is given by

$$\frac{\omega'}{\omega} = 1 - \frac{GM}{Rc^2}. \tag{5.12.14}$$

If

$$\frac{GM}{Rc^2} \to 1$$

[†] T. Gold, *Nature* **218**, 731 (1968)

Self-sustaining Nuclear Reactions and Nuclear Energy Sources

then the photon can never leave its parent body which consequently will appear as a black hole in space, absorbing energy but never radiating it and manifesting itself only by its long range gravitational field. Applying (5.12.14) to a neutron star with an assumed density of $\sim 10^{15}$ g cm^{-3} yields a limiting radius $\sim 10^6$ cm which is dangerously close to the conditions we have inferred for a neutron star.

This simple argument relating the mass and radius of a black hole is wrong by a factor 2. The equations of general relativity yield a result

$$R = 2\frac{GM}{c^2}$$

which is known as Schwartzschild's limit. While a spherically symmetric collapse of a sufficiently massive object seems certain to lead to a black hole, we do not know whether this occurs in nature. The collapsing body will almost certainly be rotating and aspherical[†] and it is not clear whether or not there is a corresponding horizon over which nothing can escape in this case.

It has recently been suggested that one component of the peculiar binary system ε Aurigae may be such a collapsar.[‡]

5.13. Thermonuclear power plants? Introduction

We conclude this chapter with a brief discussion of the problems of extracting power in a controlled way from thermonuclear reactions in a terrestrial power plant (any hydroelectric scheme extracts power in a controlled way from thermonuclear reactions taking place in the sun). Thermonuclear power is the subject of big research programmes, and has been for the last twenty years. The motivation for this effort is that high-grade deposits of fissionable or fertile materials are limited, and if fission reactors are to meet the increasing demand for power in the future, then eventually granite will have to be mined and processed for its uranium content just as coal has been mined during the past centuries. Deuterium, however, is plentiful and easily extracted from water.

Suppose we have a fusion plant with a power output of 100 MW—10^{15} erg sec^{-1} $\sim 10^{21}$ MeV sec^{-1}. If the plant can only usefully extract ~ 2 MeV per reaction then the consumption of deuterium would be 10^{21} nuclei/sec. One cm^3 of water contains $\sim 10^{19}$ deuterium nuclei. The deuterium nuclei contained in water flow of a 100 cm^3/sec could run the 100-MW generator! This is the rate at which water runs out of a tap. The supply of deuterium to a fusion power plant is thus no problem.

A further attractive feature of fusion plants—if they can be made to work—is that the residual radioactivity would be negligible compared with fission power plants with the same power rating. The neutron-rich fragments in fission result in a highly radioactive ash with a long lifetime and the safe disposal of this debris is an accumulating problem. The thermonuclear reactions envisaged for a fusion power plant do not in themselves produce radioactive nuclei.[§]

[†] In which case it is expected to be a potent source of gravitational waves.
[‡] A. G. W. Cameron, *Nature*, **229**, 178 (1971).
[§] Except for tritium which is itself a potential fuel and would be fed back in.

Nuclear Physics

Let us now consider the conditions under which we would hope to generate power in a fusion reactor. Suppose that we inject deuterium nuclei with energies 0.1–1 MeV into deuterium gas. This kinetic energy will be lost in ionizing the gas atoms long before the nuclear reaction takes place. Suppose that the target gas is ionized before the energetic deuterium nuclei are injected. This still will not help. The kinetic energy of the injected nuclei will be lost through elastic collisions with both the electrons and the gas nuclei so that the probability of a nuclear reaction becomes negligible. Because the reaction cross-sections are very small in comparison with the elastic scattering cross-sections, the only way of keeping the kinetic energy of the injected nuclei up until a reaction takes place is for them to gain as much energy in a collision as they lose, on average. This means that the ionized gas in which the reaction takes place must have a temperature $kT \sim 0.1$–1 MeV. Under these conditions we can, of course, dispense with injection of special energetic nuclei and a thermonuclear reactor would run on reactions taking place in a plasma at a temperature $\sim 10^8$ °K.

To make a thermonuclear reactor work, then, we need to raise a plasma containing light nuclei to a temperature $\sim 10^8$ °K and then hold these conditions for long enough to get out more energy than we originally put in to heat and compress the plasma, and in addition make good the radiation losses.

The nuclear physics problems have long been solved—the problems of making a reactor are those of heating and containing the plasma.

5.14. Nuclear physics of a fusion reactor

In order to make a reactor, we must choose reactions which have (1) a large Q value, (2) a large cross-section. Both criteria demand that we work with isotopes of hydrogen, which have the minimum value of $Z_1 Z_1$ ($= 1$) in the barrier factor, and build them up to helium, thus releasing ~ 25 MeV per helium nucleus formed.

It is clear that we cannot run a reactor on protons, since the PP cross-section is controlled by the weak interaction coupling. The isotopes ^{2}H and ^{3}H, deuterium and tritium, are the natural candidates.

Now $d+d$ lies about 23 MeV above the ground state of ^{4}He and at so high an excitation the compound nucleus decays by particle emission rather than electromagnetically:

$$d+d \rightarrow {}^3\text{He}+n \quad Q = 3.27 \text{ MeV}, \tag{5.14.1}$$
$$d+d \rightarrow {}^3\text{H}+p \quad Q = 4.03 \text{ MeV}. \tag{5.14.2}$$

These reactions have low cross-sections in addition to low Q values, and are of most interest as a source of tritium, for the reaction

$$d+t \rightarrow {}^4\text{He}+n \quad Q = 17.58 \text{ MeV} \tag{5.14.3}$$

has not only a large Q value but also a cross-section ~ 200 times that of the d–d reactions in the energy range of interest. Finally, the reaction

$$d+{}^3\text{He} \rightarrow {}^4\text{He}+p \quad Q = 18.34 \text{ MeV} \tag{5.14.4}$$

Self-sustaining Nuclear Reactions and Nuclear Energy Sources

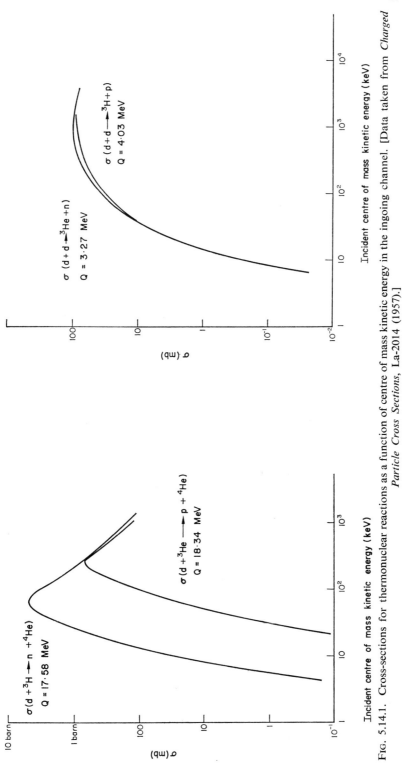

FIG. 5.14.1. Cross-sections for thermonuclear reactions as a function of centre of mass kinetic energy in the ingoing channel. [Data taken from *Charged Particle Cross Sections*, La-2014 (1957).]

Nuclear Physics

is attractive in that it has not only a high Q value but both product nuclei are charged and therefore readily communicate their energy to the plasma. However, there is an incoming and outgoing coulomb barrier with $Z_1 Z_2 = 2$ and the cross-section is only comparable with the dd reactions. The cross-sections for reactions (5.14.1) to (5.14.4) are shown in Fig. 5.14.1 as a function of centre of mass kinetic energy.

If we write the cross-sections for these exothermic processes as

$$\sigma(v) = \frac{\alpha}{v} e^{-G(v)} \left(\frac{E_B}{E}\right)^{1/2} \qquad (5.14.5,\ 5.9.7)$$

then the reaction rate at a temperature T is given by

$$R = N_1 N_2 \langle \sigma v \rangle$$
$$= N_1 N_2 \alpha C(T) \quad \text{cm}^{-3}\ \text{sec}^{-1} \qquad (5.14.6)$$

where the function $C(T)$ was defined in eq. (5.9.16):

$$C(T) = \sqrt{8/3}\ \frac{E_{\max}}{kT} \left(\frac{E_B}{kT}\right)^{1/2} \left(\frac{E_{\max}^{1/2}}{\beta}\right)^{1/2} e^{-E_{\max}/kT} e^{-\beta/E_{\max}^{1/2}},$$

$$E_{\max} = \left(\beta\frac{kT}{2}\right)^{2/3},$$

$$\beta = \sqrt{2m\pi}\ \frac{Z_1 Z_2 e^2}{\hbar}$$

where m is the reduced mass of the nuclei with charges Z_1 and Z_2. If the useful energy released in each interaction is E, then the rate of production of energy is

$$\varepsilon = ER = EN_1 N_2 \alpha C(T)\ \text{erg cm}^{-3}\ \text{sec}^{-1}.$$

If $N_1 = N_2$ and the particle density, including electrons, is n, then $N_1 = N_2 = n/4$ and

$$\varepsilon = \frac{En^2}{16} \alpha\ C(T) \qquad (5.14.7)$$

and the optimal temperature for operation of the reactor is determined by maximizing this quantity (the particle density n is in general also a function of T).

Now the work necessary to raise n particles to a temperature T is

$$\sim \tfrac{3}{2} nkT$$

and so the criterion for reactor feasibility is

$$\varepsilon \tau > \tfrac{3}{2} nkT$$

or

$$n\tau > \frac{24kT}{E\alpha C(T)} \qquad (5.14.8)$$

where τ is the confinement time of the plasma.

Self-sustaining Nuclear Reactions and Nuclear Energy Sources

5.15. Optimal conditions for fusion reactor operation

We already know that the hot plasma must be at a temperature $\sim 10^8$ °K and so must clearly be dilute so as not to melt the reactor by radiation alone. This hot dilute plasma must be kept off the walls of the containing vessel because if it touches the heat will be conducted out. In a star the plasma is confined by the stellar gravitational field. In a bomb the temperature is raised to $\sim 10^9$ °K by a fission explosion, so rapidly that inertia confines the thermonuclear plasma until explosive conditions have already been reached.

For controlled reactions all we have available to contain the plasma are electric and magnetic fields. The plasma is neutral on a large scale (this is essential if the density is not to be space charge limited) but since there are currents in the plasma (either macroscopic or microscopic) a magnetic field can be used to confine the material.

The force acting on an elementary volume of plasma can be written approximately as

$$\mathbf{F} = \tfrac{1}{c}\, \mathbf{J} \times \mathbf{B} - \nabla P \qquad (5.15.1)$$

where P is the fluid pressure and $\mathbf{J}$ the current density. The magnetic field $\mathbf{B}$ will in general be generated partly by external currents and partly by the plasma currents themselves.

Now at low frequency

$$\nabla \times \mathbf{B} = \frac{4\pi}{c}\, \mathbf{J}$$

whence

$$\mathbf{F} = \frac{1}{4\pi} (\nabla \times \mathbf{B}) \times \mathbf{B} - \nabla P$$

$$= -\frac{1}{4\pi}\left\{ \nabla \left(\frac{B^2}{2}\right) - (\mathbf{B}\cdot\nabla)\mathbf{B}\right\} - \nabla P$$

$$= \frac{1}{4\pi}\{(\mathbf{B}\cdot\nabla)\mathbf{B}\} - \nabla\left\{\frac{B^2}{8\pi} + P\right\}. \qquad (5.15.2)$$

Now $B^2/8\pi$ is the magnetic energy density, which acts as a magnetic pressure. The first term acts like a surface tension, and for many configurations may be neglected. The minimal requirement for containment of a hot plasma is then that the magnetic and fluid pressures are about equal:

$$B^2/8\pi \gtrsim P = nkT. \qquad (5.15.3)$$

Thus

$$n \sim \frac{B^2}{8\pi} \frac{1}{kT}. \qquad (5.15.4)$$

We may now use this relation with equation (5.9.5) to calculate the optimum temperature for a thermonuclear reactor. On substituting for $E_{\max}$ in eq. (5.9.16) the quantity $C(T)$ becomes

$$C(T) = \frac{4}{\sqrt{3}} \left(\frac{\beta^2}{4kT}\right)^{2/3} e^{-3(\beta^2/4kT)^{1/3}} \left(\frac{E_B}{\beta^2}\right)^{1/2}$$

Nuclear Physics

which has a maximum at

$$3\left(\frac{\beta^2}{4kT}\right)^{1/3} = 1$$

(this maximum arises from the flux factor in the cross-section).

Setting

$$n \sim \frac{B^2}{8\pi kT}$$

gives for the rate of energy production

$$\begin{aligned}\varepsilon &= \frac{EB^4}{16(8\pi)^2}\frac{1}{(kT)^2}\alpha\frac{4}{\sqrt{3}}\left(\frac{\beta^2}{4kT}\right)^{2/3}e^{-3(\beta^2/4kT)^{1/3}}\left(\frac{E_B}{\beta^2}\right)^{1/2} \\ &= \frac{EB^4}{(8\pi)^2\beta^4}\left(\frac{\beta^2}{4kT}\right)^{8/3}\frac{4}{\sqrt{3}}\alpha e^{-3(\beta^2/4kT)^{1/3}}\left(\frac{E_B}{\beta^2}\right)^{1/2}\end{aligned} \quad (5.15.5)$$

which has a maximum at

$$3\left(\frac{\beta^2}{4kT}\right)^{1/3} = 8$$

or

$$kT \sim 1.3\times 10^{-2}\beta^2 \quad (5.15.6)$$

and the value of this maximum is independent of the magnetic field. For the reaction

$$d+t \to {}^4\text{He}+n \quad (5.14.3)$$

the constant β is 1.13×10^3 (eV)$^{1/2}$ so

$$kT \sim 17 \text{ keV} \quad \text{or} \quad T = 2\times 10^8 \text{ °K}.$$

We can now work out the particle density in the plasma and the minimum confinement time.

For

$$kT = 17 \text{ keV},$$

$$C(T) \sim \left(\frac{E_B}{\beta^2}\right)^{1/2} 10^{-2} \sim 10^{-2}.$$

At a centre of mass energy of 17 keV, the approach velocity of the nuclei is 1.64×10^8 cm sec^{-1} and at this energy the cross-section for reaction (5.14.3) is 0.25 barn. The constant

$$\alpha = \sigma_S v_S e^{\beta/E_S}\left(\frac{E_S}{E_B}\right)^{1/2},$$

$$\sigma_S = 0.25 \text{ barn}, \quad v_S = 1.6\times 10^8 \text{ cm sec}^{-1}, \quad E_S = 17 \text{ keV}$$

is thus

$$\alpha \simeq 2.5\times 10^{-13}\left(\frac{E_S}{E_B}\right)^{1/2} \text{ cm}^3 \text{ sec}^{-1}.$$

Self-sustaining Nuclear Reactions and Nuclear Energy Sources

and

$$\langle \sigma v \rangle = \alpha C(T) = 2.5 \times 10^{-15} \left(\frac{E_S}{\beta^2}\right)^{1/2} = 3 \times 10^{-16} \text{ cm}^3 \text{ sec}^{-1}.$$

Finally, the useful energy E is $\sim \frac{1}{5} Q = 3.5$ MeV the remaining energy being carried off by the neutron, which we will suppose irretrievably lost. Then

$$n\tau \gtrsim 5 \times 10^{14} \tag{5.15.7}$$

for a reactor fuelled on a deuterium–tritium mixture.

The density n and hence the minimum confinement time τ depends on the fields that can be achieved. If we assume a moderate field $\sim 2 \times 10^4$ gauss

$$n \lesssim \frac{B^2}{8\pi kT} = 10^{15} \text{ cm}^{-3}.$$

Thus our prescription for a working thermonuclear reactor is: heat a deuterium–tritium mixture to $\sim 2 \times 10^8$ °K at a density $\sim 10^{15}$ particles cm^{-3}, and hold this condition for a time $\tau \gtrsim 0.1$ sec.

The power output is given by eq. (5.14.7) and under these conditions is

$$\varepsilon \sim 10^7 \text{ erg sec}^{-1} \text{ cm}^{-3}.$$

For our 100-MW generator, a volume of $\sim 10^8$ cm^3 is required.

If the problem of heating and confining the plasma can be solved, it is likely that the energy escaping from the plasma as neutron kinetic energy can be utilized to breed further supplies of tritium. The reactor would be surrounded by a blanket of material in which $(n, 2n)$ reactions would multiply and moderate the neutrons, and the emergent neutron flux would then dissociate ^{6}Li through the reaction ^{6}Li (n, α) ^{3}H which has a resonance at 250 keV.

5.16. Magnetic containment of charged particles

It has not yet proved possible to get a plasma hot enough, dense enough and hold it for long enough to achieve thermonuclear reactions under quasi-equilibrium conditions. In some machines the ignition temperature for the d–t reaction has been achieved, while in others adequate densities and confinement times have been reached. These two achievements must now be brought together in the same machine. While there are technical difficulties in providing enough power for the initial heating and compression of the plasma, the biggest limitation has been in the confinement of the hot plasma. We therefore conclude this chapter with a very brief discussion of the problems of magnetic confinement.

It is easy enough to confine a charged particle of specified momentum in a magnetic field—if it were not, no cyclic accelerator would work. The problems begin when the density of particles is sufficiently high that they may change direction and momentum by scattering.

Nuclear Physics

Consider a single charged particle moving in a uniform magnetic field. It executes a helical motion along the field lines with constant pitch and amplitude. A constant field will thus only confine a particle moving normal to the field (Fig. 5.16.1).

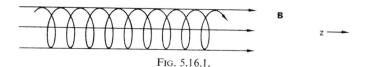

Fig. 5.16.1.

Now suppose that the field lines get closer together—that is the field strength increases with z. A radial component of the field develops and increases with increasing z, and this radial field exerts an increasing force on the orbiting particle, pushing in the direction of decreasing field strength. As the z-component of the particle velocity is reduced,

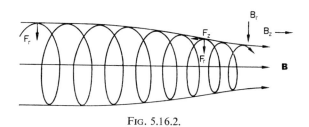

Fig. 5.16.2.

the circular component speeds up (because the total energy is constant) and so the force along the field increases. If the initial circular component is big enough, the particle will be reflected from the region of increasing field. It is this reflection mechanism that confines both protons and electrons in the Van Allen radiation belts—the approximately dipolar field of the earth increases near the magnetic poles. However, it is clear that particles with sufficiently small circular motion will always escape along the axis of the field. This is one of the major difficulties with magnetic mirror machines. Regardless of the configuration of the mirror, particles moving parallel to the field escape. A single particle may be injected at a variety of energies and angles and remain trapped, but as soon as collisions occur there will always be a leakage through the regions of strong field.

If the linear device with mirrors has its ends linked to form a toroid the mirrors may be dispensed with and no end leakage can occur. However, in a toroid, the magnetic field gets weaker with increasing radius, $\nabla(B^2/8\pi)$ is outward directed and while spiralling around the field lines a single particle tends to drift outwards into the region of weaker field, and so eventually ends up at the wall of the container (Fig. 5.16.3).

These considerations illustrate the difficulties of containing a very dilute ionized gas in a magnetic field. At the kind of densities needed for a thermonuclear reactor, the interactions between particles can no longer be treated just as isolated scattering processes tending to randomize the momenta. Any current flowing in a plasma generates its own magnetic field and this field in turn reacts back on the plasma. These collective effects lead to the growth of small departures from equilibrium and until a machine is devised in which plasma in-

Self-sustaining Nuclear Reactions and Nuclear Energy Sources

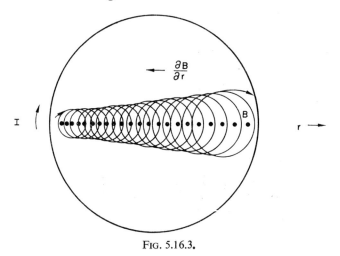

Fig. 5.16.3.

stabilities are sufficiently damped out a thermonuclear reactor is no more than a carrot suspended before the noses of the plasma physicists, who, however, seem now very close to the carrot.

A simple low-frequency example of such instabilities is provided by the pinch effect. If a high current is passed through a column (or a ring) of gas, a circulating magnetic field is generated. Outside the region of conduction this field falls off linearly with distance. Inside the column the situation is different.

Suppose that the current density **J** within the column of plasma is constant. Then at a radius r from the centre of the column, the field **B** is given by

$$2\pi r B(r) = \pi r^2 J,$$

$$B(r) = J \frac{r}{2},$$

$B(r)$ grows linearly with radius until the surface of the column of plasma is reached. The gradient of the magnetic pressures is outward, and so the force due to the self-generated magnetic pressure is inward. The column tends to collapse radially under this pressure (or if you prefer it, the attraction between the current elements collapses the column).

If the current is pulsed and the conductivity of the plasma is high, then the current density will not be uniform, rather all the current will flow in a surface layer. There will then be no magnetic field inside the column of plasma, and a magnetic field

$$B(r) = \frac{I}{2\pi r}$$

outside. The column will collapse to radius R such that

$$\left(\frac{I}{2\pi R}\right)^2 = 8\pi nkT$$

Nuclear Physics

where nkT is the final fluid pressure. This adiabatic compression is one of the ways that have been tried for achieving reactor conditions. Unfortunately, this pinch is inherently unstable. Suppose a neck develops in the column of plasma (Fig. 5.16.4). The current is constant, so that just outside the surface of the plasma column, $B \sim I/R$ is greatest in the region of the neck. The excess magnetic pressure collapses the neck, the sausage instability grows and breaks up the column of plasma.

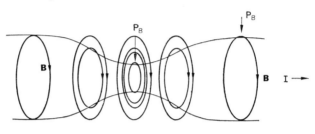

Fig. 5.16.4.

Alternatively, suppose a kink develops in the column (Fig. 5.16.5), on the inside of the kink the field B is compressed while on the outside it is extended. The magnetic pressure thus tends to make the kink grow, breaking up the column of plasma or bringing it into contact with the walls.

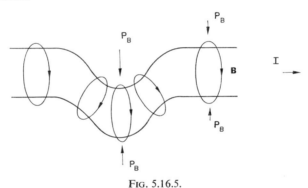

Fig. 5.16.5.

While these examples of low-frequency instabilities in a pinch can be controlled by applying external fields, new forms of instability in plasmas have been discovered by the plasma physicist at a rate comparable with that of the discovery of new particles by the high-energy physicist. These hydromagnetic turbulences tend to increase the rate at which plasma diffuses through the fields and are outside the scope of this book.

Finally, we should mention the exciting possibility that if a thermonuclear reactor can be made to work the energy may be extracted directly rather than via one of the variants of steam. As an example, suppose a pulsed pinch effect machine were feasible. The initial discharge produces the confining magnetic field and heats the plasma to thermonuclear temperatures. The energy produced by the reactions further heats the plasma, causing a re-expansion driving the magnetic field before it. The energy in this expanding magnetic field

Self-sustaining Nuclear Reactions and Nuclear Energy Sources

could, at least in principle, be tapped directly by induction. It appears more promising at present to utilize the leakage of particles out of the ends of a mirror machine, separate the charges electromagnetically and collect the positive charges on one electrode and the negative charges on another, thus supplying d.c. at a constant voltage. A scheme devised for tapping the energy directly in this way has been estimated as $\sim 90\%$ efficient and comparing favourably in cost with the traditional apparatus of heat exchanger and steam engine.[†]

[†] R. F. Post, Controlled fusion research and high temperature plasmas, *Ann. Rev. Nucl. Sci.* **20**, 509 (1970).

CHAPTER 6

Isospin

6.1. Introduction

Throughout this book we have casual references to isotopic multiplets and isospin (or isotopic spin or isobaric spin—they are all the same). Since this concept has not at any point been essential in the material we have developed, we have left the discussion of it until the end. In nuclear physics the concept of isospin provides us with a convenient framework in which to discuss the consequences of the observed charge independence of nuclear forces, but it contains nothing which is not already contained in the statement of charge independence. It is only when the ideas of isospin are extended from nuclear physics into elementary particle physics that they become fruitful in the sense of allowing predictions stretching far beyond the original input to be developed. Towards the end of this chapter we shall therefore forsake the rather classical problems of the motion of clusters of nucleons non-relativistically under their mutual interactions, and reach into the relativistic domain of high-energy physics.

6.2. Exchange symmetry: identical and non-identical nucleons

The proton and neutron have almost identical masses and the same spin. Experimentally they are distinguished one from the other by their different electromagnetic interactions. The neutron has zero charge and a magnetic moment $-1.91\ \mu_N$, while the proton has charge $+e$ and magnetic moment $2.79\ \mu_N$. The energy levels of proton and neutron in an electromagnetic field are thus very different. However, the close equality of their masses suggests that in a strong interaction field the energy levels might be the same. The reasoning behind this is that if the mass of a particle is due to the energy stored in the field around it, then two particles with the same spin and the same mass very much greater than an electron mass have maybe the same strength of strong interaction field around them. An exact test of this idea would involve turning off all electromagnetic interactions and putting first a proton and then a neutron into the same strong field. The electromagnetic interactions cannot be turned off, but are sufficiently well understood that their effects may be calculated and hence the contribution of the strong interactions alone to the interaction of nucleons with a source of both

Isospin

strong and electromagnetic fields may be inferred. The simplest such system is a system of two nucleons, which we discussed in detail in Section 4.8. The two-nucleon system has only one bound state, the deuteron, in which a proton and neutron are bound by 2.225 MeV in a triplet spin state, with zero orbital angular momentum between them (apart from the small amount of d-wave mixed in by the tensor forces). The absence of a bound state in the proton–proton system could be attributed to the electromagnetic effects, but such an explanation will not work for two neutrons. Thus a proton in an s-wave with respect to a neutron has a bound state, a neutron in an s-wave with respect to a neutron does not. This might at first sight suggest that protons and neutrons do not have the same coupling to the strong interactions, but this is illusory. The deuteron is formed from two distinguishable particles and it is the triplet s-wave that is bound. The singlet s-wave is unbound. Now any state of two identical fermions must be totally antisymmetric and so an s-wave triplet state of two protons or two neutrons does not exist. The singlet s-waves have the same strong interaction as the singlet proton–neutron s-wave: you will recall that after subtracting out electromagnetic effects the scattering lengths and effective ranges in s wave pp, nn, and singlet np scattering are the same down to about the 1% level. Thus studies of the two-nucleon system show that the strong interactions are charge independent, provided that we understand by this that two nucleons, *in a given spatial and spin configuration*, have an interaction independent of the charge on either nucleon.

If proton and neutron have very nearly the same mass, and almost identical strong interactions, can we regard them as being the same from the point of view of the strong interactions and distinguished only by their electromagnetic interactions? The answer is that we cannot, but we may regard the proton and neutron as the two almost degenerate states of a single entity, the nucleon, which consequently must possess an internal degree of freedom which is two valued. As an analogy, consider an electron in a weak magnetic field. This system has two states, corresponding to electron spin up, and electron spin down. In a weak field these two states are almost degenerate and the two-valued internal degree of freedom is the spin of the electron. If we want to talk about the proton and the neutron as two states of the nucleon, then we must introduce as the extra degree of freedom the isospin of the nucleon. However, before discussing the way in which this is done, let us first show in more detail why, even if we suppose the electromagnetic interactions to be turned off, we must still let the nucleon have an internal degree of freedom which is quite distinct from spin.

Consider two nucleons in a general strong interaction field, which may be generated by a cluster of other nucleons. Let the two nucleons be situated at r_1, r_2 and have spin coordinates s_1, s_2. The part of the wave function describing these two nucleons may then be written

$$\psi(r_1, s_1; r_2, s_2).$$

If we interchange these two particles with some operator I_{12}

$$I_{12}\psi(r_1, s_1; r_2, s_1) = \psi(r_2, s_2; r_1, s_1). \qquad (6.2.1)$$

If the two particles are completely indistinguishable, then the energy of the whole system will not be changed by this thought operation, so if

$$H\psi = E\psi$$

Nuclear Physics

then
$$I_{12}H\psi = EI_{12}\psi = HI_{12}\psi \tag{6.2.2}$$

so that
$$[H, I_{12}] = 0 \tag{6.2.3}$$

and so
$$\frac{d}{dt}I_{12} = 0. \tag{6.2.4}$$

Applying the operation twice
$$I_{12}[I_{12}\psi(\mathbf{r}_1, \mathbf{s}_1; \mathbf{r}_2, \mathbf{s}_2)] = I_{12}\psi(\mathbf{r}_2, \mathbf{s}_2; \mathbf{r}_1 \mathbf{s}_1) = \psi(\mathbf{r}_1, \mathbf{s}_1; \mathbf{r}_2, \mathbf{s}_2). \tag{6.2.5}$$

I_{12} has eigenstates which are also eigenstates of the Hamiltonian H; and the corresponding eigenvalues are thus ± 1. The operator I_{12} is in some respects analogous to the parity operator, but you should note that it is quite distinct. The similarity is most marked in a two-particle system but even in this case the parity operator does NOT interchange spins. When the system is more complicated than a two-particle system, even the effect on the coordinates is different:

$$P \left\{ \begin{array}{c} \mathbf{r}_2 \nearrow \\ \mathbf{r}_1 \end{array} \right\} = \left\{ \begin{array}{c} \mathbf{r}_1 \swarrow \\ \mathbf{r}_2 \end{array} \right\}$$

while

$$I_{12} \left\{ \begin{array}{c} \mathbf{r}_2 \nearrow \\ \mathbf{r}_1 \end{array} \right\} = \left\{ \begin{array}{c} \mathbf{r}_1 \nearrow \\ \mathbf{r}_2 \end{array} \right\}$$

Thus a wave function describing two identical particles is either even (symmetric) or odd (antisymmetric) under interchange of the two particles. This degree of freedom is, of course, due to our measuring the square of wave functions rather than the wave functions themselves. Furthermore, because $dI_{12}/dt = 0$ once a pair of particles are placed in a state of given exchange symmetry, this symmetry is maintained.

It appears that the symmetry appropriate to the wave function of two identical particles is a characteristic of the particles themselves rather than the system they are placed in: particles with half integral spins are described by wave functions antisymmetric under exchange of any identical pair, and particles with integral spin by wave functions symmetric under exchange of any identical pair. It is this property of the wave functions that gives rise to the Pauli exclusion principle for fermions. (The association of antisymmetric wave functions with fermions and symmetric wave functions with bosons is in the framework of non-relativistic quantum mechanics an empirical fact with no explanation at a deeper level. In relativistic theory, however, an explanation has been found, but this is beyond the scope of

Isospin

this book.) Now nucleons are particles with spin $\frac{1}{2}$, and so two protons or two neutrons will obey the Pauli principle and be described by antisymmetric wave functions. What about a proton and a neutron?

Because of the electromagnetic interactions, the Hamiltonian for this pair will differ from that of two protons or two neutrons. The masses are slightly different, and so is the interaction with any electromagnetic field that happens to be present. However, in the limit as $e \to 0$ we suppose the remaining strong interaction part of the Hamiltonian to affect both a proton and a neutron in the same way. Thus *in this limit* the wave function of a proton and neutron pair in a strong field will again be either symmetric or antisymmetric. If we wish to regard the proton and neutron as the same particle in the limit $e \to 0$ then these wave functions must be antisymmetric under interchange of the proton and the neutron.

Now consider the effect of turning up e. The particles are now distinguishable, and the effect of this is that the Hamiltonian is no longer unchanged under interchange of the two particles. However, it is only a little piece of the Hamiltonian that is affected (since we are turning up e from zero, leaving the strong interactions unchanged) and the effect of this little piece will be to mix into the predominantly antisymmetric eigenstates a small amount of a symmetric term, the amount governed by the size of e through first-order perturbation theory. Thus if as $e \to 0$ the proton and neutron become completely indistinguishable, and have no extra degree of freedom, then we would expect the eigenstates of neutron–proton systems to be almost completely antisymmetric under exchange. BUT we are well acquainted with the deuteron, a triplet s neutron–proton state, which is symmetric under exchange of neutron and proton. Therefore we must allow the nucleon an extra internal degree of freedom *even in the limit* $e \to 0$.

This is, of course, a rather obvious result—something must tell the electromagnetic interactions which nucleon is charged as we turn up the electronic charge e from zero. It is none the less worth pointing out why not quite identical nucleons will still have wave functions antisymmetric under exchange: it should be clear that the same considerations also apply to any thought operation for distinguishing particles, like painting the proton red and the neutron green!

So if two nucleons are to be regarded as the same in the limit $e \to 0$ regardless of whether the pair is pp, pn or nn then the wave functions describing such a pair must be antisymmetric. Given that states symmetric under interchange of the external position coordinate and the internal spin coordinate simultaneously exist, then the nucleon must be allotted an extra internal two-valued degree of freedom: the isospin. While none of the foregoing considerations really constitutes a proof, we will suppose this to be the case and see how well this hypothesis works out in practice.

6.3. The nucleon and isospin

We have concluded that the single entity known as the nucleon has two almost degenerate states, the proton state and the neutron state. The two states are ascribed to a two-valued internal degree of freedom and so the well-known formalism for handing spin $\frac{1}{2}$ would seem suitable for this new degree of freedom. We will represent the proton and neutron states by

Nuclear Physics

two component column matrices

$$\left.\begin{array}{l}\psi_p = \begin{pmatrix}1\\0\end{pmatrix},\\ \psi_n = \begin{pmatrix}0\\1\end{pmatrix},\end{array}\right\} \quad (6.3.1)$$

which are normalized to 1 and are eigenstates of the operator

$$T_z = \tfrac{1}{2}\begin{pmatrix}1 & 0\\0 & -1\end{pmatrix}.$$

T_z is analogous to the third component of spin and

$$\left.\begin{array}{l}T_z\psi_p = \tfrac{1}{2}\psi_p,\\ T_z\psi_n = -\tfrac{1}{2}\psi_n.\end{array}\right\} \quad (6.3.2)$$

The proton state is thus an eigenstate of the operator T_z with eigenvalue $+\tfrac{1}{2}$, the neutron eigenvalue $-\tfrac{1}{2}$.† The reason for the introduction of the numerical factor $\tfrac{1}{2}$ will become clear shortly.

We can also define operators in the abstract isospin space

$$T_x = \tfrac{1}{2}\begin{pmatrix}0 & 1\\1 & 0\end{pmatrix} \quad T_y = \tfrac{1}{2}\begin{pmatrix}0 & -i\\i & 0\end{pmatrix}$$

such that

$$T^2 = T_x^2 + T_y^2 + T_z^2 = \tfrac{3}{4}\begin{pmatrix}1 & 0\\0 & 1\end{pmatrix}$$

and so

$$\left.\begin{array}{l}T^2\psi_p = \tfrac{3}{4}\psi_p,\\ T^2\psi_n = \tfrac{3}{4}\psi_n.\end{array}\right\} \quad (6.3.3)$$

If the maximum value of T_z is equal to T, then $T^2 = T(T+1)$ and the number of states is $2T+1$. It is to maintain this multiplicity rule already very familiar from angular momentum that the factor $\tfrac{1}{2}$ was introduced. Thus given that the nucleon has two states and only two states we find that these states are eigenstates of an operator T_z with eigenvalues $\pm\tfrac{1}{2}$ and simultaneously eigenstates of the operator T^2 with eigenvalue $T(T+1)$. Making a formal analogy with spin, we call T_z the third component of isospin (and its value determines the charge on the nucleon) and T the isospin of the nucleon, equal to one half, which determines the multiplicity of nucleon states: the nucleon is an isospin doublet. This formalism also

---

† Another convention is also used in which $\psi_p = \begin{pmatrix}0\\1\end{pmatrix}$ and $\psi_n = \begin{pmatrix}1\\0\end{pmatrix}$ when

$$T_z\psi_p = -\tfrac{1}{2}\psi_p,$$
$$T_z\psi_p = +\tfrac{1}{2}\psi_n.$$

Which convention is used is entirely arbitrary, but they must not be mixed!

gives us a convenient way of writing into the weak interaction matrix element operators that change protons into neutrons and vice versa. Define

$$T_+ = (T_x + iT_y) = \begin{pmatrix} 0 & 1 \\ 0 & 0 \end{pmatrix},$$
$$T_- = (T_x - iT_y) = \begin{pmatrix} 0 & 0 \\ 1 & 0 \end{pmatrix}.$$
(6.3.4)

Then

$$T_+\psi_p = 0 \quad T_-\psi_p = \psi_n,$$
$$T_+\psi_n = \psi_p \quad T_-\psi_n = 0.$$
(6.3.5)

These operators change the value of T_z and are analogous to the spin flip operators of angular momentum.

Thus using the isospin formalism we would write the matrix element for neutron β-decay in the form

$$\{\bar{\psi}_e(1-\gamma_5)\gamma_\mu\psi_\nu\}\{\bar{\Psi}_p(g_V - g_A\gamma_5)\gamma_\mu T_+ \Psi_n\}$$
(6.3.6)

where the function of T_+ is to explicitly turn a neutron into a proton and the remainder of the nucleon bracket reorganizes the spin.

This merely describes the isolated nucleon in terms of the concept of isospin that we have introduced. If we want to see how isospin affects the interactions of nucleons, we must consider a system of at least two nucleons. The two-nucleon system has four states, singlet pp, nn and pn, and triplet pn. Charge independence means that the three spin singlets have the same strong interactions and the triplet pn state is on its own. We have described the nucleon as an isospin doublet with two degenerate states: an obvious generalization is to describe a system of two nucleons as an isospin triplet, with three degenerate states, and an isospin singlet with only one state. This is precisely analogous to the singlet and triplet states constructed out of two spin $\frac{1}{2}$ objects, and so is indeed contained in the isospin formalism.

The isospin state function for two protons is

$$\psi_{pp} = \begin{pmatrix} 1 \\ 0 \end{pmatrix}_1 \begin{pmatrix} 1 \\ 0 \end{pmatrix}_2$$

meaning that both nucleons 1 and 2 are in the proton state. Then

$$T_z\psi_{pp} = \left[T_z\begin{pmatrix} 1 \\ 0 \end{pmatrix}_1\right]\begin{pmatrix} 1 \\ 0 \end{pmatrix}_2 + \begin{pmatrix} 1 \\ 0 \end{pmatrix}_1 T_z\begin{pmatrix} 1 \\ 0 \end{pmatrix}_2 = +\psi_{pp}.$$
(6.3.7)

Similarly

$$\psi_{nn} = \begin{pmatrix} 0 \\ 1 \end{pmatrix}_1 \begin{pmatrix} 0 \\ 1 \end{pmatrix}_2,$$
$$T_z\psi_{nn} = -\psi_{nn}.$$
(6.3.8)

The third component of the isospin of a pp system is $+1$, the third component of the isospin of an nn state is -1.

Nuclear Physics

We may now find the value of T appropriate to a pp state by applying the operator T^2. We must remember that this is not a simple one off operator but corresponds to two applications of each of the operators T_x, T_y, T_z:

$$T^2\psi_{pp} = T_x[T_x\psi_{pp}] + T_y[T_y\psi_{pp}] + T_z[T_z\psi_{pp}]$$

$$= T_x\left\{\left[T_x\begin{pmatrix}1\\0\end{pmatrix}_1\right]\begin{pmatrix}1\\0\end{pmatrix}_2 + \begin{pmatrix}1\\0\end{pmatrix}_1 T_x\begin{pmatrix}1\\0\end{pmatrix}_2\right\}$$

$$+ T_y\left\{\left[T_y\begin{pmatrix}1\\0\end{pmatrix}_1\right]\begin{pmatrix}1\\0\end{pmatrix}_2 + \begin{pmatrix}1\\0\end{pmatrix}_1 T_y\begin{pmatrix}1\\0\end{pmatrix}_2\right\}$$

$$+ T_z\left\{\left[T_z\begin{pmatrix}1\\0\end{pmatrix}_1\right]\begin{pmatrix}1\\0\end{pmatrix}_2 + \begin{pmatrix}1\\0\end{pmatrix}_1 T_z\begin{pmatrix}1\\0\end{pmatrix}_2\right\}$$

$$= \frac{1}{2}T_x\left\{\begin{pmatrix}0\\1\end{pmatrix}_1\begin{pmatrix}1\\0\end{pmatrix}_2 + \begin{pmatrix}1\\0\end{pmatrix}_1\begin{pmatrix}0\\1\end{pmatrix}_2\right\}$$

$$+ \frac{i}{2}T_y\left\{\begin{pmatrix}0\\1\end{pmatrix}_1\begin{pmatrix}1\\0\end{pmatrix}_2 + \begin{pmatrix}1\\0\end{pmatrix}_1\begin{pmatrix}0\\1\end{pmatrix}_2\right\}$$

$$+ \frac{1}{2}T_z\left\{\begin{pmatrix}1\\0\end{pmatrix}_1\begin{pmatrix}1\\0\end{pmatrix}_2 + \begin{pmatrix}1\\0\end{pmatrix}_1\begin{pmatrix}1\\0\end{pmatrix}_2\right\}$$

$$= 2\psi_{pp},$$

so

$$T^2\psi_{pp} = T(T+1)\,\psi_{pp} = 2\psi_{pp}, \tag{6.3.9}$$

therefore

$$T = 1$$

and, of course,

$$T^2\psi_{nn} = 2\psi_{nn}. \tag{6.3.10}$$

We already know that the state with $T = 1$, $T_z = 0$ is

$$\psi_{pn}^1 = \frac{1}{\sqrt{2}}\left\{\begin{pmatrix}1\\0\end{pmatrix}_1\begin{pmatrix}0\\1\end{pmatrix}_2 + \begin{pmatrix}0\\1\end{pmatrix}_1\begin{pmatrix}1\\0\end{pmatrix}_2\right\}.$$

We get this result by applying the T_z lowering operator T_- to ψ_{pp}:

$$T_-\psi_{pp} = \left[T_-\begin{pmatrix}1\\0\end{pmatrix}_1\right]\begin{pmatrix}1\\0\end{pmatrix}_2 + \begin{pmatrix}1\\0\end{pmatrix}_1 T_-\begin{pmatrix}1\\0\end{pmatrix}_2$$

$$= \left\{\begin{pmatrix}0\\1\end{pmatrix}_1\begin{pmatrix}1\\0\end{pmatrix}_2 + \begin{pmatrix}1\\0\end{pmatrix}_1\begin{pmatrix}0\\1\end{pmatrix}_2\right\} \tag{6.3.11}$$

and normalizing to unity. The other state with the variable $T_z = 0$ is

$$\psi_{pn}^0 = \frac{1}{\sqrt{2}}\left\{\begin{pmatrix}1\\0\end{pmatrix}_1\begin{pmatrix}0\\1\end{pmatrix}_2 - \begin{pmatrix}0\\1\end{pmatrix}_1\begin{pmatrix}1\\0\end{pmatrix}_2\right\}$$

which is orthogonal to $T_-\psi_{pp}^0$ and is the isospin singlet state of a proton neutron pair. It is trivial to show that ψ_{pn}^1 and ψ_{pn}^0 have. $T_z = 0$ and application of T^2 to ψ_{pn}^1 and ψ_{pn}^0 confirms at once that

$$\left.\begin{array}{ll} T^2\psi_{pn}^1 = 2\psi_{pn}^1 & T = 1, \\ T^2\psi_{pn}^0 = 0 & T = 0. \end{array}\right\} \qquad (6.3.12)$$

We may also note that T_+ or T_- applied to ψ_{pn}^0 gives zero. Thus we have

$$\begin{array}{cccc} & T_z = 1 & T_z = 0 & T_z = -1 \\ T = 1 & \begin{pmatrix}1\\0\end{pmatrix}_1 \begin{pmatrix}1\\0\end{pmatrix}_2 & \dfrac{1}{\sqrt{2}}\left\{\begin{pmatrix}1\\0\end{pmatrix}_1\begin{pmatrix}0\\1\end{pmatrix}_2 + \begin{pmatrix}0\\1\end{pmatrix}_1\begin{pmatrix}1\\0\end{pmatrix}_2\right\} & \begin{pmatrix}0\\1\end{pmatrix}_1\begin{pmatrix}0\\1\end{pmatrix}_2, \\ T = 0 & & \dfrac{1}{\sqrt{2}}\left\{\begin{pmatrix}1\\0\end{pmatrix}_1\begin{pmatrix}0\\1\end{pmatrix}_2 - \begin{pmatrix}0\\1\end{pmatrix}_1\begin{pmatrix}0\\1\end{pmatrix}_2\right\} & \end{array}$$

and we identify the three members of the isotriplet state with the singlet pp, pn and nn systems and the isosinglet with the triplet s-wave pn system.

Now since we have introduced the new variables T, T_z into our description of the two-nucleon system, in general the interactions between the two nucleons must depend on these new variables. Let us write the general two-nucleon Hamiltonian as

$$H_{NN} = H_0 + \mathbf{H}_1 \cdot \mathbf{T} + H_2 T^2 \qquad (6.3.13)$$

where H_0, $\mathbf{H}_1$ and H_2 are independent of the isospin states and just depend on the relative space and spin coordinates. The operator $\mathbf{T}$ consists of T_x, T_y and T_z. First we note that since the states we have constructed are eigenstates of T_z, the terms H_{1x} and H_{1y} must be zero because T_x and T_y mix together different values of T_z. Electromagnetism does not alter this conclusion, for different values of T_z correspond to different charge and states with different charge are orthogonal. We are therefore restricted on very general grounds to

$$H_{NN} = H_0 + H_{1z} T_z + H_2 T^2 \qquad (6.3.14)$$

plus perhaps higher powers. Now if for a given value of T all substates are to have the same energy, then the term in T_z must also be absent in the strong interaction Hamiltonian, for

$$\begin{aligned} H_{NN}\psi_{NN} &= \{H_0 + H_{1z} T_z + H_2 T^2\}\psi_{NN} \\ &= \{E_0 + E_{1z} t_{NN} + E_2 T(T+1)\}\psi_{NN} \end{aligned}$$

where t_{NN} is the eigenvalue of T_z applied to ψ_{NN} and for a given value of T is not unique. If $T = 0$ then t_{NN} can only be zero, but if $T = 1$ then $t_{NN} = \pm 1, 0$ and the different members of a given multiplet would have different energies. Thus we are restricted to the form

$$H_{NN} = H_0 + H_2 T^2 \qquad (6.3.15)$$

by the condition of charge independence. The isosinglet and isotriplet nucleon–nucleon states are *not* degenerate and so H_2 is non zero—and because the isotriplet state has higher energy than the isosinglet, H_2 is positive.

Since the interaction H_{NN} can only depend on the isospin variables through T^2, and the isospin states are eigenstates of T^2, this means not only that the energy is independent of T_z

Nuclear Physics

but also that if we start a system off in a specified state of isospin and then apply to it a strong interaction the system will continue in the same isospin state whatever else happens to it, for there is nothing in the interaction that we have adopted that can change the isospin of the system. Thus the assumption that charge independence holds is translated in the isospin formalism into the statement that the isospin is a conserved quantity. This tells us nothing new in the context of nuclear physics and this proposition only takes on its full power when extended to all strongly interacting particles, whether or not they involve nucleons. In this wider field conservation of isospin emerges as an immensely important principle, and the charge independence of the nucleon–nucleon interaction is seen as being merely a special case.

If we now write the wave function describing two nucleons as

$$\psi_{\text{space}} \, \psi_{\text{spin}} \, \psi_{\text{isospin}}$$

we see that ψ_{space} is either symmetric or antisymmetric, ψ_{spin} is either symmetric or antisymmetric and ψ_{isospin} is either symmetric or antisymmetric. Thus for two nucleons in an s-wave the spatial part is symmetric. In the singlet states the spin part is antisymmetric, but it is the singlet s-states that we identified with an isotopic triplet, and so if we exchange position, spin and isospin coordinates the generalized wave function is antisymmetric. The triplet state has a symmetric spin part, but is identified with an isotopic singlet and so has an antisymmetric isospin part. Again the generalized wave function is antisymmetric under interchange of all coordinates. By treating the nucleon through the isospin formalism we have a generalized Pauli principle but again this is really nothing new. Consider how it works for a p-wave state of two nucleons. A p-wave state of two nucleons is antisymmetric under interchange of position. The symmetric spin function must therefore be accompanied by the symmetric isospin function and the triplet p-state is an isotriplet while the singlet p-state is an isosinglet. But this conclusion could have been reached directly. Forgetting about isospin, a p-state of two protons or two neutrons can only exist as a spin triplet, applying the ordinary Pauli principle. A p-state of a proton and a neutron can exist, in either a triplet or a singlet state. Charge independence then requires the triplet pn state to have the same energy as the triplet pp and nn states, and the singlet state is on its own. The isospin formalism and the generalized Pauli principle contain no more information about the p-wave nucleon–nucleon states.

Now let us see where electromagnetism fits into the picture we have derived by ignoring it.

$$T_z \psi_p = +\tfrac{1}{2} \psi_p,$$
$$T_z \psi_n = -\tfrac{1}{2} \psi_n.$$

Both proton and neutron have a nucleon number $N = 1$ (antiproton and antineutron have a nucleon number -1). This number is a conserved quantity since protons and neutrons can transform into one another via the weak interactions, but never disappear. We can therefore write for the nucleon

$$Q = \frac{N}{2} + T_z$$

where Qe is the charge and it is easy to see that this works for systems of nucleons.

Isospin

The coulomb interaction between two nucleons can then be written as

$$C_{12} = Q_1 Q_2 \frac{e^2}{r_{12}} = \frac{e^2}{r_{12}} \left[\frac{N_1}{2} + T_{1z}\right]\left[\frac{N_2}{2} + T_{2z}\right] \quad (6.3.16)$$

where nucleon 1 has nucleon number N_1 and third component of the isospin T_{1z} and nucleon 2 has N_2, T_{2z}. N_1, T_{1z} operate only nucleon 1 and N_2, T_{2z} operate only on nucleon 2.

The contribution of the coulomb interaction to the energy of a system of two nucleons with wave function ψ_{12} is

$$\int \psi_{12}^* C_{12} \psi_{12} \, dV \quad (6.3.17)$$

when treated as a first order perturbation acting on eigenstates of isospin. If ψ_{12} is a system of two protons then the isospin part of the wave function is

$$\begin{pmatrix}1\\0\end{pmatrix}_1 \begin{pmatrix}1\\0\end{pmatrix}_1$$

and so the coulomb energy is proportional to a factor

$$(1\ 0)_1 (1\ 0)_2 \left[\frac{N_1}{2}+T_{1z}\right]\left[\frac{N_2}{2}+T_{2z}\right]\begin{pmatrix}1\\0\end{pmatrix}_1\begin{pmatrix}1\\0\end{pmatrix}_2$$

$$= (1\ 0)_1 (1\ 0)_2 \left\{\left[\frac{N_1}{2}+T_{1z}\right]\begin{pmatrix}1\\0\end{pmatrix}_1\right\}\left\{\left[\frac{N_2}{2}+T_{2z}\right]\begin{pmatrix}1\\0\end{pmatrix}_2\right\}$$

$$= 1.$$

If ψ_{12} is a system of two neutrons then the isospin part of the wave function is

$$\begin{pmatrix}0\\1\end{pmatrix}_1 \begin{pmatrix}0\\1\end{pmatrix}_2$$

and since

$$\left[\frac{N_1}{2}+T_{1z}\right]\begin{pmatrix}0\\1\end{pmatrix}_1 \left[\frac{N_2}{2}+T_{2z}\right]\begin{pmatrix}0\\1\end{pmatrix}_2 = 0$$

there is no coulomb energy term. Similarly

$$\left[\frac{N_1}{2}+T_{1z}\right]\left[\frac{N_2}{2}+T_{2z}\right]\psi_{12} = 0$$

if ψ_{12} is the $T_z = 0$ member of an isotriplet state of two nucleons, or an isosinglet state. This also shows that the coulomb interaction introduces no mixing between the isosinglet and isotriplet members of a two-nucleon system, for

$$\psi_{12}^*(T=1, T_z=0)\left[\frac{N_1}{2}+T_{1z}\right]\left[\frac{N_2}{2}+T_{2z}\right]\psi_{12}(T=0, T_z=0) = 0. \quad (6.3.18)$$

Nuclear Physics

6.4. Isospin and nuclear structure

Let us now go to a system of three nucleons. The maximum value of T_z is $\frac{3}{2}$ corresponding to charge 3, and the minimum value is $-\frac{3}{2}$, corresponding to charge 0. A multiplet with $T = \frac{3}{2}$ has four states:

$$T = \tfrac{3}{2} \quad \begin{array}{ll} T_z = +\tfrac{3}{2} & ppp, \\ T_z = +\tfrac{1}{2} & ppn, \\ T_z = -\tfrac{1}{2} & nnp, \\ T_z = -\tfrac{3}{2} & nnn \end{array}$$

and the total number of states of three nucleons is $2^3 = 8$ because there are two kinds of nucleon. Since there is only one ppp or nnn state, the remaining four states cannot make up a quartet, and because T_z is half integral they cannot be a triplet and a singlet. The only possibility is two doublets, like

$$T = \tfrac{1}{2} \quad \begin{array}{ll} T_z = +\tfrac{1}{2} & ppn, \\ T_z = -\tfrac{1}{2} & nnp.\end{array}$$

Now the ground states of ³H and ³He will be part of an unfilled s-shell of nucleons (one missing). The two like nucleons are therefore forced to have spins opposite and the nucleus has spin $\frac{1}{2}$. Since neither three protons nor three neutrons can lie in an s-shell, there are only two charge configurations of three nucleons in the same relative space and spin states corresponding to an s-shell nucleus, and therefore the pair of mirror nuclei ³H, ³He are members of an isospin doublet just as are the proton and neutron. The excited states are unbound and will correspond to the $T = \frac{1}{2}$ excited states of the doublet and the $T = \frac{3}{2}$ quadruplet excited states. In ³H and ³He the isospin triplet pp (or nn) couples with the isospin $\frac{1}{2}$ nucleon to produce total isospin $\frac{1}{2}$.

Now add one more nucleon to the s-shell. The only possible s-shell for four nucleons is two protons and two neutrons. If a proton is to be replaced by a neutron, or vice versa, the p-shell must be used for it. Therefore the ground state of ⁴He is an isosinglet. The excited states are unbound, and will be members of further isosinglets, triplets and quintets. If we now leave the isosinglet ⁴He core undisturbed and add a nucleon the only states made are the isodoublet ⁵Li, ⁵He. Neither is bound. On adding two nucleons to a ⁴He core we form a triplet and a singlet (Fig. 6.4.1). All three levels of the triplet have their degeneracy removed by the coulomb interaction, in contrast to the nucleon–nucleon triplet. The reason is, of

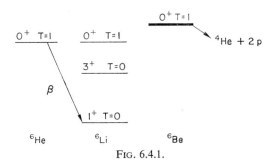

Fig. 6.4.1.

Isospin

course, that while there is no coulomb interaction between proton and neutron, a proton and neutron moving in orbit about a ^{4}He core are in the coulomb field of this core and so the energy of the proton is shifted upwards. If the core remains unexcited, the excited states of these nuclei can only be isotriplet (^{6}He, ^{6}Li and ^{6}Be) and isosinglet (^{6}Li). (Figure 6.4.1 is drawn on an atomic mass scale, and on such a scale the mass excess of the neutron makes the ground state of ^{6}He almost degenerate with the 0^+ excited state of ^{6}Li.)

Three nucleons beyond a ^{4}He core and we have once more quartets and doublets, among the nuclei ^{7}He, ^{7}Li, ^{7}Be and ^{7}B. Only the doublet of ^{7}Be and ^{7}Li is bound, ^{7}Be decaying to ^{7}Li via K-shell electron capture.

If we add four nucleons beyond ^{4}He we complete the first p-orbital in an independent particle model. The ground state of ^{8}Be decays into two α-particles, and consequently has isospin 0, as do all the excited states that decay into two α's. If the helium core is left undisturbed, the other four nucleons can be so arranged as to give quintets or triplets as well as the singlets. In the system with mass number 8, there are the singlet states of ^{8}Be, the triplet states of ^{8}Li, ^{8}Be and ^{8}B, and ^{8}He has been observed, which must be a member of a quintet. Among the light nuclei are a very large number of identified singlet, doublet and triplet states, some completed families of quartets and a number of members of quintets are known.

Purely on the basis of the Pauli principle we can see that the larger the isospin of a family, the larger we expect its energy to be. A singlet has an equal number of protons and neutrons, which can fill up the available levels. If we remove degeneracies, each level has an occupation number of four nucleons only, and if, as in an even–even $T = 0$ nucleus, all these levels are filled, then if we wish to convert a proton into a neutron or vice versa to go to a $T_z = -1$ or $T_z = +1$ member of a triplet, then the extra nucleon must go into a higher level. In an odd–odd nucleus the triplet will be reached by turning a proton into a neutron or vice versa but without raising the nucleon to a different spatial wave function. Whether the singlet or triplet is lower lying will depend on the spin dependence of the interaction. A quintet is reached from a singlet by the excitation of two nucleons and so must lie higher than the singlet.

If we take a singlet and add a single nucleon to it to make a basic doublet, then the quartet with the same A is reached by exciting another nucleon from out of the singlet core to at least the same spatial state as the added nucleon: the quartet will therefore lie higher than the doublet. These conclusions are illustrated in Fig. 6.4.2.

So in light nuclei at least, for a given A we expect ladders of states. If A is even, the lowest-lying state of mass number A is expected to be a singlet with $Z = N$. Above it we expect the three members of a triplet with

$$Z = \frac{A}{2}-1, \quad \frac{A}{2}, \quad \frac{A}{2}+1.$$

The quintet ground state will lie higher still with

$$Z = \frac{A}{2}-2, \quad \frac{A}{2}-1, \quad \frac{A}{2}, \quad \frac{A}{2}+1, \quad \frac{A}{2}+2.$$

Nuclear Physics

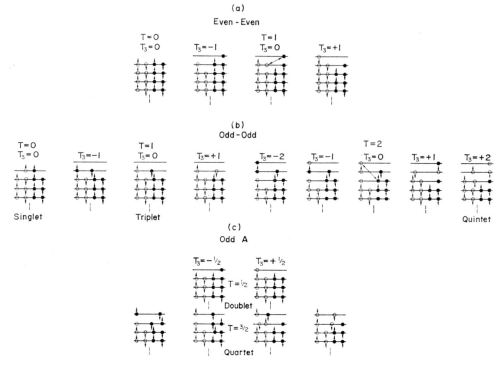

FIG. 6.4.2. (a) Schematic configuration for even–even nuclei with $T = 0$ and $T = 1$. The lowest isotriplet state lies higher than the lowest singlet. (b) Schematic configuration for odd–odd nuclei with $T = 0$ and $T = 1$. Whether the singlet or triplet levels lie higher is governed by the spin dependent properties of the interactions. A $T = 2$ configuration is also illustrated. (c) An isodoublet and quartet for odd A. The quartet will in general lie higher than the doublet unless the spin-dependent forces can close the gap.

If, on the other hand, A is odd, the lowest-lying states are a doublet with

$$Z = \frac{A}{2} - \frac{1}{2}, \quad \frac{A}{2} + \frac{1}{2}.$$

The quartet

$$Z = \frac{A}{2} - \frac{3}{2}, \quad \frac{A}{2} - \frac{1}{2}, \quad \frac{A}{2} + \frac{1}{2}, \quad \frac{A}{2} + \frac{3}{2}$$

will lie higher. All allowed isospins will recur among the highly excited states of a system with given A.

As a result, we expect for even A the nucleus with $T_z = 0$ to be richest in excited states, for the maximum allowed isospin is $A/2$ and all integral isospin are represented by a $T_z = 0$ member. The nuclei with $T_z = 1$ can only have triplet or higher states, and so on. This again is just the Pauli principle. If we forget about isospin, there are far more ways of stacking $A/2$ protons and $A/2$ neutrons than there are of stacking $(A/2)-1$ protons and $(A/2)+1$ neutrons without violating the Pauli principle.

If the mass number A is odd, then the lowest-lying states of A nucleons will be the doublet with

$$Z = \frac{A}{2} + \frac{1}{2}, \quad \frac{A}{2} - \frac{1}{2}.$$

These nuclei will be richest in excited states. The lowest-lying quartet may be expected to lie higher, and so on. All these statements apply to a situation in which the coulomb interaction has not shifted energies too greatly from the values expected only on the basis of the strong interactions—^{238}U is an even–even nucleus with $T_z = -27$. The isosinglet state of 238 nucleons is certainly NOT lower lying!

In nuclear physics, then, the concept of isospin is useful in classifying nuclear states in multiplets, and charge independence, expressed as conservation of isotopic spin, then requires that the wave functions of the various members of these multiplets are identical apart from coulomb effects. If we regard the coulomb interaction as a perturbation acting on the states of unmixed isospin, the first-order effect is to shift the energies of the otherwise degenerate levels. We expect the coulomb energy to increase as the square of the charge Q and since

$$Q = \frac{N}{2} + T_z$$

then the shift in energy is proportional to

$$\left\langle \psi_T^* \left| \left(\frac{N}{2} + T_z \right)^2 \right| \psi_T \right\rangle$$

or

$$E_c = a + bT_z + cT_z^2, \tag{6.4.1}$$

an expression which works quite well—it has been tested in identified quartets and among members of quintets.

As the charge on the nucleus increases with increasing A, the levels of a given multiplet will be expected to split further and further apart. In addition to this increasing first-order effect, we would expect the coulomb interaction acting as a perturbation in second order to mix together states of different isospin so that the nuclear levels were no longer states of unique isospin. It is therefore remarkable that the states of massive nuclei retain a high degree of isospin purity: the major effect of the coulomb interaction is to shift energy levels rather than to mix states, even in the heaviest nuclei.

The evidence that has led to this conclusion comes mostly from the study of yields in direct reactions on medium weight and heavy nuclei. The original work was done with (p, n) reactions at (for nuclear physics) fairly high bombarding energies, ~ 10 MeV or more. It is observed that when (p, n) reactions on an enormous variety of nuclei are studied at a fixed angle and a fixed bombarding energy as a function of the energy of the emitted neutron, then one or more very narrow peaks stand out above the general charge exchange background: examples of such data are shown in Fig. 6.4.3.

The replacement of a neutron by a proton in a nucleus (Z, A) leads to excited states of the nucleus $(Z+1, A)$. If the bombarding energy is ~ 10 MeV the excitation energy of the nucleus formed is ~ 8–18 MeV. At such an excitation the density of excited states is very great —in a heavy nucleus the separation of single-particle excited states is ~ 0.1 MeV and so the number of many particle modes that can be excited is enormous. The continuum neutron spectrum thus corresponds to the excited nucleus being left in these very closely spaced, and unresolved, excited states. The superimposed lines means that one particular excited state is

Nuclear Physics

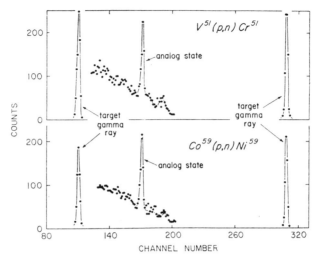

FIG. 6.4.3. Examples of analogue states strongly excited in (p, n) reactions. Neutron yield is plotted against neutron time of flight, which increases to the left. [From J. D. Anderson and C. Wong, *Phys. Rev. Lett.* **8**, 442 (1962).]

produced with very much higher probability than its neighbours: for such a state the proton fits in much more easily than in the neighbouring states. This implies that the wave function of the proton in the excited nucleus is very similar to the wave function of the neutron in the ground state target nucleus, and that there is a minimal amount of rearrangement of the rest of the nucleons. If we merely replace a neutron by a proton in a nucleus and do not change anything else, then the process takes us from the $T_z = Z - A/2$ member of an isomultiplet to the $T_z = Z + 1 - A/2$ member of the same isomultiplet. The occurrence of neutron spikes in reactions on heavy nuclei thus implies that the wave functions of neutron and proton in adjacent members of the isomultiplet are very similar in spite of the large coulomb effects which might be expected to grossly change the wave functions with increasing Z by mixing into the neighbouring excited states of different isospin. Furthermore, the strongly excited states will correspond to single particle levels rather than the compound nuclear states involving many particle excitations.

The ground state of the nucleus (Z, A) has $T_z = Z - A/2$ and probably the minimum isospin $T = |Z - A/2|$. On replacing a neutron by a proton an excited state of the nucleus $(Z+1, A)$ is formed. If the ground state of (Z, A) and the excited state formed are members of the same multiplet and this multiplet is isospin pure, then the only difference in the energy will come from the coulomb term since the strong interaction contributions will be left unchanged. The coulomb energy is

$$E_c \sim \frac{3}{5} \frac{Z^2 e^2}{R_0 A^{1/3}}$$

and so the change on replacing a neutron by a proton in a multiplet is given by

$$\Delta E_c = \frac{6}{5} \frac{Ze^2}{R_0 A^{1/3}} = \frac{1.4Z}{A^{1/3}} \text{ MeV}$$

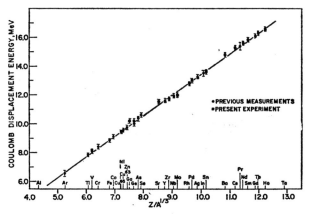

FIG. 6.4.4. The coulomb displacement energies for isobaric analogue states excited in (p, n) reactions are plotted against $Z/A^{1/3}$. [From J. D. Anderson et al., Phys. Rev. 138 B, 615 (1965).]

which is ~ 10 MeV for $\quad Z = 27 \quad (^{59}\text{Co})$
$\qquad\qquad\qquad\qquad\quad A = 59$

~ 19 MeV for $\quad Z = 83$
$\qquad\qquad\qquad A = 209 \quad (^{209}\text{Bi})$

(see Fig. 6.4.4).

On replacing a neutron by a proton in a medium or heavy nucleus the state $(Z+1, A)$ with isospin $T = |Z - A/2|$ and $T_z = Z + 1 - A/2$ is expected to lie ~ 10–20 MeV higher (depending on Z) than the ground state of $[Z, A]$. This puts it among the hash of excited states of $(Z+1, A)$ which do not have isospin $T = |Z - A/2|$ but predominantly the minimum isospin appropriate to a nucleus $(Z+1, A)$, $T = |Z+1-A/2|$ and the nucleus $(Z+1, A)$, having lower T_z, will have more excited states than the nucleus $[Z, A]$. The admirable agreement between the relative energy of the states recoiling against the neutron spikes and the ground states from which they were formed with the calculated coulomb energy shift confirms this picture and so provided the first demonstration of the remarkable isospin purity of the states of massive nuclei. With the deployment of beams of protons of energy up to ~ 20 MeV, the separation between several peaks in the neutron spectrum could be correlated with the separation between the ground and excited states of the target nucleus. The whole process is represented schematically in Fig. 6.4.5 which also shows what would happen if the coulomb interaction badly mixed states of different isospin.

Those states of isospin $T = |Z - A/2|$ which are readily excited over the background of states with $T = |Z+1-A/2|$ "natural" to a nucleus $(Z+1, A)$ are known as isobaric analogue states among the professional nuclear physicists. This term does not mean that only these states are of high isospin purity, but rather that it is only these states that can be easily correlated with their analogues in the neighbouring nucleus (Z, A).

These isobaric analogue states have been studied in a number of ways other than the direct replacement of a neutron with a proton. If a single neutron is taken out of a nucleus without disturbing the rest of the structure the residual nucleus is left in one of a restricted series of excited states which are analogues of the states of neighbouring nuclei: taking out a

Nuclear Physics

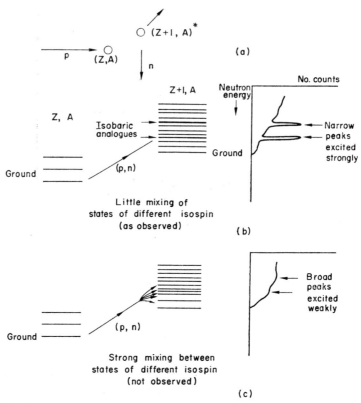

FIG. 6.4.5. (a) Excitation of isobaric analogue states by the (p, n) reaction. (b) shows what happens when the analogue states are little mixed with neighbouring states and (c) shows what would result if the analogue states were strongly mixed with the host of adjacent states of different isospin.

neutron and leaving the rest of the structure alone changes T by zero or by one unit. Thus isobaric analogues are strongly excited in pickup (p, d) reactions and also in stripping reactions where a nucleon is injected into a nucleus without disturbing the rest. In both stripping and pickup reactions the angular distribution of the product give directly the orbital angular momentum of the nucleon picked out or injected (see Section 4.15) and so provide much more detailed information than just the energy differences accessible in direct replacement reactions.

In high-energy elastic scattering of protons, strong resonance scattering is observed at definite energies. This strong scattering corresponds to formation of a compound nucleus into which the protons fit with little rearrangement of the other nucleons and again such simple states are analogues of states of neighbouring nuclei in which the only difference is injection of a neutron rather than a proton. For heavy nuclei the strong coulomb scattering interferes with resonance scattering of the protons and the variation of the scattering cross-section with energy at a few angles gives enough information to extract the orbital angular momentum of the resonating proton. Thus the strongly excited resonances of the compound nucleus (Z, A) formed in the process

$$(Z-1, A-1)\,(p, p)\,(Z-1, A-1)$$

may be compared directly and in detail with the excited states of the nucleus $(Z-1, A)$ recoiling from a proton in the reaction

$$(Z-1, A-1)\,(d, p)\,(Z-1, A)$$

and the detailed agreement of results from the two kinds of study make it quite clear that the strongly excited resonances in elastic proton scattering are indeed the isobaric analogues of the lower lying strongly excited states in the deuteron stripping reactions on the same target (Fig. 6.4.6). The single-particle input channels couple most strongly with single particle excitations of the intermediate nucleus formed, and so the analogues in the (Z, A) system of the low-lying states in the $(Z-1, A)$ system will be a close approximation to optical model shape resonances rather than to the many-particle excitations characteristic of most compound nuclear states. However, all shape resonances in $(Z-1, A)(p, p)(Z-1, A)$ reactions are not classified as isobaric analogous states. On adding a proton to $(Z-1, A)$, assumed to have isospin $T = |Z-1-A/2|$ states with isospin $T = |Z-1-A/2|$ and $T = |Z-A/2|$ may be formed. The former states are classified as isobaric analogue states and show up as sharp resonances, while the latter have the isospin "natural" to the hash of compound nuclear states in the system (Z, A). The residual interactions can thus mix the shape states with neighbouring compound nuclear states without violating isospin conservation if $T = |Z-1-A/2|$ and so a situation like that shown in Fig 6.4.5(c) obtains. Those states with $T = |Z-1-A/2|$ cannot be mixed with neighbouring $T = |Z-A/2|$ states except by violating conservation of isospin. The sharpness of the isobaric analogue states is another indication of isospin purity even in excited heavy nuclei. (In fact there is some mixing and the analogue states do not correspond to pure shape resonances and have decay modes other than the emission of the injected nucleon. Their structure is still simpler than the background states of "natural" isospin and so they are excited more strongly.)

6.5. Isospin and nuclear decay: electromagnetic decay

We complete our discussion of isospin in nuclear physics by applying the formalism to electromagnetic and weak decay processes. We first discuss electromagnetic decay, then β-decay and finally tie the two together in the conserved vector current model which we introduced briefly in Chapter 3.

In Section 3.3, we considered the general coupling between the electromagnetic field and a charged particle and wrote

$$H_{\text{int}} = q\,\frac{\mathbf{p}\cdot\mathbf{A}}{Mc} \qquad (3.3.5)$$

for a particle of momentum $\mathbf{p}$, mass M and charge q coupling with the radiation field $\mathbf{A}$. This gave rise to a matrix element

$$\frac{1}{2}\frac{e}{Mc}\int \Psi_b^*(\mathbf{r})\,e^{-i\mathbf{k}\cdot\mathbf{r}}\,\mathbf{A}_0^*\cdot\mathbf{p}\,\Psi_a(\mathbf{r})\,d^3r$$

Nuclear Physics

for a transition between the nuclear states Ψ_a and Ψ_b with emission of a photon described by

$$\mathbf{A} = \mathbf{A}_0\, e^{i(\mathbf{k}\cdot\mathbf{r}-\omega t)}$$

through the coupling with a nucleon of charge e. Thus it was necessary to specify the protons in summing over all contributions to the matrix element. We may use the isospin formalism to treat protons and neutrons on the same footing by writing

$$q = e\,\{\tfrac{1}{2} + T_z\} \tag{6.5.1}$$

and hence a matrix element

$$\frac{1}{2}\frac{e}{Mc}\int \Psi_b^*(\mathbf{r})\, e^{-i\mathbf{k}\cdot\mathbf{r}}\, \mathbf{A}_0^*\cdot\mathbf{p}\{\tfrac{1}{2}+T_z\}\,\Psi_a(\mathbf{r})\, d^3r$$

to be summed over all nucleons and not just over the protons: for a neutron the operator $[\tfrac{1}{2}+T_z]$ gives nothing.

Since the T_z operators are the analogues of J_z operators in real space, the operator $\tfrac{1}{2}+T_z$ behaves like a scalar and the third component of a vector in isospin space.

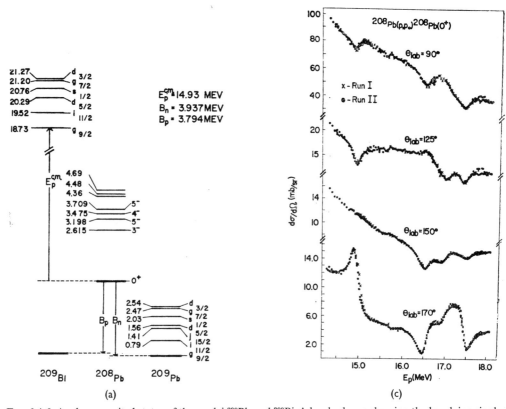

FIG. 6.4.6. Analogue excited states of the nuclei ^{209}Pb and ^{209}Bi. A level scheme showing the low-lying single-particle states in ^{209}Pb and their isobaric analogues in ^{209}Bi is shown in Fig. (a). Figure (b) shows the proton spectrum at 75° from the reaction ^{208}Pb(d, p) ^{209}Pb. The numbers above the peaks are the excitation in MeV of the corresponding excited states in ^{209}Pb. Peaks due to contamination are indicated by a

The inclusion of the intrinsic magnetic moments of nucleons does not change this picture. We may write

$$\mu_p = \mu_p \{\tfrac{1}{2}+T_z\},$$
$$\mu_n = \mu_n \{\tfrac{1}{2}-T_z\},$$

and a general nucleon magnetic moment as

$$\mu_N = \tfrac{1}{2}(\mu_p+\mu_n)+T_z\,(\mu_p-\mu_n). \tag{6.5.2}$$

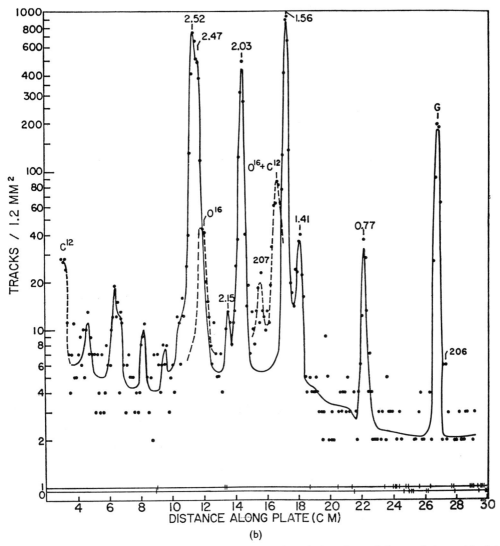

(b)

broken curve. The proton groups were separated by magnetic analysis and recorded on a photographic plate. Fig. (c) shows anomalies in the scattering process ^{208}Pb(p,p) ^{208}Pb due to the analogues of the ^{209}Pb states in ^{209}Bi. [Figs. (a) and (c) from S. A. A. Zaidi et al., *Phys. Rev.* **165**, 1312 (1968); Fig. (b) from P. Mukherjee and B. L. Cohen, *Phys. Rev.* **127**, 1284 (1962).]

Nuclear Physics

If we regard the term

$$\mathbf{J}_c = \frac{e\mathbf{p}}{M} \tag{6.5.3}$$

as the electric current operator for spinless particles, magnetic moments due to orbital motion are automatically included but magnetic moments due to spin are not. The interaction energy of a general current with an electromagnetic field generated by that current is

$$H_c = \frac{1}{2}\frac{1}{c}\mathbf{A}\cdot\mathbf{J}_c = \frac{1}{2}\frac{e}{c}\frac{\mathbf{p}\cdot\mathbf{A}}{M}. \tag{6.5.4}$$

The interaction energy of a magnetic moment $\mathbf{M}$ with an electromagnetic field is

$$H_M = \mathbf{M}\cdot\mathbf{B} = \mathbf{M}\cdot(\nabla\times\mathbf{A})$$
$$= (\mathbf{M}\times\nabla)\cdot\mathbf{A}.$$

For a spinning nucleon we write

$$\mathbf{M} = \mu\frac{e\hbar}{2Mc}\boldsymbol{\sigma} \tag{6.5.5}$$

where μ is measured in nuclear magnetons and is 2.79 for a proton, -1.91 for a neutron. Then if we write

$$H = H_E + H_M = \frac{1}{2}\frac{1}{c}\{\mathbf{J}_c + \mathbf{J}_M\}\cdot\mathbf{A} \tag{6.5.6}$$

then

$$\mathbf{J}_M = c\mu\frac{e\hbar(\boldsymbol{\sigma}\times\nabla)}{2Mc} \tag{6.5.7}$$

where the ∇ operator operates only on the position variables of the electromagnetic vector potential $\mathbf{A}$: if

$$\mathbf{A} = \mathbf{A}_0 e^{i(\mathbf{k}_1\cdot\mathbf{r}-\omega t)}$$

then

$$(\boldsymbol{\sigma}\times\nabla)\cdot\mathbf{A}^* = -i(\boldsymbol{\sigma}\times\mathbf{k})\cdot\mathbf{A}^*.$$

This inclusion of spin can only be done properly by going to the Dirac equation in which intrinsic spin appears naturally; here we have just botched it on a non-relativistic picture and the eventual justification is the emergence of such a form in the non-relativistic limit of the relativistic equations. With this result, however, we can write the general matrix element as

$$\frac{1}{2}\frac{1}{c}\int \Psi_b^*(\mathbf{J}^0 + \mathbf{J}_z^1)\cdot\mathbf{A}^*\Psi_a\, d^3r \tag{6.5.8}$$

where

$$\mathbf{J}_0 = \sum_i \frac{1}{2}\left[\frac{e}{M}\mathbf{p}_i + \frac{e\hbar}{2M}(\mu_p + \mu_n)(\boldsymbol{\sigma}_i\times\nabla)\right],$$

$$\mathbf{J}_z^1 = \sum_i \left[\frac{e}{M}\mathbf{p}_i + \frac{e\hbar}{2M}(\mu_p - \mu_n)(\boldsymbol{\sigma}_i\times\nabla)\right]T_z^i. \tag{6.5.9}$$

Isospin

The current operator J_0 is a scalar in isospin space while the operator J_z^1 is the third component of a vector in isospin space.

The utility of such a representation, if correct, is that in the limit of pure isospin states the formalism alone yields general selection rules which would otherwise have to be worked out piecemeal. These rules are called isospin selection rules and this term can be a source of some confusion, for it may be taken to imply that the rules follow from conservation of isospin, whereas the electromagnetic interaction violates conservation of isospin. The isospin selection rules for electromagnetic decay indeed depend upon the strong interactions conserving isospin, for they are derived under the assumption that the initial and final nuclear states have pure isospin. Beyond this, however, the isospin selection rules derive from the way in which we have assumed the electromagnetic interaction to depend on the isospin T, namely through one term which does not contain T and is therefore isospin conserving, and a second which contains only T_z and is therefore in general an isospin violating term.

Given that these assumptions are valid, we may derive the very general selection rule for any electromagnetic decay

$$\Delta T = 0, \pm 1 \qquad \Delta T_z = 0.$$

The first term in the current operator, J^0, contains no component of $\mathbf{T}$ and therefore cannot induce transitions between states with different T (or T_z). The second, J_z^1, contains T_z and so does not induce transitions between states of different T_z but can induce transitions between states of different T and the same T_z. The rule $\Delta T_z = 0$ is, of course, no more than the statement that charge is conserved. The rule that the maximum change of isospin is one unit follows from the fact that T_z is the third component of a vector in the abstract isospin space. To be less abstract, suppose we have a nucleus the wave function of which can be represented by a set of products of single particle functions,

$$\Psi \sim \psi_1 \psi_2 \ldots \psi_A. \tag{6.5.10}$$

In an electromagnetic transition we will suppose only a few of the outer nucleons to be involved and represent the rest by an inert core with wave function Ψ_c. If the core has an equal number of protons and neutrons, the core isospin is zero. If outside the core is one valence nucleon which transfers from one state to another, the only states of one nucleon outside a core with $T = 0$ have $T = \frac{1}{2}$. All such transitions therefore go from a $T = \frac{1}{2}$ state to another $T = \frac{1}{2}$ state, $\Delta T = 0$. Such transitions can be induced by both the isoscalar and isovector terms in the electromagnetic interaction. Let us consider nucleon 1 as the valence nucleon and nucleons 2 through A as core nucleons. (The proper wave function must of course be fully antisymmetrized.)

$$\Psi_{i,f}(1 \ldots A) = \psi_{1i,f} \Psi_c(2 \ldots A). \tag{6.5.11}$$

The scalar term is

$$\int \Psi_c^*(2 \ldots A) \psi_{1f}^* \frac{1}{2} \left[\frac{e\mathbf{p}}{M} + \frac{e\hbar}{2M}(\mu_p + \mu_n)(\boldsymbol{\sigma} \times \nabla) \right] \psi_{1i} \Psi_c(2 \ldots A) d^3r \tag{6.5.12}$$

where

$$\mathbf{p} = \sum_1^A \mathbf{p}_i,$$

$$\boldsymbol{\sigma} = \sum_1^A \boldsymbol{\sigma}_i.$$

Nuclear Physics

Since the core is inert, only the components $\mathbf{p}_1$ (and $\boldsymbol{\sigma}_1$) for the valence nucleon contribute to the matrix element.

The isovector term is similarly

$$\int \Psi_c^*(2\ldots A)\,\psi_{1f}^*\frac{1}{2}\left[\frac{e\mathbf{p}_i}{M}T_z^i + \frac{e\hbar}{2M}(\mu_p - \mu_n)(\boldsymbol{\sigma}_i \times \nabla)T_z^i\right]$$
$$\psi_{1i}\,\Psi_c(2\ldots A)\,d^3r. \tag{6.5.13}$$

Only the term $\mathbf{p}_1$ for the valence nucleon comes through, and therefore the only individual nucleon T_z operator is that appropriate to nucleon 1. For this case of $\Delta T = 0$ the isovector term differs from the isoscalar term in sign, for $T_z^{(1)}$ operating on $\psi_1\Psi_c$ gives $+\frac{1}{2}$ if the valence nucleon is a proton, and $-\frac{1}{2}$ if it is a neutron. The other difference is just the standard replacement of $\mu_p + \mu_n$ in the isoscalar case by $\mu_p - \mu_n$ in the isovector case. We do not need to consider all other possible terms in the antisymmetrized function: for this situation they will all emerge the same.

A more complicated situation is encountered for two valence nucleons with $T_z = 0$ outside an inert $T = 0$ core. We must write the isospin part of the wave function of the two-valence nucleons as either an isospin singlet or triplet

$$\frac{1}{\sqrt{2}}\left\{\begin{pmatrix}1\\0\end{pmatrix}_1\begin{pmatrix}0\\1\end{pmatrix}_2 \pm \begin{pmatrix}0\\1\end{pmatrix}_1\begin{pmatrix}1\\0\end{pmatrix}_2\right\}_T \Psi_c(3\ldots A)_T$$

where the T suffixes denote the isospin functions.

Only nucleons 1 and 2 can shift their states so we again lump the rest in the core. The iso-singlet part of the interaction can shift orbits or flip spins, but does not change the isospin-function above. The operator $T_z^{(1)}$ gives

$$\frac{1}{2}\frac{1}{\sqrt{2}}\left\{\begin{pmatrix}1\\0\end{pmatrix}_1\begin{pmatrix}0\\1\end{pmatrix}_2 \mp \begin{pmatrix}0\\1\end{pmatrix}_1\begin{pmatrix}1\\0\end{pmatrix}_2\right\}_T \Psi_c(3\ldots A)_T \tag{6.5.14}$$

and so for this particular case the final nucleus differs in isospin by 1 unit. The operator $T_z^{(2)}$ gives the same apart from a change in sign, and so in this case even where two valence nucleons are involved, the isospin can only change by zero or 1 unit. Considering only singlets or triplets, it is obvious that this is the case without doing any isospin algebra. If, however, we have three valence nucleons outside a $T = 0$ core, it is not instantly obvious that a $\Delta T = 2$ transition may not occur. It becomes obvious on applying the algebra, or from the following simple physical argument. For each nucleon that may change its state, we apply either the singlet operator 1, in which case there is no change of isospin, or alternatively the triplet interaction operator $T_z^{(i)}$, the effect of which is that a nucleon is taken out of the nucleus, a process which changes the isospin by $\pm\frac{1}{2}$ and then put back in again, changing the isospin by $\pm\frac{1}{2}$. Thus for any initial isospin, applying the operator $T_z^{(i)}$ to the ith nucleon changes the isospin by ± 1 or 0. Although many nucleons may be involved in a transition, the operators $T_z^{(i)}$ are each applied to the initial state and so each application can only lead to a final state with $\Delta T = \pm 1, 0$. There is no term of the kind $T_z^{(1)}T_z^{(2)}$ in the electromagnetic interaction we have written, and it is only such a term that could change the isospin by two units.

Isospin

The selection rule for γ-transitions

$$\Delta T = \pm 1, 0$$

thus follows from our writing the electromagnetic interaction as simply as possible, depending on the isospin only through terms linear in the $T_z^{(i)}$. This is, of course, an assumption, but is borne out in practice, although the tests are few because of the paucity of identified $T = 2, T_z = 0$ members of quintets. Finally, this rule will be violated at some level because the isospin states will not be quite pure.

There are a large number of other selection rules, of varying degrees of rigour. Here we will only mention two, of which the first is: $\Delta T = 0$ E1 transitions in self-conjugate nuclei are forbidden.

A self-conjugate nucleus has an equal number of protons and neutrons. For an E1 transition the centre of mass must move relative to the centre of charge. If, however, we try and shift the orbit of a proton we must shift a neutron with the same characteristics to the same new orbit if the isospin T is not to change. But this means the centre of charge cannot shift with respect to the centre of mass and so no electric dipole transition:

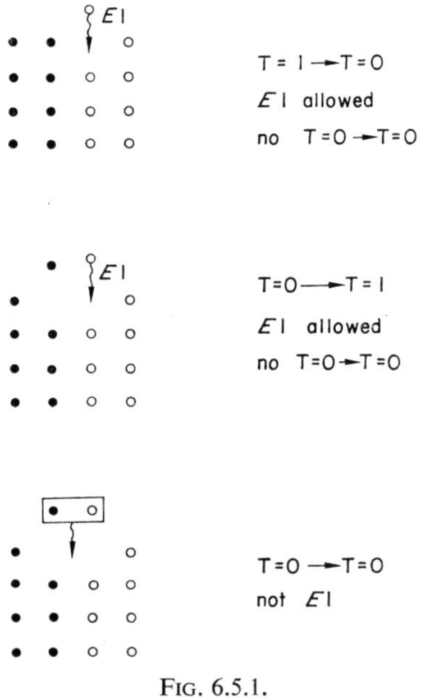

FIG. 6.5.1.

Apart from isospin impurity of the participating nuclear states, this rule is exact in the long wavelength limit. This selection rule can be derived much more formally than through the argument we have given, which has, however, the virtue that the rule is seen to be just a consequence of charge independence of nuclear forces. (It is in fact even broader based, requiring only charge symmetry.)

The second rule is much looser. In magnetic dipole transitions the isoscalar operator is proportional to

$$\sim \sum_i (\mu_p + \mu_n)_i = \sum_i 0 \cdot 88 \qquad (6.5.15)$$

Nuclear Physics

and the isovector term to

$$\sim \sum_i (\mu_p - \mu_n)_i = \sum_i 4.70. \quad (6.5.16)$$

The coupling strength of the isovector M1 term is thus ~ 5 times the coupling strength of the isoscalar M1 term, and inclusion of orbital motion increases the divergence. The M1 transition rates when only the isoscalar term can drive the transition are thus expected to be $\approx \frac{1}{100}$ of the rates when the isovector term can contribute, that is in the majority of M1 transitions. Thus a $T = 0 \to T = 0$ M1 transition will be weaker than would otherwise be expected. In addition, if $T_z = 0$ as in self-conjugate nuclei the isovector term cannot give $\Delta T = 0$, only $\Delta T = \pm 1$ [we saw an example of this in eq. (6.5.14)] and so the more general statement of this rule is that $\Delta T = 0$ M1 transitions in nuclei with $T_z = 0$ are inhibited by a factor ≈ 100.

These two rules, for particular classes of E1 and M1 transitions, are not very fundamental and are restricted in their application. They can be very useful, however, in sorting out the details of nuclear structure among the richly populated excited states of self-conjugate nuclei. It must always be remembered that if a particular γ-transition is suppressed this may be a reflection of the gross nuclear structure through an isospin selection rule, or it may just be an accidental suppression owing nothing to such general features.

6.6. Isospin and nuclear decay: β-decay

We already touched briefly on this subject in Section 6.3 in which we pointed out that the isospin formalism gives us a natural way of including explicitly the conversion of a neutron into a proton, or vice versa, without specifying from which we start. We may write the fundamental weak interaction as

$$\{\bar{\psi}_e(1-\gamma_5)\gamma_\mu\psi_\nu\}\{\bar{\psi}_N(g_V - g_A\gamma_5)\gamma_\mu T_+\psi_N\} \quad (6.6.1)$$

which turns a neutron into a proton with emission of an electron neutrino pair, and its complex conjugate similarly turns a proton into a neutron. If all particles could be non-relativistic, parity-violating terms would disappear and we would have

$$\left.\begin{array}{ll} g_V(\psi_e^* \psi_\nu^*)(\psi_N^* T_+ \psi_N) & \text{Fermi interaction,} \\ g_A(\psi_e^* \sigma \psi_\nu^*)\cdot(\psi_N^* \sigma T_+ \psi_N) & \text{Gamow–Teller interaction.} \end{array}\right\} \quad (6.6.2)$$

In a nuclear transition we would write for the matrix element

$$\{\bar{\psi}_e(1-\gamma_5)\gamma_\mu\psi_\nu\}\sum_i \{\bar{\Psi}_f(g_V - g_A\gamma_5)\gamma_\mu^{(i)} T_+^{(i)}\Psi_i\} \quad (6.6.3)$$

where the summation extends over all nucleons but only yields a non-zero term for nucleons which can transform. Once more each of the terms, being linear in T_+, corresponds to taking out one nucleon, and putting it back after changing its value of T_z. The isospin selection rules for β-decay are immediately obvious

$$\Delta T = \pm 1, 0 \quad (0 \not\to 0),$$
$$\Delta T_z = \pm 1.$$

Isospin

The weak interaction as written here is thus an isovector interaction. The isovector part of the electromagnetic interaction goes through the third component of the isovector **T**, namely T_z, while the β-interactions couple through the components

$$T_+ = T_x + iT_y, \qquad (6.6.4)$$
$$T_- = T_x - iT_y,$$

which are related to the standard components of **T** in the same way that circular polarization is related to linear polarization.

This notation is particularly nice for Fermi transitions between members of the same multiplet, where we can write the nuclear part of the matrix element as

$$g_V \sum_i \int (\Psi_f^* T_\pm^{(i)} \Psi_i) \, dV. \qquad (6.6.5)$$

The matrix element only changes charge and does not otherwise rearrange the nucleons in any way. Then

$$g_V \sum_i \int \Psi_f^*(T, T_z \pm 1) \, T_\pm^{(i)} \Psi_i(T, T_z) \, dV = g_V \int \Psi_f^*(T, T_z \pm 1) \, T_\pm \Psi_i(T, T_z) \, dV$$

where

$$T_\pm = \sum_i T_\pm^{(i)} \qquad (6.6.6)$$

and the value of the integral is quite independent of the wave functions themselves and is given by the properties of the raising and lowering operators only. (On the other hand, $\sum_i \int \Psi_f^* \sigma^{(i)} T_\pm^{(i)} \Psi_i \, dV$ is NOT independent of the wave functions because of the rearrangement of spins governed by the additional factor $\sigma^{(i)}$.)

The matrix elements for vector operators† give

$$\langle T, T_z \pm 1 | T_\pm | T, T_z \rangle = \sqrt{(T \mp T_z)(T \pm T_z + 1)}, \qquad (6.6.7)$$

thus if $T = \frac{1}{2}$

$$\langle \tfrac{1}{2}, +\tfrac{1}{2} | T_+ | \tfrac{1}{2}, -\tfrac{1}{2} \rangle = 1, \qquad (6.6.8)$$
$$\langle \tfrac{1}{2}, -\tfrac{1}{2} | T_- | \tfrac{1}{2}, +\tfrac{1}{2} \rangle = 1$$

(relations which you can easily check for a nucleon by using the raising and lowering operators we have employed before) and if $T = 1$

$$\langle 1, 0 | T_- | 1, +1 \rangle = \sqrt{2}, \qquad (6.6.9)$$
$$\langle 1, -1 | T_- | 1, 0 \rangle = \sqrt{2}$$

(which can easily be checked for two nucleons in a $T = 1$ state). Thus the strength of the Fermi part of neutron decay is just g_V^2 and the pure Fermi $^{14}O \rightarrow {}^{14}N^*$ decay has strength $2g_V^2$. The isospin formalism is not, of course, telling us something new in this application. In ^{14}O decay there are two identical protons either of which can decay into a neutron giving the excited state of $^{14}N^*$ and therefore it is obvious that the strength is $2g_V^2$. The other mem-

---

† See, for example, L. D. Landau and E. M. Lifshitz, *Quantum Mechanics*, 2nd ed. (Pergamon Press, 1965).

Nuclear Physics

ber of the $T = 1$ multiplet, ^{14}C, decays via a Gamow–Teller transition to the ground state of ^{14}N and although there are two equivalent neutrons in ^{14}C the spin flip involved means that we cannot predict the strength of the transition from isospin considerations alone.

The fact that the nuclear matrix element for allowed Fermi transitions does not depend on the details of nuclear structure but is governed only by the matrix elements of the isospin raising and lowering operators suggests that this might be a general feature of the Fermi interactions of strongly interacting particles. The pion is also an isospin triplet, and if the decay

$$\pi^+ \to \pi^0 e^+ \nu$$

can be written analogously to that for ^{14}O decay

$$g_\pi = g_V \langle \pi^0 | T_- | \pi^+ \rangle = \sqrt{2}\, g_V \tag{6.6.10}$$

then we have an unambiguous prediction for the β-decay rate of the charged pion, which is in admirable accord with experiment. This generalization arises from the conserved vector current hypothesis which we touched on briefly in Chapter 3. The quantitative connection between this result from the isospin algebra and the concept of a conserved weak interaction Fermi charge will be discussed in the next section. The hypothesis that is known as the conserved vector current hypothesis (CVC) is in fact more specific than the title implies. A general conserved vector current hypothesis implies that there exists a weak charge g_V which is conserved just as electric charge is conserved. The more restrictive version of CVC depends on noting that the electromagnetic interaction depends linearly on T_z and the β-interactions on $T_\pm$. For point particles the electric current can be written

$$J_{z\mu} = e\,(\bar{\psi} T_z \gamma_\mu \psi) \tag{6.6.11}$$

and the weak currents for the Fermi interaction as

$$J_{\pm\mu} = g_V (\bar{\psi} T_\pm \gamma_\mu \psi). \tag{6.6.12}$$

The appropriate charge conservation is expressed through

$$\sum_\mu \frac{\partial J_\mu}{\partial x_\mu} = 0 \quad \left(\text{i.e.} \quad \nabla \cdot \mathbf{J} + \frac{\partial \rho}{\partial t} = 0 \right) \tag{6.6.13}$$

and for point particles the analogy is obvious.

The nucleons, however, have considerable structure, manifested by their anomalous magnetic moments and form factors. This structure is traditionally attributed to the meson fields surrounding the nucleon but whatever the virtual processes are the electric charge is always conserved. The heart of the conserved vector current hypothesis lies in the assertion that if the electromagnetic interaction has a vector part

$$e \sum_\mu A_\mu J_{z\mu} \tag{6.6.14}$$

where $J_{z\mu}$ is a conserved current then the Fermi part of the weak interaction is

$$\frac{g_V}{\sqrt{2}} \sum_\mu [\bar{\psi}_e (1 - \gamma_5)\, \gamma_\mu \psi_\nu]\, J_{\pm\mu} \tag{6.6.15}$$

where
$$J_{z\mu} = J_\mu T_z,$$
$$J_{\pm\mu} = J_\mu T_\pm$$

so that the anomalous magnetic moment of the nucleon is reproduced in a term

$$\frac{g_V \hbar}{2Mc}(\mu_p - \mu_n) T_\pm^{(i)} \tag{6.6.16}$$

and the vector weak interaction form factors are the same as the electromagnetic form factors. As we pointed out in Chapter 3, weak magnetism has been detected and agrees within errors with this prediction, while the weak form factors studied in the process

$$\nu_\mu + n \to p + \mu^-$$

agree within (considerable) errors with the electromagnetic form factors of the nucleon. While we do not wish to dig further into this subject here, it is the isospin formalism that makes manifest the elegance of the CVC hypothesis.

While relations between various strong reaction rates in nuclear physics can be obtained using isospin, these relations will be central to our discussion of isospin and particle physics and so are dealt with and illustrated in the next section.

6.7. The pion and isospin

The particle known as the pion (or π-meson) was postulated by Yukawa in 1935 in order to explain the observed short range of nuclear forces. The argument may be constructed by analogy with electromagnetism:

A free electromagnetic field is described by the equations

$$\nabla^2 \phi - \frac{1}{c^2}\frac{\partial^2 \phi}{\partial t^2} = 0, \tag{6.7.1}$$

$$\nabla^2 \mathbf{A} - \frac{1}{c^2}\frac{\partial^2 \mathbf{A}}{\partial t^2} = 0. \tag{6.7.2}$$

In Chapter 3 we interpreted the vector potential $\mathbf{A}$ as the wave function of a photon and consequently we could write eq. (6.7.2) as

$$(p^2 - E^2/c^2)\mathbf{A} = 0 \tag{6.7.3}$$

which is just the relation between energy and momentum for a massless particle, and demonstrates that the photons corresponding to a free electromagnetic field have zero rest mass. For a free massive particle we would write

$$(p^2 - E^2/c^2 + m^2 c^2)\psi = 0 \tag{6.7.4}$$

or a wave equation

$$\left(\nabla^2 - \frac{1}{c^2}\frac{\partial^2}{\partial t^2}\right)\psi = -\frac{m^2 c^2}{\hbar^2}\psi. \tag{6.7.5}$$

Nuclear Physics

The static potential in empty space for electromagnetism is given by the equations

$$\nabla^2 \phi = 0,$$
$$\nabla^2 \mathbf{A} = 0,$$

and the corresponding static potential of a meson field is given by

$$\nabla^2 \psi = -\frac{m^2 c^2}{\hbar^2} \psi. \tag{6.7.6}$$

If the source of the static field is a point charge of some kind, then ψ will be spherically symmetric and eq. (6.7.6) can be directly integrated to give

$$\psi = G \frac{e^{-(mc/\hbar)r}}{r} \tag{6.7.7}$$

where G is a measure of the source strength. We may now interpret

$$\psi = G \frac{e^{-(mc/\hbar)r}}{r}$$

as the potential due to exchange of mesons of mass m, just as we interpret

$$\phi = \frac{q}{r}$$

as the electrostatic potential due to the exchange of photons of the coulomb field, of rest-mass zero. In first-order perturbation theory, the matrix element connecting an initial plane wave state with a final plane wave state is given by

$$\mathscr{M}_\pi = G^2 \int e^{-i\mathbf{k}'\cdot\mathbf{r}} \frac{e^{-(mc/\hbar)r}}{r} e^{i\mathbf{k}\cdot\mathbf{r}} d^3r$$

$$= G^2 \int e^{-i\mathbf{q}\cdot\mathbf{r}} \frac{e^{-(mc/\hbar)r}}{r} d^3r,$$

$$\mathscr{M}_\pi = \frac{4\pi G^2}{q^2 + m^2 c^2/\hbar^2} \tag{6.7.8}$$

to be compared with the coulomb scattering matrix element (see Chapter 4),

$$\mathscr{M}_c = \frac{4\pi e^2}{q^2}. \tag{6.7.9, 4.4.14}$$

Thus while the probability of a photon in the coulomb field surrounding a charged particle having momentum q is $\propto 1/q^4$, the probability of a pion in the field surrounding a source of pions, such as a nucleon, having momentum q is

$$\propto \left[\frac{1}{q^2 + m^2 c^2/\hbar^2}\right]^2$$

Isospin

which is fairly constant until $q^2 \gg m^2c^2/\hbar^2$. The corresponding diagrams are shown in Fig. 6.7.1.

Fig. 6.7.1.

If we write eq. (6.7.7) as

$$\psi = G\frac{e^{-r/r_0}}{r}$$

then the range of the potential is given by

$$r_0 = \frac{\hbar}{mc}$$

and varies inversely with the mass. If we set $r_0 \sim 10^{-13}$ cm then

$$m = \frac{\hbar}{r_0 c} \sim 3 \times 10^{-25} \text{ g} \sim 0.2\, M_p.$$

The mass of the neutral pion is in fact ~ 135 MeV/c^2, $\sim 0.14\, M_p$, corresponding to a range of

$$r_{\pi^0} = 1.46 \times 10^{-13} \text{ cm}.$$

(The charged pions have a mass of 139.6 MeV/c^2.) We must emphasize that the pion is only one of a large number of mesons which can be exchanged between nucleons to generate the nuclear force. The pion is, however, by far the least massive and one pion exchange seems to dominate the long-range part of the nuclear field:

Meson	Mass, MeV/c^2	Range (10^{-13} cm)
$\pi^\pm$	139.6	1.41
π^0	135.0	1.46
$\eta^0_\pm$	548.8	0.36
ϱ^0	~ 765	~ 0.26
ω^0	783.4	0.25
f^0	~ 1264	~ 0.16

and so on.

We will suppose, however, that each individual mechanism independently generates charge independent forces, for otherwise the observed charge independence would be the result of an extraordinary cancellation of terms which is hardly credible. The one-pion exchange contribution to the nucleon–nucleon interaction can be represented by the diagrams shown in Fig. 6.7.2. (If we included a π^- coming from the upper vertex with the last diagram, we should also have to include a π^0 coming from the upper vertex in the other three, and so

Nuclear Physics

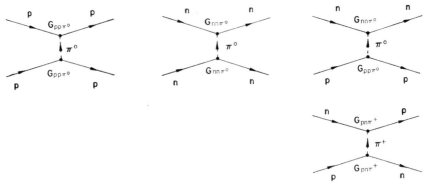

FIG. 6.7.2.

would get nothing new.) There are three arbitrary coupling strengths in the problem which are related by the requirement that the scattering should be the same in each charge state. Thus in the approximation that the charged and neutral pions have equal masses we must require

$$G_{pp\pi^0}G_{pp\pi^0} = G_{nn\pi^0}G_{nn\pi^0} = G_{pp\pi^0}G_{nn\pi_0}+G_{pn\pi+}G_{pn\pi+} \qquad (6.7.10)$$

if the scattering amplitudes are to be the same in all three states. One obvious solution is

$$G_{pp\pi^0} = G_{nn\pi^0}$$

in which case $G_{pn\pi+} = 0$. This implies that there is not a charged pion (and there is) or that the charged pion does not couple with a nucleon, and if the neutral pion does couple with a nucleon, so must the charged pion. We therefore discard this solution and adopt the alternative

$$\begin{aligned} G_{pp\pi^0} &= -G_{nn\pi^0}, \\ G_{pn\pi+} &= \pm\sqrt{2}G_{pp\pi^0}, \end{aligned} \qquad (6.7.11)$$

which determines the coupling of nucleons to pions down to unobservable signs.

We may now introduce isospin into this picture. If we have a nucleon dissociating into a nucleon and a pion, then our hypothesis that isospin is conserved in all strong interactions requires that:

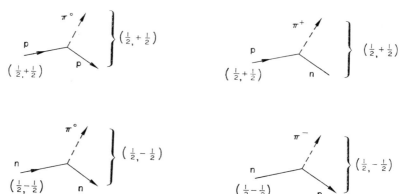

FIG. 6.7.3.

Isospin

The amplitude for $p \to n\pi^+$ is $\sqrt{2}$ times the amplitude for $p \to p\pi^0$. The amplitude for $n \to p\pi^-$ is $\sqrt{2}$ times the amplitude for $n \to n\pi^0$: a $T=(\frac{1}{2},+\frac{1}{2})$ state couples $\sqrt{2}$ times as strongly to $n\pi^+$ as $p\pi^0$. Thus

$$p \to -\sqrt{\tfrac{1}{3}}p\pi^0 + \sqrt{\tfrac{2}{3}}n\pi^+,$$
$$n \to -\sqrt{\tfrac{2}{3}}p\pi^- + \sqrt{\tfrac{1}{3}}n\pi^0,$$

or in more general terms, if we couple a triplet state to a doublet to form a doublet

$$\begin{aligned}(\tfrac{1}{2},+\tfrac{1}{2}) &= -\sqrt{\tfrac{1}{3}}(\tfrac{1}{2},+\tfrac{1}{2};1,0) + \sqrt{\tfrac{2}{3}}(\tfrac{1}{2},-\tfrac{1}{2};1,+1),\\ (\tfrac{1}{2},-\tfrac{1}{2}) &= -\sqrt{\tfrac{2}{3}}(\tfrac{1}{2},+\tfrac{1}{2};1,-1) + \sqrt{\tfrac{1}{3}}(\tfrac{1}{2},-\tfrac{1}{2};1,0),\end{aligned} \quad (6.7.12)$$

where we have chosen the arbitrary phases so as to conform with the convention for the coupling of angular momentum.

The relationship between the coupling constants $G_{pp\pi^0}$, $G_{nn\pi^0}$, $G_{pn\pi^+}$ follows just from the charge independence of nuclear forces. Combining these coupling constants with the hypothesis that isospin is conserved at every strong interaction vertex gives us relations eq. (6.7.12), which we could have obtained very formally from the hypothesis that isospin is conserved, plus the identification of the nucleon as an isospin doublet and the pion as an isospin triplet. Such a derivation relies on the algebra developed for coupling angular momentum and we have preferred to get the same results in a more physical way. It should by now be clear that the choice

$$G_{pp``\pi^0"} = G_{nn``\pi^0"}$$

is appropriate to the coupling of an isoscalar meson with a nucleon and leads to trivial relationships—the η^0 is such an isoscalar meson, as is the f^0.

Now if a pion and a nucleon are coupled together, from a doublet and a triplet we can make a doublet $T=\tfrac{1}{2}$ and a quartet $T=\tfrac{3}{2}$.

T_z		T_z	
$+1$	π^+	$+\tfrac{3}{2}$	π^+p,
$+\tfrac{1}{2}$	p	$+\tfrac{1}{2}$	$\pi^0p,\ \pi^+n$,
0	π^0		
$-\tfrac{1}{2}$	n	$-\tfrac{1}{2}$	$\pi^-p,\ \pi^0n$,
-1	π^-	$-\tfrac{3}{2}$	$\pi^-n.$

The π^+p and π^-n states have pure isotopic spin, while the π^0p, π^+n, π^-p and π^0n states are mixtures of $T=\tfrac{3}{2}$ and $T=\tfrac{1}{2}$ states. Now the assumption that isospin is conserved requires that the amplitudes for pion–nucleon scattering are independent of the value of T_z and can only depend on the value of T. Thus π^+p and π^-n scattering must be identical. This conclusion is independent of the pion–nucleon couplings derived in (6.7.11), for the diagrams of Fig. 6.7.3 contain no π^+p or π^-n vertices. At last isospin conservation has told us something we did not know already! But we can get more. For the pion–nucleon system write

$$\left.\begin{aligned}(\tfrac{3}{2},+\tfrac{3}{2}) &= (p\pi^+),\\ (\tfrac{3}{2},+\tfrac{1}{2}) &= A(p\pi^0) + B(n\pi^+),\\ (\tfrac{3}{2},-\tfrac{1}{2}) &= C(p\pi^-) + D(n\pi^0),\\ (\tfrac{3}{2},-\tfrac{3}{2}) &= (n\pi^-)\end{aligned}\right\} \quad (6.7.13)$$

Nuclear Physics

and we already have

$$(\tfrac{1}{2}, +\tfrac{1}{2}) = -\sqrt{\tfrac{1}{3}}(p\pi^0) + \sqrt{\tfrac{2}{3}}(n\pi^+),$$
$$(\tfrac{1}{2}, -\tfrac{1}{2}) = -\sqrt{\tfrac{2}{3}}(p\pi^-) + \sqrt{\tfrac{1}{3}}(n\pi^0). \tag{6.7.14}$$

We can now find A and B, C and D. First, if we normalize our isospin states to 1,

$$\left.\begin{array}{l}|A|^2 + |B|^2 = 1, \\ |C|^2 + |D|^2 = 1,\end{array}\right\} \tag{6.7.15}$$

requiring the $(\tfrac{3}{2}, +\tfrac{1}{2})$ state to be orthogonal to the $(\tfrac{1}{2}, +\tfrac{1}{2})$ state gives

$$-\sqrt{\tfrac{1}{3}}A + \sqrt{\tfrac{2}{3}}B = 0 \quad A = \sqrt{\tfrac{2}{3}} \quad B = \sqrt{\tfrac{1}{3}} \tag{6.7.16}$$

and similarly for the $T_3 = -\tfrac{1}{2}$ state

$$-\sqrt{\tfrac{2}{3}}C + \sqrt{\tfrac{1}{3}}D = 0 \quad C = \sqrt{\tfrac{1}{3}} \quad D = \sqrt{\tfrac{2}{3}}. \tag{6.7.17}$$

We can now write down a complete table of couplings between isospin doublets and isospin triplets: these numbers are identical with the Clebsch–Gordan coefficients for coupling spin $\tfrac{1}{2}$ with spin 1 (see Table 6.7.1).

TABLE 6.7.1

	$\tfrac{3}{2}, +\tfrac{3}{2}$	$\tfrac{3}{2}, +\tfrac{1}{2}$	$\tfrac{1}{2}, +\tfrac{1}{2}$	$\tfrac{1}{2}, -\tfrac{1}{2}$	$\tfrac{3}{2}, -\tfrac{1}{2}$	$\tfrac{3}{2}, -\tfrac{3}{2}$
$1, +1; \tfrac{1}{2}, +\tfrac{1}{2}$	1					
$1, +1; \tfrac{1}{2}, -\tfrac{1}{2}$		$\sqrt{\tfrac{1}{3}}$	$\sqrt{\tfrac{2}{3}}$			
$1, 0; \tfrac{1}{2}, +\tfrac{1}{2}$		$\sqrt{\tfrac{2}{3}}$	$-\sqrt{\tfrac{1}{3}}$			
$1, 0; \tfrac{1}{2}, -\tfrac{1}{2}$				$\sqrt{\tfrac{1}{3}}$	$\sqrt{\tfrac{2}{3}}$	
$1, -1; \tfrac{1}{2}, +\tfrac{1}{2}$				$-\sqrt{\tfrac{2}{3}}$	$\sqrt{\tfrac{1}{3}}$	
$1, -1; \tfrac{1}{2}, -\tfrac{1}{2}$						1

The content of such a table is difficult to appreciate as it stands, but let us see how it works out in practice, for the case of pion–nucleon scattering. The following processes can be studied:

$$\begin{aligned}\pi^+ p &\to \pi^+ p, \\ \pi^- p &\to \pi^- p, \\ \pi^- p &\to \pi^0 n, \\ \pi^+ n &\to \pi^+ n, \\ \pi^+ n &\to \pi^0 p, \\ \pi^- n &\to \pi^- n\end{aligned} \tag{6.7.18}$$

Isospin

(since the π^0 has a lifetime $\sim 10^{-16}$ sec processes with an incident π^0 are not amenable to study).

The assumption that isospin is conserved in all strong interactions implies that the strength of the interaction depends only on the total isospin T and not on its third component T_z. The possible isospins in the problem are $\tfrac{1}{2}$ and $\tfrac{3}{2}$ and so we must describe all six of the reactions (6.7.18) by only two scattering amplitudes:

$A_{3/2}$ when the pion and nucleon are in the state with $T = \tfrac{3}{2}$,

$A_{1/2}$ when the pion and nucleon are in the state with $T = \tfrac{1}{2}$.

The scattering amplitudes A are in general functions of the centre of mass energy and the centre of mass scattering angle (or two equivalent variables).
Then

$$\left.\begin{aligned}\frac{d\sigma}{d\Omega}(\pi^+ p \to \pi^+ p) &= K\,|A_{3/2}|^2, \\ \frac{d\sigma}{d\Omega}(\pi^- n \to \pi^- n) &= K\,|A_{3/2}|^2,\end{aligned}\right\} \tag{6.7.19}$$

where K is a number which may contain kinematic and normalization terms, depending on the precise definition of the scattering amplitude A.

The initial state $\pi^- p$ is not a state of pure isospin: on referring to the table we see that it is a mixture of $(\tfrac{3}{2}, -\tfrac{1}{2})$ and $(\tfrac{1}{2}, -\tfrac{1}{2})$ given by

$$\pi^- p = -\sqrt{\tfrac{2}{3}}(\tfrac{1}{2}, -\tfrac{1}{2}) + \sqrt{\tfrac{1}{3}}(\tfrac{3}{2}, -\tfrac{1}{2}).$$

Each isospin state has its own scattering amplitude and so the final state is described by

$$-\sqrt{\tfrac{2}{3}}A_{1/2}(\tfrac{1}{2}, -\tfrac{1}{2}) + \sqrt{\tfrac{1}{3}}A_{3/2}(\tfrac{3}{2}, -\tfrac{1}{2}). \tag{6.7.20}$$

We do not observe the pure isospin states $(\tfrac{1}{2}, -\tfrac{1}{2})$ and $(\tfrac{3}{2}, -\tfrac{1}{2})$ and so must enquire how they break up into the observable states:

$$(\tfrac{1}{2}, -\tfrac{1}{2}) \to \sqrt{\tfrac{1}{3}}\pi^0 n - \sqrt{\tfrac{2}{3}}\pi^- p,$$
$$(\tfrac{3}{2}, -\tfrac{1}{2}) \to \sqrt{\tfrac{2}{3}}\pi^0 n + \sqrt{\tfrac{1}{3}}\pi^- p,$$

and so the final state function after scattering is given by

$$-\sqrt{\tfrac{2}{3}}A_{1/2}\{\sqrt{\tfrac{1}{3}}\pi^0 n - \sqrt{\tfrac{2}{3}}\pi^- p\} + \sqrt{\tfrac{1}{3}}A_{3/2}\{\sqrt{\tfrac{2}{3}}\pi^0 n + \sqrt{\tfrac{1}{3}}\pi^- p\}$$
$$= \pi^0 n \{A_{3/2} - A_{1/2}\}\frac{\sqrt{2}}{3} + \pi^- p\{A_{3/2} + 2A_{1/2}\}\tfrac{1}{3} \tag{6.7.21}$$

and so

$$\left.\begin{aligned}\frac{d\sigma}{d\Omega}(\pi^- p \to \pi^- p) &= \frac{K}{9}\,|A_{3/2} + 2A_{1/2}|^2, \\ \frac{d\sigma}{d\Omega}(\pi^- p \to \pi^0 n) &= \frac{2K}{9}\,|A_{3/2} - A_{1/2}|^2.\end{aligned}\right\} \tag{6.7.22}$$

Nuclear Physics

Similarly

$$\frac{d\sigma}{d\Omega}(\pi^+ n \to \pi^+ n) = \frac{K}{9}|A_{3/2}+2A_{1/2}|^2,$$

$$\frac{d\sigma}{d\Omega}(\pi^+ n \to \pi^0 p) = \frac{2K}{9}|A_{3/2}-A_{1/2}|^2. \quad (6.7.23)$$

Thus

$$\frac{d\sigma}{d\Omega}(\pi^+ p \to \pi^+ p) = \frac{d\sigma}{d\Omega}(\pi^- n \to \pi^- n),$$

$$\frac{d\sigma}{d\Omega}(\pi^- p \to \pi^- p) = \frac{d\sigma}{d\Omega}(\pi^+ n \to \pi^+ n), \quad (6.7.24)$$

$$\frac{d\sigma}{d\Omega}(\pi^- p \to \pi^0 n) = \frac{d\sigma}{d\Omega}(\pi^+ n \to \pi^0 p)$$

and there are only three independent cross-sections among the six of (6.7.18). The remaining degrees of freedom correspond to the magnitudes of the scattering amplitudes $A_{3/2}$, $A_{1/2}$ and the relative phase between the two. The pion–nucleon interaction is NOT charge independent but DOES conserve isospin. The nucleon–nucleon interaction is charge independent because the two nucleons must be treated as identical fermions and any state of given angular momentum and parity contains only one isospin. Any pion–nucleon state of given angular momentum and parity may, however, contain two isospin states.

In a region where either $A_{3/2}$ or $A_{1/2}$ is very small, all the pion–nucleon cross-sections are approximately specified by one amplitude alone. For example, at low energy $A_{1/2} \ll A_{3/2}$ and so in the region of the famous $N^*(1236)$ pion–nucleon resonance, which has $T = \frac{3}{2}$, $\sigma(\pi^- p) = \frac{1}{3}\sigma(\pi^+ p)$. This is illustrated in Fig. 6.7.4.

Having derived the relative strengths of the various pion–nucleon couplings it is now quite easy to see in a very physical way why the conserved vector current hypothesis predicts a coupling constant for the process $\pi^+ \to \pi^0 e^+ \nu$ equal to $\sqrt{2}g_V$ and implies that the presence of the strong interactions does not modify the strength of the Fermi coupling.

A neutron with no strong interactions would decay only via the diagram of Fig. 6.7.5. But a neutron which couples to pions can some of the time be represented by Fig. 6.7.6 with a total amplitude G^2 where from eq. (6.7.12)

$$G_{nn\pi^0} = \sqrt{\tfrac{1}{3}}G \quad G_{np\pi^-} = -\sqrt{\tfrac{2}{3}}G$$

which suggests that if the pion is incapable of β-decay, these processes would reduce the effective coupling, for if $\pi^- \to \pi^0 e^- \bar{\nu}$ then virtual dissociation of the neutron into $p\pi^-$ would prevent decay for the period of the dissociation. Given that the coupling constant g_V is equal to the weak interaction coupling constant g for purely leptonic processes we must consider the β-decay diagrams of Fig. 6.7.7. These diagrams represent amplitudes for what happens following a virtual dissociation $n \to N\pi$ and must be added and the sum of the contributions of these diagrams must be equal to $g_V G^2$:

$$g_V G_{nn\pi^0} G_{pp\pi^0} + g_\pi G_{nn\pi^0} G_{pn\pi^+} + g_\pi G_{np\pi^-} G_{pp\pi^0} = g_V G^2 \quad (6.7.25)$$

if the weak charge is to be conserved in the virtual dissociation $N \to N\pi$.

Isospin

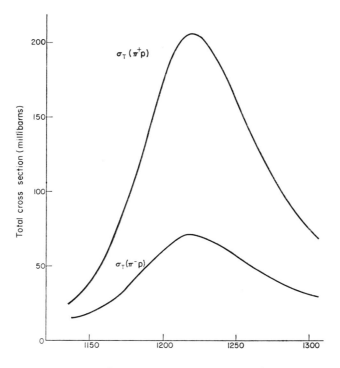

FIG. 6.7.4. The figure shows the total cross-sections for $\pi^+ p$ and $\pi^- p$ in the region of the $N^*(1236)$ resonance. The two cross-sections are closely in the ratio 3:1, indicating only a small amount of $T = \frac{1}{2}$ background to the dominant $T = \frac{3}{2}$ interaction. [Data from A. A. Carter et al., *Nuclear Physics*, B **26**, 445 (1971).]

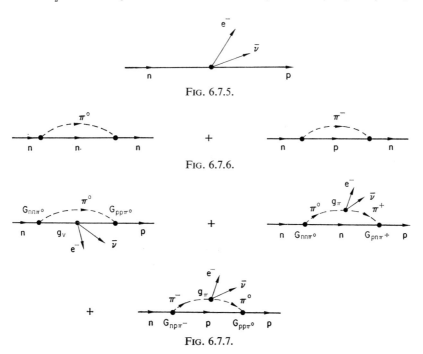

FIG. 6.7.5.

FIG. 6.7.6.

FIG. 6.7.7.

Nuclear Physics

Now with

$$G_{nn\pi^0} = \sqrt{\tfrac{1}{3}}G \quad G_{pp\pi^0} = -\sqrt{\tfrac{1}{3}}G,$$
$$G_{pn\pi^+} = \sqrt{\tfrac{2}{3}}G \quad G_{np\pi^-} = -\sqrt{\tfrac{2}{3}}G$$

[from eq. (6.7.12)],

$$\therefore \quad -\frac{1}{3}g_V + \frac{\sqrt{2}}{3}g_\pi + \frac{\sqrt{2}}{3}g_\pi = g_V$$

whence at once

$$g_\pi = \sqrt{2}g_V \tag{6.7.26}$$

which is the same result we obtained in the previous section by assuming that

$$g_\pi = \langle \pi^0 | T_+ | \pi^- \rangle g_V.$$

6.8. Strange particles

In this book the only strongly interacting fermions we have so far considered are the two nucleons (although we have mentioned πN resonant states) which have isospin $T = \tfrac{1}{2}$. Among the strongly interacting bosons, we have singled out the pion, with $T = 1$ and have only mentioned other mesons which can directly couple with the nucleon, that is, mesons with $T = 1$ or 0. There also exist fermions with integral values of T and bosons with half integral values of T: these are the strange particles that are the subject of this section. (Strange is a technical term.)

The first evidence for the existence of strongly interacting particles other than the nucleon and the pion was obtained in photographs of a cloud chamber exposed to the cosmic radiation: in 1947 Rochester and Butler observed the decay of a neutral particle into two charged particles, and also the decay of a charged particle into a charged and neutral secondary. While a great deal of quantitative work on these new particles was done with cloud chambers and photographic emulsions exposed to the cosmic radiation the real advances came after proton synchrotrons sufficiently energetic to produce these particles were constructed: the Cosmotron (at Brookhaven) and the Bevatron (at Berkeley).

The neutral decays observed in the early cosmic radiation work represent decays of two distinct particles: a boson

$$K^0 \to \pi^+ \pi^-$$

and a fermion

$$\Lambda^0 \to p\pi^-.$$

The Λ^0 also has a decay mode $\Lambda^0 \to n\pi^0$ and the K^0 a variety of other decay modes. There are charged K-mesons (or kaons) with decays

$$K^+ \to \mu^+ \nu_\mu,$$
$$\to \pi^+ \pi^0,$$
$$\to \pi^+ \pi^- \pi^+$$

and many others, and there is a family of charged fermions

$$\Sigma^+ \to p\pi^0$$
$$n\pi^+,$$
$$\Sigma^0 \to \Lambda^0 \gamma,$$
$$\Sigma^- \to n\pi^-.$$

These families, K, Λ, Σ are the strange particles studied in the late 1940s, and early 1950s. The great problem about these particles was that while their lifetimes are characteristic of weak decay processes ($\sim 10^{-8}$–10^{-10} sec) they are produced copiously in the strong interactions with cross-sections $\sim \frac{1}{10}$ of the cross-sections for pion production. This constitutes an apparent paradox: the strange particles are produced strongly by strongly interacting particles and decay weakly into strongly interacting particles.

The characteristic time for the interactions responsible for production is easily estimated. A nucleon has a radius $\sim 10^{-13}$ cm and so if any strongly interacting particle passing within $\sim 10^{-13}$ cm has a high probability of interacting, then the characteristic time for the interaction must be $\lesssim 3 \times 10^{-22}$ sec and the cross-section $\sim 10^{-26}$ cm^2, 10 millibarns. This cross-section is typical of pion–nucleon interactions. If a pion interacting with a nucleon has a cross-section ~ 1 mb for producing strange particles, then the interactions coupling the strange particles to the pion–nucleon system have a characteristic time $\sim 10^{-21}$ sec which is ~ 10 orders of magnitude down on the decay time. If the reaction

$$\pi^- p \to \Lambda^0 \pi^0$$

occurred with a time scale of $\sim 10^{-21}$ sec then we would expect the decay

$$\Lambda^0 \to p\pi^-$$

to occur with the same time scale, but

$$\tau_{\Lambda^0} = 2.6 \times 10^{-10} \text{ sec.}$$

The solution of this apparent paradox is that the strange particles are produced by the strong interactions only in pairs

$$\pi^- p \not\to \Lambda^0 \pi^0,$$
but
$$\pi^- p \to \Lambda^0 K^0,$$
$$\to \Sigma^- K^+, \text{ etc.}$$

In the decay of a Λ^0 or a K^0 there is only one strange particle involved, and so no strong decay.

The observation of both production and decay of strange particles confirmed this hypothesis of associated production beyond any doubt. Initial observations were made with cloud chambers, which were soon replaced at the accelerator laboratories by bubble chambers. In such a detector, which records the tracks of individual charged particles, the whole history may be followed. For example, in a liquid-hydrogen bubble chamber, which has the advantage of providing protons as the target, events such as those sketched in Fig. 6.8.1

Nuclear Physics

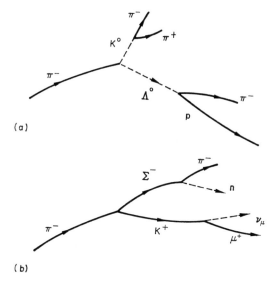

FIG. 6.8.1. Two examples of the appearance of strange particle production in a hydrogen bubble chamber are sketched

(a) $\pi^- p \to \Lambda^0 K^0$; $\quad \Lambda^0 \to p\pi^-,\quad K^0 \to \pi^+\pi^-,$

(b) $\pi^- p \to \Sigma^- K^+$; $\quad \Sigma^- \to n\pi^-,\quad K^+ \to \mu^+ \nu_\mu.$

The neutral particles, which are not bent by the applied magnetic field, and do not leave a visible track, are indicated by broken lines.

are observed. In the many millions of bubble chamber photographs which have now been obtained, reactions like

$$\pi^- p \to \Lambda^0 \pi^0,$$

or

$$np \to \Lambda^0 p$$

have never been observed. On the other hand, reactions like

$$p\bar{p} \to \Lambda^0 \bar{\Lambda}^0,$$
$$\Lambda^0 \to p\pi^-,$$
$$\bar{\Lambda}^0 \to \bar{p}\pi^+$$

have been observed: here again, we have associated production of strange particles.

The phenomenon of associated production may be expressed quantitatively by introducing a new additive quantum number, the *strangeness*, which is conserved in strong interactions, but not in the weak interactions. The nucleons and pions are allotted strangeness zero. Then

$$\underset{S=0}{\pi^-} + \underset{S=0}{p} \to \underset{S=-1}{\Lambda^0} + \underset{S=+1}{K^0} \quad \Delta S = 0 \quad \text{strong interaction}$$

$$\underset{S=-1}{\Lambda^0} \to \underset{S=0}{\pi^-} + \underset{S=0}{p} \quad \Delta S = +1 \quad \text{weak interaction}$$

$$\underset{S=+1}{K^0} \to \underset{S=0}{\pi^+} + \underset{S=0}{\pi^-} \quad \Delta S = -1 \quad \text{weak interaction}$$

$$\underset{S=0}{\pi^-} + \underset{S=0}{p} \to \underset{S=-1}{\Sigma^-} + \underset{S=+1}{K^+} \quad \Delta S = 0 \quad \text{strong interaction}$$

Isospin

The reaction

$$\pi^+ + p \rightarrow \Sigma^+ + K^+$$
$$S=0 \quad S=0 \qquad \quad S=+$$

is observed, so Σ^+ has $S=-1$, and the reaction

$$\pi^- + p \rightarrow \Sigma^+ + K^-$$
$$S=0 \quad S=0 \qquad \quad S=-1$$

is not observed, so K^- has $S=-1$.

There is yet another category of strongly interacting particles which only decay weakly, the cascade hyperons with $S=-2$, produced in reactions like

$$\pi^+ + n \rightarrow \Xi^- + K^+ + K^+$$
$$S=0 \quad S=0 \qquad S=+1 \quad S=+1$$

From the decay modes of the strange particles, it is clear that the K-mesons have integral zero spin, and the Λ, Σ and Ξ have half integral spin $\tfrac{1}{2}$. All strongly interacting particles are known as *hadrons*. Strongly interacting fermions are called *baryons*, and strange baryons are called *hyperons*. Strongly interacting bosons are called *mesons*. The idea of conservation of nucleon number is extended to reactions involving strange particles in an obvious way: so far as we know the baryon number is absolutely conserved. Some of the properties of those strange particles stable against strong interactions are summarized in Table 6.8.1.

Note that:

The antiparticle of K^+ $(B=0, S=+1)$ is K^- $(B=0, S=-1)$,
K^0 $(B=0, S=+1)$ is $\overline{K^0}$ $(B=0, S=-1)$,
Σ^+ $(B=1, S=-1)$ is $\overline{\Sigma}^-$ $(B=-1, S=+1)$.

Σ^- is NOT the antiparticle of Σ^+.

While the introduction of the new quantum number S makes quantitative the observed associated production it does not explain the phenomenon. Gell-Mann and Nishijima (1953) pointed out that it is not necessary to introduce this new quantum number: conservation of isospin in the strong interactions (together with the already known violation of isospin in the weak interactions) is sufficient to explain associated production.

If we are to allocate isospin to the strange particles, it seems plausible to identify (K^+, K^0) with an isospin doublet, Λ^0 with a singlet, $(\Sigma^+, \Sigma^0, \Sigma^-)$ with a triplet and (Ξ^0, Ξ^-) with a doublet. This works very nicely:

$$\pi^- + p \rightarrow \Lambda^0 + K^0$$

	π^-	p	Λ^0	K^0
T	1	$\tfrac{1}{2}$	0	
T_z	-1	$+\tfrac{1}{2}$	0	

T_z must be $-\tfrac{1}{2}$ for the K^0 produced with the Λ^0 and T must have a value $\tfrac{3}{2}$ or $\tfrac{1}{2}$: the observed multiplicity of kaon states fixes $T=\tfrac{1}{2}$. Reactions like

$$p + p \rightarrow \Lambda^0 + K^+ + p$$

	p	p	Λ^0	K^+	p
T	$\tfrac{1}{2}$	$\tfrac{1}{2}$	0		$\tfrac{1}{2}$
T_z	$+\tfrac{1}{2}$	$+\tfrac{1}{2}$	0		$+\tfrac{1}{2}$

$$\pi^+ + n \rightarrow \Lambda^0 + K^+$$

	π^+	n	Λ^0	K^+
T	1	$\tfrac{1}{2}$	0	
T_z	$+1$	$-\tfrac{1}{2}$	0	

Nuclear Physics

TABLE 6.8.1. STRANGE PARTICLES STABLE

Particle	Mass (MeV/c²)	Mean life (sec)	Principal decay modes		J^P	Q	S	B	T_z	T
K^+	493.8	1.24×10^{-8}	$\mu^+ \nu_\mu$ $\pi^+ \pi^0$	64% 21%	0^-	$+1$	$+1$	0	$+\frac{1}{2}$	
K^0	497.8	$K_S^0\, 0.86 \times 10^{-10}$	$\pi^+ \pi^-$ $\pi^0 \pi^0$	69% 31%	0^-	0	$+1$	0	$-\frac{1}{2}$	$\frac{1}{2}$
		$K_L^0\, 5.17 \times 10^{-8}$	$\pi e \nu$ $\pi \mu \nu$ $\pi \pi \pi$	40% 27% 34%						
Λ^0	1115.6	2.5×10^{-10}	$p \pi^-$ $n \pi$	64% 36%	$\frac{1}{2}^+$	0	-1	1	0	0
Σ^+	1189.4	0.8×10^{-10}	$p \pi^0$ $n \pi^+$	52% 48%	$\frac{1}{2}^+$	$+1$	-1	1	$+1$	
Σ^0	1192.5	$< 10^{-14}$	$\Lambda^0 \gamma$		$\frac{1}{2}^+$	0	-1	1	0	1
Σ^-	1197.4	1.49×10^{-10}	$n \pi^-$		$\frac{1}{2}^+$	-1	-1	1	-1	
Ξ^0	1314.7	3.0×10^{-10}	$\Lambda^0 \pi^0$		$\frac{1}{2}^+$	0	-2	1	$+\frac{1}{2}$	$\frac{1}{2}$
Ξ^-	1321.3	1.66×10^{-10}	$\Lambda^0 \pi^-$		$\frac{1}{2}^+$	-1	-2	1	$-\frac{1}{2}$	
Ω^-	1672	10^{-10}	$\Xi^0 \pi^-$ $\Xi^- \pi^0$ $\Lambda^0 K^-$		$\frac{3}{2}^+$	-1	-3	1	0	0

Notes: 1. Fermion and antifermion have opposite intrinsic parity.
2. K^0 and $\bar{K}^0$ do not decay exponentially but are different mixtures of two exponentially decaying states K_S and K_L.

fix the value of T_z for the K^+ as $+\frac{1}{2}$. Then

$$\pi^- + p \to \Sigma^- + K^+,$$
$$\pi^+ + p \to \Sigma^+ + K^+$$

yield $T_z = +1$ for Σ^+ and -1 for Σ^-. In the absence of any observed doubly charged Σ hyperons, this implies a triplet of Σ's and this scheme predicted the Σ^0 before it was observed experimentally.

Similarly the isospin of the Ξ hyperons is fixed by reactions like

$$\pi^+ + n \to \Xi^- + K^+ + K^+$$

which requires $T_z = -\frac{1}{2}$ for Ξ^-—the Gell-Mann–Nishijima scheme predicted the Ξ^0 before it was observed.

No reactions inconsistent with these isospin assignments have ever been observed and given these assignments conservation of strangeness appears as no more than conservation of the third component of isospin.

Isospin

AGAINST THE STRONG INTERACTIONS

							Antiparticle			
T		T_z	B	S	Q	J^P	Principal decay modes		Mean life (sec)	Mass (MeV/c^2)	
$\frac{1}{2}$	⎧	$-\frac{1}{2}$	0	-1	-1	0^-	$\mu^-\bar{\nu}_\mu$ $\pi^-\pi^0$	64% 21%	1.24×10^{-8}	493.8	K^-
	⎨	$+\frac{1}{2}$	0	-1	0	0^-	$\pi^+\pi^-$ $\pi^0\pi^0$ $\pi e\nu$ $\pi\mu\nu$ $\pi\pi\pi$	69% 31% 40% 27% 34%	$K_S^0\,0.86\times10^{-10}$ $K_L^0\,5.17\times10^{-8}$	497.8	$\bar{K}^0$
0		0	-1	$+1$	0	$\frac{1}{2}^-$	$\bar{p}\pi^+$ $n\pi^+$	64% 36%	2.5×10^{-10}	1115.6	$\bar{\Lambda}^0$
1	⎧	-1	-1	$+1$	-1	$\frac{1}{2}^-$	$\bar{p}\gamma^0$ $\bar{n}\pi^-$	52% 48%	0.8×10^{-10}	1189.4	$\bar{\Sigma}^+$
	⎨	0	-1	$+1$	0	$\frac{1}{2}^-$	$\bar{\Lambda}^0\gamma$		$<10^{-14}$	1192.5	$\bar{\Sigma}^0$
	⎩	$+1$	-1	$+1$	$+1$	$\frac{1}{2}^-$	$\bar{n}\pi^+$		1.49×10^{-10}	1197.4	$\bar{\Sigma}^-$
$\frac{1}{2}$	⎧	$-\frac{1}{2}$	-1	$+2$	0	$\frac{1}{2}^-$	$\bar{\Lambda}^0\pi^0$		3.0×10^{-10}	1314.7	$\bar{\Xi}^0$
	⎩	$+\frac{1}{2}$	-1	$+2$	$+1$	$\frac{1}{2}^-$	$\bar{\Lambda}\pi^+$		1.66×10^{-10}	1321.3	$\bar{\Xi}^-$
0		0	-1	$+3$	$+1$	$\frac{3}{2}^-$	$\bar{\Xi}^0\pi^+$ $\bar{\Xi}^-\pi^0$ $\bar{\Lambda}K^+$		10^{-10}	1672	$\bar{\Omega}^-$

For nucleons we wrote

$$q = \tfrac{1}{2} + T_z$$

and for pions

$$q = T_z$$

the two relations being combined in the form

$$q = \frac{N}{2} + T_z$$

where N is the nucleon number of the particle. This relation is generalized to embrace the strange particles by writing

$$q = \frac{S+B}{2} + T_z$$

where S is the strangeness and B the baryon number. Since q and B are conserved, conservation of strangeness is implied by conservation of T_z alone. The quantity

$$S + B = Y$$

is known as the hypercharge.

It should now be clear that the only reason why the strange particles we have listed decay weakly is that their masses are too low for decay without violation of S to be energetically

403

Nuclear Physics

possible. In this respect they are analogous to the neutron and the pions, which being among the lightest strongly interacting particles have not enough energy to decay via the strong interactions. However, just as a massive state such as the $N^*(1236)$ which is observed as a resonance in πN scattering can decay via the strong interactions

$$N^{*++}(1236) \to \pi^+ p \qquad \text{mass } 1236 \text{ MeV}/c^2,$$
$$\text{width } 120 \text{ MeV}/c^2,$$
$$N^{*+}(1236) \to \pi^+ n$$
$$\pi^0 p,$$
$$N^{*0}(1236) \to \pi^- p$$
$$\pi^0 n,$$
$$N^{*-}(1236) \to \pi^- n$$

with a lifetime $\sim 10^{-23}$ sec, massive states with non-zero strangeness may also decay via the strong interactions:

$$Y^*(1385) \to \Lambda^0 \pi^0 \qquad \text{mass } 1385 \text{ MeV}/c^2,$$
$$\text{width } 44 \text{ MeV}/c^2.$$

This state is below threshold for $K^- p$ scattering but would show up as a resonance in $\Lambda \pi$ scattering if this were experimentally feasible. Since it is not, this state is only observed through a two-stage process like

$$K^- p \to Y^* \pi,$$
$$Y^* \to \Lambda \pi.$$

The ϱ-meson, which would appear in $\pi \pi$ scattering in the $T = 1$ channel, can only be observed in processes like

$$\pi^- p \to \varrho^0 n \qquad \text{mass } 765 \text{ MeV}/c^2,$$
$$\varrho^0 \to \pi^+ \pi^- \qquad \text{width } 120 \text{ MeV}/c^2$$

and has a strange analogue, the $K^*(890)$

$$K^+ p \to K^{*+}(890) p,$$
$$K^{*+}(890) \to K^0 \pi^+ \qquad \text{mass } 890 \text{ MeV}/c^2$$
$$K^+ \pi^0 \qquad \text{width } 50 \text{ MeV}/c^2.$$

The isospin of the $K^*(890)$ is $\frac{1}{2}$. Processes like

$$K^+ p \to K^{*++} n$$

do not occur, and the assignment can be checked from the branching ratio

$$R_{K^{*+}} = \frac{K^{*+} \to K^0 \pi^+}{K^{*+} \to K^+ \pi^0}.$$

The isospin can only be $\frac{3}{2}$ or $\frac{1}{2}$. Reference to Table 6.7.1 shows that the coupling of an isospin $\frac{3}{2}$ object to an isospin 1 and an isospin $\frac{1}{2}$ object gives

$$R_{3/2} = \tfrac{1}{2}$$

while
$$R_{1/2} = 2.$$
It is the latter that is in accord with experiment.

There appears to be nothing special about these hadrons, strange or non-strange, which only decay via the weak interactions (or in a few cases the electromagnetic interactions). They are merely the least massive hadrons. Almost all the vast number of hadronic states which have been unearthed in the last ten years are sufficiently massive to decay strongly into other hadrons.

While the classification of hadrons through isospin has been enormously successful, it is by no means the whole story. Isospin multiplets themselves are grouped together to form supermultiplets reflecting restrictions in hadronic physics of greater scope than isospin alone. The best established of these supermultiplets are:

An octet of $J^P = \frac{1}{2}^+$ baryons

				Mass	S
	p	n		~939	0
Σ^+	Σ^0		Σ^-	~1190	−1
	Λ^0			~1115	−1
	Ξ^0		Ξ^-	~1315	−2

A decuplet of $J^P = \frac{3}{2}^+$ baryons

				Mass	S
N^{*++}	N^{*+}	N^{*0}	N^{*-}	~1236	0
	Y^{*+}	Y^{*0}	Y^{*-}	~1385	−1
	Ξ^{*0}		Ξ^{*-}	~1530	−2
		Ω^-		1675	−3

An octet of 0^- mesons

			Mass	S
	K^+	K^0	~495	+1
π^+	π^0	π^-	~139	0
	η^0		~545	0
	$\bar{K}^0$	K^-	~495	−1

An octet of 1^- mesons

			Mass	
	K^{*+}	K^{*0}	~890	+1
ϱ^+	ϱ^0	ϱ^-	~765	0
	ω^0		~783	0
	$\bar{K}^{*0}$	K^{*-}	~890	−1

The proton seems to be the only stable hadron. The other members of the $\frac{1}{2}^+$ octet decay via the weak interactions, except for the Σ^0 which decays electromagnetically into the Λ^0: $\Sigma^0 \to \Lambda^0 + \gamma$. All members of the $\frac{3}{2}^+$ decuplet decay strongly except for the $S = -3$, $T = 0$ Ω^-. This particle is accidentally stable against the strong interactions, being just too light to decay via

$$\Omega^- \to \Xi^0 K^-$$

and only decaying weakly through the strangeness violating modes

$$\Omega^- \to \Xi^0 \pi^-$$
$$\Lambda^0 K^-, \text{etc.}$$

Nuclear Physics

The octet of 0^- mesons is stable against strong decay (the pion is the lightest hadron) and decays weakly except for the π^0 and η^0 members which decay via the electromagnetic interactions. All members of the 1^- octet decay strongly.

We conclude this section by remarking on a curious fact. Of the order of 100 hadronic states are now known and among them are found no baryon states with $T > \frac{3}{2}$ and no meson states with $T > 1$ and there are no $S = +1$ baryon (as opposed to antibaryon) states at all.

6.9. Concluding remarks

In this chapter we have been concerned with isospin as a means of classifying both nuclear and hadronic states as members of multiplets identical as far as the strong interactions are concerned and distinguished only by the electromagnetic and weak interactions. While we derived selection rules in terms of allotted isospin for these isospin breaking interactions, we have not discussed explicitly the selection rules for strong interaction processes. It should be clear to you that the isospin selection rule for all strong interactions is

$$\Delta T = 0 \qquad \Delta T_z = 0$$

which is another way of expressing conservation of isospin. At energies high enough for electromagnetic effects to be unimportant, these rules place enormous constraints on cross-sections and branching ratios which are part of the tools of the trade of the particle physicist. They are, of course, of most use when dealing with a state of unique isospin: we saw in Section 6.7 that in the region of πN scattering where the $T = \frac{1}{2}$ amplitude is unimportant

$$\sigma(\pi^- p \to \pi^- p) = \tfrac{1}{2}\sigma(\pi^- p \to \pi^0 n) \tag{6.9.1}$$

which could be expressed as a branching ratio

$$\frac{N^{*0} \to \pi^- p}{N^{*0} \to \pi^0 n} = \frac{1}{2}$$

and in the previous section we similarly found

$$\frac{K^{*+} \to K^+ \pi^0}{K^{*+} \to K^0 \pi^+} = \frac{1}{2}.$$

The energies involved in these strong interactions are typically hundreds of MeV or greater and corrections due to electromagnetic interactions are at the 1 MeV level. In nuclear physics energies are typically ~ 1 MeV and corrections due to electromagnetism may be of the same order. Under such conditions cross-sections and branching ratios may depend strongly on T_z as well as on T: we encountered an extreme example at the end of Chapter 5. Consider the reactions

$$d + t \to {}^4\text{He} + n,$$
$$d + {}^3\text{He} \to {}^4\text{He} + p.$$

The triton (^{3}H) and ^{3}He are members of the same isospin multiplet with $T = \frac{1}{2}$. Both the deuteron and ^{4}He have isospin 0. These two reactions, then, involve the same isospin

Isospin

multiplets on both sides of the equations and both initial and final states have unique isospin $\frac{1}{2}$. If the cross-sections depended only on T and not on T_z, then these two cross-sections should be the same—but at the energies of interest in thermonuclear reactor work the extra coulomb barriers in the $d + {}^3\text{He} \to {}^4\text{He} + p$ reaction suppress it relative to $d + t \to {}^4\text{He} + n$ by a factor ~ 200.

Finally we should point out the kind of thing that isospin conservation will NOT do. Isospin conservation gives the relative couplings of two specified isospin multiplets to a third specified multiplet, but does not give the couplings of other multiplets to the same third multiplet. This is best made clear by an example. The hadronic state $N^*(1924)$ has isospin $T = \frac{3}{2}$ and $J^P = \frac{7}{2}^+$. It couples to πN and is sufficiently massive also to decay into ΣK. In both cases we are coupling an isospin doublet and a triplet to form a quartet. Using Table 6.7.1 we may write the relations (assuming a pure initial state)

$$\frac{d\sigma}{d\Omega}(\pi^+ p \to \pi^+ p) = K_1 |A_{3/2}(\pi p \to \pi p)|^2,$$

$$\frac{d\sigma}{d\Omega}(\pi^- p \to \pi^- p) = \frac{K_1}{9} |A_{3/2}(\pi p \to \pi p)|^2,$$

$$\frac{d\sigma}{d\Omega}(\pi^- p \to \pi^0 n) = \frac{2K_1}{9} |A_{3/2}(\pi p \to \pi p)|^2$$

and

$$\frac{d\sigma}{d\Omega}(\pi^+ p \to \Sigma^+ K^+) = K_2 |A_{3/2}(\pi p \to \Sigma K)|^2,$$

$$\frac{d\sigma}{d\Omega}(\pi^- p \to \Sigma^- K^+) = \frac{K_2}{9} |A_{3/2}(\pi p \to \Sigma K)|^2,$$

$$\frac{d\sigma}{d\Omega}(\pi^- p \to \Sigma^0 K^0) = \frac{2K_2}{9} |A_{3/2}(\pi p \to \Sigma K)|^2$$

which we could also express through branching ratios such as

$$\frac{N^{*0}(1924) \to \pi^- p}{N^{*0}(1924) \to \pi^0 n} = \frac{1}{2},$$

$$\frac{N^{*0}(1924) \to \Sigma^- K^+}{N^{*0}(1924) \to \Sigma^0 K^0} = \frac{1}{2}, \quad \text{etc.}$$

Conservation of isospin, however, gives us no relations between the two $T = \frac{3}{2}$ amplitudes

$$A_{3/2}(\pi p \to \pi p)$$

and

$$A_{3/2}(\pi p \to \Sigma K)$$

In order to obtain any such relations we need to go beyond the isospin formalism and use the fact that N and Σ, π and K are members of the same supermultiplet. The SU3 classification indeed provides such coupling schemes which are identical in spirit to isospin coupling schemes, but much more complicated in practice because of the extra degree of freedom involved—and well outside the scope of this book.[†]

[†] See, for example, B. T. Feld, *Models of Elementary Particles* (Blaisdell, 1969), and D. H. Perkins, *Introduction to High Energy Physics* (Addison-Wesley, 1972).

APPENDIX 1

Energy Units and Constants

1. The electron volt

The basic unit of energy in nuclear physics, as in atomic physics, is the electron volt (eV), which is the energy acquired when 1 electronic charge e falls through a potential difference of 1 volt. The electronic charge is

$$e \simeq 4.8 \times 10^{-10} \quad \text{esu}$$

and

$$1 \text{ V} \simeq \tfrac{1}{300} \text{ esu},$$

so

$$1 \text{ eV} \simeq 1.6 \times 10^{-12} \text{ erg}.$$

(The precise value is given in the list of values on p. 409.)

The electron volt is inconveniently small for most applications in nuclear physics and the following multiples are used:

$$1 \text{ keV} = 10^3 \text{ eV},$$
$$1 \text{ MeV} = 10^6 \text{ eV},$$

and in high-energy physics

$$1 \text{ GeV} = 10^3 \text{ MeV} = 10^9 \text{ eV}.$$

2. Momentum and mass units

Momentum and mass are frequently quoted in units of energy, on the basis of the relativistic formula

$$p^2 c^2 + m^2 c^4 = E^2.$$

The quantities pc and mc^2 have the dimensions of energy and

$$p = \frac{[pc]}{c} \quad m = \frac{[mc^2]}{c^2}.$$

If pc and mc^2 are expressed in MeV then p can be measured in units of MeV/c and m in units of MeV/c² (frequently the denominators are dropped).

Thus the mass of the proton is $M_p \sim 1.67 \times 10^{-24}$ g

so

$$M_p c^2 \sim 1.5 \times 10^{-4} \text{ erg} \sim 938 \text{ MeV}.$$

The rest mass energy of the proton

$$M_p c^2 \sim 938 \text{ MeV}$$

and the mass of the proton is

$$M_p \sim 938 \; \frac{\text{MeV}}{c^2}.$$

408

Energy Units and Constants

3. *Values of physical constants*

Velocity of light	c	$2.9979250(10) \times 10^{10}$ cm sec^{-1}
Electron charge	e	$4.803250(21) \times 10^{-10}$ esu
Atomic mass unit (^{12}C = 12)	amu	$1.660531(11) \times 10^{-24}$ g
Energy unit	1 MeV	$1.6021917(70) \times 10^{-6}$ erg
Electron mass	m_e	$9.109558(54) \times 10^{-28}$ g
		$5.485930(34) \times 10^{-4}$ amu
		$0.5110041(16)$ MeV/c^2
Proton mass	M_p	$1.672614(11) \times 10^{-24}$ g
		$1.00727661(8)$ amu
		$938.2592(52)$ MeV/c^2
Neutron mass	M_n	$1.674920(11) \times 10^{-24}$ g
		$1.00866520(10)$ amu
		$939.5527(52)$ MeV/c^2
Planck's constant/2π	$\hbar$	$1.0545919(80) \times 10^{-27}$ erg sec
Bohr magneton	$e\hbar/2m_e c$	$9.274096(65) \times 10^{-21}$ erg gauss^{-1}
Nuclear magneton	$e\hbar/2M_p c$	$5.050951(50) \times 10^{-24}$ erg gauss^{-1}
Electron magnetic moment	μ_e	$1.0011596389(31)$ Bohr magnetons
Proton magnetic moment	μ_p	$2.792709(17)$ nuclear magnetons
Neutron magnetic moment	μ_n	$-1.913148(66)$ nuclear magnetons
Avogadro's number	N	$6.022169(40) \times 10^{23}$ mole^{-1}
Gas constant	R	$8.31434(35) \times 10^7$ erg mole^{-1} °K^{-1}
Boltzmann's constant	k	$1.380622(59) \times 10^{-16}$ erg °K^{-1}
Gravitational constant	G	$6.6732(31) \times 10^{-8}$ dyne cm^2 g^{-2}
Stefan's constant	σ	$5.66961(96) \times 10^{-5}$ erg sec^{-1} cm^{-2} °K^{-4}
Fine structure constant	$e^2/\hbar c$	$7.297351(11) \times 10^{-3}$
Bohr radius	a_0	$5.2917715(81) \times 10^{-9}$ cm

Errors on the last two digits are appended in parentheses. The numbers are taken from B. N. Taylor, W. H. Parker and D. N. Langenberg, *Rev. Mod. Phys.* **41**, 375 (1969), with the exception of the value for neutron magnetic moment, which is from the compilation of the Particle Data Group, *Rev. Mod. Phys.* **43**, S1 (1971).

APPENDIX 2

Angular Momentum Coupling

In the table on p. 411 we give spherical harmonics up to Y_2^2 and the angular momentum coupling (Clebsch–Gordan) coefficients. The spherical harmonics are orthonormal solutions of the angular part of the Schrödinger equation for a central potential and each Y_L^M corresponds to a well-defined value of the orbital angular momentum L and the third component M. The spherical harmonics are related to the associated Legendre polynomials by a normalization factor, and have phases chosen according to the generally used Condon and Shortley convention.

As an example of the use of the table, consider $1 \times \frac{1}{2}$. A state that is a product of states with angular momentum 1 and angular momentum $\frac{1}{2}$ is in general a mixture of states with angular momentum $\frac{3}{2}$ and angular momentum $\frac{1}{2}$, and only specific mixtures give well-defined angular momentum. Referring to the table we see that a product of the states

$$|1, +1\rangle |\tfrac{1}{2}, +\tfrac{1}{2}\rangle$$

is a pure $|\tfrac{3}{2}, +\tfrac{3}{2}\rangle$ state. However, the product

$$|1, +1\rangle |\tfrac{1}{2}, -\tfrac{1}{2}\rangle$$

is a mixture given by reading horizontally across the table

$$\sqrt{\tfrac{1}{3}}\,|\tfrac{3}{2}, +\tfrac{1}{2}\rangle + \sqrt{\tfrac{2}{3}}\,|\tfrac{1}{2}, +\tfrac{1}{2}\rangle.$$

Conversely, if we have a state of angular momentum $J = \tfrac{3}{2}$ and $M = +\tfrac{1}{2}$ we read vertically down the table and find

$$|\tfrac{3}{2}, +\tfrac{1}{2}\rangle = \sqrt{\tfrac{1}{3}}\,|1, +1\rangle |\tfrac{1}{2}, -\tfrac{1}{2}\rangle + \sqrt{\tfrac{2}{3}}\,|1, 0\rangle |\tfrac{1}{2}, +\tfrac{1}{2}\rangle$$

and if we wish to write the orbital parts in terms of θ and ϕ we have

$$\left|\tfrac{3}{2}, \tfrac{1}{2}\right\rangle = \sqrt{\tfrac{1}{3}}\, Y_1^{+1} \left|\tfrac{1}{2}, -\tfrac{1}{2}\right\rangle + \sqrt{\tfrac{2}{3}}\, Y_1^0 \left|\tfrac{1}{2}, \tfrac{1}{2}\right\rangle$$

$$= \sqrt{\tfrac{1}{3}} \left\{-\sqrt{\tfrac{3}{8\pi}} \sin\theta\, e^{i\phi}\right\} \left|\tfrac{1}{2}, \tfrac{1}{2}\right\rangle$$

$$+ \sqrt{\tfrac{2}{3}} \left\{\sqrt{\tfrac{3}{4\pi}} \cos\theta\right\} \left|\tfrac{1}{2}, \tfrac{1}{2}\right\rangle.$$

The s-wave angular factor is $Y_0^0 = 1/\sqrt{4\pi}$.

The table is taken from the compilation of the Particle Data Group, Review of particle properties, *Rev. Mod. Phys.* **43**, S1 (1971).

CLEBSCH-GORDAN COEFFICIENTS AND SPHERICAL HARMONICS

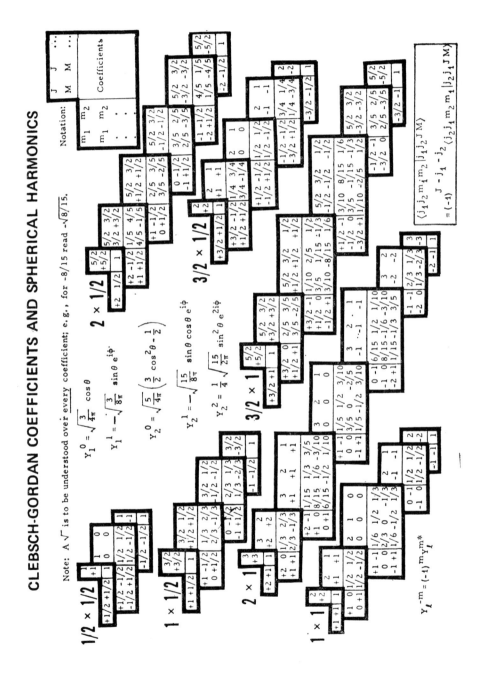

Angular Momentum Coupling

Index

α-decay 2, 175–187
 and Fermi Golden Rule 184–187
 fine structure 178
 Geiger–Nuttall rule 178
 long range α's 178
 theory 180–184
Angular correlations
 e–ν, in forbidden β-decay 147
  relativistic effects in allowed β-decay 158
 in successive decays 136
 γ–γ 136–139
Angular momentum 60–65
 definition in quantum mechanics 63–65
 orbital 60
  conservation in a central field 61
  non-conservation 62, 63
 spin, intrinsic 63
Asymmetry energy 36–38, 41
Atomic mass unit (amu) 11
Atomic number 1

Barn 191
Baryon 401
 number 401
 and strangeness 403
β-decay 2, 139–175
 allowed 145
 charge conjugation 172–174
 conserved vector current (CVC) 171, 388, 396
 electron capture 140, 153
 energetics 140–141
 favoured 148
 forbidden 145–148
 muon 170
 neutron 170
 non-conservation of parity 161–174
  lepton polarization 165–166
  V–A theory 169, 172
 pion 171, 173, 388, 396
 relativistic effects
  angular correlations 158
  Fermi fraction 160
  Fierz interference terms 159
 possible interaction forms 157
 V–A theory 161
 simple theory 142–145
 spin effects 149–152
  Fermi transitions 150, 151
  Gamow–Teller transitions 150, 151
  (*see also* relativistic effects)
 super-allowed 149
 time reversal 172–174
 unfavoured 148
Binding energy 13–15
Black hole 351
Branching ratio 6
Breeding (in nuclear reactors) 309, 321
Breit–Wigner formula 8, 103, 108, 260, 270, 271
 connection with causality 263
 p-wave 250, 260
 s-wave 260
 scattering of light 244

Carbon burning (in stars) 347–348
Carbon–nitrogen cycle 331–333
Causality 208, 245–247, 250
 Breit–Wigner formula 263
Centrifugal barrier 52, 180, 249, 252, 255, 267, 271
Chain reaction 312
Chandrasekhar limit 342–345, 349, 350
Charge conjugation 172–174
Charge independence (of nuclear forces) 29, 30, 40, 41, 220, 362
 isospin formalism 365–370
 pion–nucleon coupling 391–393
 two-nucleon system 233–235, 362–364
Clebsch–Gordan coefficients 216, 394, Appendix 2
Coherent scattering, of neutrons by protons 229–230
 coherent scattering length 230–231
Compound nucleus 30, 191, 247, 261, 264–268
Configuration mixing 42
Conserved vector current (weak interactions) 170–172, 175, 388, 389, 396–398
Coulomb barrier 17, 30, 206, 272–287
 α-decay 180, 279
 fission 303, 304, 307

413

Index

Coulomb barrier (*cont.*)
  resonance width 283
  thermonuclear reactions 285, 321, 325, 332, 339
Coulomb energy 29, 34, 43–45, 376–377
  fission 301
Crab nebula 349
Critical mass (^{235}U) 319
Critical size (fission reactor) 317, 319
Cross-section 15, 16, 201, 210–213
  general discussion and definition 192–198
  spin effects 213–217

Decay rate 5, 101, 102
  Fermi Golden Rule 103–107
Degeneracy pressure (of electrons) 337
  Chandrasekhar limit 342–344
Deuteron 62, 63, 70, 220, 227
  formation by neutron capture 237–242
  formation in hydrogen burning 324–329
  pickup and stripping reactions 292–294
  thermonuclear reactions 352–354
Dirac equation 154
  free field solutions 155
Direct reactions 191, 287
  neutrino interactions 294–299
  peripheral 287–292
    elastic scattering 289
    inelastic scattering 289–291
    knock-on 292
    pickup and stripping 292–294

e (equilibrium) process (nucleosynthesis) 348
Effective range 226
  parameters, for two-nucleon system 228, 233
Electromagnetic decay 101, 109
  angular correlations 136–139
  basic theory 109–114
  determination of multipolarity 135–139
  electric dipole 114–118
  electric quadrupole 121–122
  internal conversion 124–129
  magnetic dipole 118–121
  measurement of lifetimes 132–135, 268
  rates in atoms and nuclei 123, 124
  0→0 transitions 129–132
Electron capture 140, 141, 152, 153
Exchange forces 57–59
Explosion
  fission 319
  supernova 344, 349
  thermonuclear 355

Factorization (of cross-sections) 261, 268, 270, 271
Fermi gas
  of electrons in stars 343
  of nucleons 46, 264

Fermi Golden Rule
  basic derivation
    decay processes 103–108
    extension to reaction cross-sections 198–201
    resonance reactions 269–272
  1st application (electromagnetic processes) 109–114
  2nd application (internal conversion) 124–128
  3rd application (β-decay) 141–144
  4th application (α-decay) 184–187
  5th application (cross-sections) 198–201
  6th application (neutron capture by protons) 237–242
  7th application (peripheral reactions) 287–292
  8th application (ν interactions) 294, 298
Fermi statistics (neutron–proton ratio in nuclei) 36
Fission 300–302
  fragments 309–311, 317, 319
  induced 301, 308–309
    activation energy 308
  secondary features 309–311
  spontaneous 301–307
    conditions for stability 302, 305, 306
Fission reactor 311–321
  breeding 309, 321
  moderation of neutrons 313–314
  time constant 318–320
Flux factor (cross-sections) 200, 202, 204
Form factors
  of nucleon
    electromagnetic 23, 299
    weak 299
  of nucleus
    electric 20
    peripheral scattering 289
Fourier transform
  energy structure of unstable states 8
  of coulomb field, in scattering 20
  wave packets 197, 245
ft value (β-decay) 148

γ-decay *see* Electromagnetic decay
γ-rays 2
Gamow factor (coulomb barriers) 183, 206, 279, 286, 322, 325

Half-life 6
Helicity states 156
  projection operators 162, 172
Helium burning (in stars) 189, 338–342
Hertzsprung–Russell diagram (star clusters) 334, 335, 341
Hydrogen burning (in stars) 324–334
  CN cycle 331–333
  PP chains 324–330

Index

Impact parameter 15, 192
  and uncertainty principle 195
Independent particle model 45–47
Intermediate boson *see* Weak interactions
Internal conversion 124–128
  coefficients 128
Isobaric analogue states *see* Isospin; Resonances
Isobaric spin *see* Isospin
Isobars 35, 39, 45
Isomerism 49, 56, 124, 128
Isospin
  analogue states 375–379
  β-decay 386–389
    conserved vector current 388
    pion 388, 396–398
    selection rules 386
  conservation 370
  coulomb interaction 371, 375
  electromagnetic decay 379–386
    selection rules 383–386
  identical and non-identical particles 362–365
  nuclear structure 372–379
  nucleon 365–367
  pion 391–398
    pion–nucleon coupling 391–394
    pion–nucleon scattering 394–396
  selection rules for strong interactions 369, 370, 406
  strange particles 398–405
  two-nucleon system 363, 367–371
Isotopes 1, 2, 35
Isotopic doublet 29, 149
Isotopic multiplets 175, 372–379
Isotopic shift (atomic spectra) 26
Isotopic spin *see* Isospin
Isotopic triplet 30, 149, 175

j–j coupling 40, 53–56

K-meson (Kaon) 398–403
  CP violation 174
  parity violation 161
Kurie plot (β-decay) 145–147

Lamb shift 26
Legendre polynomials 61, 69, 70, 207
Leptons 139
  conservation 140, 168, 172
Level density
  compound nucleus 266, 267, 375
  single-particle 38, 266, 375
Liquid drop model 34, 35, 45
  fission 305

Magic numbers 48–53, 56

Main sequence (stars) 333–335
Mass defect 13
Mass number 1
Mass spectrometry 10–12
Maxwell–Boltzmann distribution 326
Mean life 6
Mirror nuclei 29, 149
Moderation of neutrons 312–315
Molecular scattering of neutrons
  hydrocarbon mirror 231–232
  ortho- and para-hydrogen 229–231
Molecular spectra (nuclear spin) 96–99
Multipole moments (static) 72, 82
  electric dipole 73–74
    and parity 73–74
  electric quadrupole 74–82
    collective effects 81–82
  magnetic dipole 82–96
    independent particle model and Schmidt limits 89–93
Multipole radiation *see* Electromagnetic decay
Muon (μ) 28
  β-decay 170
  muonic atoms 28–29

Neutrino 2, 139–140
  astronomy 297, 332
  charge conjugation violation 172–173
  helicity 166
  mass 145
  muon neutrino 154, 298–299
  parity violation 172–173
  radiation (by stars) 342, 345–347
  reactions 140, 294–299
  two-component theory 166, 168
Neutron emission (fission)
  delayed 311
    time constant of reactor 319
  prompt 310
Neutron number 1
Neutron resonances 253, 267
  in uranium 313–315
Neutron stars 349
Neutrons, thermal 312
Nuclear forces x, 56–59
  charge independence 30, 220, 233–235, 363, 370
  exchange 57–59
  hard core 59, 235
  isospin 369–370
  meson exchange 57, 390–392
  three-body x
Nuclear mass 9–15
Nuclear size x, 15–33
Nucleon x
  number 370, 403

Optical model 30–31, 218–219, 294

Index

Pairing energy 39–41, 308
Parity 66–71
  non-conservation (in weak interactions) 70–71, 161–175
  selection rules (in nuclear decay)
    β-decay 150
    electromagnetic decay 123
Partial wave expansion 118, 206–208
Pauli exclusion principle
  and electron degeneracy pressure 337, 342–343
  and nuclear structure 36–38, 46, 48, 56–57, 265–267, 373–374
Pauli spin matrices 151, 156, 239
Peripheral reactions *see* Direct reactions
Phase shift 209–212, 244–246, 247–253
Phase space 108, 112–113, 202, 204
Photons 109–111
  virtual 57, 125, 132, 390
Pinch effect (plasmas) 359
Pion (π-meson) 57, 188, 389–398
  β-decay 172, 388, 396–398
  decay 71, 165, 173
  isospin 392–398
Polarization
  longitudinal (in β-decay) 161–174
  transverse (in scattering) 54–55
Polarization vectors (of photons) 109–111, 114–115, 119–120, 122, 240
  correlations 139
  virtual photons 57, 132
Pregnant nuclei (among fission fragments) 319
Proton–proton chains (in hydrogen burning)
  PP I 324
  PP II 329–330
  PP III 329–330
Pulsars (and neutron stars) 349–350

Q value (nuclear reactions) 12, 205

r (rapid) process (nucleosynthesis) 348
Radioactivity 1–6
Reactors
  fission 300, 311–321
    breeder 309, 321
    time constant 318–320
  thermonuclear (fusion) 300, 351–361
Red giants 335–337, 340
Repulsive core (in nucleon–nucleon interaction) 59, 235
Residual interactions 39–42, 46, 48
  configuration mixing 42
  magnetic moments 93
  pairing 39–41
  quadrupole moments 81
Resonances 243–272
  and coulomb barrier 282–284
  and Fermi Golden Rule 269–272
  γ-induced resonance 268
  in π–p scattering 261, 396–397
  in square well scattering 247–252
  isobaric analogue resonances 375–379
  shape resonances 247, 252, 264
  structure resonances 247, 252, 264–267
Rutherford formula (coulomb scattering) 16, 193, 203

s (slow) process (nucleosynthesis) 348
Scattering 206–213, 244–260
  and nuclear size 15–23, 30–33
  coulomb 15–21, 201–203, 213
  elastic and inelastic 189
  in two-nucleon system 220–237
  of light 227, 243–244
  optical model 218–219
  potential and nuclear 257, 267
  resonant 17, 267
Scattering amplitude 17–18
  and spin 213–217
Scattering length 222–227
  in two-nucleon systems 228–233
  coherent 230–232
Schmidt limits 91–93
Schwartzschild's limit 351
Selection rules
  in β-decay
    Fermi transitions 150
    Gamow–Teller transitions 150
    isospin selection rules 386
  in electromagnetic decay 123
    electric dipole transitions 118
    electric quadrupole transitions 122
    isospin selection rules 384–386
    magnetic dipole transitions 121
Semiempirical mass formula 34–45
  and stability against fission 301–307
Shape independent approximation (in scattering) 225–227
Shell structure 36–37, 48–54
Slowing-down time (of nuclei in matter) 135, 310
Solar constant 328
Spin *see* Angular momentum
Spin–orbit coupling 53–55, 236
Stability curve 9, 43–45
Star clusters 334–335, 341
Stark effect (in atomic hydrogen) 73–74
Statistics of nuclei 99
Strange particles 398–407
  weak decay 161, 172
Sun
  central temperature and other physical characteristics 323–324
  radius of hydrogen burning core 329, 330
  solar constant 328
Supernova 15, 349–350
Surface energy 34–36, 47
  in fission 301–306

Index

Tensor interaction 59, 237
Thermonuclear reactions 300, 321–323
   for reactors 351–354
   in stars
      advanced evolutionary stages 347–348
      helium burning 338–341
      hydrogen burning 324–334
Thomas precession 156
Thomas–Fermi model of atom 14
Time reversal 74, 170, 174
Transition rate 5–6, 101–102
   and cross-section 201
   Fermi Golden Rule 102, 103–108, 199
21-cm line 89, 94
Two-nucleon system 220–237
   charge independence 233–235
   isospin formalism 363, 367–371

Uncertainty principle
   and atomic structure ix
   and nuclear structure ix
   for energy 8
   in scattering 195–197
Unitarity limit 245, 258
   and weak interactions 298–299
Urca process 349

Van Allen radiation belts 358
Variable stars 341–342

W.B.K. method (and coulomb barrier) 181, 273–277
Wave packets
   and reactions 196–197
   and resonances 245–246
   and causality 209, 245–246
   and decaying states 6–8
Weak interactions 101, 153, 172–174, 299
   intermediate boson 153, 299
   see also β-decay; Neutrino
White dwarfs 338, 347–350
Width of state (in energy) 8, 108, 257–263, 269–271
   relation to lifetime 8, 108, 262–263
Woods–Saxon potential 32, 33, 219

X-ray spectra 25–26
   of muonic atoms 28

Zero range approximation (to deuteron wave function) 241, 325

OTHER TITLES IN THE SERIES IN NATURAL PHILOSOPHY

- Vol. 1. DAVYDOV—Quantum Mechanics
- Vol. 2. FOKKER—Time and Space, Weight and Inertia
- Vol. 3. KAPLAN—Interstellar Gas Dynamics
- Vol. 4. ABRIKOSOV, GOR'KOV and DZYALOSHINSKII—Quantum Field Theoretical Methods in Statistical Physics
- Vol. 5. OKUN'—Weak Interaction of Elementary Particles
- Vol. 6. SHKLOVSKII—Physics of the Solar Corona
- Vol. 7. AKHIEZER et al.—Collective Oscillations in a Plasma
- Vol. 8. KIRZHNITS—Field Theoretical Methods in Many-body Systems
- Vol. 9. KLIMONTOVICH—The Statistical Theory of Non-equilibrium Processes in a Plasma
- Vol. 10. KURTH—Introduction to Stellar Statistics
- Vol. 11. CHALMERS—Atmospheric Electricity (2nd Edition)
- Vol. 12. RENNER—Current Algebras and their Applications
- Vol. 13. FAIN and KHANIN—Quantum Electronics, Volume 1—Basic Theory
- Vol. 14. FAIN and KHANIN—Quantum Electronics, Volume 2—Maser Amplifiers and Oscillators
- Vol. 15. MARCH—Liquid Metals
- Vol. 16. HORI—Spectral Properties of Disordered Chains and Lattices
- Vol. 17. SAINT JAMES, THOMAS and SARMA—Type II Superconductivity
- Vol. 18. MARGENAU and KESTNER—Theory of Intermolecular Forces (2nd Edition)
- Vol. 19. JANCEL—Foundations of Classical and Quantum Statistical Mechanics
- Vol. 20. TAKAHASHI—An Introduction to Field Quantization
- Vol. 21. YVON—Correlations and Entropy in Classical Statistical Mechanics
- Vol. 22. PENROSE—Foundations of Statistical Mechanics
- Vol. 23. VISCONTI—Quantum Field Theory, Volume 1
- Vol. 24. FURTH—Fundamental Principles of Modern Theoretical Physics
- Vol. 25. ZHELEZNYAKOV—Radioemission of the Sun and Planets
- Vol. 26. GRINDLAY—An Introduction to the Phenomenological Theory of Ferroelectricity
- Vol. 27. UNGER—Introduction to Quantum Electronics
- Vol. 28. KOGA—Introduction to Kinetic Theory: Stochastic Processes in Gaseous Systems
- Vol. 29. GALASIEWICZ—Superconductivity and Quantum Fluids
- Vol. 30. CONSTANTINESCU and MAGYARI—Problems in Quantum Mechanics
- Vol. 31. KOTKIN and SERBO—Collection of Problems in Classical Mechanics
- Vol. 32. PANCHEV—Random Functions and Turbulence
- Vol. 33. TALPE—Theory of Experiments in Paramagnetic Resonance
- Vol. 34. TER HAAR—Elements of Hamiltonian Mechanics (2nd Edition)
- Vol. 35. CLARKE and GRAINGER—Polarized Light and Optical Measurement
- Vol. 36. HAUG—Theoretical Solid State Physics, Volume 1
- Vol. 37. JORDAN and BEER—The Expanding Earth
- Vol. 38. TODOROV—Analytical Properties of Feynman Diagrams in Quantum Field Theory
- Vol. 39. SITENKO—Lectures in Scattering Theory
- Vol. 40. SOBEL'MAN—An Introduction to the Theory of Atomic Spectra
- Vol. 41. ARMSTRONG and NICHOLLS—Emission, Absorption and Transfer of Radiation in Heated Atmospheres
- Vol. 42. BRUSH—Kinetic Theory, Volume 3
- Vol. 43. BOGOLYUBOV—A Method for Studying Model Hamiltonians
- Vol. 44. TSYTOVICH—An Introduction to the Theory of Plasma Turbulence
- Vol. 45. PATHRIA—Statistical Mechanics
- Vol. 46. HAUG—Theoretical Solid State Physics, Volume 2

Other Titles in the Series

Vol. 47. NIETO—The Titius–Bode Law of Planetary Distances: Its History and Theory
Vol. 48. WAGNER—Introduction to the Theory of Magnetism
Vol. 49. IRVINE—Nuclear Structure Theory
Vol. 50. STROHMEIER—Variable Stars
Vol. 51. BATTEN—Binary and Multiple Systems of Stars
Vol. 52. ROUSSEAU and MATHIEU: Problems in Optics